Excited States and Photochemistry of Organic Molecules

Excited States and Photochemistry of Organic Molecules

Martin Klessinger
Westfälische Wilhelms-Universität Münster
Josef Michl
University of Colorado

Martin Klessinger
Organisch-Chemisches Institut
Westfälische Wilhelms-Universität
D-48149 Münster
Germany

Josef Michl
Department of Chemistry and
Biochemistry
University of Colorado
Boulder, CO 80309-0215

Library of Congress Cataloging-in-Publication Data
Klessinger, Martin.
Excited states and photochemistry of organic molecules / Martin Klessinger, Josef Michl.
p. cm.
Includes index.
ISBN 1-56081-588-4
1. Chemistry, Physical organic. 2. Photochemistry. 3. Excited state chemistry. I. Michl, Josef, 1939– . II. Title.
QD476.K53 1994
547.1′35—dc20 92-46464
CIP

Printed in the United States of America

ISBN 1-56081-588-4 VCH Publishers, Inc.

Printing History:
10 9 8 7 6 5 4 3 2 1

Published jointly by

VCH Publishers, Inc.
220 East 23rd Street
New York, New York 10010

VCH Verlagsgesellschaft mbH
P.O. Box 10 11 61
D-69451 Weinheim
Federal Republic of Germany

VCH Publishers (UK) Ltd.
8 Wellington Court
Cambridge CB1 1HZ
United Kingdom

To our teachers

WOLFGANG LÜTTKE
AND RUDOLF ZAHRADNÍK

Preface

This graduate textbook is meant primarily for those interested in physical organic chemistry and in organic photochemistry. It is a significantly updated translation of *Lichtabsorption und Photochemie organischer Moleküle,* published by VCH in 1989. It provides a qualitative description of electronic excitation in organic molecules and of the associated spectroscopy, photophysics, and photochemistry. The text is nonmathematical and only assumes the knowledge of basic organic chemistry and spectroscopy, and rudimentary knowledge of quantum chemistry, particularly molecular orbital theory. A suitable introduction to quantum chemistry for a German-reading neophyte is *Elektronenstruktur organischer Moleküle,* by Martin Klessinger, published by VCH in 1982 as a volume of the series, *Physikalische Organische Chemie*. The present textbook emphasizes the use of simple qualitative models for developing an intuitive feeling for the course of photophysical and photochemical processes in terms of potential energy hypersurfaces. Special attention is paid to recent developments, particularly to the role of conical intersections. In emphasizing the qualitative aspects of photochemical theory, the present text is complementary to the more mathematical specialized monograph by Josef Michl and Vlasta Bonačić-Koutecký, *Electronic Aspects of Organic Photochemistry,* published by Wiley in 1990.

Chapter 1 describes the basics of electronic spectroscopy at a level suitable for nonspecialists. Specialized topics such as the use of polarized light are mentioned only briefly and the reader is referred to the monograph by Josef Michl and Erik Thulstrup, *Spectroscopy with Polarized Light,* published by VCH in 1986 and reprinted as a paperback in 1995. Spectra of the most important classes of organic molecules are discussed in Chapter 2. A unified view of the electronic states of cyclic π-electron systems is based on the classic perimeter model, which is formulated in simple terms. Chapter 3 completes the discussion of spectroscopy by examining the interaction of circularly polarized light with chiral molecules (i.e., natural optical activity), and with molecules held in a magnetic field (i.e., magnetic optical activity). An understanding of the perimeter model for aromatics comes in very handy for the latter.

Chapter 4 introduces the fundamental concepts needed for a discussion of photophysical and photochemical phenomena. Here, the section on biradicals and biradicaloids has been particularly expanded relative to the German original. The last three chapters deal with the physical and chemical transformations of excited states. The photophysical processes of radiative and radiationless deactivation, as well as energy and electron transfer, are treated in Chapter 5. A qualitative model for the description of photochemical reactions in condensed media is described in Chapter 6, and then used in Chapter 7 to examine numerous examples of phototransformations of organic molecules. All of these chapters incorporate the recent advances in the understanding of the role of conical intersections ("funnels") in singlet photochemical reactions.

Worked examples are provided throughout the text, mostly from the recent literature, and these are meant to illustrate the practical application of theory. Although they can be skipped during a first reading of a chapter, it is strongly recommended that the reader work them through in full detail sooner or later. The textbook is meant to be self-contained, but provides numerous references to original literature at the end. Moreover, each chapter concludes with a list of additional recommended reading.

We are grateful to several friends who offered helpful comments upon reading sections of the book: Professors F. Bernardi, R. A. Caldwell, C. E. Doubleday, M. Olivucci, M. A. Robb, J. C. Scaiano, P. J. Wagner, M. C. Zerner, and the late G. L. Closs. The criticism of the German version provided by Professors W. Adam, G. Hohlneicher and W. Rettig was very helpful and guided us in the preparation of the updated translation. We thank Dr. Edeline Wentrup-Byrne for editing the translation of the German original prepared by one of us (M. K.), and to Ms. Ingrid Denker for a superb typing job and for drawing numerous chemical structures for the English version. It was a pleasure to work with Dr. Barbara Goldman of VCH and her editorial staff, and we appreciate very much their cooperation and willingness to follow our suggestions. Much of the work of one of us (J. M.) was done during the tenure of a BASF professorship at the University of Kaiserslautern; thanks are due to Professor H.-G. Kuball for his outstanding hospitality. We are much indebted to our respective families for patient support and understanding during what must have seemed to be interminable hours, days and weeks spent with the manuscript. Last but not least, we wish to acknowledge the many years of generous support for our work in photochemistry that has been provided by the Deutsche Forschungsgemeinschaft and the U.S. National Science Foundation.

Many fine books on organic excited states, photophysics, and photochemistry are already available. Ours attempts to offer a different perspective by placing primary emphasis on qualitative theoretical concepts in a way that we hope will be useful to students of physical organic chemistry.

Münster M. K.
Boulder J. M.
March 1995

Acknowledgments

The authors wish to thank the following for permission to use their figures in this book.

Academic Press, Orlando (USA)
Figures 2.9, 2.10, 7.4, 7.5 and 7.53
Academic Press, London (UK)
Figure 7.15
American Chemical Society, Washington (USA)
Figures 2.11, 2.37, 3.7, 3.9, 3.10, 3.13, 3.17, 3.21, 4.8, 5.14, 5.17, 5.34, 5.39, 5.40, 6.1, 6.16, 6.17, 6.21, 6.27, 7.2, 7.18, 7.24, 7.34, 7.38, 7.39, 7.42 and 7.43
American Institut of Physics, New York (USA)
Figures 2.6, 2.17 and 2.18
The Benjamin/Cummings Publishing Company, Menlo Park (USA)
Figures 1.11, 5.4, 5.11, 7.19 and 7.57
Bunsengesellschaft für Physikalische Chemie, Darmstadt (G)
Figures 1.15 and 1.17
Elsevier Science Publishers B.V., Amsterdam (NL)
Figures 1.20, 5.32, 5.33, 7.3, 7.6, and 7.13
Gordon and Breach Science Publishers, Yverdon (CH)
Figure 6.8
Hevetica Chimica Acta, Basel (CH)
Figures 2.35 and 7.50
International Union of Pure and Applied Chemistry, Oxford (UK)
Figures 2.30, 3.15, 3.16, 5.24 and 5.25
Kluwer Academic Publishers, Dordrecht (NL)
Figures 1.25, 4.12, 4.13, 7.20 and 7.21
R. Oldenbourg Verlag GmbH, München (G)
Figure 5.38
Pergamon Press, Oxford (UK)
Figures 2.15, 2.29, 3.11, 3.14, 3.18, 3.19, 4.27, 4.28, 5.15, 5.16, 5.30 and 5.36
Plenum Publishing Corp, New York (USA)
Figure 5.10
Royal Society of Chemistry, Cambridge (UK)
Figure 2.28

The Royal Society, London (UK)
Figure 1.14
Springer-Verlag, Heidelberg (G)
Figures 1.23, 1.24, 6.3, 6.20 and 7.8
VCH Publishers, Inc., New York (USA)
Figure 1.16
VCH Verlagsgesellschaft mbH, Weinheim (G)
Figures 1.8, 2.7, 2.25, 2.27, 2.34, 2.38, 2.42, 3.3, 3.6, 4.21, 5.9, 5.18, 5.19, 5.20, 6.5, 6.9, 6.13, 6.23, 6.25, 7.28, 7.33 and 7.51
Weizmann Science Press of Israel, Jerusalem
Figure 7.22
John Wiley & Sons, Inc., New York (USA)
Figures 1.3, 2.2, 2.3, 2.45, 4.5, 4.6, 4.10, 4.11, 4.16, 4.20, 4.22, 4.23, 4.24, 6.19 and 6.28
John Wiley & Sons, Ltd., West Sussex (UK)
Figures 7.12 and 7.14

Contents

7. Organic Photochemistry 361

Notation

Operators

One-electron	$\hat{a}$
Many-electron	$\hat{A}$

Vectors $\boldsymbol{a}, \boldsymbol{A}$

Matrices $\mathbf{A}$

Wave functions

Electronic configuration	$\Phi_K(1, \ldots, n)$
Electronic state	$\Psi_j = \sum_K C_{Kj}\Phi_K$
Nuclear	$\chi^j_v(\boldsymbol{Q})$
Vibronic	$\Psi^{\mathrm{T}} \equiv \Psi^{\mathrm{T}}_{j,v}(\boldsymbol{q},\boldsymbol{Q}) = \Psi^{\mathrm{Q}}_j(\boldsymbol{q})\chi^j_v(\boldsymbol{Q})$

Orbitals

General	$\varphi_k(i)$
Atomic	$\chi_\mu(i)$
Molecular	$\phi_k(i)$
Spin orbital	$\psi_k(i)$ $[\phi_k(i)\alpha(i) \equiv \phi_k,\ \phi_k(i)\beta(i) \equiv \bar{\phi}_k]$

Universal Constants

$c_0 = 2.9979 \times 10^{10}$ cm/s	speed of light in vacuum
$e = -1.6022 \times 10^{-19}$ C	electron charge

$h = 6.626 \times 10^{-27}$ erg s	Planck constant
$\hbar = h/2\pi = 1.0546 \times 10^{-27}$ erg s	
$m_e = 9.109 \times 10^{-28}$ g	electron rest mass
$N_L = 6.022 \times 10^{23}$ mol^{-1}	Avogadro constant
$k = 1.3805 \times 10^{-16}$ erg K^{-1}	Boltzmann constant

Throughout this book, we use energy units common among U.S. chemists. Their relation to SI units is as follows:

$$1 \text{ cal} = 4.194 \text{ J}$$
$$1 \text{ eV} = 1.602 \times 10^{-19} \text{ J}$$
$$1 \text{ erg} = 10^{-17} \text{ J}$$

Frequently Used Symbols
(Section of first appearance or definition is given in parentheses)

$a = -\langle\phi_N\|\hat{h}'\|\phi_{-N}\rangle$	interaction matrix element between perimeter MOs ϕ_N and ϕ_{-N} (2.2.3)
$A = \log(I_0/I) = \varepsilon cd$	absorbance or optical density (1.1.2)
A_i, B_i, C_i	A, B, and C term of the i-th transition of the MCD spectrum (3.3.1)
$\boldsymbol{A}(\boldsymbol{r}_j, t)$	vector potential (1.3.1)
$b = \langle\phi_{-N-1}\|\hat{h}'\|\phi_{N+1}\rangle$	interaction matrix element between perimeter MOs ϕ_{-N-1} and ϕ_{N+1} (2.2.3)
^{1,3}B	dot–dot states of biradicals (4.3.1)
$\boldsymbol{B}(\boldsymbol{r}, t)$	magnetic flux density (1.3.1)
$\boldsymbol{B}(x)$	magnetic field vector (1.1.1)
B_a, B_b	excited states of ($4N + 2$)-electron perimeter (2.2.1)
B_1, B_2	excited states of ($4N + 2$)-electron perimeter (2.2.3)
c	concentration (1.1.2)
$c_{\mu i}$	LCAO coefficient of AO χ_μ in MO ϕ_i (1.2.2)
d	path length (1.1.2)
D	doubly excited state of the 3×3 model (4.4.1) and of $4N$-electron perimeter (2.2.7)
D_i	dipole strength of transition i (3.3.1)
$\boldsymbol{e}_U$	unit vector in direction U (1.3.1)
E, E_i	energy (1.1.2)
$\boldsymbol{E}$, $\boldsymbol{E}(\boldsymbol{r}, t)$	electric field vector (1.1.1)
EA_k	electron affinity (1.2.3)
$\Delta E_{i\to k}$	excitation energy (1.2.3)
f	oscillator strength (1.3.1)

F	fluorescence (5.1.1)
F	Fock matrix (1.5.1)
$\boldsymbol{F}$	reaction field (2.7.2)
$F_{\mu\nu} = <\chi_\mu\|\hat{F}\|\chi_\nu>$	matrix element of the Fock operator between AOs χ_μ and χ_ν (1.5.1)
$\hat{g}(i, j)$	electron repulsion operator (1.2.3)
G	ground state of the 3 × 3 model (4.4.1) and of 4*N*-electron perimeter (2.2.7)
ΔG, ΔG^0, $\Delta G^\ddagger$	free energy (1.4.3)
$\hat{h}(j)$	one-electron operator of kinetic and potential energy (1.2.3)
$\hat{H}(1, \ldots, n)$	Hamiltonian (1.2.3)
$H_{KL} = <\Phi_K\|\hat{H}\|\Phi_L>$	matrix element of the Hamiltonian between configurational functions Φ_K and Φ_L (1.2.4)
$\hat{H}_{el}^{(Q)}(\boldsymbol{q})$	electronic Hamiltonian (1.2.1)
$\hat{H}_{vib}(j, \boldsymbol{Q})$	nuclear Hamiltonian (1.2.1)
$\hat{H}_{SO}$	spin-orbit coupling operator (1.3.2)
$\boldsymbol{H}^{SO}$	spin-orbit coupling vector (4.3.4)
ΔH, ΔH^0	enthalpy (1.4.3)
ΔHL, ΔHSL	parameters for MCD spectra of systems derived from 4*N*-electron perimeter (3.3.4)
ΔHOMO	energy splitting of the pair of highest occupied perimeter MOs (2.2.3)
$I = I_0 e^{-\alpha d}$	intensity (1.1.2)
IC	internal conversion (5.1.1)
ISC	intersystem crossing (5.1.1)
IP_i	ionization potential (1.2.3)
J_{ik}	Coulomb integral (1.2.3)
k_j	rate constant of process *j* (5.1.2)
$\boldsymbol{K}$	wave vector (1.3.1)
K_{ik}	exchange integral (1.2.3)
K'_{ab}, K'_{12}	electron repulsion integrals in biradicals and biradicaloids (4.3.2)
$l = \pm k$	angular momentum quantum number (1.3.2)
$\hat{\boldsymbol{l}}_j^\mu = \boldsymbol{r}_j^\mu \times \hat{\boldsymbol{p}}_j$	orbital angular momentum operator, origin of coordinates at atom μ (1.3.2)
L_a, L_b	lowest excited states of (4*N* + 2)-electron perimeter (2.2.1)
L_1, L_2	lowest excited states of (4*N* + 2)-electron perimeter (2.2.3)
ΔLUMO	energy splitting of the pair of lowest unoccupied perimeter MOs (2.2.3)

m_j	mass of particle j (1.3.1)
$\hat{\boldsymbol{m}}_j$	one-electron electric dipole operator (1.3.1)
$\hat{\mathbf{m}}_j$	one-electron magnetic dipole operator (3.2.2)
$\hat{\boldsymbol{M}}$	electric dipole operator (1.1.2)
$\hat{\mathbf{M}}$	magnetic dipole operator (1.1.2)
M_S	z component of total spin (4.2.2)
$\bar{\boldsymbol{M}}_{0v\to fv'} = <\Psi^{T}_{fv'}\|\hat{\boldsymbol{M}}\|\Psi^{T}_{0v}> = \boldsymbol{M}_{0\to f} <\chi^{f}_{v'} \mid \chi^{0}_{v}>$	vibrionic transition moment (1.3.1)
$\boldsymbol{M}_{0\to f} = <\Psi^{Q}_{f}\|\hat{\boldsymbol{M}}\|\Psi^{Q}_{0}>$	electronic transition moment (1.3.1)
$\boldsymbol{M}_{i\to k} = \sqrt{2} <\phi_k\|\hat{\boldsymbol{m}}\|\phi_i>$	configurational electronic transition moment (1.3.1)
n	number of electrons (1.1.2)
n	number of atoms in the perimeter (2.2.2)
n	refractive index (1.1.1)
$n_L - n_R$	circular birefringence (3.1.2)
n_i	occupation number of MO ϕ_i (1.2.3)
N	number of π electrons in the perimeter is $4N + 2$ or $4N$ (2.2.1)
N_1, N_2, P_1, P_2	excited states of $4N$-electron perimeter (2.2.7)
P	phosphorescence (5.1.1)
P	degree of polarization (5.3.3)
$\hat{\boldsymbol{p}}_j = -i\hbar\nabla_j$	linear momentum operator of particle j (1.3.1)
q_j	charge of particle j (1.3.1)
$\boldsymbol{q}, \boldsymbol{q}_j$	electronic coordinates (1.2.1)
$\hat{\mathbf{Q}}$	electric quadrupole operator (1.1.2)
$\boldsymbol{Q}, \boldsymbol{Q}_k$	nuclear coordinates (1.2.1)
Q	quencher (5.4.1)
$\boldsymbol{r}_j$	position vector of electron j (1.3.1)
$\boldsymbol{r}^{\mu}_j = \boldsymbol{r}_j - \boldsymbol{R}_\mu$	vector pointing from nucleus μ to electron j (1.3.1)
R	degree of anisotropy (5.3.3)
$\boldsymbol{R}_\mu, \boldsymbol{R}_A$	position vector of nucleus μ or A (1.3.1)
$\mathcal{R}$	rotational strength (3.2.2)
$\|s\| = \|<\phi_N\|\hat{h}'\|\phi_{-N}>\|$	interaction matrix element between MOs ϕ_N and ϕ_{-N} of a $4N$-electron perimeter (2.2.7)
$\hat{\boldsymbol{s}}_j$	spin angular momentum operator (1.3.2)
S	singly excited state of the 3×3 model (4.4.1) and of $4N$-electron perimeter (2.2.7)

$S_0, S_1, S_2, \ldots$	singlet states (1.4.3)
$\mathbf{S}_{0\rightarrow f}$	two-photon transition tensor (1.3.6)
ST	singlet-triplet intersystem crossing (5.1.2)
t	time (1.1.1)
T	temperature (1.1.2)
T	triplet state of the 3 × 3 model (4.4.1)
$T_1, T_2, \ldots$	triplet states (1.4.3)
TS	triplet-singlet intersystem crossing (5.1.2)
U, V, X, Y	real (4N + 2)-electron perimeter configurations (2.2.3)
$\boldsymbol{U}_{i\rightarrow k} = \langle\Psi_k\vert\hat{\boldsymbol{U}}\vert\Psi_i\rangle$	transition moment (1.1.2)
$U_{\mu\mu}$	one-center valence-state energy (1.5.1)
$\boldsymbol{x}_1, \boldsymbol{x}_2$	gradient difference and nonadiabatic coupling vectors that define the branching space at a conical intersection (4.1.2)
Z	partition function (1.1.2)
Z_A, Z_μ	charge of nucleus A or μ (1.3.1)
Z_1, Z_2	hole–pair states of biradical (4.3.1)
$\alpha\ [\alpha]_M$	rotation angle, molar rotation (3.1.2)
$\alpha\ (\tilde{\nu})$	absorption coefficient (1.1.2)
$\delta\alpha_\mu$	perturbation of Coulomb integral α_μ (2.4.1)
α, β	phase factor in complex interaction matrix element a or b (2.2.3)
$\beta_{\mu\nu}$	resonance integral (1.5.1)
$\delta\beta_{\mu\nu}$	perturbation of resonance integral (2.4.1)
γ	covalent perturbation in homosymmetric and nonsymmetric biradicaloids (4.3.3)
$\gamma_{\mu\nu}$	Coulomb repulsion integral (1.5.1)
δ	polarization perturbation in heterosymmetric and nonsymmetric biradicaloids (4.3.3)
δ_0	polarization perturbation in critically heterosymmetric biradicaloids (4.3.3)
$\delta_{\uparrow\uparrow}, \delta_{\circlearrowright\circlearrowright}$	two-photon absorption cross sections (1.3.6)
$\varepsilon = \varepsilon\ (\tilde{\nu})$	decadic molar extinction coefficient (1.1.2)
$\Delta\varepsilon = \varepsilon_L - \varepsilon_R$	circular dichroism (3.1.2)
ε_i	orbital energy of MO ϕ_i (1.2.3)
ζ_μ	atomic spin-orbit coupling parameter (4.3.4)
η_j	efficiency (5.1.3)
$\theta, [\theta]_M$	ellipticity, molar ellipticity (3.1.2)
$\Theta_x, \Theta_y, \Theta_z$	triplet spin functions (4.3.4)

λ	reorganization energy (5.4.4)
$\lambda = c/\nu$	wavelength (1.1.1)
$\boldsymbol{\mu}$	dipole moment vector (2.7.2)
ν	frequency (1.1.1)
$\tilde{\nu} = 1/\lambda$	wave number (1.1.1)
ϱ_E	density of states (5.2.3)
σ	phase factor in complex interaction matrix element of 4*N*-electron perimeter (2.2.7)
Σ	singlet spin function (4.3.4)
τ	lifetime (5.1.2)
$\phi_i = \sum_{\mu} c_{\mu i}\chi_\mu$	molecular orbital (MO) (1.2.2)
ϕ_a and ϕ_s	highest occupied real perimeter MOs (2.2.5)
$\phi_{a'}$ and $\phi_{s'}$	lowest unoccupied real perimeter MOs (2.2.5)
ϕ_i and $\phi_{i'}$	paired MOs of alternant hydrocarbon (1.2.4)
ϕ_k and ϕ_{-k}	complex perimeter MOs (2.2.2)
Φ_0	ground configuration (1.2.2)
$\Phi_{i \to k}$	singly excited configuration (1.2.2)
$\Phi_K(1, \ldots, n)$	configurational wave function (1.2.2)
Φ_j	quantum yield (5.1.3)
χ	character of irreducible representation (1.2.4)
χ_μ	atomic orbital (AO) (1.2.2)
$\chi_v^j(\boldsymbol{Q})$	vibrational wave function (1.2.1)
$\lvert(<\chi_{v'}^{\mathrm{f}} \lvert \chi_v^0>)\rvert^2$	Franck-Condon factor (1.3.3)
$\psi_k(j)$	spin orbital (1.2.2)
$\Psi_j^{\varrho}(\boldsymbol{q})$	electronic wave function (1.2.1)
Ψ_0, Ψ_{f}	wave functions of ground and final state (1.2.2)
$\Psi_j^{\varrho} = \sum_K C_{Kj}\Phi_K$	CI (configuration interaction) wave function (1.2.2)
$\Psi^{\mathrm{T}} = \Psi_{j,v}^{\mathrm{T}}(\boldsymbol{q}, \boldsymbol{Q}) = \Psi_j^{\varrho}(\boldsymbol{q})\chi_v^j(\boldsymbol{Q})$	Born-Oppenheimer wave function (1.2.1)
$\Omega = \delta_{\circlearrowright\circlearrowleft}/\delta_{\uparrow\uparrow}$	polarization degree for two-photon absorption (1.3.6)

CHAPTER

1

Spectroscopy in the Visible and UV Regions

1.1 Introduction and Theoretical Background

1.1.1 Electromagnetic Radiation

Ultraviolet (UV) and visible (VIS) light constitute a small region of the electromagnetic spectrum which also comprises infrared (IR) radiation, radio waves, X-rays, etc. A diagrammatic representation of the electromagnetic spectrum is shown in Figure 1.1. Electromagnetic radiation can be envisaged in terms of an oscillating electric field and an oscillating magnetic field that are perpendicular to each other and to the direction of propagation. The case of linearly polarized light where the planes of both the electric and the magnetic field are fixed is shown in Figure 1.2.

In vacuum, the electric vector of a linearly polarized electromagnetic wave at any point in space is given by

$$\boldsymbol{E}(t) = \boldsymbol{E}_0 \sin (2\pi\nu t + \theta) \tag{1.1}$$

where $\boldsymbol{E}_0$ is a constant vector in a plane perpendicular to x, the direction of propagation of the light; $(2\pi\nu t + \theta)$ is the phase at time t; θ is the phase at time $t = 0$; and ν is the frequency in Hz. The direction of $\boldsymbol{E}$ is referred to as the polarization direction of the light. As a function of position along the x axis, the electric vector $\boldsymbol{E}(x)$ and the magnetic vector $\boldsymbol{B}(x)$ are given by

$$\begin{aligned} \boldsymbol{E}(x,t) &= \boldsymbol{E}_0 \sin [2\pi(\nu t - x/\lambda) + \theta] \\ \boldsymbol{B}(x,t) &= \boldsymbol{B}_0 \sin [2\pi(\nu t - x/\lambda) + \theta] \end{aligned} \tag{1.2}$$

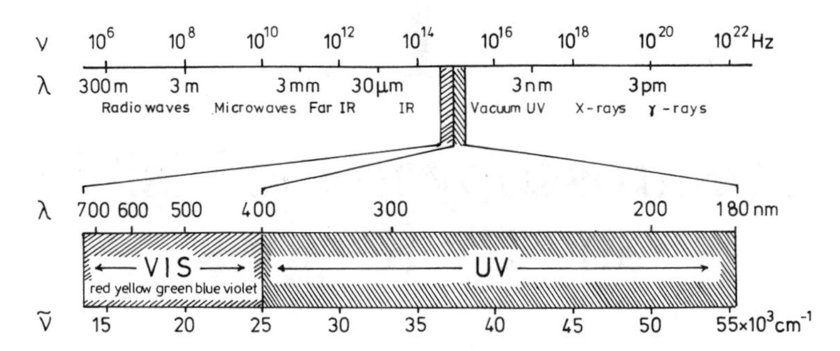

Figure 1.1. Frequencies ν and wavelengths λ for various regions of the electromagnetic spectrum. In the UV/VIS region, which is of special interest in this book, nm is the commonly used unit of wavelength. Wave numbers $\tilde{\nu}$, which are proportional to frequencies, are expressed in cm^{-1}.

$\boldsymbol{B}_0$ is a constant vector perpendicular to $\boldsymbol{E}_0$ (Figure 1.2),

$$\lambda = c/\nu \tag{1.3}$$

is the wavelength, and c is the speed of light, whereas θ is the phase for $x = 0$, $t = 0$. In vacuum, $c = c_0 = 2.9979 \times 10^{10}$ cm/s, and in a medium of refractive index n, $c = c_0/n$. If light passes from one medium to another, the frequency ν remains constant, whereas the wavelength λ changes according to Equation (1.3) with the speed of light.

If the polarization directions of two linearly polarized light waves, 1 and 2, with identical amplitudes $|\boldsymbol{E}_0^{(1)}| = |\boldsymbol{E}_0^{(2)}|$, frequencies $\nu_1 = \nu_2$, and directions of propagation x, are mutually orthogonal, and if the phases of the two waves are identical, $\theta_1 = \theta_2 = \theta$, their superposition will produce a new linearly polarized wave

$$\boldsymbol{E}(x) = (\boldsymbol{E}_0^{(1)} + \boldsymbol{E}_0^{(2)}) \sin [2\pi(\nu t - x/\lambda) + \theta] \tag{1.4}$$

The amplitude of the new wave is $\sqrt{2}$ times larger than that of either of the original waves, and its direction of polarization forms an angle of 45° with the polarization directions of either of the two waves. If the phases of wave 1 and wave 2 differ by $\pi/2$, which according to Equation (1.2) is equivalent to a difference in optical path lengths of $x = \lambda/4$, the superposition of the two waves results in a circularly polarized wave

$$\begin{aligned} \boldsymbol{E}(x) = {} & \boldsymbol{E}_0^{(1)} \sin [2\pi(\nu t - x/\lambda) + \theta] \\ & + \boldsymbol{E}_0^{(2)} \cos [2\pi(\nu t - x/\lambda) + \theta] \end{aligned} \tag{1.5}$$

that has a constant amplitude $|\boldsymbol{E}_0^{(1)}| = |\boldsymbol{E}_0^{(2)}|$. The direction of its electric vector at a given point x_0 rotates with frequency ν about the x axis. The general result of a superposition of linearly polarized waves with phases that do not

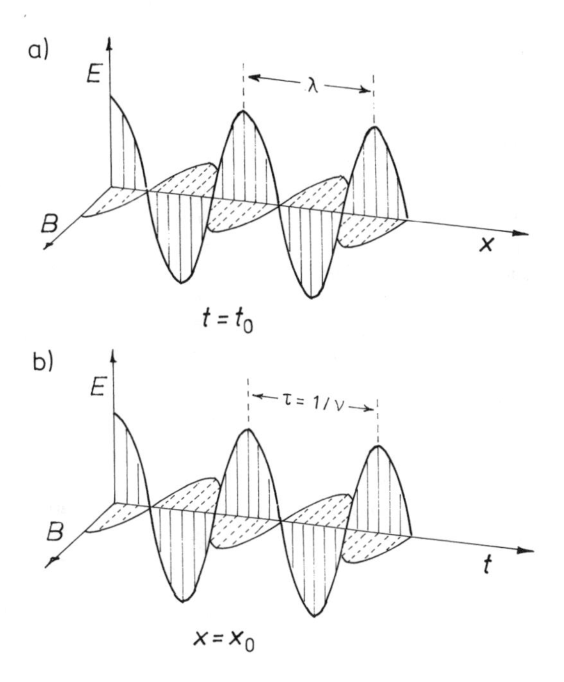

Figure 1.2. The variation of the electric field (***E***) and the magnetic flux (***B***) of linearly polarized light of wavelength λ a) in space (at time $t = t_0$) and b) in time (at point $x = x_0$). The vectors ***E*** and ***B*** and the wave vector ***K***, whose direction coincides with the propagation direction x, are mutually orthogonal.

differ by an integral multiple of $\pi/2$ is a wave of elliptical polarization. (Cf. Chapter 3.)

In changing from the classical to the quantum mechanical description of light, one of the principal results is that light is emitted or absorbed in discrete *quanta* known as *photons*, with an energy of

$$E = h\nu = hc\tilde{\nu} \tag{1.6}$$

where h is Planck's constant, $h = 6.626 \times 10^{-27}$ erg s, and $\tilde{\nu}$ is the wave number defined as

$$\tilde{\nu} = 1/\lambda \tag{1.7}$$

The common wave-number unit is cm^{-1}. Since $\tilde{\nu}$ is linearly related to the energy according to Equation (1.6), in spectroscopy "energies" are frequently expressed in wave numbers; that is, E/hc is used instead of E. Table 1.1 shows the numerical relationship between wavelengths, wave numbers, and energies for the visible and the adjacent regions of the spectrum; the values in the last columns have been converted into molar energies by multiplication with Avogadro's number. (Cf. Example 1.1.)

Table 1.1 Conversion of Wavelength λ, Wave Number $\tilde{\nu}$, and Energy ΔE[a]

Spectral Region	λ (nm)	$\tilde{\nu}$ (cm^{-1})	ΔE eV	ΔE kJ/mol	ΔE kcal/mol
UV	200	50,000	6.20	598	142.9
	250	40,000	4.96	479	114.3
	300	33,333	4.13	399	95.2
	350	28,571	3.54	342	81.6
VIS	400	25,000	3.10	299	71.4
	450	22,222	2.76	266	63.5
	500	20,000	2.48	239	57.1
	550	18,182	2.25	218	51.9
	600	16,666	2.07	199	47.6
	650	15,385	1.91	184	44.0
	700	14,826	1.77	171	40.8
IR	800	12,500	1.55	150	35.7
	1,000	10,000	1.24	120	28.6
	5,000	2,000	0.25	24	5.7

[a] Conversion factors: 1 eV = 8,066 cm^{-1} = 96.485 kJ/mol = 23.045 kcal/mol; 1 kcal/mol = 4.1868 kJ/mol.

Example 1.1:
From Equation (1.6)

$$\Delta E = h\nu = hc\tilde{\nu} = (6.626 \times 10^{-27}) \times (2.998 \times 10^{10}) \times \tilde{\nu}$$

which yields ΔE in erg if $\tilde{\nu}$ is given in cm^{-1}; after multiplication with Avogadro's number $N_L = 6.022 \times 10^{23}$ mol^{-1} and taking into account the appropriate conversion factor (1.4383×10^{13}), the molar energy in kcal/mol is found to be

$$\Delta E = 0.0029\ \tilde{\nu}$$

An absorption at 50 cm^{-1}, 1,500 cm^{-1}, or 33,333 cm^{-1} therefore corresponds to an energy gain of 0.14 kcal/mol, 4.3 kcal/mol, and 95 kcal/mol, respectively. The average bond energy of a C—C bond is roughly 85 kcal/mol; that is, the energies corresponding to the absorption of visible light are of the same order of magnitude as bond energies. They transfer the molecule from the ground state into an electronically excited state. On the other hand, the amount of energy corresponding to an absorption in the IR region is considerably smaller and is in the range of energy required to excite molecular vibrations.

The intensity of radiation is measured in erg s^{-1} cm^{-2} as the energy of radiation falling on unit area of the system in unit time; this energy is related directly through Planck's constant to the number of quanta and their associated frequency. On the other hand, in the classical description of light as an electromagnetic wave the intensity is proportional to the squared amplitude of the electric field vector (or the magnetic field vector).

1.1.2 Light Absorption

Light is absorbed if the molecule accepts energy from the electromagnetic field, and spontaneous or stimulated emission occurs if it provides energy to the field. The latter is the basis of laser action but will not be treated here.

A molecule in a state i of energy E_i can change into a state k of energy E_k by absorption of light of frequency ν, if the relation

$$E_i - E_k = h\nu \tag{1.8}$$

is fulfilled. A photon can be absorbed only if its energy corresponds to the difference in energy between two stationary states of the molecule.

Absorption occurs only if the light can interact with a transient molecular charge or current distribution characterized by the quantity

$$U_{i \to k} = \langle \Psi_k | \hat{U} | \Psi_i \rangle \tag{1.9}$$

referred to as the transition moment between molecular states i and k, described by the wave functions Ψ_i and Ψ_k, respectively.*

$\hat{U}$ is an operator that corresponds either to the electric dipole moment ($\hat{M}$), to the magnetic dipole moment ($\hat{\boldsymbol{M}}$), or to the electric quadrupole moment ($\hat{\boldsymbol{Q}}$). Accordingly electric dipole transitions, magnetic dipole transitions, and electric quadrupole transitions are distinguished. (Higher-order transitions can normally be ignored.) For an allowed transition in the visible region the transition moments of the electric and the magnetic dipole operator and of the electric quadrupole operator are roughly in the ratio $10^7 : 10^2 : 1$. It is therefore quite common to confine the discussion to electric dipole transitions. However, magnetic dipole transition moments cannot be neglected in magnetic resonance spectroscopy and in the treatment of optical activity. (Cf. Chapter 3.)

Since the electric dipole moment operator is a vector operator, the electric dipole transition moment will also be a vector quantity. The probability of an electric dipole transition is given by the square of the scalar product between the transition moment vector in the molecule and the electric field vector of the light, and is therefore proportional to the squared cosine of the angle between these two vectors. Thus, an orientational dependence results for the absorption and emission of linearly polarized light. The orientation of the transition moment with respect to the molecular system of axes is

---

* The bracket notation $\langle \Psi_i | \hat{U} | \Psi_k \rangle$ introduced by Dirac for the matrix element $\int \Psi_i^* (1, \ldots, n) \hat{U}(1, \ldots, n) \Psi_k(1, \ldots, n) d\tau_1 \ldots d\tau_n$ of the operator $\hat{U}$ is advantageous, particularly if the integration is not carried out explicitly.

frequently called the *absolute polarization direction,* whereas the *relative polarization direction* of two distinct transitions refers to the angle between their two transition moments.

Samples used in spectroscopic measurements usually consist of a very large number of molecules. According to *Boltzmann's law,* in thermal equilibrium at temperature T the number N_j of molecules in a state of energy E_j is given by

$$N_j = (N_0/Z)\, e^{-E_j/kT} \tag{1.10}$$

where N_0 is the total number of molecules, k is the Boltzmann constant, and Z is the partition function, that is, the sum over the Boltzmann factors $e^{-E_j/kT}$ for all possible quantum states of the molecule. The average number N_j of molecules in a state of energy E_j is thus larger than the number N_l of molecules in a state of higher energy E_l. Since absorption and stimulated emission are intrinsically equally probable, more molecules are raised from the lower state j into the higher state l than the reverse. This perturbs the thermal equilibrium distribution, but due to interactions with the environment, transitions to other energy states are possible and the equilibrium distribution can be restored. For excitations in the UV/VIS region the return to equilibrium, which is referred to as *relaxation,* is so fast that at ordinary light intensities the thermal equilibrium in the irradiated sample is hardly perturbed at all. Saturation, which corresponds to identical populations in the ground and the excited state and inhibits further absorption of energy, a well-known phenomenon in NMR spectroscopy, is therefore hard to achieve in optical spectroscopy except temporarily in the case of laser excitation.

Example 1.2:
From Equation (1.10) the ratio N_l/N_j of the number of molecules in two states of energy E_l and E_j, respectively, is given by

$$N_l/N_j = e^{-(E_l - E_j)/kT}$$

If wave numbers are used the energies have to be replaced by $\tilde{E}_i = E_i/hc$. At room temperature, given $k = 1.3805 \times 10^{-16}$ erg K^{-1} and the conversion factor 1 erg/hc = 5.034×10^{15} cm^{-1}, we have $kT \approx 200$ cm^{-1}. Thus, for an energy difference $\tilde{E}_l - \tilde{E}_j = 50$ cm^{-1}

$$N_l/N_j = e^{-50/200} = 0.78$$

or

$$N_l = 0.78\, N_j$$

With $N_j + N_l = 100\%$, we obtain $N_j = 56\%$ and $N_l = 44\%$. In the same way, for $\tilde{E}_l - \tilde{E}_j = 1{,}500$ cm^{-1} and for $\tilde{E}_l - \tilde{E}_j = 33{,}333$ cm^{-1},

$$N_l/N_j = e^{-7.5} = 0.006$$

or

$$N_j > 99.9\%$$

and

$$N_l < 0.1\%$$

and

$$N_l/N_j = e^{-167} = 4 \times 10^{-73}$$

or

$$N_j = 100\%$$

In the case of molecular vibrations with excitation energies of about 50 cm^{-1}, exemplified by restricted rotations about single bonds, nearly one-half of all molecules reside in the energetically higher state. For excitation energies of 1,500 cm^{-1}, which are typical for a stretching vibration, there are less than 0.1% of the molecules in the upper state. Finally, for electronically excited states the population at room temperature is so small that it can be ignored, and practically all the molecules are in the ground state.

Emission in the UV/VIS region is observed at room temperature only if the equilibrium of the molecular system with its surroundings has been disturbed by external effects, for example, by radiation, heat, collision with an electron, or chemical reaction.

When a collimated monochromatic light beam of intensity I passes through an absorbing homogeneous isotropic sample, it is attenuated. The loss of intensity dI is proportional to the incident intensity I and to the thickness dx of the absorbing material, that is,

$$\mathrm{d}I = -\alpha I \mathrm{d}x$$

where $\alpha = \alpha(\tilde{\nu})$ is an absorption coefficient characteristic of the absorbing medium and dependent on the wave number of the light. It is proportional to the difference $(N_j - N_l)$ of the number of molecules in the ground state and the excited state. With $I = I_0$ for $x = 0$, integration over the thickness d of the sample yields

$$\int_{I_0}^{I} \mathrm{d}I/I = -\int_{0}^{d} \alpha \mathrm{d}x$$

or

$$I = I_0 e^{-\alpha d}$$

Setting $\alpha = 2.303\,\varepsilon c$, where c is the concentration of the absorbing species, yields the *Lambert-Beer law*:

$$A = \log(I_0/I) = \varepsilon c d \tag{1.11}$$

The dimensionless quantity A is called the *absorbance* or *optical density* of the sample. The concentration c is traditionally given in mol/L and the thickness of the sample d in cm. $\varepsilon = \varepsilon(\tilde{\nu})$ is then the decadic molar extinction coefficient; its unit is L mol^{-1} cm^{-1} and is understood but not explicitly stated on spectra.

In general the Lambert-Beer law is obeyed quite well. Exceptions can be attributed, for example, to interactions between the solute molecules (changes of the composition of the system with concentration), to perturbations of the thermal equilibrium by very intense radiation, or to the population of a very long-lived state.

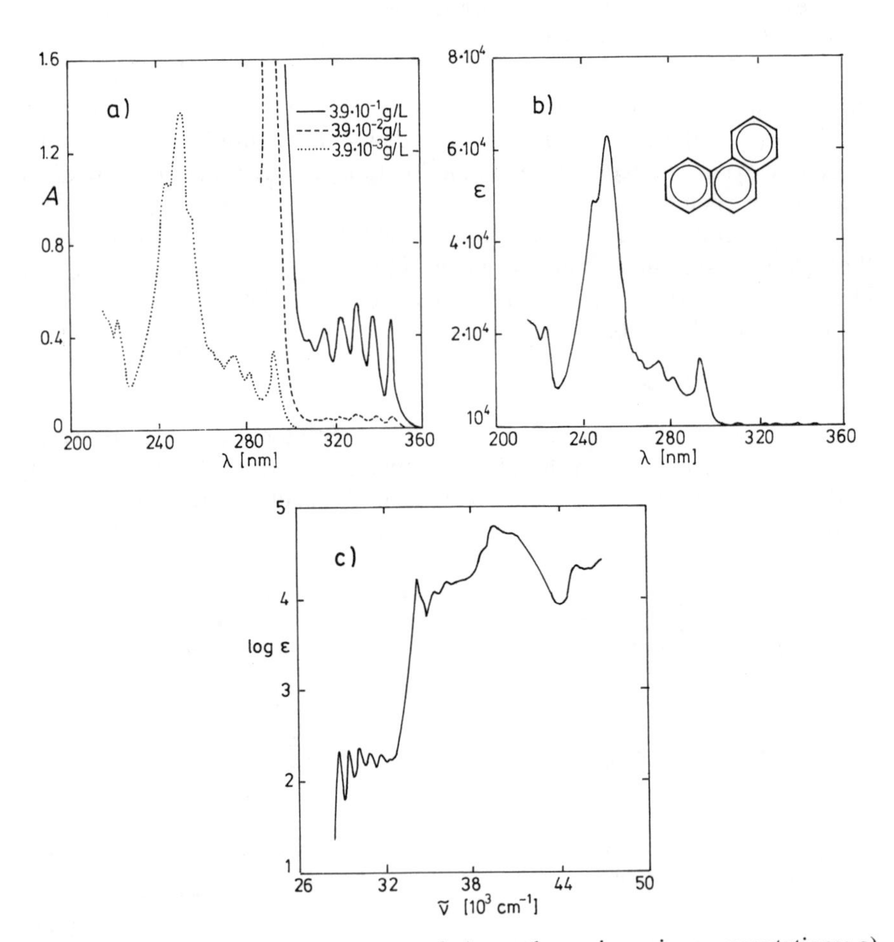

Figure 1.3. The absorption spectrum of phenanthrene in various presentations: a) absorbance A versus λ, b) ε versus λ, and c) log ε versus $\tilde{\nu}$ (by permission from Jaffé and Orchin, 1962).

The example of the UV spectrum of phenanthrene (**1**) in Figure 1.3 shows that the molar extinction coefficient $\varepsilon = \varepsilon(\lambda)$ or $\varepsilon = \varepsilon(\tilde{\nu})$, expressed as a function of wavelength or wave numbers varies over several orders of magnitude. It is therefore common to sacrifice some detail and to use a logarithmic scale or to plot log ε versus $\tilde{\nu}$, as shown in Figure 1.3c. Spectra are usually measured down to 200 nm. (Most often solutions of a concentration of about 10^{-4} mol/L are used in cells of 1-cm thickness.) The region of shorter wavelengths, which is sometimes referred to as the far-UV region, is experimentally less accessible, because solvents and even air tend to absorb strongly. The term vacuum UV is applied to the region below 180 nm since an evacuated spectrometer is required.

1.2 MO Models of Electronic Excitation

1.2.1 Energy Levels and Molecular Spectra

Absorption spectra of atoms consist of sharp lines, whereas absorption spectra of molecules show broad bands in the UV/VIS region. These may exhibit some vibrational structure, particularly in the case of rigid molecules (Figure 1.4b). Rotational fine structure can be observed only with very high resolution in the gas phase and will not be considered here. (See, however, Section 1.3.6.) Polyatomic molecules possessing a large number of normal vibrational modes of varying frequencies have very closely spaced energy levels. As a result of line broadening due to the inhomogeneity of the interactions

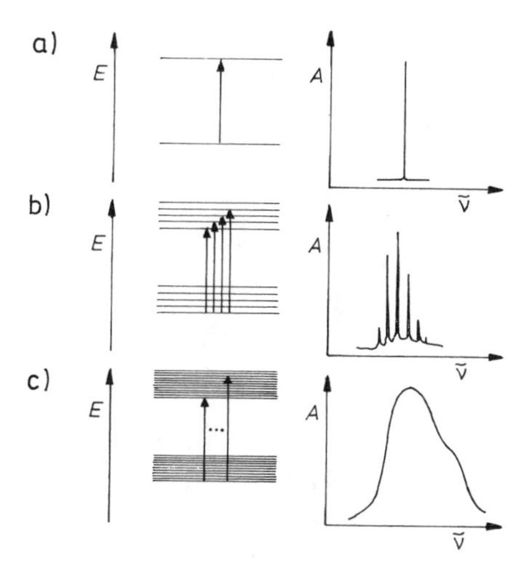

Figure 1.4. Atomic and molecular spectra: a) sharp-line absorption typical for isolated atoms in the gas phase, b) absorption band with vibrational structure typical for small or rigid molecules, and c) structureless broad absorption typical for large molecules in solution (adapted from Turro, 1978).

between solute molecules and solvent, to hindered rotations, and to the short lifetimes of the higher excited states, the vibrational structure may be either unresolved or only partly resolved, so that in general only broad unstructured bands can be observed in condensed phase (Figure 1.4c).

The vibrational structure may be explained as follows: For each state of a molecule there is a wave function that depends on time, as well as on the internal space and spin coordinates of all electrons and all nuclei, assuming that the overall translational and rotational motions of the molecule have been separated from internal motion. A set of stationary states exists whose observable properties, such as energy, charge density, etc., do not change in time. These states may be described by the time-independent part of their wave functions alone. Their wave functions are the solutions of the time-independent Schrödinger equation and depend only on the internal coordinates $\boldsymbol{q} = q_1, q_2, \ldots$ of all electrons and the internal coordinates $\boldsymbol{Q} = Q_1, Q_2, \ldots$ of all nuclei.

Within the Born-Oppenheimer approximation (cf. McWeeny, 1989: Section 1.1) the total wave function Ψ^{T} of a stationary state is written as

$$\Psi^{\mathrm{T}}_{j,v}(\boldsymbol{q},\boldsymbol{Q}) = \Psi^{Q}_{j}(\boldsymbol{q})\chi^{j}_{v}(\boldsymbol{Q}) \tag{1.12}$$

where j characterizes the electronic state and v the vibrational sublevel of that state. (Cf. Figure 1.5.) The electronic wave function $\Psi^{Q}_{j}(\boldsymbol{q})$ is an eigenfunction of the electronic Hamiltonian $\hat{H}^{Q}_{\mathrm{el}}(\boldsymbol{q})$ defined for a particular geom-

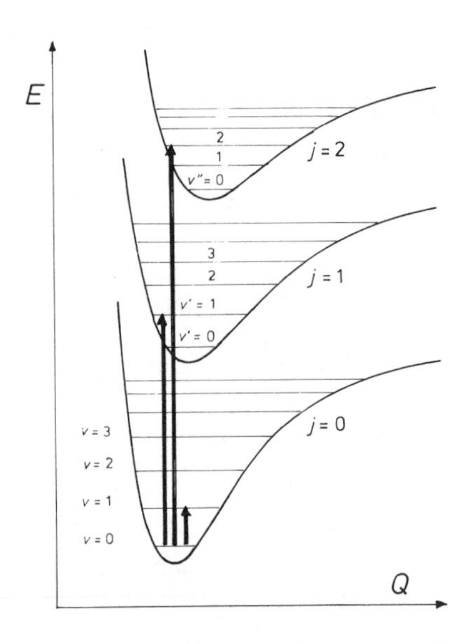

Figure 1.5. Schematic representation of potential energy curves and vibrational levels of a molecule. (For reasons of clarity rotational sublevels are not shown.)

etry $\boldsymbol{Q}$. There is a different electronic wave function $\Psi_j^Q(\boldsymbol{q})$ with a different energy E_j^Q for a given (j-th) state for each value of the parameter $\boldsymbol{Q}$.

The vibrational wave functions $\chi_v^j(\boldsymbol{Q})$ are eigenfunctions of the vibrational Hamiltonian $\hat{H}_{\text{vib}}(j,\boldsymbol{Q})$, which is defined for a particular electronic state j as an operator containing E_j^Q, the electronic plus nuclear repulsion energy of state j, as the potential energy of the nuclear motions. For every electronic state j, there is a different potential energy and therefore a different vibrational Hamiltonian $\hat{H}_{\text{vib}}(j,\boldsymbol{Q})$.

Due to the product form of the total wave function in Equation (1.12) the energy of a stationary state can be written as

$$E = E^{(\text{el})} + E^{(\text{vib})}$$

As a result, each electronic state of a molecule with energy $E^{(\text{el})} = E_j^Q$ carries a manifold of vibrational sublevels, and the energy of an electronic excitation may be separated into an electronic component and a vibrational component (cf. Figure 1.5) according to

$$\Delta E = \Delta E^{(\text{el})} + \Delta E^{(\text{vib})}$$

Similarly, a rotational component $E^{(\text{rot})}$ and a translational component $E^{(\text{trans})}$ are obtained when all $3N$ displacement coordinates of the N nuclei are used rather than the internal coordinates, which are obtained by separating the motion of center-of-mass and the rotational motions.

1.2.2 MO Models for the Description of Light Absorption

The determination of state energies and transition moments requires the knowledge of the wave functions $\Psi_0^Q(\boldsymbol{q})$ and $\Psi_\text{f}^Q(\boldsymbol{q})$ of the ground state 0 and the excited (final) state f. In general, the exact wave functions are not known, but nowadays approximate semiempirical or even ab initio LCAO-MO-SCF-CI wave functions are fairly easily available for most molecules.

These wave functions are obtained starting with *atomic orbitals* (AOs) χ_μ, from which *molecular orbitals* (MOs) ϕ_i are formed as linear combinations by application of the self-consistent field (SCF) method:

$$\phi_i = \sum_\mu c_{\mu i}\chi_\mu \tag{1.13}$$

Multiplying each MO with one of the spin functions α and β yields the *spin orbitals* $\psi_k(j) = \phi_i(j)\alpha(j) \equiv \phi_i$ and $\psi_l(j) = \phi_i(j)\beta(j) \equiv \bar{\phi}_i$. Here, the space and spin coordinates of the electron are indicated by the number j of the electron and in the shorthand notation only the β spin is indicated by a bar. Each possible selection of occupied orbitals ϕ_i defines an *orbital configuration,* from which configurational functions may be obtained. These are antisymmetric with respect to the interchange of any pair of electrons and are spin eigenfunctions. Thus the singlet ground configurational function $^1\Phi_0$

of a closed-shell molecule, with the lowest-energy $n/2$ orbitals all doubly occupied, is given by an antisymmetrized spin-orbital product known as a *Slater determinant:*

$$^1\Phi_0 = |\phi_1\bar{\phi}_1\phi_2\bar{\phi}_2 \ldots \phi_{n/2}\bar{\phi}_{n/2}| \tag{1.14}$$

The general definition of a Slater determinant is

$$|\psi_{k_1}(1)\psi_{k_2}(2) \ldots \psi_{k_n}(n)| = 1/\sqrt{n!} \begin{vmatrix} \psi_{k_1}(1) & \psi_{k_1}(2) & \ldots & \psi_{k_1}(n) \\ \psi_{k_2}(1) & \psi_{k_2}(2) & \ldots & \psi_{k_2}(n) \\ \ldots & & & \\ \psi_{k_n}(1) & \psi_{k_n}(2) & \ldots & \psi_{k_n}(n) \end{vmatrix} \tag{1.15}$$

Other configurations are referred to as *excited configurations*. Singly and multiply excited configurations differ from the ground configuration in that one or several electrons, respectively, are in orbitals that are not occupied in the ground configuration. Consider the following example of singly excited singlet and triplet configurations:

$$^1\Phi_{i\to k}(1, \ldots, n) = 1/\sqrt{2}\,(|\phi_1\bar{\phi}_1 \ldots \phi_i\bar{\phi}_k \ldots| + |\phi_1\bar{\phi}_1 \ldots \phi_k\bar{\phi}_i \ldots|) \tag{1.16}$$

$$^3\Phi_{i\to k}(1, \ldots, n) = \begin{cases} |\phi_1\bar{\phi}_1 \ldots \phi_i\phi_k \ldots| \\ 1/\sqrt{2}\,(|\phi_1\bar{\phi}_1 \ldots \phi_i\bar{\phi}_k \ldots| - |\phi_1\bar{\phi}_1 \ldots \phi_k\bar{\phi}_i \ldots|) \\ |\phi_1\bar{\phi}_1 \ldots \bar{\phi}_i\bar{\phi}_k \ldots| \end{cases} \tag{1.17}$$

The configurational functions of all three components of the triplet state are listed on the three lines of Equation (1.17). From top to bottom, they correspond to the occupation of the MOs ϕ_i and ϕ_k with two electrons with an α spin, with one electron each with α and β spin, and with two electrons with a β spin. The z component of the total spin is equal to $M_S = 1$, 0, and -1, respectively. The three triplet functions are degenerate (i.e., have the same energy) in the absence of external fields and ignoring relativistic effects (i.e., with a spin-free Hamiltonian). For our purposes, it is therefore sufficient to consider only one of the components, e.g., the one corresponding to $M_S = 0$.

Finally, states of given multiplicity $M = 2S + 1$, e.g., singlet ($S = 0$) and triplet ($S = 1$) states, may be described by a linear combination of configurational functions of appropriate multiplicity and symmetry:

$$^M\Psi_j^Q(1, \ldots, n) = \sum_K C_{Kj}\ {}^M\Phi_K(1, \ldots, n) \tag{1.18}$$

Here Φ_K can be either the ground configuration Φ_0 or any of the singly or multiply excited configurational functions $\Phi_{i\to k}$, etc. The coefficients C_{Kj} are determined from the variational principle by solving a secular problem. (Cf.

Michl and Bonačić-Koutecký, 1990: Appendix III.) This procedure is referred to as *configuration interaction* (CI).

Once wave functions of the type shown in Equation (1.18) are known, the electronic excitation energies $\Delta E^{(el)}$ may be calculated from Equation (1.8) as the energy difference between the ground state described by Ψ_0^Q and the excited state described by Ψ_f^Q:

$$\Delta E_{0\rightarrow f}^{(el)} = {}^{M'}E_f - {}^{M}E_0 = \langle {}^{M'}\Psi_f^Q|\hat{H}|{}^{M'}\Psi_f^Q\rangle - \langle {}^{M}\Psi_0^Q|\hat{H}|{}^{M}\Psi_0^Q\rangle \qquad (1.19)$$

According to Equation (1.9), the transition moment is given by

$$U_{0\rightarrow f} = \langle {}^{M'}\Psi_f^Q|\hat{U}|{}^{M}\Psi_0^Q\rangle \qquad (1.20)$$

$\hat{U}$ may be the electric dipole operator $\hat{M}$, the magnetic dipole operator $\hat{\mathbf{M}}$, or the electric quadrupole operator $\hat{\mathbf{Q}}$.

1.2.3 One-Electron MO Models

The singlet ground state of most organic molecules is reasonably described by the ground configuration Φ_0; that is, C_0 in Equation (1.18) is nearly 1. The lowest singlet and triplet excited states are frequently characterized by singly excited configurations, but in some cases doubly excited configurations may also be of vital importance. (Cf. Section 2.1.2.)

If, however, an excited state can be described by just one singly excited configuration ${}^{M}\Phi_{i\rightarrow k}$, as is frequently possible to a good approximation for transitions from the highest occupied MO (HOMO) into the lowest unoccupied MO (LUMO), the formulae for the excitation energies and the transition moments can be simplified considerably. This is particularly true for simple one-electron models, such as the Hückel MO method (HMO), that do not take electron repulsion into account explicitly.

The Hamiltonian of the HMO model,

$$\hat{H}(1, \ldots, n) = \sum_j \hat{h}(j)$$

is a sum of one-electron operators. The energy E_K of the electron configuration Φ_K is given by

$$E_K = \sum_i n_i \varepsilon_i$$

where ε_i is the orbital energy and $n_i = 0, 1,$ or 2 is the occupation number of MO ϕ_i. It follows that the excitation energy is

$$\Delta E_{i\rightarrow k} = E_{i\rightarrow k} - E_0 = \varepsilon_k - \varepsilon_i \qquad (1.21)$$

so the excitation energy is equal to the difference of orbital energies at the Hückel level. In practical applications, it can be useful to replace Equation (1.21) by

$$\Delta E_{i\rightarrow k} = E_{i\rightarrow k} - E_0 = \varepsilon_k - \varepsilon_i + C \qquad (1.22)$$

which allows for the fact that electron interaction is implicitly present in the HMO operator $\hat{h}(j)$. [Cf. Equation (1.23) and Equation (1.24).] The additive constant C has different values for different classes of compounds. (Cf. Section 2.2.1.)

If electron interaction is taken explicitly into account by writing the Hamiltonian in the form

$$\hat{H}(1, \ldots, n) = \sum_j \hat{h}(j) + \sum_{i<j}\sum \hat{g}(i, j)$$

where $\hat{h}(j)$ is a one-electron operator for the kinetic and potential energy of an electron j in the field of all atoms or atomic cores, whereas $\hat{g}(i, j)$ represents the Coulomb repulsion between electrons i and j, the excitation energy is calculated to be

$$^1\Delta E_{i\rightarrow k} = \langle {}^1\Phi_{i\rightarrow k}|\hat{H}|\, {}^1\Phi_{i\rightarrow k}\rangle - \langle {}^1\Phi_0|\hat{H}|{}^1\Phi_0\rangle$$
$$= \varepsilon_k - \varepsilon_i - J_{ik} + 2K_{ik} \quad (1.23)$$

$$^3\Delta E_{i\rightarrow k} = \langle {}^3\Phi_{i\rightarrow k}|\hat{H}|{}^3\Phi_{i\rightarrow k}\rangle - \langle {}^1\Phi_0|\hat{H}|{}^1\Phi_0\rangle = \varepsilon_k - \varepsilon_i - J_{ik} \quad (1.24)$$

(See, e.g., Michl and Bonačić-Koutecký, 1990: Section 1.3.) The Coulomb integral J_{ik}, which represents the Coulomb interaction between the charge distributions $|\phi_i|^2$ and $|\phi_k|^2$, and the exchange integral K_{ik}, which is given by the electrostatic self-interaction of the overlap density $\phi_i^*\phi_k$, are both positive, and the singlet as well as the triplet excitation energy is smaller than the difference in orbital energies. Within this approximation, the difference between the singlet and triplet excitation energies is just twice the exchange integral K_{ik}. As a rule, K_{ik} is much smaller than J_{ik}, and the energies of the singlet and triplet levels resulting from the same orbital occupancy are not vastly different.

Example 1.3:
The ionization potential and electron affinity of naphthalene were determined experimentally as IP = 8.2 eV and EA ≈ 0.0 eV. According to Koopmans' theorem it is possible to equate minus the orbital energies of the occupied or unoccupied MOs with molecular ionization potentials and electron affinities, respectively ($\text{IP}_i = -\varepsilon_i$ and $\text{EA}_k = -\varepsilon_k$). Thus, in the simple one-electron model, the excitation energy of the HOMO→LUMO transition in naphthalene may be written according to Equation (1.22) as

$$\Delta E_{i\rightarrow k} = \varepsilon_k - \varepsilon_i + C = \text{IP}_i - \text{EA}_k + C = 8.2 \text{ eV} + C$$

Experimental values for the singlet and triplet excitations corresponding to the HOMO→LUMO transition are 4.3 and 2.6 eV, respectively. If the value of C in the above expression is equated to the electron repulsion terms in Equation (1.23) and Equation (1.24),

$$^1\Delta E_{i\rightarrow k} = 8.2 \text{ eV} + C = 4.3 \text{ eV}$$

so

$$C = -J_{ik} + 2K_{ik} = -3.9 \text{ eV}$$

and

$$K_{ik} = \tfrac{1}{2}({}^1\Delta E_{i\to k} - {}^3\Delta E_{i\to k}) = \tfrac{1}{2}(4.3 - 2.6) = 0.85 \text{ eV}$$

Thus, due to the existence of the electron repulsion term $J_{ik} - 2K_{ik}$ the singlet excitation energy is only about half the orbital energy difference, and the exchange interaction $2K_{ik}$ is about one third of the Coulomb interaction J_{ik}.

The dipole moment operator is a one-electron operator, and within the independent particle model, with explicit treatment of electron repulsion as well as without, the transition moment becomes

$$\begin{aligned} \boldsymbol{M}_{i\to k} &= \langle \Phi_{i\to k}|\hat{\boldsymbol{M}}|\Phi_0\rangle \\ &= \frac{1}{\sqrt{2}}(\langle \phi_i|\hat{\boldsymbol{m}}|\phi_k\rangle + \langle \phi_k|\hat{\boldsymbol{m}}|\phi_i\rangle) \\ &= \sqrt{2}\langle \phi_i|\hat{\boldsymbol{m}}|\phi_k\rangle \end{aligned} \tag{1.25}$$

where we have used the rules for matrix elements between Slater determinants (Slater rules; see, e.g., McWeeny, 1989: Section 3.3). Thus, if the excited state is described by a single configuration, the excitation energy and transition moment are completely determined by the two MOs, ϕ_i and ϕ_k. This is rather an oversimplification, and configuration interaction is indispensable for a more realistic description of electronic excitations. However, in Chapter 2 it will become clear under which favorable conditions qualitative predictions and even rough quantitative estimates based on this simple model are possible.

Example 1.4:
Particularly illuminating is the free-electron MO model (FEMO) based on the assumption that π electrons can move freely along a one-dimensional molecular framework. Stationary states are then characterized by standing waves, and using the de Broglie relationship

$$\lambda = h/m_e v$$

for the wavelength of an electron of mass m_e and velocity v one obtains standing waves only if there is an integral number of half-wavelengths between the ends of the potential well of length L—that is, if

$$L = k \times (\lambda/2)$$

Eliminating λ by means of the de Broglie relationship yields

$$\varepsilon_k = \tfrac{1}{2}m_e v^2 = \frac{k^2h^2}{8m_eL^2}$$

for the orbital energy. The energy needed to excite an electron from the HOMO, the highest-occupied MO ϕ_k ($k = n/2$), into the LUMO, the lowest-unoccupied MO ϕ_{k+1}, is then

$$\Delta E_{k \to k+1} = \frac{h^2}{8m_e L^2}[(k + 1)^2 - k^2]$$

Thus, for the trimethinecyanine **2** with 6 π electrons,

$$\Delta E = \frac{h^2}{8m_e L^2}[4^2 - 3^2] = \frac{7h^2}{8m_e L^2}$$

and insertion of the values of the physical constants $h = 6.626 \times 10^{-27}$ erg s and $m_e = 9.109 \times 10^{-28}$ g and of the plausible value $L = 6 \times 140$ pm for the length of the potential well yields an energy that corresponds to a wavelength of approximately 333 nm. This is in excellent agreement with the experimental value $\lambda_{max} = 313$ nm for this cyanine in methanol solvent.

$R_2N{-}CH{=}CH{-}CH{=}\overset{\oplus}{N}R_2$

2

1.2.4 Electronic Configurations and States

It has already been mentioned that the ground configuration Φ_0 is in many cases quite sufficient to characterize organic molecules in their singlet ground states, whereas a single configuration constructed from ground-state SCF MOs is generally not suited for the description of an excited state; that is, an excited-state wave function can in general be improved considerably by configuration interaction.

The interaction between two configurations Φ_K and Φ_L increases with the increasing absolute value of the matrix element H_{KL} and the decreasing absolute value of the difference $H_{KK} - H_{LL}$ of the energies of the two configurations. If two configurations Φ_K and Φ_L belong to different irreducible representations of the point group of the molecule, $H_{KL} = 0$, and the two configurations cannot interact. Therefore, configurations can contribute to the wave function of a given state only if they are of the appropriate symmetry, that is, if they belong to the same irreducible representation as the state under consideration.

In approximate models such as the PPP method (cf. Section 1.5.1), degeneracies of orbital energy differences may carry over to the corresponding many-electron excitation energies, leading to degenerate configurations of the same symmetry. In these cases, configuration interaction is of paramount importance, since it determines the energy-level scheme. It has therefore been termed *first-order configuration interaction,* in contrast to configuration interaction among nondegenerate configurations, which in general affects the results only to a smaller and less fundamental degree, and which is therefore referred to as *second-order configuration interaction* (Moffitt, 1954b). Hence, in the case of first-order configuration interaction, two or more configurational functions are needed to construct the wave function of a state in a given MO basis. Such a state can no longer be characterized by specifying a single electron configuration. In the case of second-order configuration interaction, one of the configurational functions may predominate to such an extent that the specification of a state in terms of a single electron configuration may still be justified, at least qualitatively.

The excited states of alternant hydrocarbons may serve as an example of the importance of first-order configuration interaction. Due to the pairing theorem that is valid for both the HMO and the PPP methods (cf. Koutecký, 1966), configurational functions for the excitation of an electron from ϕ_i into $\phi_{k'}$ and from ϕ_k into $\phi_{i'}$ are degenerate, if the MO ϕ_i is paired with $\phi_{i'}$, and the MO ϕ_k is paired with $\phi_{k'}$. The configurational functions $\Phi_{i\rightarrow k'}$ and $\Phi_{k\rightarrow i'}$, are of the same symmetry, and their energies are split by configuration interaction. Figure 1.6 shows the splitting for the lowest excited configurations of an alternant hydrocarbon. If the interaction matrix element H_{KL} is sufficiently large, the splitting may be sufficient to bring one of the two states that result from the splitting of the degenerate configurations below the state that corresponds primarily to the excitation of an electron from the HOMO to the LUMO.

In this text orbitals that are occupied in the ground configuration will frequently be numbered 1, 2, 3 . . . starting from the HOMO, and the unoccupied MOs by 1′, 2′, 3′ . . . starting from the LUMO. The advantage of this numbering system is that the frontier orbitals responsible for light absorption will be denoted in the same way (ϕ_1 and ϕ_2 for the two highest occupied MOs and $\phi_{1'}$ and $\phi_{2'}$ for the lowest unoccupied MOs) for all mol-

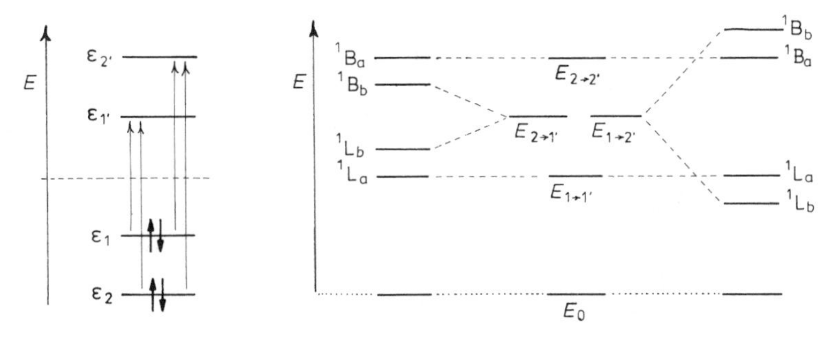

Figure 1.6. Schematic representation of first-order configuration interaction for alternant hydrocarbons. Within the PPP approximation, configurations corresponding to electronic excitation from MO ϕ_i into $\phi_{k'}$ and from MO ϕ_k into $\phi_{i'}$ are degenerate. The two highest occupied MOs ($i = 1$, $k = 2$) and the two lowest unoccupied MOs ($i' = 1'$ and $k' = 2'$) are shown. Depending on the magnitude of the interaction, the HOMO→LUMO transition $\phi_i \rightarrow \phi_{i'}$ corresponds approximately to the lowest or to the second-lowest excited state.

ecules irrespective of which electrons (π electrons, valence electrons, or even inner shells) are taken into account.

Within the PPP approximation the singlet and triplet CI matrices for an alternant hydrocarbon each factor into two separate matrices for the "plus" states and the "minus" states corresponding to the linear combinations

$$^{1,3}\Phi^{\pm}_{i\to k} = [^{1,3}\Phi_{i\to k'} \pm {}^{1,3}\Phi_{k\to i'}]/\sqrt{2} \tag{1.26}$$

The ground state behaves like a minus state since it interacts only with minus states; excited configurations of the type $\Phi_{i\to i'}$, however, behave like plus states. In this approximation the transition moments between two plus states or two minus states vanish, and such transitions are forbidden. Electric dipole transitions are allowed only between plus and minus states (Pariser, 1956).

Example 1.5:
In Figure 1.7 the HMO orbital energy levels of anthracene and phenanthrene are given together with the labels of the irreducible representations of the point groups D_{2h} and C_{2v}. The ground configuration with fully occupied orbitals is totally symmetric and belongs to the irreducible representation A_g or A_1. The symmetry of a singly excited configuration $\Phi_{i\to k}$ is given by the direct product of the characters χ of the irreducible representation of the singly occupied MOs ϕ_i and ϕ_k. For phenanthrene,

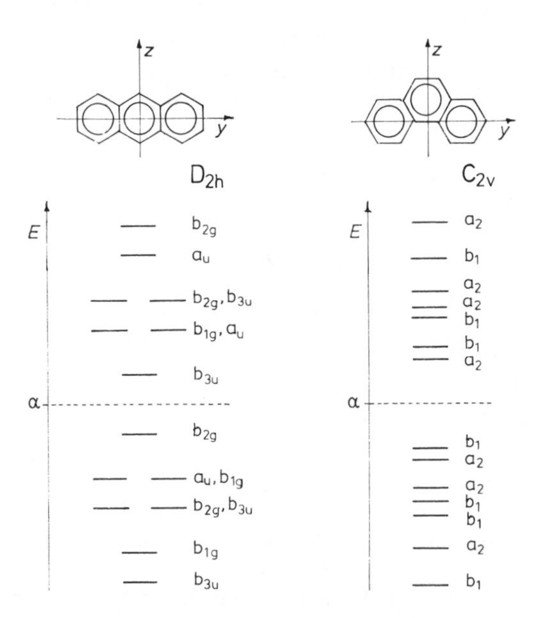

Figure 1.7. Orbital energy diagrams of anthracene and phenanthrene. For each HMO energy level the irreducible representation of the π MO is given.

$$\chi(\Phi_{1\to1'}) = \chi(\phi_1) \times \chi(\phi_{1'}) = b_1 \times a_2 = B_2$$

and making allowance for the classification into plus and minus states, the HOMO→LUMO transition may be denoted as a transition from an $^1A_1^-$ ground state into an excited state of symmetry $^1B_2^+$. The degenerate configurations $\Phi_{1\to2'}$ and $\Phi_{2\to1'}$ are of the same symmetry:

$$\chi(\Phi_{1\to2'}) = \chi(\phi_1) \times \chi(\phi_{2'}) = b_1 \times b_1 = A_1$$
$$\chi(\Phi_{2\to1'}) = \chi(\phi_2) \times \chi(\phi_{1'}) = a_2 \times a_2 = A_1$$

Using typical PPP parameters it is found that $\langle\Phi_{1\to2'}|\hat{H}|\Phi_{2\to1'}\rangle$ is larger than the difference $E(\Phi_{1\to2'}) - E(\Phi_{1\to1'})$, which corresponds to the situation shown in Figure 1.6 on the right, with the $^1A_1^-$ state below the $^1B_2^+$ state. Thus, the lowest-energy transition is forbidden and the next one allowed.

For anthracene, for the state that is characterized by the HOMO→LUMO excitation, we have

$$\chi(\Phi_{1\to1'}) = \chi(\phi_1) \times \chi(\phi_{1'}) = b_{2g} \times b_{3u} = B_{1u}$$

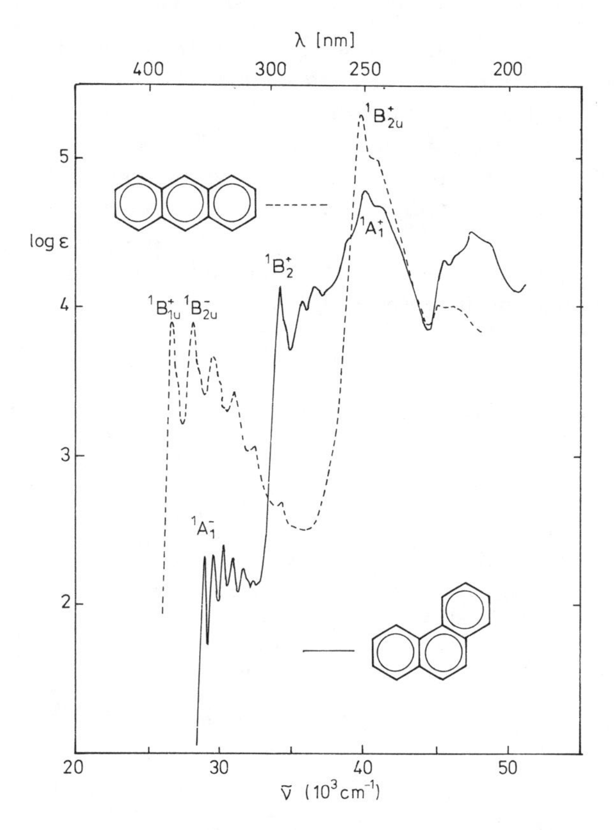

Figure 1.8. Absorption spectra of anthracene (---) and phenanthrene (—) (by permission from DMS-UV-Atlas 1966–71).

whereas the orbitals below the HOMO and above the LUMO are accidentally degenerate, so four configurations have to be considered. Only the following two have u symmetry:

$$\chi(\Phi_{2\rightarrow 1'}) = \chi(\phi_2) \times \chi(\phi_{1'}) = b_{1g} \times b_{3u} = B_{2u}$$
$$\chi(\Phi_{1\rightarrow 2'}) = \chi(\phi_1) \times \chi(\phi_{2'}) = b_{2g} \times a_u = B_{2u}$$

For anthracene $<\Phi_{2\rightarrow 1'}|\hat{H}|\Phi_{1\rightarrow 2'}>$ is smaller than the difference $E(\Phi_{1\rightarrow 2'}) - E(\Phi_{1\rightarrow 1'})$ and the allowed HOMO→LUMO transition ($^1A_g^- \rightarrow {}^1B_{1u}^+$) is lower in energy than both the $^1B_{2u}^-$ and $^1B_{2u}^+$ states split by configuration interaction. This is shown on the left of Figure 1.6. Now, the lowest-energy transition is allowed and the next one is forbidden. This contrast of phenanthrene and anthracene is obvious in the absorption spectra shown in Figure 1.8. The forbidden second transition in anthracene is hard to discern under the intense first one.

1.2.5 Notation Schemes for Electronic Transitions

Various schemes have been proposed to denote the states of a molecule and the absorption bands that correspond to transitions between these states. Some of these schemes are collected in Table 1.2.

The discussion in the previous section revealed the advantages as well as the disadvantages of the group theoretical nomenclature.

For cyclic π-electron hydrocarbons two more schemes, introduced by Clar (1952) and Platt (1949), are widely used. Clar's empirical scheme is based on the appearance of the absorption bands and designates the first three bands in the spectrum of phenanthrene as the α, p, and β bands, re-

Table 1.2 Labeling of Electronic Transitions

System	State Symbol		Example
Enumerative	S_0	Singlet ground state	$S_0 \rightarrow S_1$
	$S_1, S_2, S_3 \ldots$	Excited singlet states	$S_0 \rightarrow S_2$
	$T_1, T_2, T_3 \ldots$	Triplet states	$T_1 \rightarrow T_2$
Group theory	A, B, E, T (with indices g, u, 1, 2, ′, ″)	Irreducible representations of point group of the molecule	$^1A_1 \rightarrow {}^1B_2$†
			$^1A_{1g} \rightarrow {}^1B_{2u}$
			$^1A_{1g} \rightarrow {}^3E_{1u}$
Kasha	σ, π, n	Ground state orbitals	$\pi \rightarrow \pi^*$
	σ^*, π^*	Excited state orbitals	$n \rightarrow \pi^*$
Platt	A	Ground state	$^1A \rightarrow {}^1B_a$†
	B, L (with indices a, b)	Excited states	$^1A \rightarrow {}^1L_b$
Mulliken	N	Ground state	V←N
	Q, V, R	Excited states	Q←N
Clar	α, p, β	Intensity and band shape	

† The upper left index indicates the multiplicity.

spectively. Platt's nomenclature is derived from the perimeter model and denotes the same bands as ^1L_a, ^1L_b, and ^1B_b. (Cf. Section 2.2.2.)

A very simple scheme is obtained using consecutive numbering of the singlet states, denoted by S, and the triplet states, denoted by T. The longest wavelength transition in the spectrum of phenanthrene or anthracene is then referred to as the $S_0 \rightarrow S_1$ transition. This nomenclature does not reveal anything about the nature of the states involved except for their multiplicity and their energy order. Another disadvantage is that the detection of a new transition automatically means that all higher transitions have to be renamed.

According to Mulliken (1939) the ground state is denoted by N and the valence excited states by V. The bands observed in the phenanthrene spectrum are called V←N transitions. In addition, Rydberg transitions and transitions that involve electron lone pairs (cf. Section 2.5) are denoted by R←N and Q←N, respectively. Finally, Kasha (1950) specifies only the nature of the orbitals involved in the transition, using $\pi \rightarrow \pi^*$ transitions, etc.

1.3 Intensity and Band Shape

1.3.1 Intensity of Electronic Transitions

A rough measure of the intensity of an electronic transition is provided by the maximum value ε_{max} of the molar extinction coefficient. A physically more meaningful quantity is the total area under the absorption band, given by the integral $\int \varepsilon d\tilde{\nu}$, or the *oscillator strength*

$$f = 4.319 \times 10^{-9} \int \varepsilon(\tilde{\nu}) d\tilde{\nu} \tag{1.27}$$

which is proportional to the integrated intensity. f is a dimensionless quantity that represents the ratio of the observed integrated absorption coefficient to that calculated classically for a single electron in a three-dimensional harmonic potential well. The maximum value of f for a fully allowed transition is of the order of unity.

In order to obtain a theoretical expression for the oscillator strength, perturbation theory may be used to treat the interaction between electromagnetic radiation and the molecule. Since an oscillating field is a perturbation that varies in time, time-dependent perturbation theory has to be used. Thus, the Hamiltonian of the perturbed system is

$$\hat{H} = \hat{H}^{(0)} + \hat{H}^{(1)}(t)$$

where $\hat{H}^{(0)}$ is the Hamiltonian of the unperturbed system, that is, of the molecule in the absence of any electromagnetic radiation, and $\hat{H}^{(1)}(t)$ describes the interaction between light and molecule.

It turns out that the probability of a transition from a state 0 with total wave function Ψ_0^T into a state with wave function Ψ_f^T is proportional to the light intensity and to the squared matrix element $|<\Psi_f^T|\hat{H}^{(1)}|\Psi_0^T>|^2$ of the perturbation operator over the wave functions of the two states involved. For light with a wave vector $\boldsymbol{K}$ and polarization characterized by the unit vector $\boldsymbol{e}_U$, it is convenient to express $\hat{H}^{(1)}(t)$ by the vector potential

$$\boldsymbol{A}(\boldsymbol{r}_j,t) = A_0\,[\boldsymbol{e}_U \mathrm{e}^{\mathrm{i}(2\pi\nu t - \boldsymbol{K}\cdot\boldsymbol{r}_j)} + \boldsymbol{e}_U^* \mathrm{e}^{-\mathrm{i}(2\pi\nu t - \boldsymbol{K}\cdot\boldsymbol{r}_j)}]/2$$

instead of the electric and magnetic field to describe the radiation.* Here, $\boldsymbol{K} = (2\pi/\lambda)\boldsymbol{e}_X$, where $\boldsymbol{e}_X$ is a unit vector in the propagation direction X, and $\boldsymbol{r}_j$ is the position vector of the electron j. If we consider light absorption, only the second of the exponential terms in the square brackets need to be kept (the other term gives analogous results for stimulated emission), and neglecting terms of higher order in $\boldsymbol{A}$, $\hat{H}^{(1)}(t)$ may be written as

$$\hat{H}^{(1)}(t) = \sum_j \frac{q_j}{m_j} \boldsymbol{A}(\boldsymbol{r}_j,t)\cdot\hat{\boldsymbol{p}}_j \approx -\mathrm{i}\hbar\boldsymbol{e}_U^* \sum_j \frac{q_j}{m_j} \mathrm{e}^{\mathrm{i}\boldsymbol{K}\cdot\boldsymbol{r}_j}\cdot\hat{\nabla}_j \tag{1.28}$$

where $\hat{\boldsymbol{p}}_j = -\mathrm{i}\hbar\hat{\nabla}_j$ is the linear momentum operator of particle j with mass m_j and charge q_j, and $\hat{\nabla}_j = \left(\frac{\partial}{\partial x_i}, \frac{\partial}{\partial y_i}, \frac{\partial}{\partial z_i}\right)$ is the gradient operator.

When the origin of coordinates is chosen to lie within a molecule of ordinary size, the length $\boldsymbol{r}_j$ is much smaller than λ for UV light and light of longer wavelengths. Therefore, $\boldsymbol{K}\cdot\boldsymbol{r}_j \ll 1$, and the expansion of the exponential $\mathrm{e}^{\mathrm{i}\boldsymbol{K}\cdot\boldsymbol{r}_j}$ into an infinite series converges rapidly

$$\mathrm{e}^{\mathrm{i}\boldsymbol{K}\cdot\boldsymbol{r}_j} = 1 + \mathrm{i}\boldsymbol{K}\cdot\boldsymbol{r}_j - \tfrac{1}{2}(\boldsymbol{K}\cdot\boldsymbol{r}_j)^2 + \ldots \tag{1.29}$$

The dominant contribution to the squared matrix element of the interaction operator $\hat{H}^{(1)}(t)$ stems from the first term in Equation (1.29), unity. The oscillator strength of the transition $\Psi_0^T \rightarrow \Psi_f^T$ is then

$$f^{(p)} = (4\pi/3he^2m_e\nu)\cdot|\boldsymbol{e}_U^*\cdot\boldsymbol{p}_{0\rightarrow f}|^2 \tag{1.30}$$

where

$$\boldsymbol{p}_{0\rightarrow f} = (\hbar/\mathrm{i})<\Psi_f^T|\sum_j(q_j/m_e)\hat{\nabla}_j|\Psi_0^T> \tag{1.31}$$

is the matrix element of the linear momentum operator. Equation (1.30) is called the dipole velocity formula for the oscillator strength.

---

* Both the electric field $\boldsymbol{E}(\boldsymbol{r},t)$ and the magnetic flux density $\boldsymbol{B}(\boldsymbol{r},t)$ may be derived from the vector potential $\boldsymbol{A}(\boldsymbol{r},t)$. Using the so-called Coulomb gauge one obtains

$$\boldsymbol{E}(\boldsymbol{r},t) = -(1/c)\partial\boldsymbol{A}(\boldsymbol{r},t)/\partial t$$

and

$$\boldsymbol{B}(\boldsymbol{r},t) = \hat{\nabla}\times\boldsymbol{A}(\boldsymbol{r},t)$$

More commonly used (cf. Example 1.6) is the related dipole length formula

$$f^{(r)} = (4\pi m_e/3he^2)\mu \cdot |e_U \cdot \tilde{M}_{0\to f}|^2 \tag{1.32}$$

where

$$\tilde{M}_{0\to f} = <\Psi_f^T|\hat{M}|\Psi_0^T> \tag{1.33}$$

is the transition moment, and

$$\begin{aligned}\hat{M} &= -|e|\sum_j r_j + |e|\sum_A Z_A R_A \\ &= \sum_j \hat{m}_j + |e|\sum_A Z_A R_A\end{aligned} \tag{1.34}$$

is the dipole moment operator. The first sum in Equation (1.34) runs over the electrons j and the second sum over the nuclei A of charge Z_A. Making use of the product form of the total wave function given in Equation (1.12), one has

$$\begin{aligned}\tilde{M}_{0v\to fv'} &= <\Psi_f^Q\chi_{v'}^f| -|e|\sum_j r_j + |e|\sum_A Z_A R_A|\Psi_0^Q\chi_v^0> \\ &= <\Psi_f^Q|\Psi_0^Q> <\chi_{v'}^f| +|e|\sum_A Z_A R_A|\chi_v^0> + <\Psi_f^Q| -|e|\sum_j r_j|\Psi_0^Q> <\chi_{v'}^f|\chi_v^0>\end{aligned}$$

where the first term vanishes for every fixed nuclear configuration Q due to the orthogonality of ground-state and excited-state wave functions. For electronically allowed transitions the geometry dependence of the electronic transition moment

$$M_{0\to f} = -|e|<\Psi_f^Q|\sum_j r_j|\Psi_0^Q> \tag{1.35}$$

may be neglected, so the overall transition dipole moment becomes

$$\tilde{M}_{0v\to fv'} = M_{0\to f}<\chi_{v'}^f|\chi_v^0> \tag{1.36}$$

The electronic transition dipole moment $M_{0\to f}$ determines the overall intensity of the transition. The overlap integrals $<\chi_{v'}^f|\chi_v^0>$ of the vibrational wave functions of the ground and excited electronic states determine the intensity of the individual vibrational components of the absorption band. (Cf. Section 1.3.3.) If only the total intensity of the electronic transition is of interest, it is sufficient to calculate the electronic dipole transition moment from Equation (1.35). Here, and frequently in the following, the index Q for molecular geometry is omitted for clarity of notation.

The transition dipole moment $M_{0\to f}$ is a vector quantity, and one has

$$M^2 = M_x^2 + M_y^2 + M_z^2 \tag{1.37}$$

with

$$M_\alpha = <\Psi_f|\sum_j q_j\alpha_j|\Psi_0>$$

where $\alpha = x, y, z$ denotes the electronic coordinates.

If the transition dipole moment is measured in Debye (1D $= 10^{-8}$ esu cm), the oscillator strength is given by

$$f = 4.702 \times 10^{-7}\, \tilde{\nu} |\boldsymbol{M}_{0\rightarrow f}|^2 \tag{1.38}$$

Example 1.6:

From the above derivation it is seen that after the series expansion of the exponential in the space part of the vector potential, the transition moment operator involves the linear momentum operator $\hat{\boldsymbol{p}}_j$ or the gradient operator $\hat{\nabla}_j$. Equation (1.32) is obtained from Equation (1.30) in the following way: From the commutation relation $[\hat{H}, \boldsymbol{r}_j] = \hat{H}\boldsymbol{r}_j - \boldsymbol{r}_j\hat{H} = \hat{\boldsymbol{p}}_j$, we have

$$<\Psi_f|\hat{\boldsymbol{p}}_j|\Psi_0> = <\Psi_f|\hat{H}\boldsymbol{r}_j - \boldsymbol{r}_j\hat{H}|\Psi_0> = <\Psi_f|\hat{H}\boldsymbol{r}_j|\Psi_0> - <\Psi_f|\boldsymbol{r}_j\hat{H}|\Psi_0>$$

From the Hermitean property of $\hat{H}$ it follows that $<\Psi_f|\hat{H}\boldsymbol{r}_j|\Psi_0> = <\hat{H}\Psi_f|\boldsymbol{r}_j|\Psi_0>$, so for exact eigenfunctions of $\hat{H}$, for which $\hat{H}\Psi_K = E_K\Psi_K$, the following relation holds:

$$<\Psi_f|\hat{\boldsymbol{p}}_j|\Psi_0> = (E_f - E_0)<\Psi_f|\boldsymbol{r}_j|\Psi_0>$$

Thus, two different expressions are obtained for the oscillator strength:

$$f^{(r)}_{0\rightarrow f} \sim (E_f - E_0)|<\Psi_f|\sum_j \boldsymbol{r}_j|\Psi_0>|^2$$

and

$$f^{(p)}_{0\rightarrow f} \sim (E_f - E_0)^{-1}|<\Psi_f|\sum_j \hat{\boldsymbol{p}}_j|\Psi_0>|^2$$

If Ψ_0 and Ψ_f are exact eigenfunctions of the molecular Hamiltonian the two expressions give identical results, but this is generally not true for approximate wave functions. (Cf. Yang, 1976, 1982.) Thus, nonempirical SCF calculations based on the ZDO approximation yield for the N→V transition of ethylene (the longest-wavelength singlet–singlet transition) the following values:

$$f^{(r)}_{NV} = 0.59$$

and

$$f^{(p)}_{NV} = 0.28$$

which differ by a factor of 2. Refinement of the wave function by including some configuration interaction leads to

$$f^{(r)}_{NV} = 0.33$$

and

$$f^{(p)}_{NV} = 0.44$$

(Hansen, 1967). Convergence to a common value of $f^{(r)}$ and $f^{(p)}$ on further refinement of the wave function is to be expected. (See also Example 1.14 and Figure 1.23, as well as Bauschlicher and Langhoff, 1991.)

The second term in the series expansion of the exponential in Equation (1.29), $i\boldsymbol{K}\cdot\boldsymbol{r}_j$, leads to an integral that may be separated into two parts. The square of the first part

$$|(\boldsymbol{e}_X \times \boldsymbol{e}_U^*) \cdot <\Psi_f|\hat{\boldsymbol{M}}|\Psi_0>|^2$$

describes the contribution of the magnetic dipole transition moment to the intensity of the transition $\Psi_0 \rightarrow \Psi_f$, whereas the square of the second part

$$|\boldsymbol{e}_X \cdot <\Psi_f|\hat{\boldsymbol{Q}}|\Psi_0> \cdot \boldsymbol{e}_U^*|^2$$

determines the value of the electric quadrupole transition moment. Here,

$$\hat{\boldsymbol{M}} = \sum_j \frac{q_j}{2m_j c}(\boldsymbol{r} \times \hat{\boldsymbol{p}}_j)$$

and

$$\hat{\boldsymbol{Q}} = \sum_j \frac{q_j}{2m_j c}(\boldsymbol{r}_j\hat{\boldsymbol{p}}_j + \hat{\boldsymbol{p}}_j\boldsymbol{r}_j)$$

are the operators of the magnetic dipole moment and of the electric quadrupole moment, respectively. We have already mentioned that contributions from these operators may be neglected for dipole-allowed transitions, and that even higher-order terms in Equation (1.29) may be safely ignored.

From Equation (1.35), the electric dipole transition moment $\boldsymbol{M}_{0\rightarrow f}$ may be thought of as the dipole moment of the transition density $\Psi_f^*\Psi_0$. The transition density is a purely quantum mechanical quantity and cannot be inferred from classical arguments. A more pictorial representation of the electric dipole transition moment equates it to the amplitude of the oscillating dipole moment of the molecule in the transient nonstationary state that results from the mixing of the initial and the final states of the transition by the time-dependent perturbation due to the electromagnetic field, and which can be written as a linear combination: $c_0\Psi_0 + c_f\Psi_f + \ldots$ This emphasizes the fact that the *absolute direction* of the transition moment vector has no physical meaning.

Using the CI expansion (1.18), Ψ_0 and Ψ_f can be written in terms of the configurations Φ_K. From Equation (1.25), the contribution $C_{00}C_{i\rightarrow k,f}\boldsymbol{M}_{i\rightarrow k}$ to the transition moment that is provided by the configurations Φ_0 and $\Phi_{i\rightarrow k}$, is proportional to

$$\boldsymbol{M}_{i\rightarrow k} = <\Phi_{i\rightarrow k}|\hat{\boldsymbol{M}}|\Phi_0> = \sqrt{2}<\phi_k|\hat{\boldsymbol{m}}|\phi_i> \tag{1.39}$$

where the total electric dipole moment operator $\hat{\boldsymbol{M}} = \Sigma\hat{\boldsymbol{m}}_j = -|e|\Sigma\boldsymbol{r}_j$ is given by a sum of contributions provided by each electron j. By substituting the LCAO expansion from Equation (1.13) one obtains

$$\boldsymbol{M}_{i\rightarrow k} = -|e|\sqrt{2}\sum_\mu\sum_\nu c_{\mu k}^* c_{\nu i}<\chi_\mu|\boldsymbol{r}|\chi_\nu> \tag{1.40}$$

Introduction of the ZDO approximation, appropriate for π-electron systems, gives $<\chi_\mu|\boldsymbol{r}|\chi_\nu> = \delta_{\mu\nu}<\chi_\mu|\boldsymbol{r}|\chi_\mu>$; the remaining integrals may be evaluated easily by writing the position vector $\boldsymbol{r}_j$ of electron j as

$$\boldsymbol{r}_j = \boldsymbol{R}_\mu + \boldsymbol{r}_j^\mu$$

where $\boldsymbol{R}_\mu$ is the position vector of the nucleus μ on which the orbital χ_μ is centered, and is independent of the electronic coordinates. The vector pointing from nucleus μ to electron j is $\boldsymbol{r}_j^\mu = \boldsymbol{r}_j - \boldsymbol{R}_\mu$. By symmetry of the AO χ_μ, $<\chi_\mu|\boldsymbol{r}_j^\mu|\chi_\mu> = 0$, and

$$\boldsymbol{M}_{i\to k} = -|e|\sqrt{2}\sum c_{\mu k}^* c_{\mu i}\boldsymbol{R}_\mu \tag{1.41}$$

Therefore, in the ZDO approximation, $\boldsymbol{M}_{i\to k}$ is given by the electric dipole moment of point charges at the centers μ. However, these point charges are not due to the orbital charge distributions $-|e|\ |\phi_i|^2$ and $-|e|\ |\phi_k|^2$ of an electron in one of the MOs ϕ_i or ϕ_k, but rather from the overlap charge distribution $-|e|\phi_k^*\phi_i$, which is positive in some regions of space and negative in others. Integration over all space yields zero, as the MOs ϕ_i and ϕ_k are mutually orthogonal.

Example 1.7:
Starting from the HMOs of butadiene

$$\begin{aligned}
\phi_1 &= 0.37\ (\chi_1 + \chi_4) + 0.60\ (\chi_2 + \chi_3)\\
\phi_2 &= 0.60\ (\chi_1 - \chi_4) + 0.37\ (\chi_2 - \chi_3)\\
\phi_3 &= 0.60\ (\chi_1 + \chi_4) - 0.37\ (\chi_2 + \chi_3)\\
\phi_4 &= 0.37\ (\chi_1 - \chi_4) - 0.60\ (\chi_2 - \chi_3)
\end{aligned}$$

the transition dipole for the HOMO→LUMO transition may be calculated within the HMO approximation as follows: From Equation (1.41),

$$\begin{aligned}
\boldsymbol{M}_{2\to 3} &= -|e|\sqrt{2}<\phi_3|\boldsymbol{r}|\phi_2>\\
&= -|e|\sqrt{2}[(0.60)^2(\boldsymbol{R}_1 - \boldsymbol{R}_4) - (0.37)^2(\boldsymbol{R}_2 - \boldsymbol{R}_3)]
\end{aligned}$$

Using Equation (1.37) for the components of $\boldsymbol{M}$ in the system of axes shown in Figure 1.9, and assuming that all bond lengths are equal to l_0 and all valence angles to 120°, one obtains for *s-trans*-butadiene

$$(\boldsymbol{M}_{2\to 3})_x = \sqrt{2}[(0.60)^2(m + m) - (0.37)^2(\tfrac{1}{2}m + \tfrac{1}{2}m)] = 0.82m$$

$$(\boldsymbol{M}_{2\to 3})_y = \sqrt{2}[(0.60)^2(\tfrac{1}{2}\sqrt{3}m + \tfrac{1}{2}\sqrt{3}m) - (0.37)^2(0 + 0)] = 0.88m$$

where $m = -|e|l_0$. The x as well as the y components are nonzero, and the transition moment is oriented along the long axis of the molecule.

The system of axes for *s-cis*-butadiene is different, and the y component is the same as the x component for *s-trans*-butadiene. The z component vanishes, and the transition moment is oriented parallel to the y axis. Since $\boldsymbol{M}^2 = M_x^2 + M_y^2$ or $\boldsymbol{M}^2 = M_y^2 + M_z^2$, the absorption band due to the

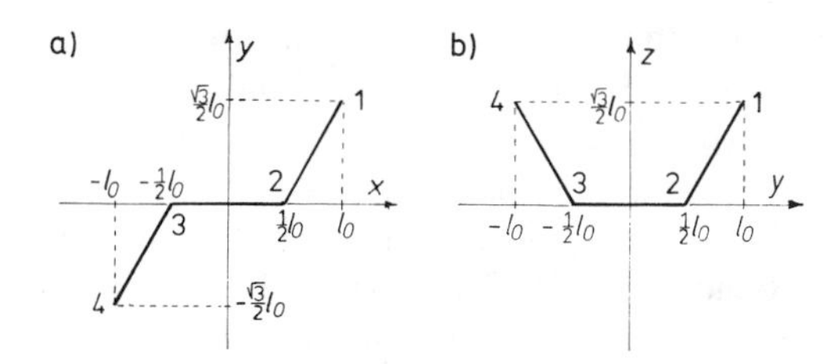

Figure 1.9. Cartesian coordinates of π centers of a) *s-trans*-butadiene and b) *s-cis*-butadiene, assuming equal lengths l_0 for all bonds.

HOMO→LUMO transition is predicted to be more intense for *s-trans*-butadiene than for *s-cis*-butadiene.

For the transitions $\phi_2 \rightarrow \phi_4$ and $\phi_1 \rightarrow \phi_3$, which are degenerate due to the pairing properties of alternant hydrocarbons, the same arguments lead to

$$\boldsymbol{M}_{2\rightarrow 4} = \boldsymbol{M}_{1\rightarrow 3} = -|e|\sqrt{2}[0.37 \times 0.60\,(\boldsymbol{R}_1 + \boldsymbol{R}_4 - \boldsymbol{R}_2 - \boldsymbol{R}_3)]$$

When first-order configuration interaction is taken into account, the contributions from these two transitions add in-phase for excitation into the plus state and out-of-phase for excitation into the minus state; cf. Equation (1.26). From the coordinates x_μ and y_μ or y_μ and z_μ depicted in Figure 1.9, it is easy to see that for *s-trans*-butadiene $M_x = M_y = 0$. Accordingly $\boldsymbol{M}^2 = 0$, the transition probability is zero and the transitions into the plus and the minus states are both forbidden. For *s-cis*-butadiene, which does not possess a center of symmetry, $M_y = 0$, but $M_z \neq 0$. The transition moment differs from zero and is oriented perpendicular to the long axis of the molecule. The transition into the plus state is therefore allowed. For excitation into the minus state the two contributions cancel each other and this transition is forbidden irrespective of the molecular geometry. (Cf. also Example 1.9.)

1.3.2 Selection Rules

Selection rules that differentiate between formally "allowed" and "forbidden" transitions can be derived from the theoretical expression for the transition moment. A transition with a vanishing transition moment is referred to as being forbidden and should have zero intensity. But it should be remembered that the transition moment of an allowed transition, although nonvanishing, can still be very small, whereas a forbidden transition may be observed in the spectrum with finite intensity if the selection rule is relaxed by an appropriate perturbation. The most important example are vibrationally induced transitions, which will be discussed later. (Cf. Section 1.3.4.) Other effects such as solvent perturbations may play a significant role also. Finally, since a series of approximations is necessary in order to derive the selection rules, they can be obeyed only within the limits of validity of these approximations.

For instance, from the approximations introduced into the theoretical treatment of the radiation field it follows that only one-photon processes are allowed. However, very intense radiation fields, especially those produced by lasers, can cause simultaneous absorption of two photons, thus making it possible to reach molecular states that are not accessible from the ground state via one-photon absorption. Quite often, the only other evidence for the existence of these states is indirect, and two-photon absorption spectroscopy is complementary to conventional one-photon spectroscopy. (Cf. Section 1.3.6.)

The *spin selection rule* is a consequence of the fact that the electric dipole and quadrupole moment operators do not operate on spin. Integration over the spin variables then always yields zero if the spin functions of the two states Ψ_0 and Ψ_f are different, and an electronic transition is spin allowed only if the multiplicities of the two states involved are identical. As a result, singlet–triplet absorptions are practically inobservable in the absorption spectra of hydrocarbons, or for that matter, other organic compounds without heavy atoms. Singlet–triplet excitations are readily observed in electron energy loss spectroscopy (EELS), which obeys different selection rules (Kuppermann et al., 1979).

Strictly speaking, however, the spin angular momentum and its components are not constants of motion in nonlinear molecules, and the classification of states by multiplicity is therefore only approximate. Spin-orbit coupling is the most important of the terms in the Hamiltonian that cause a mixing of zero-order pure multiplicity states. The interaction between the spin angular momentum of an electron and the orbital angular momentum of the same electron causes the presence of a minor term in the Hamiltonian, which may be written as

$$\hat{H}_{SO} = \frac{e^2}{2m_e^2c^2} \sum_j \sum_\mu \frac{Z_\mu}{|\boldsymbol{r}_j^\mu|^3} \hat{\boldsymbol{l}}_j^\mu \cdot \hat{\boldsymbol{s}}_j \tag{1.42}$$

Here, $\hat{\boldsymbol{l}}_j^\mu = \boldsymbol{r}_j^\mu \times \hat{\boldsymbol{p}}_j$ is the orbital angular momentum operator of the electron j, $\boldsymbol{r}_j^\mu$ is the vector pointing from nucleus μ to electron j, and the sums run over all electrons j and all nuclei μ. Spin-orbit coupling is particularly significant in the presence of atoms of high atomic number Z_μ ("heavy atom effect").

The presence of an inseparable scalar product $\hat{\boldsymbol{l}}_j \cdot \hat{\boldsymbol{s}}_j$ of one operator that acts only on the spatial part of the total electronic wave function and another one that acts only on its spin part will cause an interaction between the various pure multiplicity wave functions. The resulting eigenfunctions of $\hat{H}_{el}$ will therefore be represented by mixtures of functions that differ in multiplicity. However, since $\hat{H}_{SO}$ constitutes only a very small term in $\hat{H}_{el}$, mixing is normally not very severe, and the resulting wave functions contain predominantly functions of only one multiplicity. Commonly, such "impure singlets," "impure triplets," etc., are still referred to simply as singlets, triplets, etc. However, since such a "singlet" state contains a small triplet contribution and the "triplet" state a small singlet contribution, the transition moment for such a singlet–triplet transition

$$\begin{aligned} <^3\Psi_l|\hat{\boldsymbol{M}}|^1\Psi_k> &= <\,^3\Psi_l^{(0)} + \sum_s \lambda_s\, ^1\Psi_s^{(0)} + \ldots |\hat{\boldsymbol{M}}|^1\Psi_k^{(0)} + \sum_r \lambda_r\, ^3\Psi_r^{(0)} + \ldots > \\ &= \sum_s \lambda_s <^1\Psi_s^{(0)}|\hat{\boldsymbol{M}}|^1\Psi_k^{(0)}> + \sum_r \lambda_r <^3\Psi_l^{(0)}|\hat{\boldsymbol{M}}|^3\Psi_r^{(0)}> + \ldots \quad (1.43) \end{aligned}$$

can be different from zero, depending on the values of λ_r and λ_s.

Example 1.8:
Singlet–triplet transitions are of considerable importance not only for heavy atom compounds but also for carbonyl compounds. This can be visualized simply by reference to the principle of conservation of angular momentum, based on the fact that the spin-orbit coupling operator mixes states related by a simultaneous change of spin angular momentum and orbital angular momentum. Intersystem crossing between states represented by single configurations is therefore most favorable if they differ by promotion of an electron between molecular orbitals containing atomic p orbitals whose axes are mutually perpendicular in one and the other MO (Salem and Rowland, 1972; Michl, 1991).

In order to examine the spin-orbit interaction between a $^1(n,\pi^*)$ and a $^3(\pi,\pi^*)$ state for a carbonyl compound, it is sufficient to consider the $2p_x$ and the $2p_y$ AOs, x_O and y_O, on oxygen and the $2p_y$ AO, y_C, on carbon. The $^1(n,\pi^*)$ state and the three components T_1, T_0, and T_{-1} of the $^3(\pi,\pi^*)$ state may then be written as

$$^1(n,\pi^*) = [|x_O\bar{y}_C y_O\bar{y}_O| + |y_C\bar{x}_O y_O\bar{y}_O|]/\sqrt{2}$$

and

$$^3(\pi,\pi^*) = \begin{cases} |x_O\bar{x}_O y_C y_O| \\ [|x_O\bar{x}_O y_C\bar{y}_O| - |x_O\bar{x}_O y_O\bar{y}_C|]/\sqrt{2} \\ |x_O\bar{x}_O\bar{y}_C\bar{y}_O| \end{cases}$$

For an evaluation of the matrix element $<^1(n,\pi^*)|\hat{H}_{SO}|^3(\pi,\pi^*)>$ of the operator given in Equation (1.42), it is convenient to first determine the effect of the spin part of the three components $\hat{l}_x\hat{s}_x$, $\hat{l}_y\hat{s}_y$, and $\hat{l}_z\hat{s}_z$ of the scalar product $\hat{l}\cdot\hat{s}$ on the triplet function and to perform the spin integration, which reduces the number of terms considerably. In this way one obtains the following:

$$\begin{aligned} &<[|x_O\bar{y}_C y_O\bar{y}_O| + |y_C\bar{x}_O y_O\bar{y}_O|]/\sqrt{2}|\sum_j \hat{\boldsymbol{l}}^j \cdot \hat{\boldsymbol{s}}^j|[|x_O\bar{x}_O y_C\bar{y}_O| - |x_O\bar{x}_O y_O\bar{y}_C|]/\sqrt{2}> \\ &= \tfrac{1}{2}(<y_O|\hat{l}_z^1|x_O> + <y_O|\hat{l}_z^2|x_O>) = \hbar^2, \\ &<[|x_O\bar{y}_C y_O\bar{y}_O| + |y_C\bar{x}_O y_O\bar{y}_O|]/\sqrt{2}|\sum_j \hat{\boldsymbol{l}}^j \cdot \hat{\boldsymbol{s}}^j|\,|x_O\bar{x}_O y_C y_O|> = 0 \\ &<[|x_O\bar{y}_C y_O\bar{y}_O| + |y_C\bar{x}_O y_O\bar{y}_O|]/\sqrt{2}|\sum_j \hat{\boldsymbol{l}}^j \cdot \hat{\boldsymbol{s}}^j|\,|x_O\bar{x}_O\bar{y}_C\bar{y}_O|> = 0 \end{aligned}$$

In this case, the x and y components of the operator $\hat{l} \cdot \hat{s}$ give vanishing contributions.

If the orbitals involved are centered at the same atom, as is the case for carbonyl compounds, the spin-orbit coupling is particularly large due to the factor $|r_j^\mu|^3$ in the denominator of Equation (1.42). The $^1(\pi,\pi^*) \leftrightarrow {}^3(\pi,\pi^*)$ and $^1(n,\pi^*) \leftrightarrow {}^3(n,\pi^*)$ transitions are forbidden by the angular momentum conservation rule, since only the spin angular momentum changes, causing all one-center terms of the spin-orbit coupling to vanish. The results for a $^1(\pi,\pi^*) \leftrightarrow {}^3(n,\pi^*)$ transition are similar to those for a $^1(n,\pi^*) \leftrightarrow {}^3(\pi,\pi^*)$ transition.

This result is sometimes represented pictorially using up-and-down-directed arrows to represent electron spin. It is difficult to do this correctly, since it is the T_0 component of the triplet state that is responsible for the intersystem crossing, and not the T_1 or T_{-1}, which are the ones that are easily represented pictorially.

Spatial *symmetry selection rules* are another very important type of selection rules. They occur because the transition dipole moment in Equation (1.35) can vanish not only due to spin integration but also as a result of integration over the space coordinates of the electrons. This is always the case if the integrand is not totally symmetric or does not contain a totally symmetric component. If the integrand is antisymmetric with respect to at least one symmetry operation of the point group of the molecule, the integral vanishes, since positive and negative contributions from different regions of space defined by one or more symmetry operations cancel. Whether or not the integrand is totally symmetric can be decided easily by determining the irreducible representation χ of the point group of the molecule for the ground state Ψ_0, the excited state Ψ_f, and for the transition moment operator, and forming the direct product of all three.

The components M_α of the electric dipole transition moment are given by the Cartesian coordinates $\alpha = x, y$, and z. The α component of the transition dipole vanishes unless the direct product

$$\chi(M_\alpha) = \chi(\Psi_0) \times \chi(\alpha) \times \chi(\Psi_f) \tag{1.44}$$

forms a basis for the totally symmetric irreducible representation or a reducible representation that contains the totally symmetric irreducible representation. If all three components M_α vanish by symmetry the transition is said to be symmetry forbidden. Since the ground state Ψ_0 of closed-shell systems is totally symmetric, a transition can be symmetry allowed only if the excited state Ψ_f and at least one of the Cartesian coordinates α form bases of the same irreducible representation of the molecular point group.

In point groups with a center of symmetry, x, y, and z are of ungerade parity (u); that is, they change sign upon inversion. Therefore, one and only one of the two states Ψ_0 and Ψ_f has to be of ungerade parity if the transition moment is to be nonzero. Thus, within the electric dipole approximation only transitions g $\leftrightarrow$ u between a gerade and a ungerade state are allowed, whereas transitions g $\leftrightarrow$ g and u $\leftrightarrow$ u between states of the same parity are forbidden.

In applications of these symmetry selection rules it has to be remembered that the symmetry of a molecule can be lowered by vibrational motions so that symmetry-forbidden transitions may nevertheless be observed—for instance, the two longest-wavelength singlet–singlet transitions in benzene. Vibronic coupling and the shape of the absorption bands will be discussed in the following sections.

Example 1.9:
In applying the symmetry selection rules it is frequently quite convenient to start from Equation (1.39) and examine whether or not the integral $<\phi_k|\alpha|\phi_i>$ vanishes by symmetry. Thus, it is easy to see that for *s-trans*-butadiene, which belongs to the point group C_{2h} and has been discussed already in Example 1.7, the transition dipole of the HOMO→LUMO transition $\phi_2(b_g)\rightarrow\phi_3(a_u)$ has nonzero x and y components: both the x and y coordinates form a basis of the irreducible representation B_u and $B_g \times B_u \times A_u = A_g$. The transitions $\phi_2(b_g)\rightarrow\phi_4(b_g)$ and $\phi_1(a_u)\rightarrow\phi_3(a_u)$ are parity forbidden, as $B_g \times B_g = A_u \times A_u = A_g$ and the symmetry of the Cartesian coordinates is ungerade. (Cf. Figure 1.10.)

Since *s-cis*-butadiene belongs to the point group C_{2v}, which has no center of symmetry, it has no parity-forbidden transitions. $M_y \neq 0$ for the transition

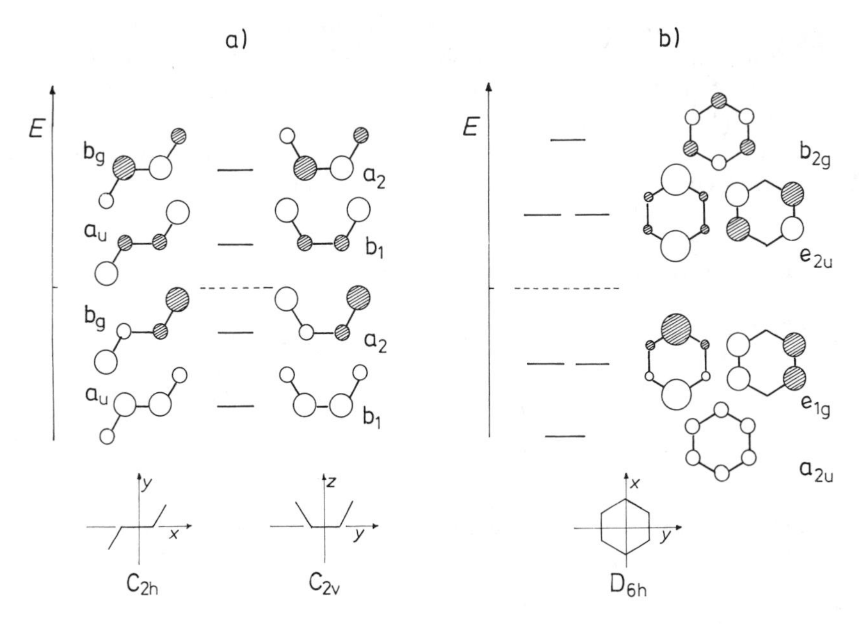

Figure 1.10. Schematic representation and symmetry labels of π MOs a) of *s-trans-* and *s-cis*-butadiene and b) of benzene.

$\phi_2(a_2) \rightarrow \phi_3(b_1)$ because $A_2 \times B_1 = B_2$, and the y coordinate transforms like B_2. However, $M_z \neq 0$ for the transitions described by $\phi_2(a_2) \rightarrow \phi_4(a_2)$ and $\phi_1(b_1) \rightarrow \phi_3(b_1)$, since the z coordinate forms a basis for the totally symmetric representation A_1.

In benzene the HOMO and LUMO are degenerate. The MO ϕ_2 and ϕ_3 form a basis for the irreducible representation E_{1g}, whereas the MOs ϕ_4 and ϕ_5 form a basis for the irreducible representation E_{2u} of the point group D_{6h}. (Cf. Figure 1.10.) From the direct product

$$E_{1g} \times E_{2u} = B_{1u} + B_{2u} + E_{1u}$$

it is seen that one allowed degenerate transition is to be expected, since the x and y coordinates form a basis for the irreducible representation E_{1u}. For the other two transitions one has $\boldsymbol{M} = 0$; hence they are electronically forbidden by symmetry and can be observed only as a result of vibronic coupling. (Cf. Section 1.3.4.)

Finally, according to Equation (1.39) the transition moment also vanishes if the differential overlap $\phi_k^*\phi_i$ is zero everywhere in space. This is not strictly possible (except within the ZDO approximation), but the product $\phi_k^*\phi_i$ may reach very small values if the amplitude of MO ϕ_k only is large in those regions of space where the amplitude of ϕ_i is very small, and vice versa. Consequently the transition dipole moment will also be very small in such a case. This is true for $n \rightarrow \pi^*$ transitions, where an electron is excited from a lone-pair orbital in the molecular plane into a π^* orbital of an unsaturated system, for which the molecular plane is also a nodal plane. Similar reasoning applies to so-called *charge transfer transitions,* that is, those in which an electron is transferred from one subsystem to another, the orbitals ϕ_i and ϕ_k are localized in different regions of space. Such transitions are *overlap forbidden.* This is not, however, a very strict selection rule, since it is not based on a vanishing but only on a small value of the transition moment. The differential overlap is never exactly zero in practice.

Table 1.3 Selection Rules for Electronic Transitions

Rule	Criteria	Validity
Spin	$\langle\theta_0\|\theta_f\rangle = 0$	Strict (however, see spin-orbit coupling)
Symmetry	$\chi(\Psi_0) \times \chi(\Psi_f) \neq \chi(\alpha)$ or $\chi(\phi_i) \times \chi(\phi_k) \neq \chi(\alpha)$, with $\alpha = x, y, z$	Strict (however, see vibronic coupling)[a]
Alternant pairing	$\Psi^+ \rightarrow \Psi^+$ or $\Psi^- \rightarrow \Psi^-$	Moderately strict
Local symmetry or overlap	$\phi_k^*\phi_i \equiv 0$	Weak

[a] Section 1.3.4.

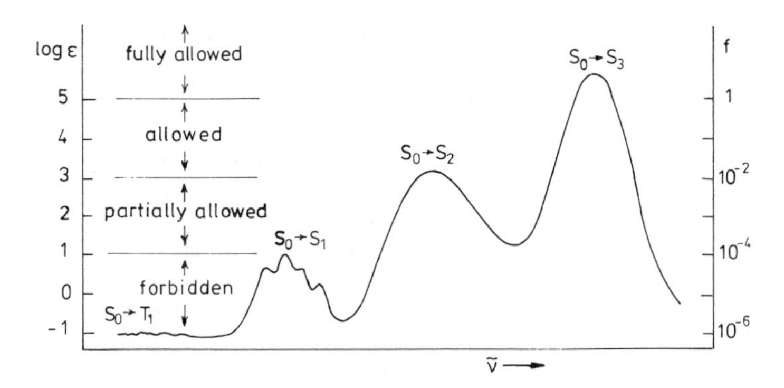

Figure 1.11. Schematic representation of the absorption spectrum of an organic molecule with some allowed transitions and some that are forbidden by spin, symmetry, or overlap selection rules (from left to right). The log ε and f values of the ordinate are only meant to provide rough orientation. In particular, according to Equation (1.27), there is no simple relation between log ε and f (by permission from Turro, 1978).

Note that overlap-forbidden transitions also tend to have small singlet–triplet splitting. This is given in the first approximation by $2K_{ik}$, that is, by twice the self-repulsion of the overlap charge density $\phi_k^*\phi_i$. If the overlap charge density is small everywhere, its self-repulsion can hardly be large.

Table 1.3 presents all the selection rules discussed so far, and Figure 1.11 represents schematically the relative intensity of various allowed and forbidden transitions.

Example 1.10:

The absorption spectrum of naphthalene is shown in Figure 1.17 (cf. also Figure 2.7): A very weak transition near 32,000 cm^{-1} is followed by a medium-intensity transition near 35,000 cm^{-1} and a very intense transition near 45,000 cm^{-1}. All three are $\pi\rightarrow\pi^*$ transitions; the first one is nearly exactly forbidden by the (approximate) alternant pairing symmetry, as it is a transition between two minus states (cf. Example 1.5); the intensity is due nearly entirely to vibronic coupling. (Cf. the first transition of benzene, Example 1.11.) The second one is predominantly the HOMO→LUMO transition, and the transition moment is determined by the transition density $\phi_{\text{LUMO}}^*\phi_{\text{HOMO}}$. In the azulene spectrum (Figure 2.28) the HOMO→LUMO transition corresponds to the longest-wavelength absorption near 14,500 cm^{-1} and is much weaker ($\varepsilon \approx 200$) than the HOMO→LUMO transition of naphthalene ($\varepsilon \approx 6{,}500$). This lower intensity is due to the fact that the HOMO and LUMO of azulene are localized largely on different atoms in the π system, so the transition density $\phi_{\text{LUMO}}^*\phi_{\text{HOMO}}$ is small, and so is its dipole moment (Section 1.3.1). This is only possible because azulene is nonalternant. The HOMO and LUMO of naphthalene are mostly localized on the same atoms, as they must be for an alternant hydrocarbon; the transition density is much larger, as is its dipole moment. It is not a coincidence that the weak HOMO→LUMO transition in azulene also has a small singlet–triplet splitting. From Equations (1.23) and (1.24), this is given by ${}^1\Delta E_{i\rightarrow k} - {}^3\Delta E_{i\rightarrow k} = 2K_{ik}$, where K_{ik} is the exchange integral that is equal to the self-repulsion of the transition density $\phi^*_{\mathrm{LUMO}}\phi_{\mathrm{HOMO}}$ (Section 1.2.3). Clearly, a small transition density is more likely to have a small self-repulsion energy than a large one. In fact, from the experimental excitation energies one finds $2K_{\mathrm{HOMO,LUMO}} = 1.70$ eV for naphthalene and 0.50 eV for azulene. (Cf. Example 1.3 and Michl and Thulstrup, 1976.)

Similar arguments apply to n→π* transitions in carbonyl compounds. The n and the π* orbitals are largely located in different parts of space and the transition density is small. These transitions are weak: $\varepsilon \approx 35$ in acrolein (see also the spectra of polyene aldehydes shown in Figure 2.37), and the singlet–triplet splitting is small: 0.2 eV in the case of the n→π* transition of acrolein (Alves et al., 1971). In the case of CT complexes, the HOMO is primarily located on the donor, the LUMO is primarily located on the acceptor, and the transition density is again small. The intensity of the CT transition is again low and the S—T splitting small.

1.3.3 The Franck-Condon Principle

When discussing symmetry selection rules it was mentioned that vibrational motion can influence both the shape and the intensity of electronic absorption bands. In the usual Born-Oppenheimer approximation with molecular wave functions written as products as in Equation (1.12) this can be understood as follows.

Electronic motion with a typical frequency of 3×10^{15} s^{-1} ($\tilde{\nu} = 10^5$ cm^{-1}) is much faster than vibrational motion with a typical frequency of 3×10^{13} s^{-1} ($\tilde{\nu} = 10^3$ cm^{-1}). As a result of this, the electric vector of light of frequencies appropriate for electronic excitation oscillates far too fast for the nuclei to follow it faithfully, so the wave function for the nuclear motion is still nearly the same immediately after the transition as before. The vibrational level of the excited state whose vibrational wave function is the most similar to this one has the largest transition moment and yields the most intense transition (is the easiest to reach). As the overlap of the vibrational wave function of a selected vibrational level of the excited state with the vibrational wave function of the initial state decreases, the transition moment into it decreases; cf. Equation (1.36). Absorption intensity is proportional to the square of the overlap of the two nuclear wave functions, and drops to zero if they are orthogonal. This statement is known as the *Franck-Condon principle* (Franck, 1926; Condon, 1928; cf. also Schwartz, 1973):

$$|\tilde{\boldsymbol{M}}_{0v\rightarrow fv'}|^2 = |\boldsymbol{M}_{0\rightarrow f}|^2 \, |\langle\chi^{f}_{v'}|\chi^{0}_{v}\rangle|^2 \tag{1.45}$$

The squared overlap integrals of the vibrational wave functions are referred to as the *Franck-Condon factors*.

If the potential governing the nuclear motion is accidentally similar in the initial and final electronic states described by Ψ_0 and Ψ_f, respectively, with a minimum at the same equilibrium geometry, the two operators $\hat{H}_{vib}(j, Q)$ for these two states as well as their vibrational wave functions are identical. The vibrational wave functions $\chi^f_{v'}$ and χ^0_v then are orthonormal. The non-vanishing factors will be $<\chi^f_{v'}|\chi^0_v> = \delta_{v'v}$, and only the 0→0, 1→1, . . ., $v \rightarrow v$ vibrational transitions will be observed in absorption or emission. The transition is then called Franck-Condon allowed. More commonly, the potentials and the equilibrium geometries of the electronic states Ψ_0 and Ψ_f will differ, $\chi^f_{v'}$ and $\chi^0_{v'}$ will be eigenfunctions of different operators $\hat{H}_{vib}(j,Q)$, and will be nonorthogonal. The Franck-Condon factor $<\chi^f_{v'}|\chi^0_v>^2$ will then be a measure of the relative intensity of the vibrational component of the absorption band that corresponds to a transition from the vibrational level v of the initial state (frequently the lowest vibrational level) into the vibrational level v' of the excited state.

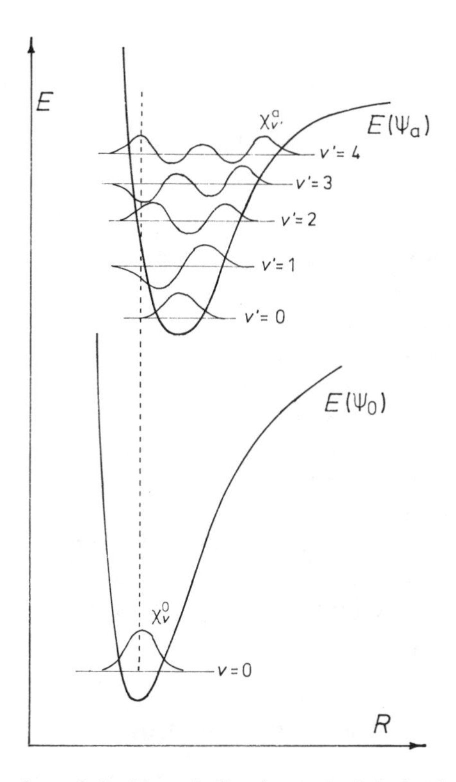

Figure 1.12. Illustration of the Franck-Condon principle in the case of a diatomic molecule: the absolute value of the integral $<\chi^f_{v'} \mid \chi^0_v>$ is largest for "vertical transitions," since the amplitudes of χ^0_v and $\chi^f_{v'}$ are largest at the equilibrium bond length and near the turning point, respectively.

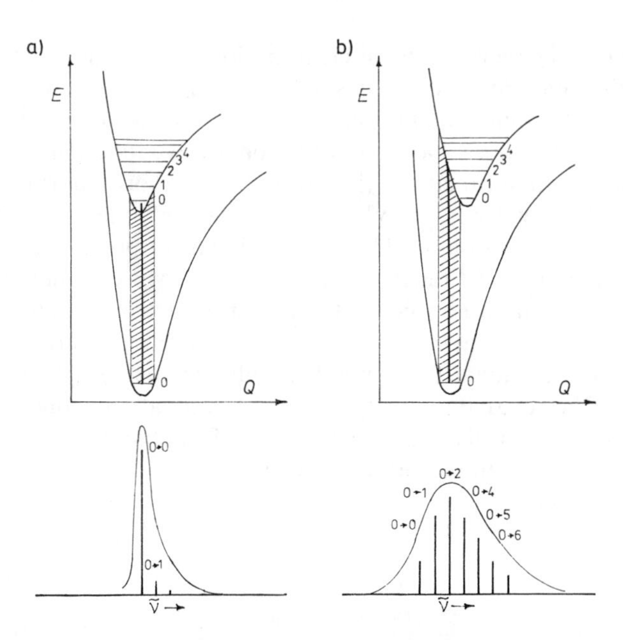

Figure 1.13. Schematic representation of the band shape to be expected from the Franck-Condon principle as a function of geometry changes on excitation: a) approximately equal equilibrium internuclear distances in the ground and excited state; b) internuclear distance in the excited state larger than in the ground state.

Figure 1.12 pictures schematically the potential energy curves of two states of a diatomic molecule together with a few vibrational energy levels and the corresponding harmonic oscillator eigenfunctions. It is seen that the amplitude of the function $\chi^{f}_{v'}$ of the excited vibrational levels is largest near the classical turning point, so the value of the integral $\langle\chi^{f}_{v'}|\chi^{0}_{v''}\rangle$ is particularly large if the classical turning point is located vertically above the equilibrium internuclear separation of the ground state. Such "vertical" transitions are most probable according to the Franck-Condon principle.

Since the Franck-Condon factors for all vibrational levels of the excited state add up to unity, the total intensity of a transition is given by the electronic dipole transition moment $\boldsymbol{M}_{0\to f}$. The resulting intensity distribution of the vibrational fine structure is depicted in Figure 1.13 for some typical cases.

1.3.4 Vibronically Induced Transitions

In Example 1.9 it was pointed out that symmetry-forbidden transitions may become observable due to symmetry-lowering vibrational motions. This cannot be explained by means of the Franck-Condon principle, which governs only the distribution of intensity due to a nonvanishing transition moment over the various vibrational components of the band. The phenomenon can be understood by taking into account the geometry dependence of the transition dipole moment $M_{0\rightarrow f} = <\Psi_f^Q|\hat{M}|\Psi_0^Q>$, which was neglected in the derivation of the Franck-Condon principle and in Equation (1.36).

Herzberg and Teller (1947) described the dependence of the electronic transition dipole moment on the nuclear geometry Q by a McLaurin series expansion of the Hamiltonian

$$\hat{H}_{el}^{(Q)}(q) = \hat{H}_{el}^{(0)} + \sum_k (\partial \hat{H}_{el}/\partial Q_k)_0 Q_k \quad (1.46)$$

around the equilibrium geometry denoted by 0, and truncation after the linear term. The use of first-order perturbation theory to obtain the electronic wave function Ψ_f^Q of the final electronic state for an arbitrary geometry Q in the vicinity of both equilibrium geometries, Q_0 and Q_f, yields

$$\tilde{M}_{0v\rightarrow fv'} = <\chi_{v'}^f(Q)|\chi_v^0(Q)> M_{0\rightarrow f} + \sum_{k=1}^{3N-6} <\chi_{v'}^f|<\Psi_b|(\partial \hat{H}_{el}/Q_k)_0|\Psi_f> Q_k|\chi_v^0> M_{0\rightarrow b}/(E_b - E_f) \quad (1.47)$$

Here the sum runs over all $3N-6$ normal modes of vibration with normal coordinates Q_k. Furthermore it is assumed that only one state b is energetically close to state f, so the perturbation expansion can be restricted to a single term.

The first term in Equation (1.47) is identical with the expression derived in the last section for electronically allowed transitions. It is presently assumed to be very small or zero. ($M_{0\rightarrow f} = 0$ for symmetry-forbidden transitions.) The second term results from vibronic mixing and represents a first-order vibronic contribution to the transition moment. It is seen that in this description the forbidden transition 0→f "steals" or "borrows" intensity from the allowed transition 0→b. If $M_{0\rightarrow f}$ is exactly zero all observed components of the electronic transition will be polarized along the direction of the transition dipole moment $M_{0\rightarrow b}$. The 0→0 transition ($v = v' = 0$) will have zero intensity and only vibrational levels of overall symmetry given by the direct product of symmetries of the states Ψ_f and Ψ_b will appear.

Example 1.11:
In benzene both the $^1B_{2u}$ and the $^1B_{1u}$ states can mix with the $^1E_{1u}$ state through vibronic coupling using an e_{2g} vibration. Since the $^1A_{1g} \rightarrow {}^1E_{1u}$ transition is symmetry allowed, both the $^1A_{1g} \rightarrow {}^1B_{2u}$ and the $^1A_{1g} \rightarrow {}^1B_{1u}$ transitions become allowed through vibronic coupling. One of the factors that make the intensity gain larger for the $^1A_{1g} \rightarrow {}^1B_{1u}$ transition near 200 nm than for the $^1A_{1g} \rightarrow {}^1B_{2u}$ transition near 260 nm is the energetic closeness of the former to the $^1A_{1g} \rightarrow {}^1E_{1u}$ transition at 180 nm, from which the intensity is borrowed. [Cf. Equation (1.47) with the energy difference between the two excited states in the denominator.] In Figure 1.14 the absorption corresponding to the $^1A_{1g} \rightarrow {}^1B_{2u}$ transition is shown: The $0 \rightarrow 0$ transition is symmetry forbidden and is not observable. Its energy can be determined if the "hot" band labeled H, whose intensity

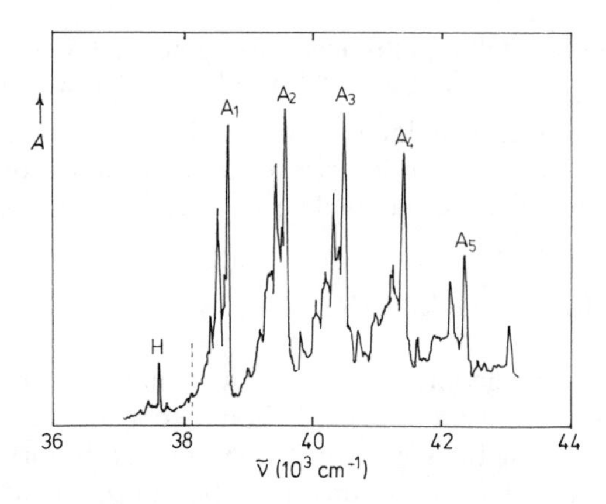

Figure 1.14. The 260-nm absorption band of benzene (by permission from Callomon et al., 1966). The vibronic progression based on the totally symmetric breathing mode $\tilde{\nu} = 920$ cm^{-1} is labeled A_1, A_2, . . . , and the "hot" band is marked H.

increases with temperature, is assigned to a transition from a vibrationally excited level of the ground state, with a quantum of an e_{2g} vibration at $\tilde{\nu} = 608$ cm^{-1}. The vibronic progression labeled by A_1, A_2 . . . may be identified as due to vibronic coupling induced by the corresponding e_{2g} vibration of the excited state, for which $\tilde{\nu} = 520$ cm^{-1} is obtained from the wave-number difference of the bands H and A_1. The progression contains multiples of the totally symmetric breathing mode, $\tilde{\nu} = 920$ cm^{-1}.

1.3.5 Polarization of Electronic Transitions

From Equation (1.33) it is seen that the transition moment is a vector quantity. The square of its absolute value determines the transition probability, whereas its direction is called the *polarization direction*. If one of the principal axes of a molecule is the long axis, the transition as well as the corresponding absorption band is often called parallel or *long axis polarized*; if the direction of the transition moment vector is perpendicular to the long axis, the transition and the absorption band are called perpendicular or *short-axis polarized*. The longest wavelength $\pi \rightarrow \pi^*$ transition of *s-cis*-butadiene in Example 1.7 is long-axis polarized, the second $\pi \rightarrow \pi^*$ transition is short-axis polarized.

The polarization direction can be measured by investigating oriented molecules, most often with polarized light. From Equation (1.32) the oscillator strength $f_{0\rightarrow f}$ is seen to be proportional to the squared scalar product of the transition moment and the electric field vector of the light. The absorption reaches its maximum if the direction of the transition moment and the polarization direction of the light coincide, whereas no light is absorbed if they are perpendicular to each other.

In the gas phase and in solution all orientations of the molecules are equally probable. Measurements of the direction of the transition moment with respect to the polarization direction of the light by polarized absorption are therefore not possible. However, if the molecules are imbedded in a single crystal, in an aligned liquid crystal, or in a stretched polyethylene or poly(vinyl alcohol) sheet, such measurements become possible. (Cf. Example 1.12.) Molecules may be frozen in dilute glassy solutions, where they are not initially oriented, but cannot rotate. When irradiated with linearly polarized light, they are excited with different probabilities according to the relative orientation of their transition dipole moment with respect to the polarization direction of the light. If the emission of these molecules can be measured, information about the polarization of the transitions involved may be obtained from the fluorescence polarization. (Cf. Section 5.3.3.) If such molecules undergo a phototransformation, both the remaining starting solute and the photoproduct may be partially oriented, again permitting polarization information to be deduced. A detailed treatment of the various spectroscopic measurements that are possible with polarized light has been given by Michl and Thulstrup (1986).

Example 1.12:
Figure 1.15 shows the absorption spectrum of a cyanine dye embedded in a stretched poly(vinyl alcohol) sheet, measured with linearly polarized light. The

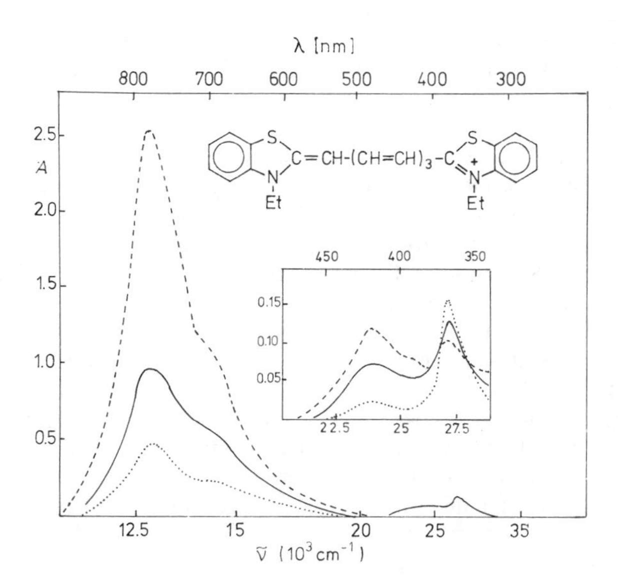

Figure 1.15. Polarization spectrum of a cyanine dye with benzothiazole end groups in a poly(vinylalcohol) sheet; (—) unstretched sheet; polarization direction of the light and stretching direction parallel (---) and perpendicular (···) (by permission from Eckert and Kuhn, 1960).

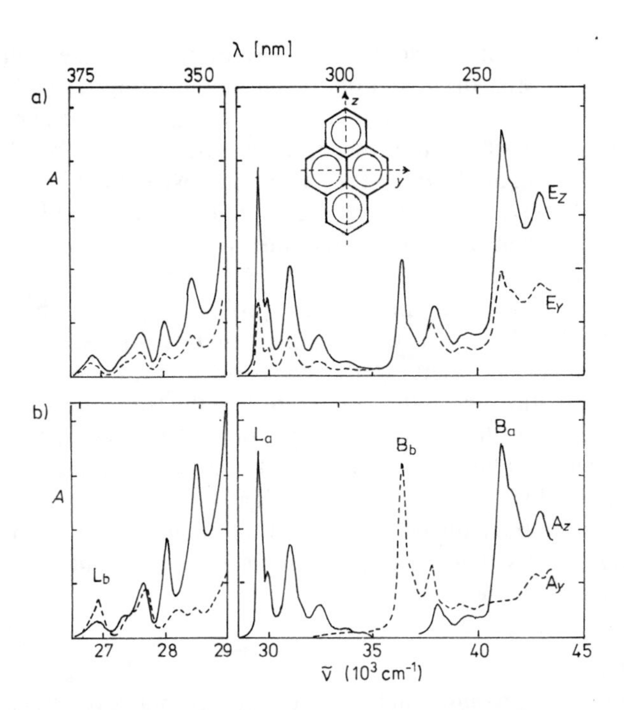

Figure 1.16. Polarized absorption spectra of pyrene in stretched polyethylene: a) observed absorbances E_Z and E_Y parallel and perpendicular to the polarization direction of the light, after baseline correction; b) reduced spectra $A_Z(\lambda) = E_Z - 1.0\, E_Y$ (—) and $A_Y(\lambda) = 0.625\,(2.8\, E_Y - E_Z)$ (---). The absorbance scale is in arbitrary units, different for the different sections of the spectrum (by permission from Michl and Thulstrup, 1986).

rod-shaped dye molecules are preferentially aligned in the direction of stretching. Since the absorption is more intense if the stretching direction and the polarization direction of the light coincide, the transition shown must be approximately long-axis polarized. As the alignment of the molecules is not perfect, the absorption does not disappear entirely if the stretching direction and the light polarization direction are mutually perpendicular.

Figure 1.16 gives the absorption spectra of pyrene, a higher-symmetry molecule whose transition moments lie in the molecular symmetry axes. One of the two absorption curves is due to transition moments oriented along the long axis of the molecule; the other is due to moments oriented along the in-plane short axis. The reduced spectra were calculated from the experimental spectra measured on stretched polyethylene sheets at different relative orientations of the stretching direction and the polarization direction of the light.

1.3.6 Two-Photon Absorption Spectroscopy

From Equation (1.32) it is seen that for the extinction coefficient, which is proportional to the absorption cross section, one obtains

$$\varepsilon \sim |e_z \cdot M_{0\to f}|^2 \quad (1.48)$$

for a linearly polarized photon with a polarization vector e_z. The corresponding expression for the simultaneous absorption of two photons with linear polarizations e_α and e_β is

$$\delta \sim |e_\alpha \cdot \mathbf{S}_{0\to f} \cdot e_\beta|^2 \quad (1.49)$$

with $\alpha,\beta = x, y, z$

$\mathbf{S}_{0\to f}$ is the two-photon transition moment tensor that can be visualized as a (3 × 3) matrix with elements $(\mathbf{S}_{0\to f})_{\alpha,\beta}$, which contain the sums of the products $\sum_i M^\alpha_{0\to i} M^\beta_{i\to f}/\Delta E_i$. If the ground state Ψ_0 is of even (g) parity the intermediate state Ψ_i has to be odd (u) and the final state Ψ_f even (g) again in order for $M_{0\to i}$ as well as $M_{i\to f}$ to have nonvanishing components. Therefore, two-photon transitions between states of identical parity (g or u) are allowed and two-photon transitions between states of different parity are forbidden. Hence two-photon spectroscopy applied to molecules with a center of symmetry complements one-photon spectroscopy in the same way as Raman and IR spectroscopies complement each other.

The elements of transition moment tensors reflect the symmetry of the states involved, so even the allowed transitions can be observed only if the polarization of the photons in the molecular frame matches the nonzero components of the tensor. A major difference with respect to one-photon spectroscopy lies in the fact that this dependence on polarization does not vanish upon averaging over all orientations, and two-photon measurements on isotropic samples such as liquid solutions provide polarization information.

Among the various methods useful for monitoring two-photon absorption by solutes, fluorescence detection has proved most popular; that is, excitation spectra (cf. Section 5.3) are measured. When both photons are taken from a single beam and linearly as well as circularly polarized light is used, two independent measurements are possible and provide the absorption cross sections $\delta_{\uparrow\uparrow}$ and $\delta_{\circlearrowright\circlearrowright}$, respectively. The quantity

$$\Omega = \delta_{\circlearrowright\circlearrowright}/\delta_{\uparrow\uparrow} \quad (1.50)$$

defines the polarization degree. It is determined by the symmetry of the excited state and can therefore be used to characterize the two-photon transitions.

As an example of the use of two-photon spectroscopy in assigning excited states that are not observed in one-photon UV spectroscopy, the two-photon absorption spectrum of naphthalene is shown in Figure 1.17. Since for the point group D_{2h} all B_g states have a theoretical polarization degree of $\Omega = 3/2$, the polarization measurement reveals immediately a B_g state near 42,000 cm^{-1}. In the two-photon absorption spectrum this shows up only as a shoulder, whereas the maximum at 44,500 cm^{-1} can be assigned to an A_g state. Neither state is prominent in the one-photon spectrum.

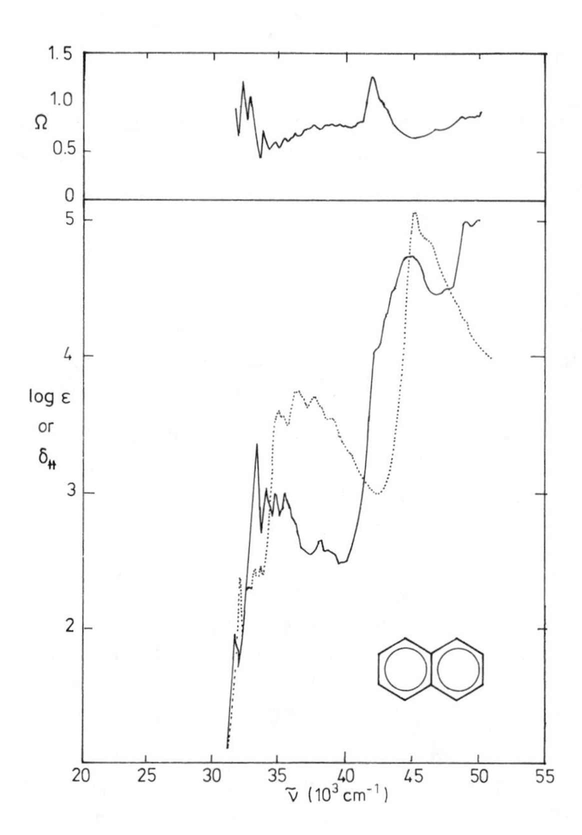

Figure 1.17. The two-photon absorption spectrum of naphthalene. The absorption cross section $\delta_{\uparrow\uparrow}$ in arbitrary units and the polarization degree Ω are shown. For comparison the one-photon absorption spectrum is given by a dotted line (by permission from Dick et al., 1981).

Polarization measurements can also be used to distinguish two-photon transitions from two-step excitations where the absorption of one photon produces an intermediate that may in turn absorb another photon. Frequently this intermediate is a triplet state. In liquid solutions $\Omega = 1$ should be found for such processes, since the absorption cross section will be given by the product of the average values for both steps and will therefore be independent of the state of polarization of the photons, if molecular rotation is rapid with respect to the intermediate and fluorescence lifetimes.

Two-photon spectroscopy may also be used to obtain highest-resolution spectra. Doppler broadening, which originates in the random motion of molecules in the gas phase, prevents individual rotational lines of a vibronic transition from being resolved in conventional spectroscopy. However, if two photons of extremely monochromatic light coming from exactly opposite directions are absorbed simultaneously, the Doppler shifts of the two photons just cancel each other and the Doppler broadening is eliminated. As an example of the resolution that can be achieved in Doppler-free spectroscopy, a part of the $S_0 \rightarrow S_1$ transition in benzene is shown in Figure 1.18. Spectrum b shows the central part of the most intense vibrational band from spectrum a, which corresponds to an excitation of the normal vibration ν_{14} from $v = 0$ to $v' = 1$, at a resolution $\delta\tilde{\nu} = 0.06\ \mathrm{cm}^{-1}$ that is limited only by the Doppler broadening. This is the highest possible resolution that may be achieved by conventional one-photon spectroscopy at room temperature. The observed structure is due to the overlap of many rotational lines. Elimination of the Doppler broadening increases the resolution by a factor of 50 and allows individual rotational lines to be observed. This is apparent from spectrum c, which only shows the short-wavelength edge of the vibrational band. In the case of benzene and its isotopomers (e.g., C_6D_6) very precise values of the rotational constants could thus be derived, which yield CC and CH bond distances of 139.7 pm and 107.9 pm, respectively, for the electronic ground state S_0, and 143.2 pm and 107.5 pm for the S_1 state. (Cf. Lin et al., 1984.)

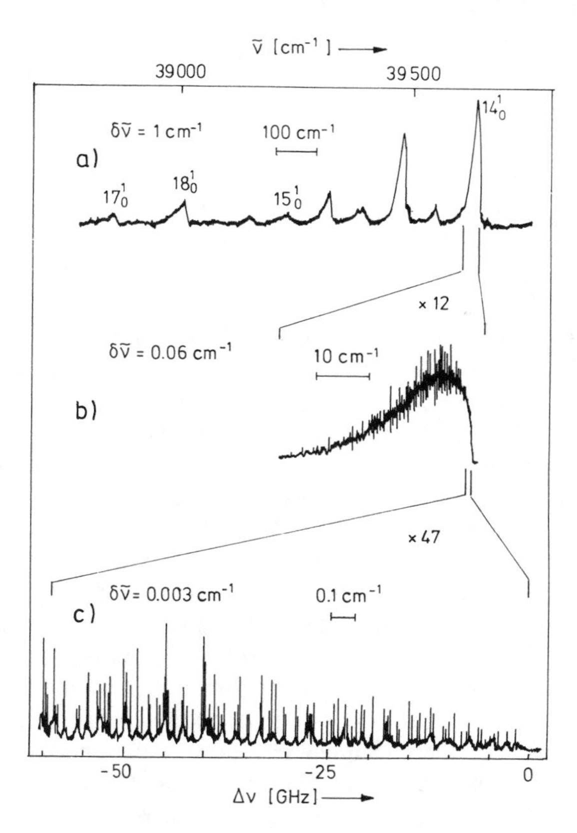

Figure 1.18. Part of the two-photon spectrum of benzene at different resolutions; a) vibrational structure of the $S_0 \rightarrow S_1$ transition, b) Q branch of the most intense vibrational line (14_0^1) with a resolution $\delta\tilde{\nu}$ limited by the Doppler broadening, and c) elimination of the Doppler broadening which yields individual rotational lines (by permission from Neusser and Schlag, 1992).

Another method for obtaining highest-resolution gas spectra is based on supersonic beam techniques. (Cf. Hayes, 1987.) A collimated molecular beam of molecules with a negligible spread in the transverse velocity distribution is produced by passing the central part of a supersonic jet beam through a small-opening aperture. This reduces the Doppler broadening from approximately 0.05 cm^{-1} for the bulk gas phase to 0.001 cm^{-1} and is therefore sufficient to resolve rotational lines in the electronic spectrum (Beck et al., 1979). Supersonic jet laser spectroscopy had a tremendous impact on the understanding of molecular coupling mechanisms such as vibronic coupling or Fermi resonance, which strongly influence the energy redistribution in polyatomic molecules. (Cf. Majewski et al., 1989; Weber et al., 1990.)

1.4 Properties of Molecules in Excited States

1.4.1 Excited-State Geometries

The excitation of an electron from a bonding orbital into a nonbonding or antibonding orbital changes the bonding situation and thus the equilibrium geometry of the molecule as well. The larger the differences in the bonding or antibonding contributions to individual bonds that are provided by the electrons located in the orbitals occupied before and after the transition, the more pronounced will be the geometry changes. Hence, for extended π systems with strongly and fairly uniformly delocalized π orbitals, only small bond-length changes are to be expected. Steric and electronic effects are frequently very delicately balanced, so it is mainly the dihedral angles that differ appreciably in the different electronic states of the molecule. (Cf. Example 1.13.)

The experimental determination of the structure of molecules in excited states is difficult. For small molecules accurate data may be obtained from the moments of inertia determined from the rotational fine structure in the electronic spectra. (Cf. Section 1.3.6.) Structural data for formaldehyde in the ground and the lowest excited singlet and triplet states are collected in Table 1.4. For larger molecules the necessary information may be retrieved from Franck-Condon factors if a sufficient number of progressions corresponding to the various normal modes of vibration show up in the UV spectrum. The first Franck-Condon analysis for a larger molecule showed that the $S_0 \rightarrow S_1$ transition in benzene is accompanied by a CC bond-length change of $\Delta R = 3.7$ pm and a very small contraction of the CH bonds ($\Delta R = -1$ pm) (Craig, 1950). Similar results were obtained for naphthalene (Figure 1.19). Since only $(\Delta R)^2$ can be derived from the spectra, supplementary experimental information or theoretical data are needed to determine the sign of the bond-length change.

Most data for geometries of excited states stem from theoretical arguments or calculations. Qualitative concepts such as the Walsh rules (cf. Buenker and Peyerimhoff, 1974) have proven very useful. Another valuable

Table 1.4 Excited-State Geometries, all Distances in pm

	CH_2O[a,b]				$CH{\equiv}CH$[c]		
	S_0	S_1	T_1		S_0	S_1	T_1
R_{CO}	120.8	132.1	130	$R_{C\equiv C}$	120	131.7	135.6
R_{CH}	111.6	109.2	111	∢HCC	180°	130.5°	128.0°
∢HCO	116.5°	121.5°	116°			(trans)	(cis)
φ[d]	0.0	21°	36°				

[a] Johnson et al., 1972.
[b] Job et al., 1969.
[c] Winkelhofer et al., 1983.
[d] Deviation from planarity.

rule was proposed by Imamura and Hoffmann (1968) for situations in which two π systems, each with q π electrons, are connected by a formal single bond. If $q = 4n + 2$, as in biphenyl (**3**), a twisted ground state and a planar excited state are predicted. If $q = 4n$ both the ground state and the excited state are predicted to be planar. If q is odd, the ground state should be planar and the excited state twisted.

Initial estimates of geometry changes on excitation of π systems may be obtained from the linear relation

$$R_{\mu\nu} = a - bp_{\mu\nu} \tag{1.51}$$

between the π bond order $p_{\mu\nu}$ and bond distance $R_{\mu\nu}$. Using this relationship, very satisfactory results for the geometry changes for the excitation of naphthalene are obtained even from HMO calculations. A symmetry-based procedure for the determination of molecular geometry changes following electronic excitation has been proposed by Bachler and Polanski (1990).

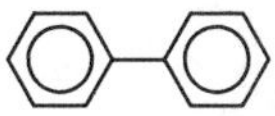

3 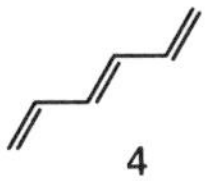 4

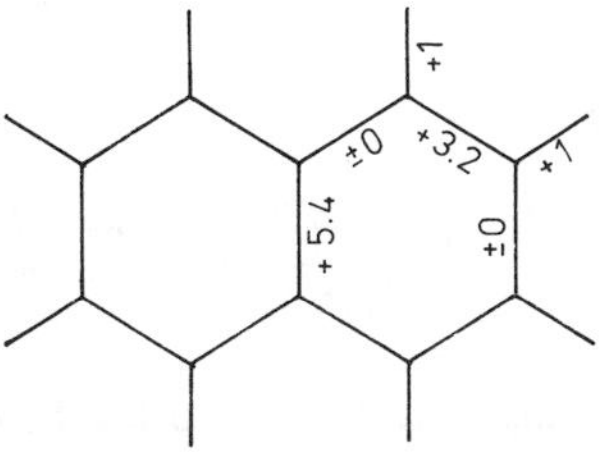

Figure 1.19. Bond-length changes of naphthalene (pm) on 310-nm excitation (adapted from Innes, 1975).

Ab initio calculations on small molecules give very detailed information. Thus, acetylene is calculated to be bent into a trans configuration in the lowest singlet state and into a cis configuration in the lowest triplet state. (Cf. Table 1.4.) Hexatriene (**4**) was shown to have three minima of comparable energy in the T_1 state, one planar and two at geometries twisted by 90° around the central or the terminal double bond. For longer polyenes, twisting around the central double bond is to be expected (Bonačić-Koutecký and Ishimura, 1977). Twisted excited states of π systems play a very important role in many photochemical reactions and will be discussed in detail in Section 7.1.

Example 1.13:
In many cases a combination of force-field methods and semiempirical calculations in the π approximation has proven very useful for calculating geome-

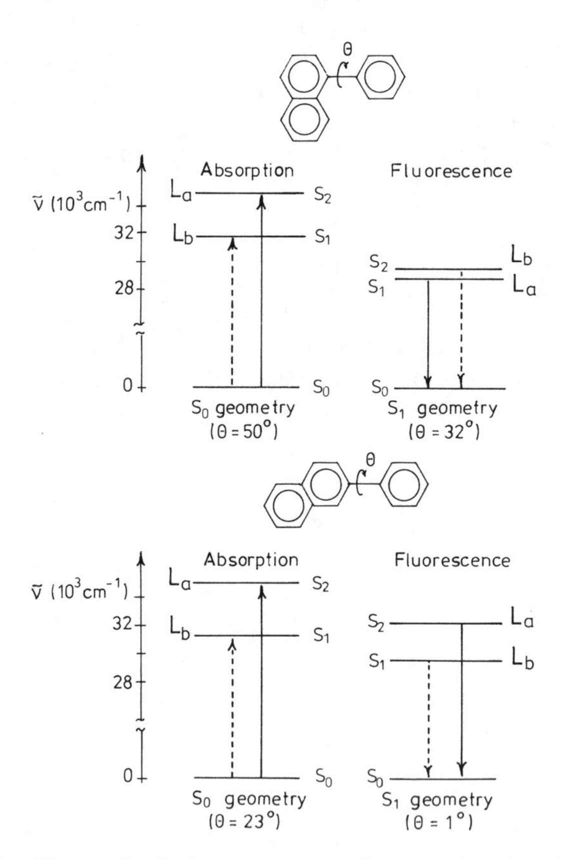

Figure 1.20. Electronic singlet term system for 1- and 2-phenylnaphthalene calculated for ground and excited-state geometries. The weakly allowed character of transitions between the ground state and the ^1L_b state is indicated by broken arrows (by permission from Gustav et al., 1980).

tries of excited states (Gustav and Sühnel, 1980). Along these lines it was shown that in going from S_0 to S_1 the torsional angle between naphthalene and the phenyl group in 1-phenylnaphthalene (**5**) changes from 50° to 32° and in 2-phenylnaphthalene from 23° to a value close to 0°. From Figure 1.20 it can be seen that calculations for these optimized excited-state geometries lead to an interchange of the first two excited states of **5** compared to the order calculated for the ground-state geometry. As a consequence the vertical fluorescence from S_1 is strongly allowed for the 1-phenyl derivative but only weakly allowed for the 2-phenyl derivative, as is the vertical absorption into S_1 for both molecules. This explains the difference in the fluorescence lifetimes of $\tau_F(S_1)$ = 35 ns (**5**) and $\tau_F(S_1)$ = 438 ns (**6**) observed for these two compounds (Gustav et al., 1980).

5 6

1.4.2 Dipole Moments of Excited-State Molecules

Molecules in an electronically excited state have chemical and physical properties that differ from those of ground-state molecules. As a result of

Table 1.5 Dipole Moments of Ground and Excited States, all Values in D, Calculated Data in Parentheses

Molecule	State				Ref.
	S_0	T_1	S_1	S_2	
$H_2C{=}O$	2.33	1.29	1.56		a
$Ph_2C{=}O$	2.98	1.72	1.23		b
C_6H_5—CH=O	2.75 (2.8)	—	—	7.6 (7.9)	c
$(CH_3)_2N$—C_6H_4—C≡N	5.6 (4.6)	—	— (10.7)	13.3 (17.7)	c
O_2N—C_6H_4—C_6H_4—C≡N	6.0	—	16.5		c
Naphthalene . . . (amine) Exciplex			10.8		d
$(CH_3)_2N$—C_6H_4—C≡N			12 (TICT)		e

[a] Buckingham et al., 1970.
[b] Hochstrasser and Noe, 1971.
[c] Labhart, 1966.
[d] Taylor, 1971.
[e] Bischof et al., 1985.

excitation, the electron distribution changes, and the dipole moment may as well. This is often apparent from the solvent dependence of absorption and luminescence (cf. Section 2.7.2 and Section 5.5.3) as well as from the chemical properties of the molecule, such as basicity and acidity.

In Table 1.5 some data for ground- and excited-state dipole moments are collected. The $n \rightarrow \pi^*$ excitation of carbonyl compounds is accompanied by a charge shift from the electronegative oxygen atom into the delocalized π^* orbital, and the dipole moments of the T_1 and the S_1 states are smaller than those of the ground state. The lowest excited states of molecules with donor and acceptor groups on the π system often exhibit intramolecular charge transfer from the donor to the π system and to the acceptor or from the π system to the acceptor. As a consequence, the excited-state dipole moment is often much higher than the ground-state value. Particularly large changes of the dipole moment are to be expected for exciplexes, since a pronounced charge transfer is characteristic for their excited states. (Cf. Section 2.6 and Section 5.4.3.) Finally, the so-called TICT states (*twisted internal charge transfer,* cf. Section 4.3.3) exhibit large dipole moments.

In solution, even centrosymmetric molecules can have large dipole moments in the excited state, as was first demonstrated in the case of 9,9′-bianthryl (**7**) (Beens and Weller, 1968). The symmetrical compound **8** is an example in which electronic excitation is localized almost entirely in one of the polar aromatic end groups, due to solvent-induced local site perturbation (Liptay et al., 1988a). In solution, polyenes may also show unsymmetrical charge distributions with nonvanishing dipole moments (Liptay et al., 1988b; see also Section 2.1.2).

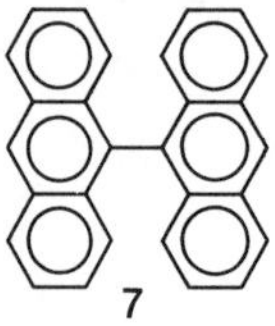

NC CN $(CH_3)_2N$ $N(CH_3)_2$ NC CN

7 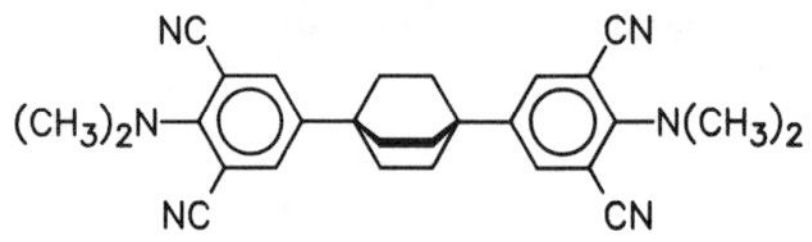 8

1.4.3 Acidity and Basicity of Molecules in Excited States

Changes in the electron distribution on excitation are connected with changes in basicity and acidity. This becomes evident if one considers the indicator equilibrium

$$BH^{\oplus} \rightleftharpoons B + H^{\oplus}$$

For the base B to act as an indicator it must absorb at a different frequency after protonation. Suppose that the excitation energy into the lowest singlet state ${}^1\Delta E = E(S_1) - E(S_0)$ or into the lowest triplet state ${}^3\Delta E = E(T_1) - E(S_0)$ is higher for $BH^{\oplus}$ than for B. Imagine a solution of pH equal to the ground-state pK, so that $[B] = [BH^{\oplus}]$. On excitation $BH^{\oplus}$ will find itself at a higher energy than excited B and will exhibit a strong tendency to change to B. Thus, it will become a stronger acid. If B absorbs at higher frequency than $BH^{\oplus}$, then B will become a stronger base on excitation. Whether a proton transfer in the excited state will actually take place depends on its lifetime and on the presence of barriers in the excited potential energy surface.

The relationship between the excitation energy change upon protonation and the change in equilibrium constant upon excitation is formalized in the Förster cycle (Förster, 1950), which is illustrated in Figure 1.21. Proceeding from the ground state $BH^{\oplus}$ to excited $B^* + H^{\oplus}$ by two different routes yields the equality

$$N_L h\nu_{BH^{\oplus}} + \Delta H^* = N_L h\nu_B + \Delta H \tag{1.52}$$

or

$$\Delta H^* - \Delta H = N_L(h\nu_B - h\nu_{HB^{\oplus}})$$

where ΔH and ΔH^* are the enthalpy changes in the ground and excited states, respectively, and N_L is the Avogadro number. If in dilute solutions ΔH can be approximated by the standard value ΔH_0 and if entropy effects may be neglected, Equation (1.52) yields, using $\Delta G_0 = -RT\ln K = -2.303$ $RTpK$

$$N_L(h\nu_B - h\nu_{BH^{\oplus}}) \approx \Delta\Delta H_0 \approx \Delta\Delta G_0 \approx 2.303\ RT\Delta pK$$

When numerical values for constants are inserted, the following relation between the change in the pK value and the shift $\Delta\tilde{\nu}$ of the absorption maxi-

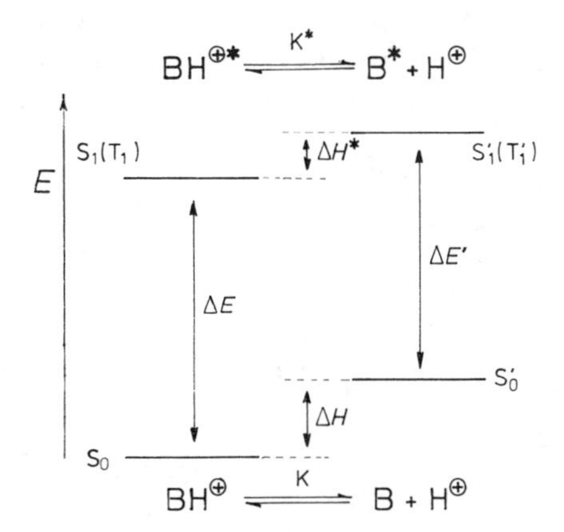

Figure 1.21. The Förster cycle. ΔE and $\Delta E'$ refer to the excitation energies and ΔH and ΔH^* to the reaction enthalpies corresponding to the equilibrium $BH^{\oplus} \rightleftharpoons B + H^{\oplus}$ in the ground and excited state, respectively.

mum for the lowest singlet or triplet transition due to protonation is obtained for $T = 298$ K:

$$\Delta pK = 0.00209\,(\tilde{\nu}_B - \tilde{\nu}_{BH\oplus})/\text{cm}^{-1} \qquad (1.53)$$

Near 300 nm a spectral shift of 30 nm corresponds approximately to a change in wave number of $\Delta\tilde{\nu} = 3{,}300\ \text{cm}^{-1}$. From Equation (1.53) this results in a pK value change of 7 units. Since shifts of this extent are quite common on protonation, changes in dissociation constants of protonated compounds by 6–10 orders of magnitude on excitation are often observed.

Depending on whether the change in ΔpK is negative or positive, i.e. depending on whether the pK value of the excited state is smaller or larger than that of the ground state, the excited-state species will be a stronger acid (the conjugate base will be weaker) or a stronger base (the conjugate acid will be weaker). Molecules with acidic substituents that are electron donors, such as 1-naphthol (**9**) with $pK(S_0) = 9.2$ and $pK(S_1) = 2.0$ (Weller, 1958) are more acidic in the excited singlet state. Molecules with acidic substituents that are electron acceptors, such as naphthalene-1-carbocylic acid (**10**) with $pK(S_0) = 3.7$ and $pK(S_1) = 10.0$ (Watkins, 1972), are less acidic in the excited singlet state. This is due to the generally much larger degree of charge transfer between the substituent and the parent π system in the excited singlet. For the same reason, a molecule with a basic electron-donor substituent, such as aniline, becomes much less basic in the excited singlet state.

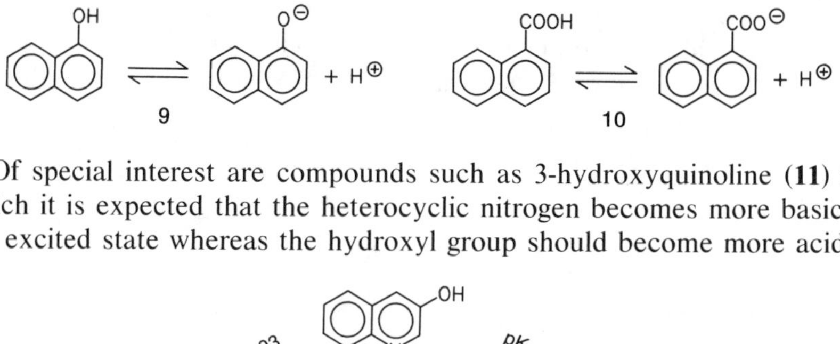

OH ⇌ O⊖ + H⊕ (9)   COOH ⇌ COO⊖ + H⊕ (10)

Of special interest are compounds such as 3-hydroxyquinoline (**11**) for which it is expected that the heterocyclic nitrogen becomes more basic in the excited state whereas the hydroxyl group should become more acidic.

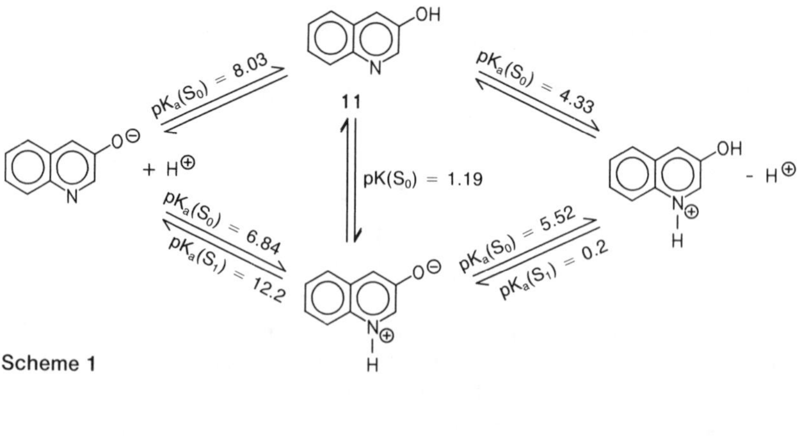

11

$pK_a(S_0) = 8.03$

$pK_a(S_0) = 4.33$

$pK(S_0) = 1.19$

+ H⊕   − H⊕

$pK_a(S_0) = 6.84$

$pK_a(S_1) = 12.2$

$pK_a(S_0) = 5.52$

$pK_a(S_1) = 0.2$

Scheme 1

As a consequence, different ionization sequences can be obtained in the S_0 and S_1 states, as illustrated in Scheme 1 (Mason et al., 1968).

At the extremes of the acidity range studied, the cation and the anion are the species found in both the ground and S_1 states. At intermediate pH values, the hydroxy tautomer of the neutral molecule is the predominant species in the ground state. On excitation, however, the phenolic group becomes more acidic and the nitrogen more basic. The zwitterion is more stable, and only the fluorescence of this species is observed.

Excitation into the T_1 state changes the pK value in the same direction as excitation into the S_1 state, but much less, and the p$K(T_1)$ value tends to be closer to p$K(S_0)$ than to p$K(S_1)$:

$$pK(S_0) > pK(T_1) \gg pK(S_1)$$
$$pK(S_0) < pK(T_1) \ll pK(S_1)$$

The trends are easy to understand in qualitative terms. For instance, acidic and basic substituents that are π-electron donors, such as OH and NH_2, become more acidic and less basic in the excited state, respectively. The ability of the substituent to transfer electron density into the arene is greatly increased upon excitation, since it then can occur into one of the bonding orbitals that were doubly occupied in the ground state. (Cf. Section 2.4.2.) The increased positive charge on the substituent functionality increases its acidity and reduces its basicity.

In order to understand the striking difference between the S_1 and T_1 states, one needs to go beyond a simple consideration of the charge densities (Constanciel, 1972). From the Förster cycle shown in Figure 1.21 it is clear that the critical quantity is the difference in the substituent effect on the

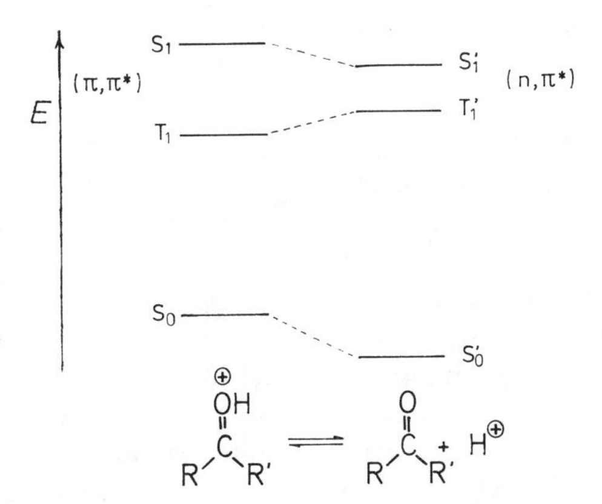

Figure 1.22. Schematic representation of the energy levels for a case in which the lowest excited states T_1 and S_1 originate from a $\pi \rightarrow \pi^*$ excitation in the protonated species, and from a $n \rightarrow \pi^*$ excitation in the nonprotonated species, with a correspondingly smaller singlet–triplet splitting $E(S_1) - E(T_1)$.

singlet versus the triplet excitation energy. Differences are to be expected, since in simple arenes such as benzene and naphthalene the two states originate in different kinds of configurations, S_1 being of the L_b and T_1 of the L_a type. (See Section 2.2.) Also, with increasing charge-transfer contributions to an excited state, the orbital ϕ_i out of which excitation occurs tends to separate in space from the orbital ϕ_k into which it occurs, and their exchange integral K_{ik}, tends to decrease. (Cf. Section 1.3.2.) Upon protonation, the charge-transfer nature of the excited states is greatly decreased and K_{ik} increases. Equations (1.23) and (1.24) show that the energy of the singlet but not the triplet excited configuration increases as a result. This is reflected in the state energies, making protonation of the S_1 species less advantageous (aniline) or its deprotonation more advantageous (phenol).

Xanthone (**12**) and several substituted benzophenones (**13**) have p$K(T_1)$ values lying outside the range defined by p$K(S_0)$ and p$K(S_1)$ (Ireland and Wyatt, 1973). In these molecules the lowest excited state is a (n, π^*) state in the ketone and a (π, π^*) state in its conjugate acid. Since the S_1–T_1 energy difference is much smaller for (n, π^*) states than for (π, π^*) states, molecules of this type, which are already stronger bases in the S_1 than in the S_0 state, can become even stronger bases in the T_1 state. The relative energies of the states are shown schematically in Figure 1.22. The pK order is therefore p$K(S_0) < pK(S_1) <$ p$K(T_1)$.

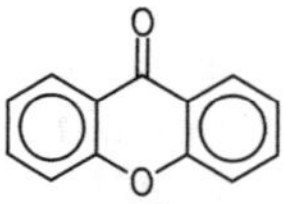

12 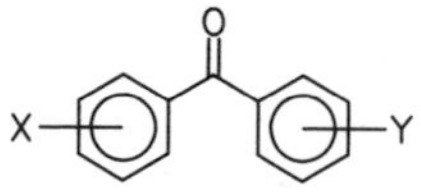 13

1.5 Quantum Chemical Calculations of Electronic Excitation

Quantum chemical calculations of excitation energies and transition dipole moments vary according to the level of sophistication. The HMO model discussed in Section 1.2.3 is the easiest to apply; excitation energies are linearly related to orbital energy differences. [Cf. Equation (1.22).]

It has already been pointed out several times that electron repulsion terms play a major part in the discussion of electronic excitation energies. Within the Hartree-Fock approximation, electron interaction in closed-shell ground states can be taken care of in a reasonable way using SCF methods. In a treatment of excited states, however, configuration interaction usually has to be taken into account. (Cf. Section 1.2.4.) This can be achieved either by semiempirical methods, especially in those cases where the π approximation is sufficient for a discussion of light absorption, or, by ab initio methods in the case of small molecules.

1.5.1 Semiempirical Calculations of Excitation Energies

Pariser and Parr (1953) and Pople (1953) developed the PPP method independently by introducing the ZDO approximation and empirical values of electron-repulsion integrals into the MO method for π-electron systems. This method has proven very useful for calculating the properties of $\pi \rightarrow \pi^*$ transitions for both unsaturated and aromatic molecules.

The procedure starts with a set of π-symmetry SCF-MOs $\phi_i = \sum_\mu c_{\mu i} \chi_\mu$.

The LCAO-MO coefficients $c_{\mu i}$ are given as components of the eigenvector $\boldsymbol{c}_i$ of the Fock matrix $\mathbf{F}$ with elements

$$\begin{aligned} F_{\mu\mu} &= U_{\mu\mu} + \tfrac{1}{2} q_\mu \gamma_{\mu\mu} + \sum_{\varrho \neq \mu} (q_\varrho - Z_\varrho) \gamma_{\mu\varrho} \\ F_{\mu\nu} &= \beta_{\mu\nu} - \tfrac{1}{2} p_{\mu\nu} \gamma_{\mu\nu} \end{aligned} \tag{1.54}$$

$U_{\mu\mu}$, $\beta_{\mu\nu}$, and $\gamma_{\mu\mu}$ are the parameters used: $U_{\mu\mu}$ is the energy of an electron occupying a π-symmetry atomic orbital appropriate to the particular valence state of the atom (e.g., sp^2 hybridized carbon); the resonance integral $\beta_{\mu\nu}$ is taken as an empirical parameter that is assumed to be nonzero only for directly bonded atoms μ and ν. The one-center Coulomb repulsion integrals $\gamma_{\mu\mu}$ are determined from the ionization potential IP_μ and the electron affinity EA_μ of the atom in its valence state,

$$\gamma_{\mu\mu} = \mathrm{IP}_\mu - \mathrm{EA}_\mu \tag{1.55}$$

and the two-center electron repulsion integrals $\gamma_{\mu\nu}$ between the π centers μ and ν are chosen empirically so as to provide a smooth interpolation between the one-center value and the theoretical value of $1/R_{\mu\nu}$ for the Coulomb repulsion of two point charges, approached asymptotically in the limit of very large distances. A commonly used expression due to Mataga and Nishimoto (1957) is

$$\gamma_{\mu\nu} = 1/(a + R_{\mu\nu}) \tag{1.56}$$

with $1/a = \mathrm{IP}_\mu - \mathrm{EA}_\mu$. There are several critical reviews of the various parameterization schemes proposed for the PPP method using singly excited configurations. (Cf. Klessinger, 1968; Scholz and Köhler, 1981.)

The excited-state wave functions are written as linear combinations of spin-adapted configurational functions Φ_K

$$\Psi = \sum_K C_K \Phi_K \tag{1.57}$$

expressed in terms of Slater determinants.

The various functions Φ_K describe the ground configuration,

$$\Phi_0 = |\phi_1\bar{\phi}_1 \ldots \phi_n\bar{\phi}_n| = |1\bar{1} \ldots n\bar{n}| \tag{1.58}$$

singly excited configurations,

$$^{1,3}\Phi_{i\to k} = 1/\sqrt{2}\{| \ldots i\bar{k} \ldots | \pm | \ldots k\bar{i} \ldots |\} \tag{1.59}$$

and more highly excited configurations. For example, the doubly excited configurations $^1\Phi_{i\to k}^{i\to k}$ and $^1\Phi_{j\to l}^{i\to k}$ are obtained by promoting two electrons from either the same or from two different orbitals occupied in the ground configuration into either the same or into two different orbitals unoccupied in the ground configuration. If multiple excitations are considered it must be remembered that there may exist different independent spin functions for the same electronic configuration. Thus, for four electrons in four different orbitals two linearly independent singlet spin functions have to be considered. Frequently the calculation is confined to single excitations (singly excited configuration interaction, SCI) or even to a selection of single excitations from the p highest occupied MOs into the q lowest unoccupied MOs. In the latter cases the ground configuration Φ_0 does not contribute to the excited-state wave functions [$C_0 = 0$ in Equation (1.57)], since according to Brillouin's theorem the matrix element $<\Phi_0|\hat{H}|\Phi_{i\to k}>$ between the SCF ground configuration and singly excited configurations vanishes. (See, e.g., McWeeny, 1989: Section 6.1.)

It has been shown that two principal factors are responsible for the ordering of the low-lying excited states:

1. First-order CI, which for alternant hydrocarbons gives rise to the distinction between "plus" and "minus" states (cf. Section 1.2.4).
2. CI with doubly excited configurations.

The effect of the introduction of doubly excited configurations into the calculation is to permit the removal of excess zwitterionic character from the calculated wave function of the lowest totally symmetric excited singlet state, thus lowering its energy so that it may even become the lowest singlet excited state. The low-energy position of this "covalent," partly "doubly excited" state, and the observability of this state by two-photon spectroscopy were predicted (Koutecký, 1967), and its importance for photochemical reactions (van der Lugt and Oosterhoff, 1969; cf. Section 7.1) was pointed out already in the sixties. It was also shown that due to the geometry dependence of the electron repulsion terms the influence of multiply excited configurations is smaller in more compact molecules (Koutecký, 1967). As a consequence, first-order CI predominates with cyclic π systems, whereas multiply excited configurations are particularly important with *all-trans*-polyenes. (Cf. Example 2.2.)

The ordering of excited states is in general not affected by more highly than doubly excited configurations, but the actual values calculated for the state energies are. In this connection it is important to note that PPP calculations that include CI with doubly or more highly excited configurations require different parameters than SCI calculations (Koutecký et al., 1964). In particular, the electron repulsion integrals $\gamma_{\mu\nu}$ should decrease more slowly with increasing $R_{\mu\nu}$ than suggested by the Mataga-Nishimoto formula, and the resonance integrals should be a little larger (Karwowski, 1973). Effective programs for large-scale CI are based on the unitary group approach (UGA), which is notably suited for complete CI (CCI) (Paldus, 1967). In this latter case it is common to build up the CI from AOs rather than from SCF-MOs since the construction of MOs corresponds just to a unitary basis transformation that does not affect the results.

Unfortunately, using CCI is rarely feasible in routine applications since the number of configurations soon becomes too large if the number of π centers increases. Therefore, criteria for truncating the CI have been established; for example, by exclusion of all those configurations Φ_K for which $\Delta E = E(\Phi_K) - E(\Phi_R)$, the energy difference with respect to a reference configuration Φ_R is larger than a preselected threshold, or for which the interaction matrix element $<\Phi_K|\hat{H}|\Phi_R>$ is smaller than a given threshold (Downing et al., 1974).

In calculating triplet states it should be noted that the lowest triplet configuration ${}^3\Phi_{i\rightarrow k}$ is already singly excited with respect to the ground configuration Φ_0. Many configurations that are singly excited with respect to ${}^3\Phi_{i\rightarrow k}$ are doubly excited with respect to Φ_0 and contain up to four open shells. It has been proposed to calculate triplet energies using parameter values different from those employed in calculating singlet energies (Pancir and Zahradnik, 1973). However, when feasible, it is more consistent to include doubly and multiply excited configurations instead and to truncate the CI by appropriate selection criteria in the same way as for the singlet states.

The CNDO/S method (Del Bene and Jaffé, 1968) has also been used successfully to calculate excitation energies. Determinations of $n\rightarrow\pi^*$ transitions and calculations for nonplanar π systems become possible using this method because all valence electrons are considered. The main difference between the CNDO/S and the CNDO/2 method is an additional parameter $\varkappa$ that reduces the weight of π overlap relative to σ overlap in calculating the resonance integrals. This is a prerequisite for calculating reasonable values of $\pi\rightarrow\pi^*$ excitation energies. When limited configuration interaction including singly and doubly excited configurations is considered, fairly good agreement between computed and experimental results is obtained for singlet as well as triplet excitation energies (Dick and Nickel, 1983). The INDO/S method, a similar variant of the INDO/2 procedure (Ridley and Zerner, 1973, 1976), has enjoyed considerable popularity. Its main advantage is the correct treatment of the splitting between the ${}^1(n, \pi^*)$ and ${}^3(n, \pi^*)$ states. In general, these all-valence electron methods still tend to give better results for (π, π^*) states than for others, whose energies are at times described very poorly.

A frequent problem in excited-state applications of semiempirical valence-electron methods including configuration interaction with multiply excited configurations is that the parametrization already may have been determined so as to include correlation effects for the ground-state results (Dewar, 1969), and these end up being counted twice. Thiel (1981) proposed the semiempirical MNDOC method, whose parameterization permits an explicit treatment of correlation effects by CI. Together with appropriate selection criteria for truncating the CI and with the possibility of using different reference configurations for different states, this method is well suited for calculating excitation energies, potential energy surfaces, and hence geometries of excited states (Klessinger et al., 1991).

1.5.2 Computation of Transition Moments

The use of the dipole length formula is based on the evaluation of the transition dipole moment

$$\boldsymbol{M}_{0\to\mathrm{f}} = <\Psi_{\mathrm{f}}|\sum_j \hat{\boldsymbol{m}}_j|\Psi_0>$$

which within the framework of the LCAO-MO method reduces to the computation of matrix elements $\langle\chi_\mu|\hat{\boldsymbol{m}}_j|\chi_\nu\rangle$ over AOs. Setting $\hat{\boldsymbol{m}}_j = -|e|\boldsymbol{r}_j$, the ZDO approximation yields

$$<\chi_\mu|\hat{\boldsymbol{m}}_j|\chi_\nu> = \delta_{\mu\nu} <\chi_\mu|\hat{\boldsymbol{m}}_j|\chi_\nu> = -|e|\delta_{\mu\nu}\boldsymbol{R}_\mu \qquad (1.60)$$

where $\boldsymbol{R}_\mu$ is the position vector of center μ.

The use of the dipole velocity formulation requires the matrix elements of the linear momentum operator $\hat{\boldsymbol{p}}$. From the commutation relation of $\hat{H}$ with $\boldsymbol{r}$, Linderberg (1967) showed that the following expression is obtained in the ZDO approximation:

$$<\chi_\mu|\hat{\boldsymbol{p}}|\chi_\nu> = \mathrm{i}m_e\beta_{\mu\nu}(\boldsymbol{R}_\nu - \boldsymbol{R}_\mu)/\hbar \qquad (1.61)$$

$\beta_{\mu\nu}$ is the usual resonance integral and m_e is the mass of the electron. This relation permits facile use of the dipole velocity formulation in semiempirical π-electron methods.

The two methods only give identical results for oscillator strengths in the complete CI limit. In SCI calculations one has in general

$$f^{(p)} \leq f_{\mathrm{exp}} \leq f^{(r)}$$

where $f^{(p)}$ and $f^{(r)}$ refer to the oscillator strength determined from the dipole velocity formula Equation (1.30) and from the dipole length formula Equation (1.32), respectively.

Multiply excited configurations are not only important for calculating excitation energies but also for calculating intensities and polarization directions, where even triply excited configurations may be needed for molecules of low symmetry. Therefore, rational methods for selecting configurations are of great importance. Energy criteria alone are not sufficient for the purpose. Rather, rough estimates of the matrix elements $<\Phi_0|\Sigma\hat{m}_j|\Phi_K>$ are needed in order to decide which configurations affect the computed value of the oscillator strength and of the polarization direction. Such a scheme was developed, for example, by Downing et al. (1974).

Example 1.14:
As an example of the influence of multiply excited configurations, Figure 1.23 shows the excitation energies and oscillator strengths calculated for the hydrocarbon **14** using different configuration selection methods. The calculated transition moment directions are depicted in Figure 1.24. The dimensions of SCI, SDT-CI (all singly, doubly, and triply excited configurations), and of complete CI are 10, 120, and 175, respectively, whereas a selection of doubly excited configurations by energy criteria, and the additional consideration of such doubly and triply excited configurations whose contribution to the transition moment exceeds a certain threshold, yields a CI with 28 and 48 configurations. It

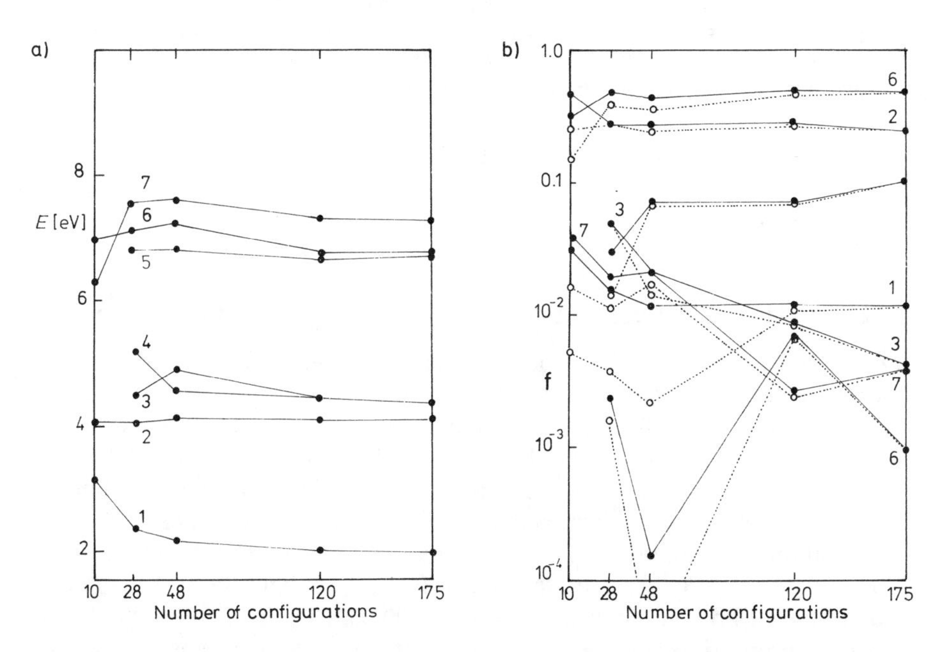

Figure 1.23. Spectral characteristics of the alkylidenecyclopropene **14** calculated as a function of the number of configurations used in the CI procedure: a) excitation energies and b) oscillator strengths, using dipole length (—) and dipole velocity (···) expressions (by permission from Downing et al., 1974).

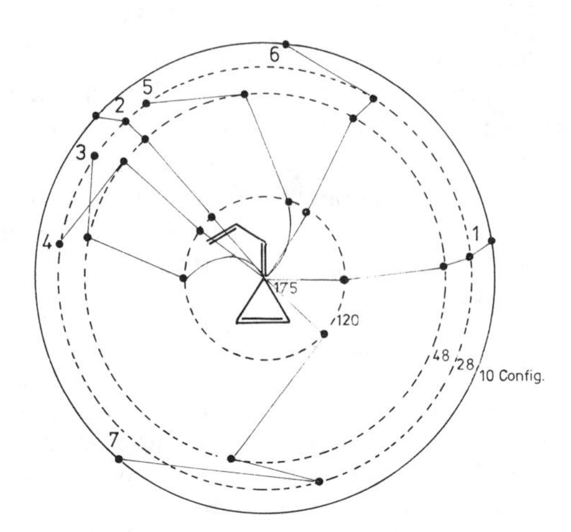

Figure 1.24. Direction of the transition moments of **14** calculated from the dipole length expression as a function of the number of configurations used in the CI procedure, plotted with respect to the molecular framework shown in the center (by permission from Downing et al., 1974).

is clear from the diagrams that the length and in particular the direction of the transition moment depend on the extent of CI considered. It is equally clear that the results converge and that dipole length and dipole velocity formulas give similar results if sufficient CI is included.

14

1.5.3 Ab Initio Calculations of Electronic Absorption Spectra

For a detailed ab initio description of excitation energies and transition moments it is necessary to use diffuse basis functions in addition to considering electron correlation effects for the ground state as well as for the open-shell excited state. A widely used scheme for such calculations is the MRD-CI method (multireference double excitation CI), which considers all doubly excited configurations with respect to a number of reference configurations. Since all those configurations that contribute appreciably to the wave function of the state of interest are taken as reference configurations, the dimension of the CI problem may easily exceed 500,000. The CI is therefore truncated by appropriate selection criteria and energy extrapolations are used to approximate the contributions from configurations omitted (Peyerimhoff and Buenker, 1975). The most accurate data for excited states are obtained from CASSCF calculations (multiconfiguration SCF calculations including all excitations within a preselected active space; Roos et al., 1980) and a

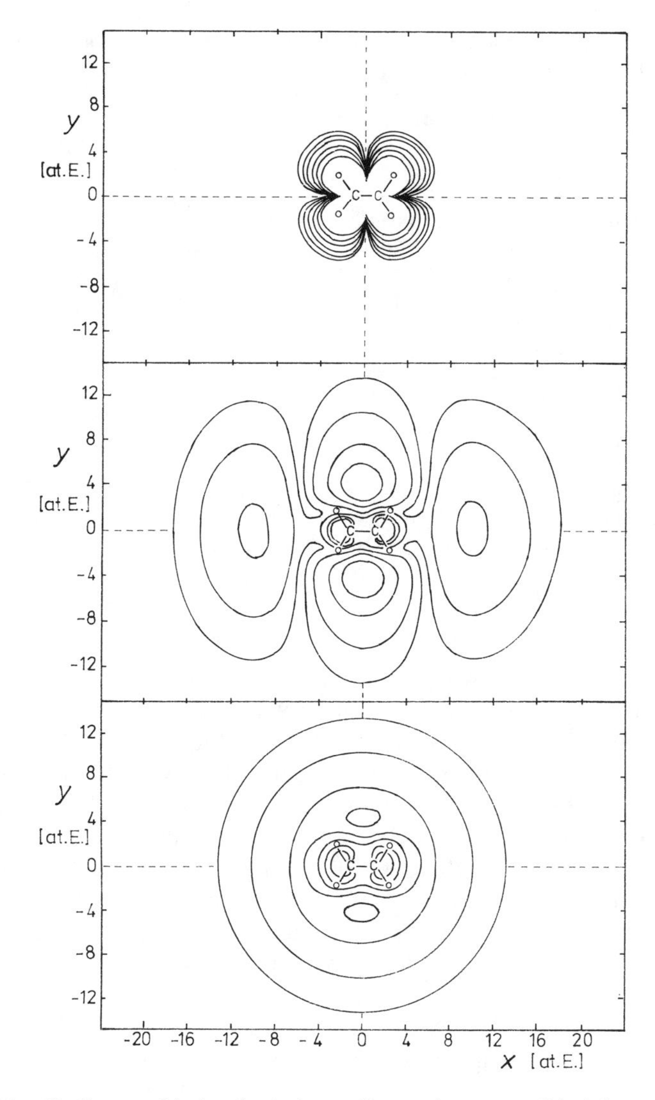

Figure 1.25. Rydberg orbitals of ethylene. Shown is an ns orbital (bottom) and an ndσ orbital (middle) as compared to the π^* orbital (top) (by permission from Peyerimhoff and Buenker, 1973).

subsequent step using second order perturbation theory with the CASSCF wave function as a reference state (Serrano-Andrés et al., 1993).

These methods lead in general to very good agreement between calculated and observed quantities. Rydberg states are also described well. They are characterized by high quantum number orbitals, which are much more extended than valence orbitals. Molecular Rydberg orbitals resemble atomic orbitals since an electron in such an orbital is so far from the nuclei of the

Table 1.6 Computed Ab Initio and Experimental State Energies (in eV) for the Lowest Triplet and Singlet Valence States of Butadiene

State	CI[a]	MRDCI[b]	Effective Valence Shell[c]	MCSCF[d]	CI[e]	CASSCF PT2[i]	Exp.
1^1A_g	0.0	0.0	0.0	0.0	0.0	0.0	0.0
1^3B_u	3.45	3.31	3.74	3.74	—	3.20	3.22[f]
1^3A_g	5.04	4.92	5.49	5.06	—	4.89	4.93[f]
2^1A_g	6.77	7.02	7.19	6.77	6.67	6.27	—
1^1B_u	7.05	7.67	6.61	6.88	6.23	6.23	5.76–6.09[g]
2^1B_u ($3p\pi$)	8.06	6.67	6.93	6.88	7.16	6.70	7.07[h]

[a] Hosteny et al., 1975.
[b] Buenker et al., 1976.
[c] Lee et al., 1983.
[d] Aoyagi et al., 1985.
[e] Cave and Davidson, 1988.
[f] Mosher et al., 1973.
[g] Innes and McDiarmid, 1978.
[h] McDiarmid, 1976.
[i] Serrano-Andrés et al., 1993.

molecule that it experiences a potential characteristic of a single point charge. This is illustrated in Figure 1.25 where an *n*s and an $nd\sigma$ Rydberg orbital of ethylene are compared with the π^* orbital.

For $\pi \rightarrow \pi^*$ excitation energies the agreement between the calculated and experimentally observed data is more difficult to achieve. This is probably related to the intricacies of the correlation of σ with π electrons in states with a highly zwitterionic character in the π part of the wave function. (Cf. Malmqvist and Roos, 1992.) Some of the difficulty is due to the limited availability of definitive and accurate experimental data. As an example, some data for butadiene are collected in Table 1.6.

Supplemental Reading

UV Spectroscopy

Jaffé, H.H. Orchin, M. (1962), *Theory and Applications of Ultraviolet Spectroscopy*; J. Wiley: New York, London.

Lamola, A.A. (1971–76), *Creation and Detection of Excited States,* Vol. **1–4**; Marcel Dekker: New York.

Mataga, N., Kubota, T. (1970), *Molecular Interactions and Electronic Spectra*; Marcel Dekker: New York.

Murrell, J.N. (1963), *The Theory of Electronic Spectra of Organic Molecules*; Methuen & Co.: London.

Perkampus, H.-H. (1992), *UV-VIS Spectroscopy and its Applications;* Springer: Berlin.

Perkampus, H.-H. (1995) *Encyclopedia of Spectroscopy;* VCH Inc.: New York.

Steinfeld, J.I. (1985), *Molecules and Radiation*; the MIT Press: Cambridge MA.

Vibronic Coupling

Fischer, G. (1984), *Vibronic Coupling*; Academic Press: New York.

Ziegler, L.D., Hudson, B.S. (1982), "The Vibronic Spectroscopy of Benzene: Old Problems and New Techniques," in *Excited States* **5**; Lim, E.C., Ed.; Academic Press: New York.

Properties of Molecules in Excited States

Innes, K.K. (1975), "Geometries of Molecules in Excited Electronic States," in *Excited States* **2**; Lim, E.C., Ed.; Academic Press: New York.

Ireland, J.F., Wyatt, P.A.H. (1976), "Acid-Base Properties of Electronically Excited States of Organic Molecules," *Adv. Phys. Org. Chem.* **12**, 131.

Klessinger, M., Pötter, T. (1991), "Properties of Molecules in Excited States" in *Theoretical Models of Chemical Bonding* Vol. 3; Maksić, Z.B., Ed.; Springer: Berlin.

Kosower, E.M. (1986), "Excited State Electron and Proton Transfer," *Ann. Rev. Phys. Chem.* **37**, 127.

Shizuka, H. (1985), "Excited-State Proton-Transfer Reactions and Proton-Induced Quenching of Aromatic Compounds," *Acc. Chem. Res.* **18**, 141.

Wan, P., Shukla, D. (1993), "Utility of Acid-Base Behavior of Excited States of Organic Molecules," *Chem. Rev.* **93** 571.

Zink, J.I., Shin, K.-S.K. (1991), "Molecular Distortions in Excited Electronic States Determined from Electronic and Resonance Raman Spectroscopy," *Adv. Photochem.* **16**, 119.

Polarization Spectra

Collett, E. (1993), *Polarized Light;* Marcel Dekker: New York.

Michl, J., Thulstrup, E.W. (1986), *Spectroscopy with Polarized Light*; VCH Inc.: New York.

Schellmann, J., Jensen, H.P. (1987), "Optical Spectroscopy of Oriented Molecules," *Chem. Rev.* **87**, 1359.

Thulstrup, E.W., Michl, J. (1989), *Elementary Polarization Spectroscopy*; VCH, Inc.: New York.

Thulstrup, E.W., Michl, J. (1982), "Orientation and Linear Dichroism of Symmetrical Aromatic Molecules Imbedded in Stretched Polyethylene," *J. Am. Chem. Soc.* **104**, 5594.

MO Methods and Calculations

Bruna, P.J., Peyerimhoff, S.D. (1987), in *Ab Initio Methods in Quantum Chemistry* **1**; Lawley, K.P., Ed.; Wiley: New York.

Clark, T. (1985), *A Handbook of Computational Chemistry*; Wiley: New York.

Davidson, E.R., McMurchie, L.E. (1982), "Ab Initio Calculations of Excited-State Potential Surfaces of Polyatomic Molecules," in *Excited States* **5**; Lim, E.C., Ed.; Academic Press: New York.

Ellis, R.L., Jaffé, H.H. (1977), "Electronic Excited States of Organic Molecules," in *Modern Theoretical Chemistry* **8**; Segal, G.A., Ed.; Plenum Press.

Hehre, W.J., Radom, L., Schleyer, P.v.R., Pople, J.A. (1986), *Ab Initio Molecular Orbital Theory*; Wiley: New York.

Klessinger, M. (1982), *Elektronenstruktur organischer Moleküle,* Verlag Chemie: Weinheim.

McWeeny, R. (1989), *Methods of Molecular Quantum Mechanics*; Academic Press: London.

Michl, J., Bonačic-Koutecký, V. (1990), *Electronic Aspects of Organic Photochemistry*; Wiley: New York.

Parr, R.G. (1963), *Quantum Theory of Molecular Electronic Structure*; Benjamin: New York, Amsterdam.

Salem, L. (1966), *The Molecular Orbital Theory of Conjugated Systems*; Benjamin: New York, Amsterdam.

Szabo, A., Ostlund, N.S. (1982), *Modern Quantum Chemistry. Introduction to Advanced Electronic Structure Theory*; New York.

Rydberg States

Robin, M.B. (1974/1985), "Highly Excited States of Polyatomic Molecules," Vol. 1–3; Academic Press: New York.

CHAPTER

2

Absorption Spectra of Organic Molecules

Saturated organic compounds are generally colorless and do not absorb in the near-UV region, either. Only the UV/VIS absorption spectra of those organic compounds that have π electrons are easy to observe experimentally. Absorptions in this region are usually due to electronic transitions from bonding π orbitals or from nonbonding n orbitals into antibonding π^* orbitals. They are referred to as $\pi \rightarrow \pi^*$ and $n \rightarrow \pi^*$ transitions, respectively. The $n \rightarrow \sigma^*$ transition of halogen compounds should also be noted. This chapter is concerned with the $\pi \rightarrow \pi^*$ transitions of the most important classes of π-electron compounds, that is, polyenes and cyclic conjugated compounds, and with the $n \rightarrow \pi^*$ transitions of carbonyl and nitrogen compounds. In addition, the influence of substituents on these electronic transitions will be discussed, and steric effects as well as solvent effects will be outlined briefly.

2.1 Linear Conjugated π Systems

The π-electron system of linear polyenes is particularly suitable for theoretical study. Even the simple free-electron model of Example 1.4 describes the most important feature of light absorption by linear π systems, that is, its shift to longer wavelengths as the length of the conjugated chain increases. Once a certain chain length is reached, the systems show absorption in the visible region and are therefore colored. Ethylene, the first member of the series, shows only absorption in the UV region. Although its spectrum is not yet understood in full detail, it may serve to explain many basic facts of UV spectroscopy.

2.1.1 Ethylene

Ethylene has two π electrons. The MO diagram of its π system therefore resembles the MO diagram of the H_2 molecule, which has a bonding σ_g MO and an antibonding σ_u MO. The electronic configurations of the H_2 molecule derived from this scheme are collected in Table 2.1 together with the Mulliken notation and the group theoretical labels. Rydberg states are not considered in the simple model. They correspond to electronic transitions from the σ_g MO into one of the diffuse MOs resulting from the use of an extended basis set, which can be best labeled as AOs of the united atom. (Cf. Figure 1.25.)

Figure 2.1 gives a qualitative MO scheme for ethylene. It also pictures the shape of the MOs, from which the relation with the AOs of the united atom becomes apparent; thus, it is possible to recognize that the $4a_g$ MO is related to a united atom 3s AO, and can therefore acquire appreciable Rydberg character.

The experimental spectrum of ethylene (cf. Mulliken, 1977) shows a broad absorption band in the vacuum UV region, at 180–145 nm. This has been assigned to a V←N or $\pi \rightarrow \pi^*$ transition with superimposed Rydberg bands. The $\pi \rightarrow \pi^*$ band extends toward 207 nm with weak vibrational structure due to the torsional mode. Vibrational analysis reveals that at the equilibrium geometry of the (π, π^*) excited state V the two methylene groups lie in mutually perpendicular planes. The extremely weak band near 270 nm ($\varepsilon = 10^{-4}$), which becomes clearly observable under O_2 pressure, confirms that this is also true at the equilibrium geometry of the triplet state. The analysis of the UV spectrum is summarized in Figure 2.2, which depicts the energies of the various states as a function of the torsional angle, and which is based on very detailed quantum chemical calculations. In fact, ethylene has been the subject of numerous theoretical investigations. (Cf. Buenker et al., 1980.) This is due not least to the fact that geometry changes of the same kind as those observed for ethylene are prototypical for photoisomerization reactions that proceed through π-bond dissociation (cf. Sec-

Table 2.1 Electronic Configurations of the H_2 Molecule

Configuration	Mulliken Notation	Group Theoretical Label
$(1\sigma_g)^2$	N	${}^1\Sigma_g^+$
$(1\sigma_g)(1\sigma_u)$	T	${}^3\Sigma_u^+$
	V	${}^1\Sigma_u^+$
$(1\sigma_u)^2$	Z	${}^1\Sigma_g^+$
$(1\sigma_g)(\mathrm{Ry})$[a]	R	

[a] Ry denotes the 2s, 3s, 3p, 3d . . . AOs of the united atom.

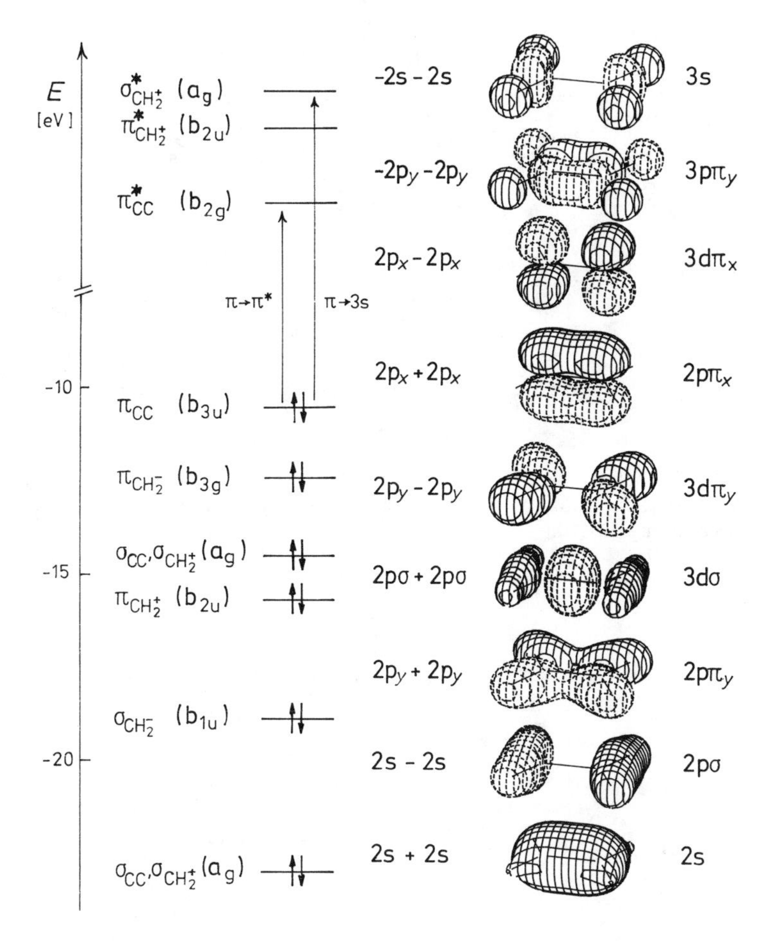

Figure 2.1. Schematic MO diagram of ethylene. Shown are orbital energy levels, qualitative shape and labeling of MOs, as well as the corresponding linear combinations of carbon AOs and labeling of united atom orbitals (adapted from Merer and Mulliken, 1969).

tion 4.1.3) and that play an important role in the chemistry of vision. (Cf. Section 7.1.5.)

2.1.2 Polyenes

Essential features of the electronic spectra of linear polyenes have already been mentioned in Chapter 1. The HOMO→LUMO transition is shifted to longer wavelengths as the number of conjugated double bonds increases, and this is easy to rationalize in the FEMO model. However, the next transition that is to be expected from the simple MO model is only allowed if the molecule does not have a center of symmetry; that is, if it is not in the all-trans configuration. (Cf. Example 1.7.) The absorption spectrum of β-caro-

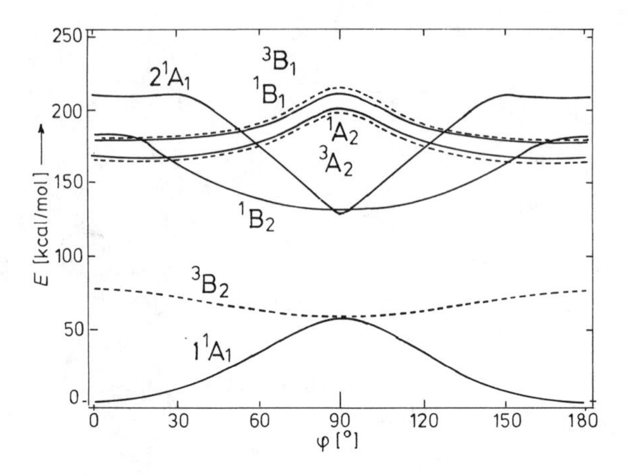

Figure 2.2. Potential energy curves of the ground state and some excited states of ethylene as a function of the torsional angle (by permission from Michl and Bonačić-Koutecký, 1990).

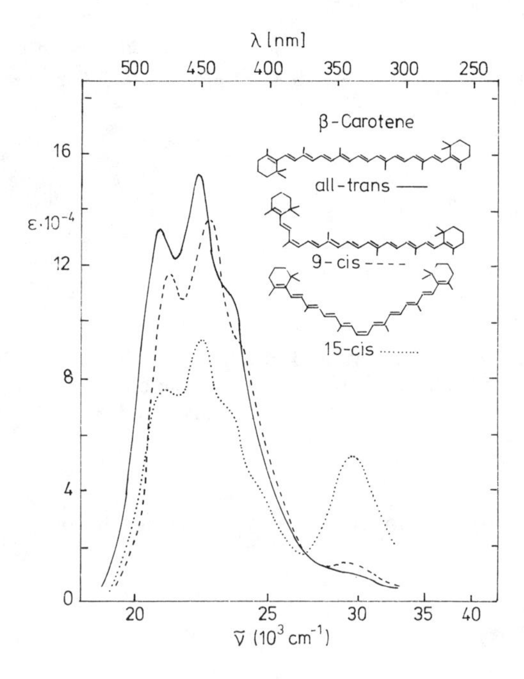

Figure 2.3. Absorption spectrum of β-carotene in the all-trans, 9-cis and 15-cis configuration (by permission from Jaffé and Orchin, 1962).

Table 2.2 Wavelength λ_{max} (in nm) and Oscillator Strength f of the Longest-Wavelength Absorption of Dimethyl and Diphenyl Polyenes (Adapted from Murrell, 1963)

	$CH_3-(CH=CH)_n-CH_3$			$C_6H_5-(CH=CH)_n-C_6H_5$		
n	λ_{max}	f	f/n	λ_{max}	f	f/n
2	227	0.74	0.37	334	0.77	0.39
3	275	1.11	0.37	358	1.26	0.42
4	310	1.66	0.42	384	1.39	0.35
5	341	2.17	0.43	403	1.41	0.28
6	380	—	—	420	1.58	0.26
7	396	—	—	435	1.90	0.27

tene in Figure 2.3 shows clearly the so-called "cis band" originating from a transition that is forbidden in the *all-trans*-polyene. (Cf. Zechmeister, 1960.)

Some data for the longest-wavelength absorption bands of polyenes and α,ω-diphenylpolyenes are collected in Table 2.2. It is seen that the intensity of this band is to a very good approximation proportional to the number n of double bonds. This is not true for the wavelength λ_{max}, which instead converges to a finite value for an infinitely long polyene, as is apparent from Figure 2.4. This contrasts with Example 1.4, where it was shown that the

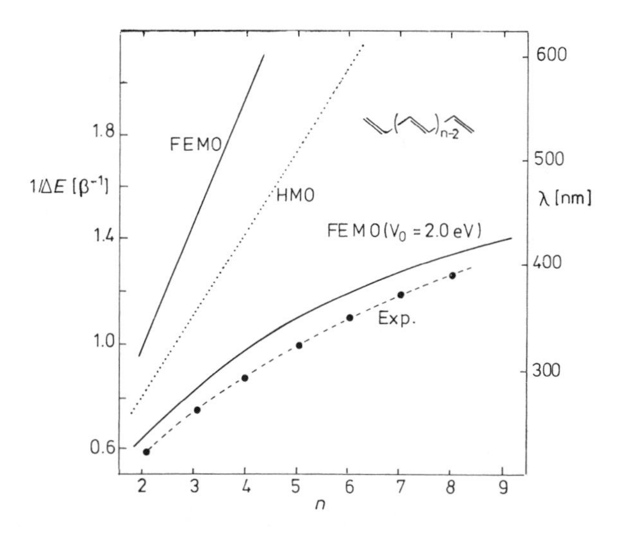

Figure 2.4. Calculated wavelength dependence of the longest wavelengths $\pi\rightarrow\pi^*$ transition of polyenes on the number n of double bonds. FEMO results (—, right-hand scale) for completely delocalized π electrons as well as for alternating double and single bonds with $V_0 = 2$ eV, experimental points (---, right-hand scale), and HMO results (···, left-hand scale).

FEMO model predicts a linear relationship between λ_{max} and n. The HMO method yields a similar linear dependence of $\lambda_{max} \sim 1/\Delta E_{1\rightarrow1'}$ on n, as is also shown in Figure 2.4.

One of the reasons simple MO models produce an incorrect dependence of the wavelength of the first absorption band on the length of the conjugated system is because of the choice of the constant $C = 0$ in calculating ΔE from Equation (1.22). However, of great importance is also the fact that the theoretical models assume that the π electrons are freely mobile, as would be the case in a π system in which all CC bonds are of equal length, whereas real polyenes are characterized by alternating double and single bonds. If the effect of bond alternation is introduced into the FEMO model by a potential that is deeper in the region of a formal double bond than in the region of a formal single bond, λ_{max} no longer increases linearly with n, but correctly converges to a finite value, as may be seen from Figure 2.4 (Kuhn, 1949).

Example 2.1:
The different manner in which the longest wavelength absorption λ_{max} depends on the chain length for systems with equal and alternating bond lengths is also revealed by HMO calculations, provided that equal β values for all bonds are used in the former case, and two distinct values in the latter: the β_2 values for double bonds are larger in absolute value than the β_1 values for single bonds. In Figure 2.5, the HMO results obtained from Equation (1.22), with all β equal and with $\beta_2/\beta_1 = 1.25$, are compared with experimental data for cyanine dyes

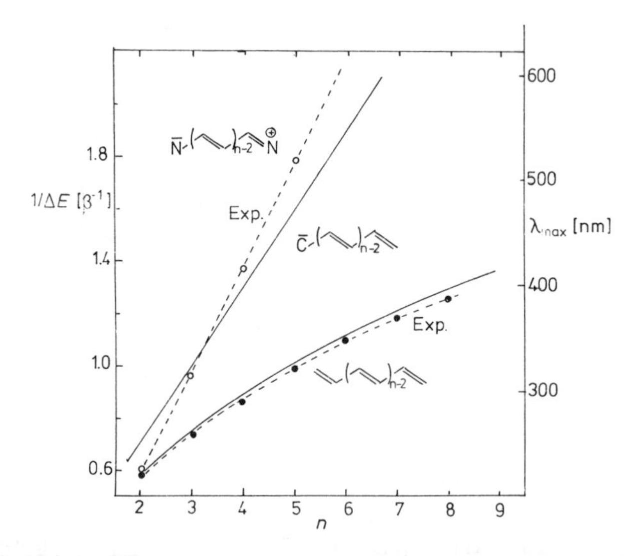

Figure 2.5. HMO results (—, left-hand scale) for the wavelength dependence of the first $\pi \rightarrow \pi^*$ transition on the chain length for odd linear polyene anions (equal β) and for even linear polyenes ($\beta_2/\beta_1 = 1.25$) compared with experimental data (---, right-hand scale) for symmetrical cyanines (Malhotra and Whiting, 1959) and polyenes (Nayler and Whiting, 1955).

of type **1** and for polyenes. It is seen that the wavelength λ_{max} of the HOMO→LUMO transition of a conjugated system with n double bonds is expected to be longer when the bond lengths are more uniform. This fact, together with the value of the constant C that is expected for cyanines from the small value of overlap density (cf. Example 1.3), explains the extraordinarily long wavelength absorption of cyanine dyes. This is utilized for producing photographic films sensitive to infrared light with dyes such as **2**, which contains equal terminal groups, and thus nearly equal bond lengths throughout.

$R_2N{-}(CH{=}CH)_n{-}CH{=}\overset{\oplus}{N}R_2$

1

=CH–(CH=CH)$_n$–

2

Quantum chemical calculations (Koutecký, 1967; Dunning et al., 1973) and experimental investigations (Hudson and Kohler, 1972; Hudson et al., 1982; Kohler, 1993) suggest that the lowest excited state of linear *all-trans*-polyenes is not the ^1B_u state that corresponds to the symmetry-allowed HOMO→LUMO transition, but a ^1A_g state. This state is characterized by an appreciable contribution from a doubly excited configuration, which is very high in energy in the simple HMO model. It is only through configu-

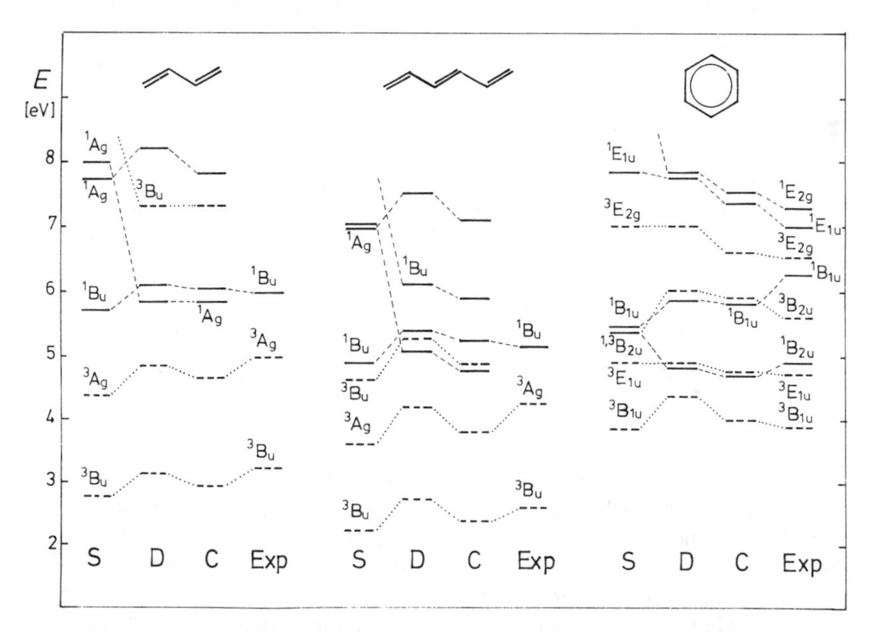

Figure 2.6. Theoretical and experimental excitation energies of butadiene, hexatriene, and benzene. The calculations include all singly excited configurations (S), all singly and doubly excited configurations (D) or complete CI (C), respectively (by permission from Schulten et al., 1976a).

ration interaction that the ^1A_g state is stabilized to such an extent that its energy is below or roughly equal to that of the singlet state assigned to the HOMO→LUMO transition. This becomes apparent from Figure 2.6, in which the results of PPP calculations with varying amounts of configuration interaction are compared with experimental data. If configuration interaction is limited to singly excited configurations, as is common practice in routine calculations, the doubly excited A_g state in question is far beyond the interesting energy range.

The forbidden $^1A_g \rightarrow {}^1A_g$ transition can be observed directly using two-photon spectroscopy, which has been developed in recent years. Measurements for octatetraene and other polyenes fully confirm the theoretical predictions (Granville et al., 1980).

Example 2.2:
Two factors are of primary importance for the energy ordering of the lowest excited states of alternant hydrocarbons: first-order CI between degenerate singly excited configurations, which produces plus and minus states (cf. Section 1.2.4), and CI with multiply excited configurations. From theoretical arguments it is known (Koutecký, 1967) that the former dominates in cyclic systems (cf. Section 2.2.2), whereas the latter is the most important for polyenes. In order to decide whether the topology or the geometry or both these factors are responsible for the dissimilarity of these two classes of molecules, so-called hairpin polyenes (**3**) were synthesized, whose geometrical structures are closely related to those of the corresponding bridged annulenes (**4**), but whose topology is that of linear polyenes:

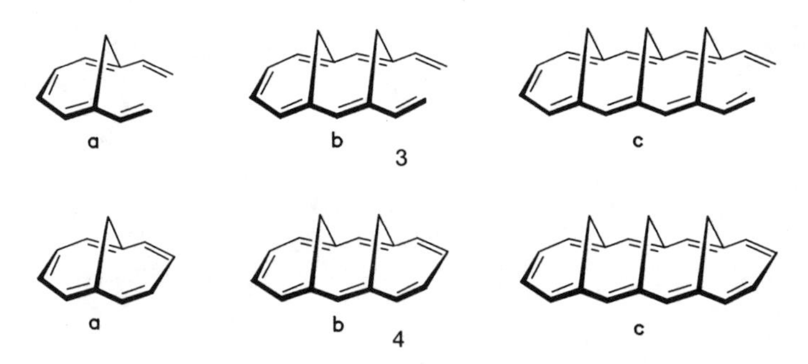

In both series the lowest A^- and B^+ states are nearly degenerate, largely due to interactions with singly excited configurations. From this comparison it may be concluded that the effect of configuration interaction on excitation energies as well as transition moments is dictated primarily by molecular geometry, and only to a minor degree by topology. These conclusions have general applicability. For instance, they provide a simple explanation for the lower energy of the B^+ transition in *s-cis*-butadiene relative to *s-trans*-butadiene (Frölich et al., 1983).

2.2 Cyclic Conjugated π Systems

The electronic spectra of cyclic conjugated π systems depend inherently on the number of π electrons. Closed-ring systems with $4N+2$ π electrons in the perimeter are aromatic compounds, of which benzene is the most important representative. Benzenoid hydrocarbons constitute a class of compounds whose UV spectra have been investigated most extensively both experimentally and theoretically. The fact that the spectra of aromatic compounds are so characteristic meant that formerly they were of considerable importance in the structure determination of organic compounds. However, these spectra cannot be explained in terms of the simple HMO model. If one seeks a theoretical basis for an understanding, one has the choice between the perimeter model and the Pariser-Parr-Pople or a more complicated numerical method. Before discussing these theoretical models, some empirical relations will be presented. Finally, cyclic systems derived from a perimeter of $4N$ π electrons will be considered.

2.2.1 The Spectra of Aromatic Hydrocarbons

Typical UV spectra of aromatic hydrocarbons are shown in Figure 2.7. They contain absorption bands that fall into three categories according to their intensity:

Bands of the first type (^1L_b) are of low intensity ($\varepsilon = 10^2$–10^3), may be hidden by the other bands, and often possess a complicated vibrational structure.
Bands of the second type (^1L_a or ^1B_a) are moderately intense ($\varepsilon \approx 10^4$); ^1L_a bands usually show a regular vibrational structure;
Bands of the third type (^1B_b) are very strong ($\varepsilon > 10^5$) and have little vibrational structure.

The ^1L_b, ^1L_a, ^1B_b, ^1B_a nomenclature originates in the perimeter model, discussed in detail in Section 2.2.2.

Figure 2.8 shows how the experimentally observed wavelengths for two series of cata-condensed benzenoid hydrocarbons, the linearly annelated *acenes* and the angularly annelated *phenes,* change with increasing number of benzene rings. The shifts in the ^1L_b and ^1B_b positions are parallel to one another in both series, whereas the bathochromic shift of the ^1L_a band of acenes upon annelation is so pronounced that even in anthracene it masks the ^1L_b band.

These findings are difficult to understand on the basis of the simple HMO model. This model suggests that in the case of benzene a fourfold degenerate transition from the degenerate highest occupied MO into the equally degenerate lowest unoccupied MO is to be expected instead of the observed three bands. In Example 1.9 it was shown that three bands of different symmetry are to be expected from group theoretical arguments, in agreement with experimental observations.

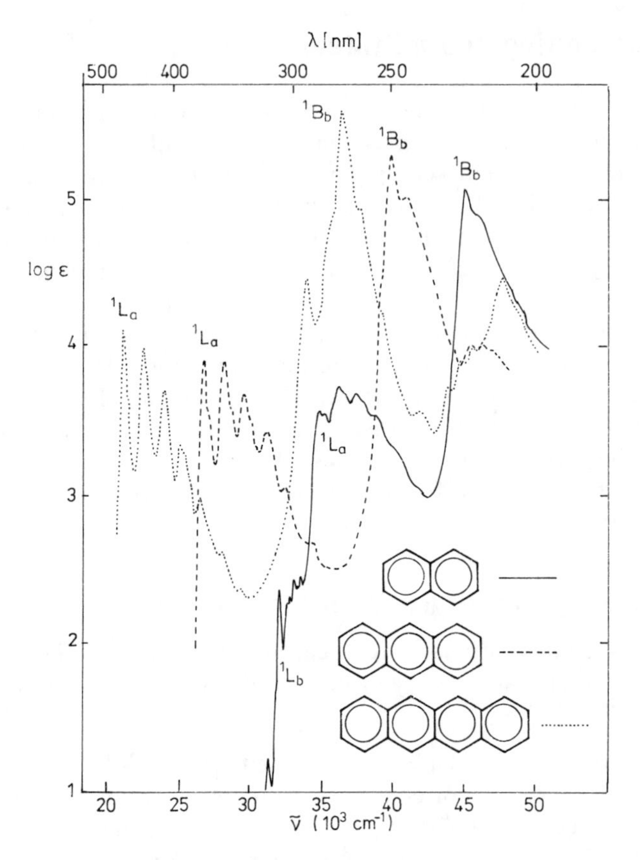

Figure 2.7. Absorption spectra of naphthalene, anthracene, and tetracene as typical examples of the UV spectra of aromatic hydrocarbons (by permission from DMS-UV-Atlas, 1966–71).

In general the HOMO and LUMO of condensed aromatic hydrocarbons are not degenerate, so simple theory gives a HOMO→LUMO transition and a degenerate transition from the HOMO ϕ_1 into the second LUMO $\phi_{2'}$, or from the second HOMO ϕ_2 into the LUMO $\phi_{1'}$.

As was shown in Section 1.2.4, the degenerate configurations $^1\Phi_{1\rightarrow 2'}$ and $^1\Phi_{2\rightarrow 1'}$ are split by configuration interaction (Pople, 1955; cf. Figure 1.6). The transition from the ground state into the lower of these states yields the ^1L_b band, and is forbidden within the PPP model. This is due to the fact that the transition moments $\boldsymbol{M}_{1\rightarrow 2'}$ and $\boldsymbol{M}_{2\rightarrow 1'}$ between the ground configuration Φ_0 and the singly excited configurations $\Phi_{1\rightarrow 2'}$ and $\Phi_{2\rightarrow 1'}$ cancel one another exactly for that transition. The transition into the higher of these two states corresponds to the ^1B_b band and, in agreement with experimental data, is expected to be very intense since the individual transition moments $\boldsymbol{M}_{1\rightarrow 2'}$ and $\boldsymbol{M}_{2\rightarrow 1'}$ provided by its two constituents enhance one another.

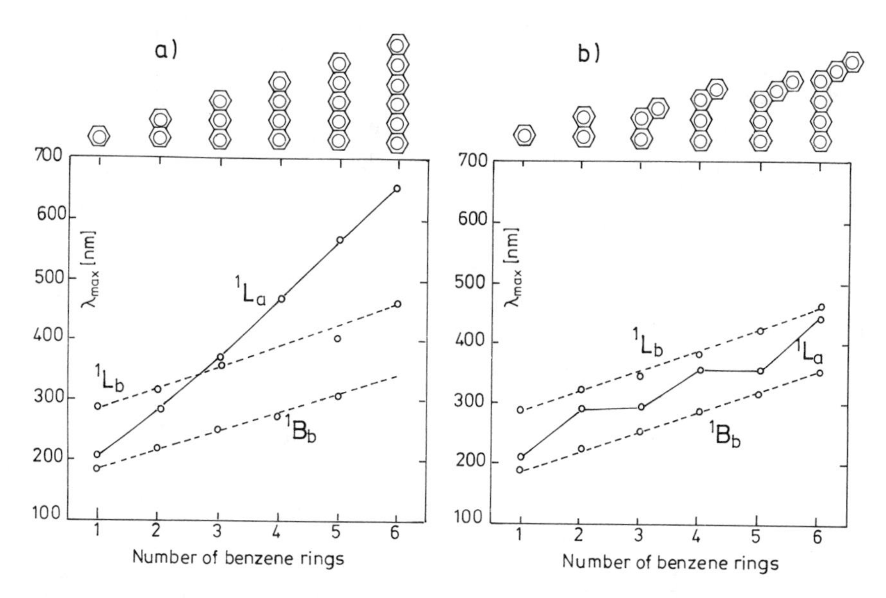

Figure 2.8. Wavelengths of ^1L_b, ^1L_a, and ^1B_b bands of condensed aromatic hydrocarbons plotted versus the number of benzene rings: a) linearly annelated aromatics (acenes); b) angularly annelated aromatics (phenes) (by permission from Badger, 1954).

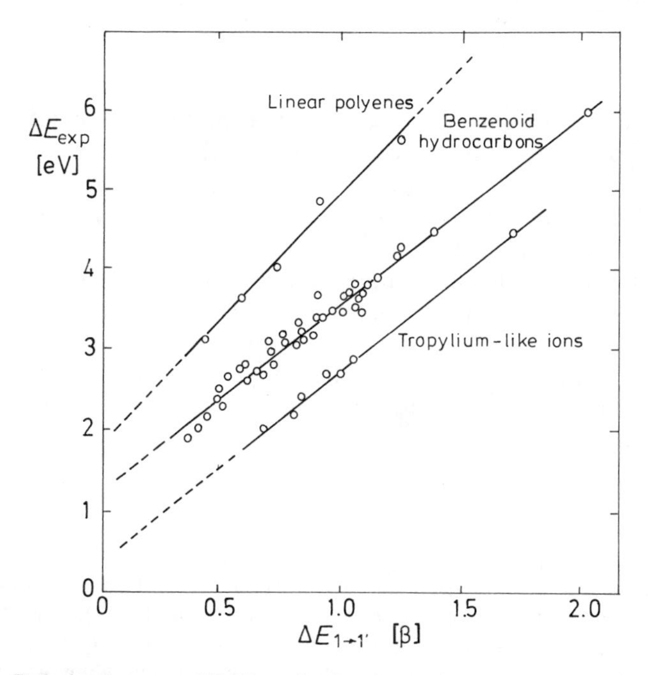

Figure 2.9. Relation between HMO excitation energies $\Delta E = \varepsilon_{1'} - \varepsilon_1$ and observed excitation energies of the ^1L_a band of various series of hydrocarbons (by permission from Koutecký, 1965).

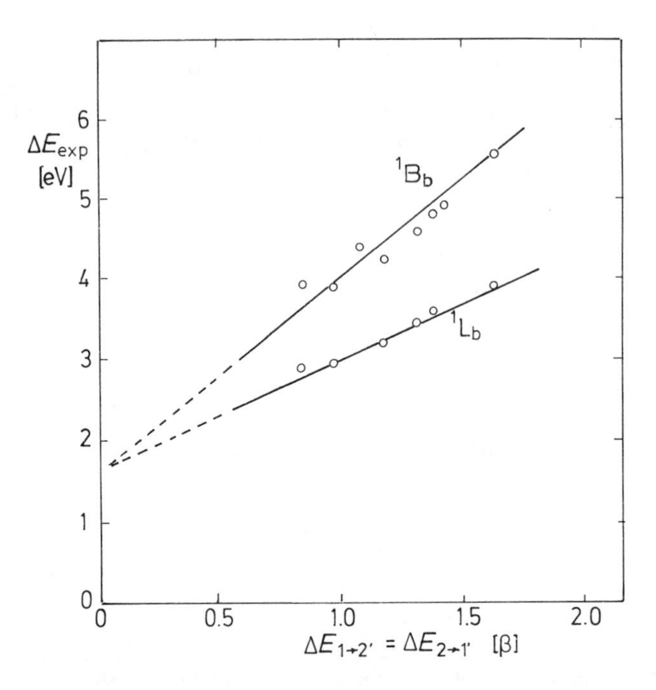

Figure 2.10. Relation between HMO excitation energies $\Delta E = \varepsilon_{2'} - \varepsilon_1 = \varepsilon_{1'} - \varepsilon_2$ and observed excitation energies of the ^1L_b and ^1B_b bands of condensed aromatic hydrocarbons (by permission from Koutecký, 1965).

From these MO arguments it becomes apparent that the excitation energy of the ^1L_a band is related to the HOMO-LUMO energy difference. In fact, the relation between the experimental transition energy of the ^1L_a band and the HMO excitation energies calculated as orbital energy differences $\varepsilon_{1'} - \varepsilon_1$ is linear to a fairly good approximation, as is seen from Figure 2.9. From Figure 2.10 a similar linear relation between the transition energies of the ^1L_b band as well as the ^1B_b band and the orbital energy difference $\varepsilon_{2'} - \varepsilon_1 = \varepsilon_{1'} - \varepsilon_2$ is seen to hold. This satisfactory proportionality between excitation energies and HMO orbital energy differences suggests that contributions of the electron repulsion terms to the excitation energies are proportional to orbital energy differences.

Example 2.3:
If the longest-wavelength transition corresponds to the ^1L_a band, its excitation energy can be estimated easily from a model proposed by Dewar (1950). An even-alternant hydrocarbon can formally be considered as composed of two odd-alternant fragments, each of which must have a nonbonding orbital. When bonds are introduced that link the two fragments to form the total system, the two nonbonding orbitals interact to give one doubly occupied bonding orbital,

which will be the HOMO, and one antibonding orbital, the LUMO. From PMO theory the first-order splitting of the two nonbonding orbitals is given by

$$\delta\varepsilon_{\text{PMO}} = 2 \sum_{\varrho-\sigma} c_\varrho^R c_\sigma^S \beta_{\varrho\sigma}$$

where c_ϱ^R and c_σ^S are the LCAO coefficients of the nonbonding MOs of fragments R and S at atoms involved in the formation of a new ϱ—σ bond, and the

Table 2.3 HOMO-LUMO Excitation Energies of Condensed Aromatic Hydrocarbons. Comparison of Values (in cm^{-1}) Calculated from the PMO Model[a] with Experimental Data for the ^1L_b Band (Adapted from Heilbronner and Bock, 1968)

Compound	$\Delta\varepsilon_{\text{PMO}}$	$\tilde{\nu}_{\text{PMO}}$[b]	$\tilde{\nu}_{\text{exp}}$	Δ
Chrysene (**5**)	−0.676	32,000	31,350	−650
3,4-Benzphenanthrene (**6**)	−0.734	34,000	33,000	−1,000
3,4-Benztetraphene (**7**)	−0.598	29,200	26,270	−2,930
1,2-Benztetraphene (**8**)	−0.648	31,000	26,880	−4,120
Pentaphene (**9**)	−0.588	28,900	28,160	−740
1,2,-5,6-Dibenzanthracene (**10**)	−0.600	29,300	28,790	−510
1,2-7,8-Dibenzanthracene (**11**)	−0.651	31,100	29,890	−1,210
Picene (**12**)	−0.650	31,100	30,740	−360
3,6-Benzchrysene (**13**)	−0.705	33,000	31,200	−1,800

[a] $2a$, $-a$, $-a$, a; $a = \frac{1}{\sqrt{7}}$ $3b$, $-b$, b, $-2b$, $-b$, $2b$; $b = \frac{1}{\sqrt{20}}$ c, $-2c$, $3c$, $-c$, c, $-c$; $c = \frac{1}{\sqrt{17}}$

[b] $\tilde{\nu}_{\text{PMO}} = 8{,}200 - 22{,}000\,(2x_m)\ [\text{cm}^{-1}]$

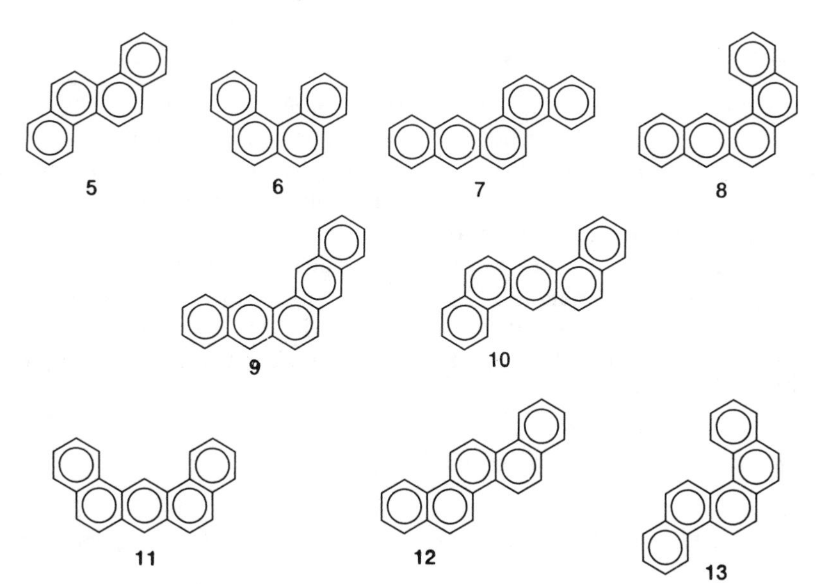

sum runs over all these new bonds. (Cf. Dewar and Dougherty, 1975.) In Table 2.3 wave numbers calculated from this relation using the regression line

$$\tilde{\nu}_{PMO}/\text{cm}^{-1} = 8{,}200 - 22{,}000\ \Delta\varepsilon/\beta$$

are collected for the condensed aromatic hydrocarbons **5–13**, which can be made up from benzyl, α-naphthyl, and β-naphthyl fragments, and compared with experimental data. Considering the very approximate nature of the model, the agreement is very good.

Triplet states of aromatics can be classified in a similar fashion as their singlet states. For instance, the ^3L_a state is well described by the $^3\Phi_{1\rightarrow 1'}$ configuration. It is the lowest triplet state, even in those benzenoid hydrocarbons in which ^1L_b rather than ^1L_a is the lowest singlet, since the ^1L_a–^3L_a energy splitting is generally large, whereas the ^1L_b–^3L_b splitting vanishes in the first approximation. (Cf. Figure 6.18.)

2.2.2 The Perimeter Model

The perimeter model introduced by Platt (1949), reformulated in the LCAO MO form by Moffitt (1954a), and extended by Gouterman (1961), Heilbronner and Murrell (1963), and Michl (1978), has been very useful in understanding trends in the electronic spectra of cyclic π systems. It applies equally to singlet and triplet states and has provided the commonly used nomenclature for both. The following discussion is limited to the singlet states, which are more important in ordinary spectroscopy.

Platt's original perimeter model was a free-electron (FEMO) model based on a one-dimensional circular potential along which the π electrons can move freely. The orbitals of an electron confined to such a circular ring are given by

$$\phi_l = (1/\sqrt{2\pi})e^{il\theta}, \quad l = 0, \pm 1, \pm 2 \ldots \tag{2.1}$$

These FEMOs are degenerate except for $l = 0$. $l\hbar$ is the angular momentum of a particle on a circular orbit, and the degeneracy may be thought of as arising from the fact that the two orbitals $\phi_{\pm|l|}$ describe an electron moving counterclockwise or clockwise along the perimeter. The original Platt model is nowadays of historical interest only.

The model becomes much more meaningful if the perimeter orbitals are described as linear combinations of n AOs χ_μ (Moffitt, 1954a). Due to the properties of the cyclic point group C_n these LCAO MOs are determined completely by symmetry for regular polygons. (Cf. Cotton, 1971.) They may be written as

$$\begin{aligned}\phi_{\pm k} &= (1/\sqrt{n}) \sum_{\mu=0}^{n-1} \exp\,(\pm 2\pi i k\mu/n)\chi_\mu \\ &= (1/\sqrt{n}) \sum_{\mu=0}^{n-1} [\cos\,(2\pi k\mu/n) \pm i \sin\,(2\pi k\mu/n)]\chi_\mu\end{aligned} \tag{2.2}$$

where k is a nonnegative integer and $l = \pm k$ may again be thought of as an *angular momentum quantum number*. At the HMO level of theory, the orbitals have energies

$$\varepsilon_{\pm k} = \alpha + 2\cos(2\pi k/n)\beta, \quad k = 0, 1, 2 \ldots \tag{2.3}$$

Due to the periodicity of the trigonometric functions one has $\phi_{k \pm jn} = \phi_k$ for an integer j. In contrast to the FEMOs the total number of LCAO-MOs is limited to the number n of perimeter AOs χ_μ.

The most stable MO ϕ_0 is real. In the order of increasing energy it is followed by complex conjugate pairs of MOs, ϕ_k and ϕ_{-k}, which are degenerate. If the number of atoms in the perimeter n is even, one has $\phi_{n/2} \equiv \phi_{-n/2}$, so that the least stable MO is nondegenerate and can be equally well labeled by $n/2$ or $-n/2$.

An electron in an orbital ϕ_k has a positive z component of orbital angular momentum and is found, on the average, to move counterclockwise along the perimeter when viewed from the positive end of the z axis. An electron in an orbital ϕ_{-k} tends to move clockwise and has a negative angular momentum of equal magnitude. Since it is negatively charged, an electron in the MO ϕ_k produces a negative z component of the magnetic moment μ_z. Similarly, the motion of an electron in MO ϕ_{-k} produces a positive μ_z. The magnitude of μ_z is a function of the number of atoms n in the perimeter and of the orbital subscript k. It is zero for $k = 0$, increases with increasing k until the latter slightly exceeds $n/4$, and then decreases again. When n is even, μ_z vanishes in the nondegenerate highest energy orbital. These features are schematically illustrated in Figure 2.11, which shows the orbital energies, the sense and magnitude of electron circulation, and the resulting orbital magnetic moments for the perimeter depicted on the left.

The physical reasons for this unusual nonmonotonic behavior, which differs from what one would expect for an atom or for the free-electron model of the perimeter, are related to the only n-fold symmetry which is present in the polygon of n perimeter atoms, instead of full cylindrical symmetry (Michl and West, 1980). The fact that the magnetic moment can either increase or decrease for a transition from k to $k+1$ is of fundamental importance for an understanding of MCD spectra. (Cf. Section 3.3.)

In the ground configuration of a perimeter with $4N+2$ π electrons all orbitals up to and including ϕ_N and ϕ_{-N} will be doubly occupied. Thus, the degenerate MOs ϕ_N and ϕ_{-N} are the HOMOs and the degenerate MOs ϕ_{N+1} and the ϕ_{-N-1} are the LUMOs of the π system. If only HOMO→LUMO excitations are considered, four singly excited configurations

$$\Phi_1 = \Phi_{N \to N+1} \tag{2.4}$$

$$\Phi_2 = \Phi_{-N \to -N-1}$$

$$\Phi_3 = \Phi_{-N \to N+1}$$

$$\Phi_4 = \Phi_{N \to -N-1}$$

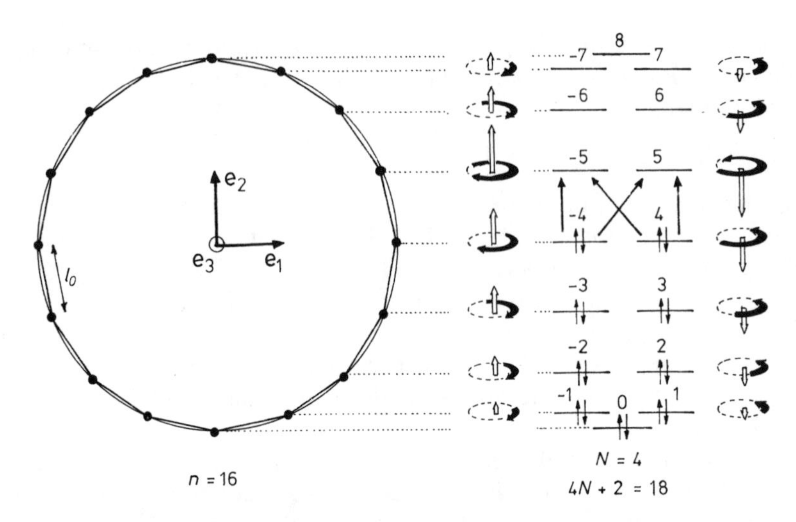

Figure 2.11. The perimeter model of an $(4N+2)$-electron [n]annulene, geometry on the left, energies of the MOs ϕ_k on the right. The angular momentum quantum number is given for each MO. The sense and magnitude of electron circulation and the resulting orbital magnetic moment are shown schematically in a perspective view. Orbital occupancy in the ground configuration and the four one-electron HOMO→LUMO excitations are indicated (by permission from Michl, 1978).

are produced in addition to the ground configuration Φ_0, where $\Phi_{i\rightarrow k}$ and $\Phi_{-i\rightarrow -k}$ are complex conjugate and therefore degenerate.

If the perimeter is charged, that is, if $n \neq 4N+2$, the five configurations belong to distinct irreducible representations a, ε_1, ε_{-1}, ε_{2N+1}, and $\varepsilon_{-(2N+1)}$ of the point group C_n; ε_k and ε_{-k} are complex conjugate and are equivalent to the real two-dimensional representation E_k. (Cf. Cotton, 1971.) Each of the five configurations has a different symmetry behavior and the Hamiltonian, which is totally symmetric, cannot mix them, that is, $\langle\Phi_i|\hat{H}|\Phi_k\rangle = 0$ for $i \neq k$. The CI matrix is diagonal and the state wave functions are identical with the configuration wave functions. At the present level of treatment, the model thus yields a ground state and four excited states that come as two degenerate pairs with Φ_1 and Φ_2 higher in energy than Φ_3 and Φ_4. For such systems the description of the excited states is qualitatively correct already in the one-electron model.

If the perimeter is uncharged, $n = 4N+2$ and the configurations Φ_3 and Φ_4 belong to the same irreducible representation b $= \varepsilon_{n/2} \equiv \varepsilon_{-n/2}$ of the point group C_n. They differ in two spin orbitals and therefore interact only through the electron repulsion part $\hat{g}(i, j)$ of the Hamiltonian. Consequently, the one-electron model fails and first-order configuration interaction has to be taken into account. If $Q = \langle\Phi_3|\hat{H}|\Phi_4\rangle$ is the interaction matrix element between configurations Φ_3 and Φ_4, the CI matrix reads

$$\begin{pmatrix} E(\Phi_1) & 0 & 0 & 0 \\ 0 & E(\Phi_2) & 0 & 0 \\ 0 & 0 & E(\Phi_3) & Q \\ 0 & 0 & Q & E(\Phi_4) \end{pmatrix} \tag{2.5}$$

with $E(\Phi_1) = E(\Phi_2)$ and $E(\Phi_3) = E(\Phi_4)$

Diagonalization of this matrix yields the states

$$\Psi(B_{2u}) = (\Phi_3 - \Phi_4)/(i\sqrt{2})$$

$$\Psi(B_{1u}) = (\Phi_3 + \Phi_4)/\sqrt{2}$$

in addition to the degenerate $\Psi(E_{1u})$ states Φ_1 and Φ_2, where the nomenclature of the full D_{nh} symmetry of the annulenes has been used. To summarize, then, the annulenes with an uncharged perimeter have a singlet ground state of symmetry A_{1g} and three excited states of symmetry B_{2u}, B_{1u}, and E_{1u}, respectively, in increasing energy order, as has been observed for benzene. (Cf. Section 2.2.1.) Following Platt, it is customary to refer to the states composed of the configurations Φ_1 and Φ_2 as the ^1B states and the states composed of the configurations Φ_3 and Φ_4 as the ^1L states. In the case of an uncharged perimeter, in which the L states are not degenerate, the lower of these two states ($^1B_{2u}$ in the full D_{nh} symmetry) is referred to as the ^1L_b state and the high-energy one ($^1B_{1u}$ in D_{nh}) as the ^1L_a state. For a ^1L_b state the transition density between the ground state and the excited state has its maxima at bonds and its nodes pass through atoms on opposite sides of the perimeter, whereas for a ^1L_a state the maxima are at atoms and nodes pass through the centers of bonds.

Example 2.4:

The transition moment can be considered as the dipole moment of the transition density. (Cf. Section 1.3.1.) The transition density between the ground configuration Φ_0 and a singly excited configuration $\Phi_{j\rightarrow k}$ is equal to the overlap density $\sqrt{2}\phi_j^*\phi_k$, as may be seen from Equation (1.39). Using the complex perimeter MOs of Equation (2.2) and applying the ZDO approximation ($\chi_\mu^*\chi_\nu \equiv 0$ for $\mu \neq \nu$) the overlap density is found to be

$$\phi_j^*\phi_k = (1/n) \sum_\mu \exp\,[2\pi i(k - j)\mu/n]\chi_\mu^*\chi_\mu$$

The complex phases of the LCAO coefficients can be visualized as a phase polygon. This is a regular polygon of unit radius with its center at the origin in the complex plane, oriented in such a way that its vertex number zero coincides with +1 on the real axis. As one proceeds from one AO of the real polygon-shaped perimeter to the next, the complex phase of the coefficient changes, in a counterclockwise fashion for ϕ_k and in a clockwise fashion for

ϕ_{-k}. The rate at which the complex phase changes from one AO χ_μ to the next AO $\chi_{\mu+1}$ is proportional to k. It jumps from one vertex of the phase polygon to the k-th, counterclockwise for ϕ_k and clockwise for ϕ_{-k}. This procedure is illustrated for the HOMOs and LUMOs of benzene in Figure 2.12a. The products of AO coefficients needed for evaluating the overlap densities are easily obtained from these diagrams by simply adding the complex phases. Thus,

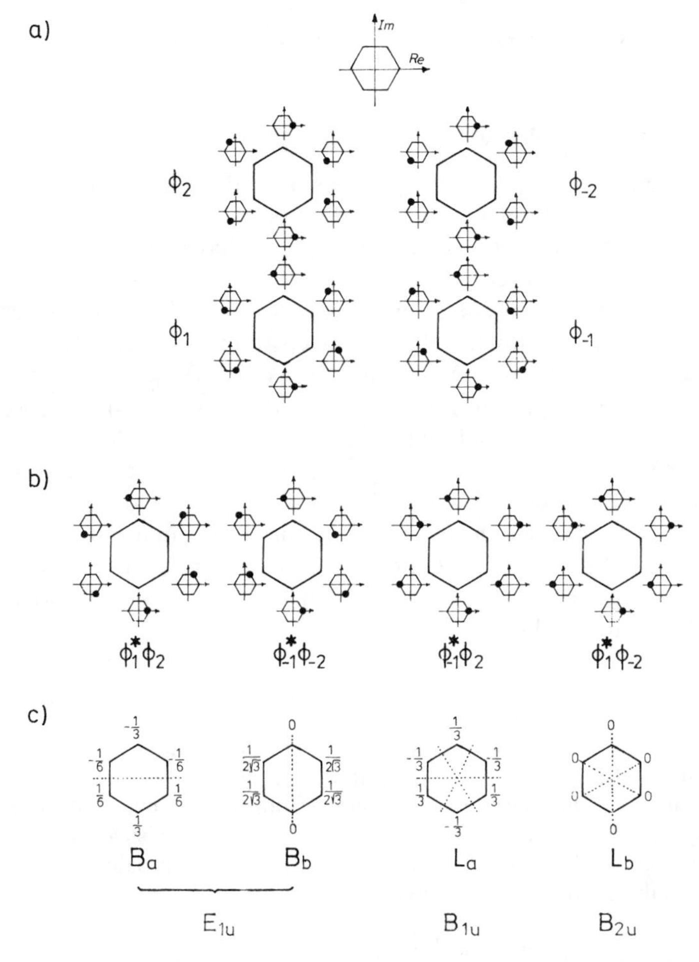

Figure 2.12. Nodal properties of the transition densities of the first four transitions in benzene. a) Representation of the complex LCAO coefficients of HOMOs ϕ_1 and ϕ_{-1} as well as LUMOs ϕ_2 and ϕ_{-2} by means of a phase polygon. Each coefficient has the absolute magnitude $n^{-1/2}$ and the complex phase shown by a dot in the complex plane of which the real and imaginary axes are abscissa and ordinate. b) Representation of the overlap densities evaluated from the complex coefficients, and c) values of the overlap densities at the vertices of the perimeter and the resulting nodal properties.

taking into account that $\phi_k^* = \phi_{-k}$, the diagrams in Figure 2.12b are obtained. From these the nodal properties of the transition densities may be derived. They are shown in Figure 2.12c for all four states considered. Use was made of the fact that $\Phi_{1\to 2}$ and $\Phi_{-1\to -2}$ are degenerate, so the real linear combinations $(\Phi_{1\to 2} + \Phi_{-1\to -2})/\sqrt{2}$ and $(\Phi_{1\to 2} - \Phi_{-1\to -2})/(i\sqrt{2})$ can be employed instead. The transition densities between the ground state Φ_0 and these excited states are then obtained by means of the Euler formulae $\cos z = (e^{iz} + e^{-iz})/2$ and $\sin z = (e^{iz} - e^{-iz})/(2i)$ as

$$\sqrt{2}(\phi_1^*\phi_2 + \phi_{-1}^*\phi_{-2})/\sqrt{2} = (1/6)\sum_\mu[\exp(2\pi i\mu/6) + \exp(-2\pi i\mu/6)]|\chi_\mu|^2$$
$$= (1/3)\sum_\mu \cos(\pi\mu/3)|\chi_\mu|^2$$

or

$$\sqrt{2}(\phi_1^*\phi_2 - \phi_{-1}^*\phi_{-2})/(i\sqrt{2}) = (1/3)\sum_\mu \sin(\pi\mu/3)|\chi_\mu|^2$$

They have dipole moments directed one along the line joining the atoms 0 and 3, and one perpendicular to this line. (Cf. Figure 2.12c.)

The transition densities between Φ_0 and the combinations $(\Phi_{-1\to 2} - \Phi_{1\to -2})/(i\sqrt{2})$ and $-(\Phi_{-1\to 2} + \Phi_{1\to -2})/\sqrt{2}$ are similarly seen to be

$$\sqrt{2}(\phi_{-1}^*\phi_2 - \phi_1^*\phi_{-2})/(i\sqrt{2}) = (1/3)\sum_\mu \sin(\pi\mu)|\chi_\mu|^2$$

and

$$-\sqrt{2}(\phi_{-1}^*\phi_2 + \phi_1^*\phi_{-2})/\sqrt{2} = -(1/3)\sum_\mu \cos(\pi\mu)|\chi_\mu|^2$$

neither of which has a dipole moment.

The subscript k on a perimeter orbital ϕ_k defined in Equation (2.2) can be viewed as a quantum number related to the z component of orbital angular momentum associated with an electron in the orbital ϕ_k. The selection rule for one-electron transitions between perimeter orbitals is similar to that familiar from atoms: the quantum number k is allowed to increase or decrease by 1. Thus promotions from Φ_0 to $\Phi_{N\to N+1}$ and to $\Phi_{-N\to -N-1}$ are allowed, and the transition from the ground state to the B state is intense. In contrast, promotions from Φ_0 to $\Phi_{N\to -N-1}$ and $\Phi_{-N\to N+1}$ change the quantum number k by $\pm(2N+1)$, and therefore are forbidden in the unperturbed perimeter. Indeed, in the parent aromatic annulenes the transitions to the L states are weak, and derive their intensity from vibronic perturbations of the perimeter.

2.2.3 The Generalization of the Perimeter Model for Systems with $4N+2$ π Electrons

The perimeter model can be extended quite easily to apply to all systems that may be derived from a regular $4N+2$ π-electron perimeter by introducing structural perturbations. Examples of such perturbations are the replacement of a carbon by a heteroatom, substitution, cross-linking or bridging, and similar structural changes (Michl, 1978).

In most cases, the structural perturbation that produces the actual molecule of interest from the pure perimeter will lower the symmetry and remove the pairwise degeneracy of the perimeter MOs.

If the perturbation can be written as a sum of one-electron perturbations,

$$\hat{H}' = \sum_i \hat{h}'(i) \tag{2.6}$$

the CI matrix Equation (2.5) for the four excited configurations Φ_1 to Φ_4 will change into

$$\begin{pmatrix} E(\mathrm{B}) & 0 & a^* & b^* \\ 0 & E(\mathrm{B}) & b & a \\ a & b^* & E(\mathrm{L}) & Q \\ b & a^* & Q & E(\Phi_4) \end{pmatrix} \tag{2.7}$$

with interaction matrix elements

$$a = \langle\Phi_3|\hat{H}'|\Phi_1\rangle = -\langle\phi_N|\hat{h}'|\phi_{-N}\rangle \tag{2.8}$$

$$b = \langle\Phi_4|\hat{H}'|\Phi_1\rangle = \langle\phi_{-N-1}|\hat{h}'|\phi_{N+1}\rangle$$

When the symmetry becomes sufficiently low due to such a perturbation, all four states can mix and produce two low-energy excited states $^1\mathrm{L}_1$ and $^1\mathrm{L}_2$ and two high-energy states $^1\mathrm{B}_1$ and $^1\mathrm{B}_2$. Crucial for the extent of mixing are the interaction matrix elements a and b defined in Equation (2.8). These complex quantities may be written in the form

$$a = |a|e^{i\alpha} \tag{2.9}$$

and

$$b = |b|e^{i\beta}$$

It follows that after applying the perturbation the energy splitting of the HOMO pair and the LUMO pair will be given to first order by

$$\Delta\mathrm{HOMO} = 2|a|$$

and (2.10)

$$\Delta\mathrm{LUMO} = 2|b|$$

The phases α and β determine the orientation of the nodes of the perturbed MOs with respect to the molecular framework. Only the relative phase $\varphi = \alpha - \beta$ is of interest; the absolute phase depends on the numbering of the perimeter atoms with respect to the perturbation.

If at least one plane of symmetry perpendicular to the molecular plane is present, the possibilities for the location of nodes are limited, and φ can only have the values 0 and π. In this case it is customary to retain the notation ^1L_b, ^1L_a, ^1B_b, and ^1B_a, introduced for the unperturbed perimeter for the four singlet states of uncharged perimeters, or for the transitions from the ground state into these states. Depending on the nature of the perturbation either the ^1L_b state or the ^1L_a state can be lower in energy. The former occurs in naphthalene, the latter in anthracene.

Example 2.5:
Starting from the perimeter of [10]annulene the introduction of a cross-link between centers $\varrho = 0$ and $\sigma = 5$ yields naphthalene (**14**). From Equation (2.8)

$$\begin{aligned} a &= -\langle\phi_N|\hat{h}'|\phi_{-N}\rangle \\ &= -(1/n)\sum_{\varrho,\sigma} \mathrm{e}^{-2\pi i N(\varrho+\sigma)/n}\langle\varrho|\hat{h}'|\sigma\rangle \\ &= -(1/10)\mathrm{e}^{-2\pi i \times 2(0+5)/10}\beta_{05} \\ &= -(1/10)\beta_{05}\mathrm{e}^{-2\pi i} = -(1/10)\beta_{05} \end{aligned}$$

and

$$\begin{aligned} b = \langle\phi_{-N-1}|\hat{h}'|\phi_{N+1}\rangle &= (1/10)\mathrm{e}^{2\pi i \times 3(0+5)/10}\beta_{05} \\ &= (1/10)\beta_{05}\mathrm{e}^{i\pi} \end{aligned}$$

Since $\beta_{05} < 0$ and $\mathrm{e}^{i\pi} = -1 = -\mathrm{e}^{i0}$, comparison with Equation (2.9) shows that $|a| = |b| = |\beta_{05}|/10$ and $\alpha = \beta = 0$; that is, the relative phase angle is $\varphi = \alpha - \beta = 0$.

A cross-link between centers $\varrho = 2$ and $\sigma = 8$ produces azulene (**15**). In this case one has

$$\begin{aligned} a &= -(1/10)\mathrm{e}^{-2\pi i \times 2(2+8)/10}\beta_{28} = -\beta_{28}/10 \\ \text{and } b &= (1/10)\mathrm{e}^{2\pi i \times 3(2+8)/10}\beta_{28} = -\beta_{28}/10 \cdot (-1) \end{aligned}$$

and from Equation (2.9) with $-1 = \mathrm{e}^{-i\pi}$ one has $|a| = |b| = |\beta_{28}|/10$ as well as $\alpha = 0$ and $\beta = -\pi$, and the relative phase angle becomes $\varphi = \alpha - \beta = \pi$.

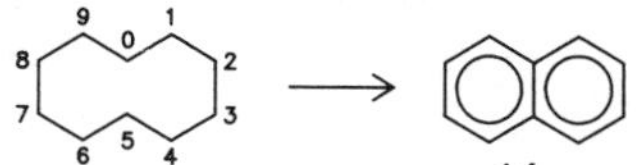

14 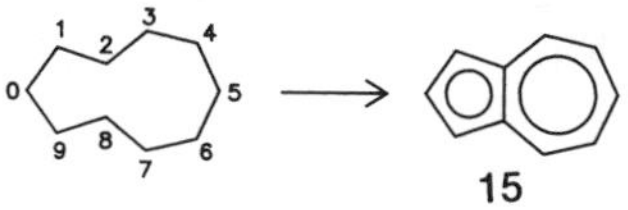 **15**

From the form of the CI matrix [Equation (2.7)] a number of general results may be derived. For this purpose it is convenient to transform the basis set of conformations $\Phi_1, \ldots, \Phi_4$ into a real one. This is accomplished by defining

$$X = (e^{-i\gamma}\Phi_1 + e^{i\gamma}\Phi_2)/\sqrt{2} \tag{2.11}$$

$$Y = (e^{-i\gamma}\Phi_1 - e^{i\gamma}\Phi_2)/(i\sqrt{2})$$

$$U = (\Phi_3 + \Phi_4)/\sqrt{2}$$

$$V = (\Phi_3 - \Phi_4)/(i\sqrt{2})$$

where $\gamma = (\alpha + \beta)/2$.* In the real basis the CI matrix can be written as follows:

$$\begin{array}{cccc} X & Y & U & V \end{array}$$

$$\begin{pmatrix} E(B) & 0 & (|a| + |b|)\cos(\varphi/2) & -(|a| + |b|)\sin(\varphi/2) \\ 0 & E(B) & (|a| - |b|)\sin(\varphi/2) & (|a| - |b|)\cos(\varphi/2) \\ (|a| + |b|)\cos(\varphi/2) & (|a| - |b|)\sin(\varphi/2) & E(L) + Q & 0 \\ -(|a| + |b|)\sin(\varphi/2) & (|a| - |b|)\cos(\varphi/2) & 0 & E(L) - Q \end{pmatrix} \tag{2.12}$$

From this matrix representation it is easy to see that for ΔHOMO = ΔLUMO Y does not mix with any of the other states, because in this case $(|a| - |b|) = 0$, and that the interaction between X and V vanishes for $\varphi = 0$ and that between X and U for $\varphi = \pi$.

In Example 2.6 it will be shown that similar arguments apply to transition dipole moments. Thus, it follows that the ^1L_b and ^1B_b transitions are both polarized along a line that bisects two opposite bonds, whereas if for symmetry reasons $\varphi = 0$ or π (cf. Figure 2.12) the ^1L_a and ^1B_a transitions are polarized along a line that passes through two opposite atoms and is perpendicular to the former. If ΔHOMO = ΔLUMO and $\varphi = 0$, ^1L_b has zero intensity, and if $\varphi = \pi$, ^1L_a has zero intensity. Even if there is no symmetry, as long as ΔHOMO = ΔLUMO and $\varphi = 0$ or π, the ^1L_b transition is polarized parallel to the ^1B_b transition and perpendicular to the ^1L_a and ^1B_a transitions and either ^1L_b ($\varphi = 0$) or ^1L_a ($\varphi = \pi$) has zero intensity. For other values of the relative phase angle, the polarization directions of the four transitions $^1L_1$, $^1L_2$, $^1B_1$, and $^1B_2$ need no longer be parallel or at right angles; indeed, when ΔHOMO = ΔLUMO, $^1L_1$ and $^1L_2$ actually may be polarized parallel to each other, for example, in 2-substituted naphthalenes (Friedrich et al., 1974).

* For $\gamma = 0$ the definition of X and Y is identical with that given by Moffitt (1954a), that is, $X = (\Phi_1 + \Phi_2)/\sqrt{2}$ and $Y = (\Phi_1 - \Phi_2)/(i\sqrt{2})$; however, $X = -(\Phi_1 - \Phi_2)/(i\sqrt{2})$ and $Y = -(\Phi_1 + \Phi_2)/\sqrt{2}$ for $\gamma = \pi/2$.

Example 2.6:
For evaluating transition dipole moments the position vector $\boldsymbol{R}_\mu$ of perimeter atom μ, that is, of vertex μ of a regular polygon, may be written in the form

$$\boldsymbol{R}_\mu = d_0/[4 \sin(\pi/n)] \{\boldsymbol{e}_1(e^{2\pi i\mu/n} + e^{-2\pi i\mu/n}) - i\boldsymbol{e}_2(e^{2\pi i\mu/n} - e^{-2\pi i\mu/n})\}$$

where $\boldsymbol{e}_1$ and $\boldsymbol{e}_2$ are orthogonal unit vectors in the molecular plane, d_0 is the bond distance between adjacent perimeter atoms, and use has been made of the Euler formulae $\cos z = (e^{iz} + e^{-iz})/2$ and $\sin z = (e^{iz} - e^{-iz})/(2i)$. This yields

$$<\phi_k|\hat{\boldsymbol{m}}|\phi_{k\pm1}> = m(\boldsymbol{e}_1 \pm i\boldsymbol{e}_2)/\sqrt{2}$$

$$<\phi_k|\hat{\boldsymbol{m}}|\phi_{k\pm l}> = 0 \text{ for } l > 1$$

$$<\phi_k|\hat{\boldsymbol{m}}|\phi_{\pm k}> = 0$$

where, neglecting overlap, m is given by

$$m = -|e|d_0/[2\sqrt{2}\sin(\pi/n)]$$

Thus, in view of Section 1.3.1, we obtain the transition moment matrix elements for the unperturbed [n]annulene,

$$<\Phi_0|\hat{\boldsymbol{M}}|\Phi_1> = \sqrt{2}<\phi_N|\hat{\boldsymbol{m}}|\phi_{N+1}> = m(\boldsymbol{e}_1 + i\boldsymbol{e}_2)$$

$$<\Phi_0|\hat{\boldsymbol{M}}|\Phi_2> = \sqrt{2}<\phi_{-N}|\hat{\boldsymbol{m}}|\phi_{-N-1}> = m(\boldsymbol{e}_1 - i\boldsymbol{e}_2)$$

whereas

$$<\Phi_0|\hat{\boldsymbol{M}}|\Phi_3> = <\Phi_0|\hat{\boldsymbol{M}}|\Phi_4> = 0$$

Thus, in agreement with the arguments in Section 2.2.2, nonzero transition moments are obtained only for transitions in which the angular momentum quantum number changes by ±1, that is, where the sense of electron circulation is preserved. Finally, the application of Equation (2.11) gives the transition moments between the ground configuration and the real excited perimeter configurations

$$<\Phi_0|\hat{\boldsymbol{M}}|X> = \sqrt{2}m(\boldsymbol{e}_1 \cos\gamma + \boldsymbol{e}_2 \sin\gamma)$$

$$<\Phi_0|\hat{\boldsymbol{M}}|Y> = \sqrt{2}m(-\boldsymbol{e}_1 \sin\gamma + \boldsymbol{e}_2 \sin\gamma)$$

$$<\Phi_0|\hat{\boldsymbol{M}}|U> = <\Phi_0|\hat{\boldsymbol{M}}|V> = 0$$

The intensities and polarization directions of the four transitions $^1L_1$, $^1L_2$, $^1B_1$, and $^1B_2$ can then be estimated by introducing configuration mixing as given by Equation (2.12). Thus it is easy to see that for $|a| - |b| = 0$ the L states can acquire their intensity only by mixing with the X state and hence have to be polarized parallel to each other.

2.2.4 Systems with Charged Perimeters

The spectra of $(4N+2)$-electron charged perimeter π systems are appreciably different from uncharged perimeter π systems, irrespective of whether the perimeter is even or odd—that is, of whether or not the ion is derived from an alternant or a nonalternant hydrocarbon. This is illustrated in Figure 2.13, which compares the absorption spectra of the tropylium ion and the cyclooctatetraene dianion with that of benzene.

At the HMO level all three systems are characterized by a fourfold degenerate transition between the degenerate HOMO and the degenerate LUMO. Of the configurations composed of the perimeter MOs according to Equation (2.4), Φ_1 and Φ_2 as well as Φ_3 and Φ_4 are pairwise degenerate. This corresponds to a splitting into the $^1B_{a,b}$ states on the one hand and the $^1L_{a,b}$ states on the other hand. In a charged perimeter with $n \neq 4N+2$, these configurations do not interact, as has been discussed already in Section

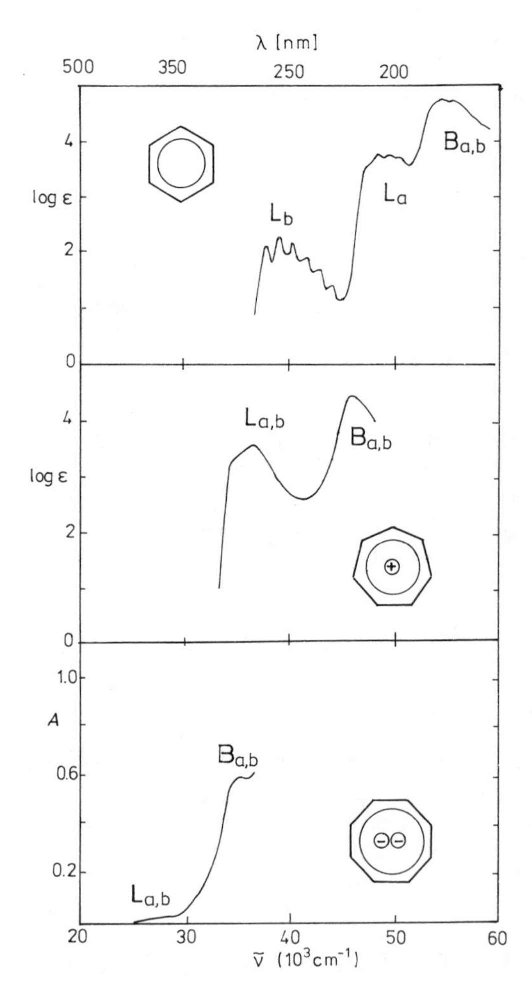

Figure 2.13. Comparison of the absorption spectra of benzene, the tropylium ion, and the cyclooctatetraene dianion (adapted from Heilbronner, 1966 and Dvorak and Michl, 1976).

2.2.2, so only two transitions are to be expected. The difference between the benzene spectrum and the spectra of the ions is therefore due to different symmetry of the excited-state configurations. This was first pointed out by Heilbronner (1966) in the case of the tropylium ion. Symmetry also affects the spectra of derivatives of charged perimeters with $4N+2$ electrons. (Cf. Section 2.2.5.)

Similar arguments apply to perimeters with $4N$ π electrons, but the situation is more complicated and a different classification of excited states applies. (Cf. Section 2.2.7.)

2.2.5 Applications of the PMO Method within the Extended Perimeter Model

The following hierarchy of perturbations has been found useful in discussing the spectra of aromatic molecules derived from a $(4N+2)$-electron perimeter through the introduction of a series of perturbations.

$(4N+2)$-Electron $[n]$Annulene
(1) ↓ Bridging
Hydrocarbon
(2) ↓ Cross-linking
Hydrocarbon
(3) ↓ 2-Electron or 0-Electron Heteroatom Replacement
Heterocycle
(4) ↓ 1-Electron Heteroatom Replacement
Heterocycle
(5) ↓ Substitution
Target Molecule

Bridging is defined as the insertion of additional π centers within the perimeter. Examples are the production of acenaphthylene (**16**) from the [11]annulenium cation and a $C^{\ominus}$ fragment or the production of pyrene (**17**) from [14]annulene and the C=C ethylene unit:

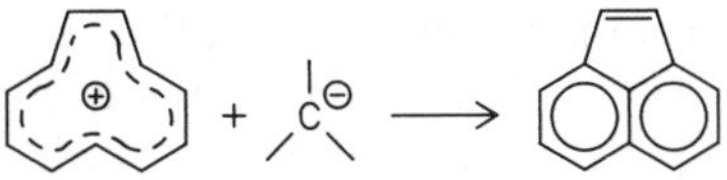

16

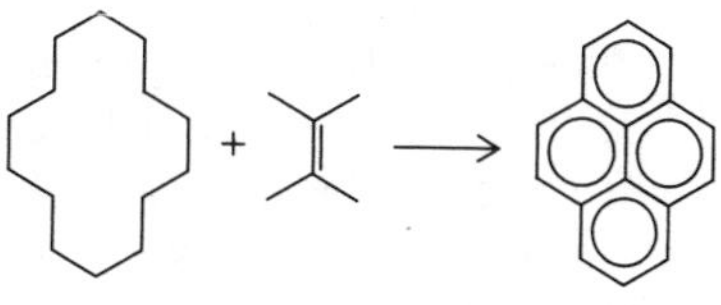

17

Bridging units tend to introduce new orbitals between the HOMOs and the LUMOs of the perimeter. This is especially true in cases such as in **16** where the bridge contains an odd number of π centers and possesses nonbonding orbitals. Excitations to or from these additional orbitals cannot of course be classified in terms of the unperturbed perimeter transitions.

Cross-linking is defined as the introduction of π bonds between nonadjacent centers in the perimeter, producing, for example, naphthalene (**14**) and azulene (**15**) from [10]annulene; 2-electron and 0-electron heteroatom replacement refers to the replacement of $C^{\ominus}$ by —O—, —NR—, —S—, or a similar group with two π electrons, and the replacement of $C^{\oplus}$ by —BH— or a similar heteroatom with no π electron in its p_z AO, respectively, as illustrated in **18** and **19**.

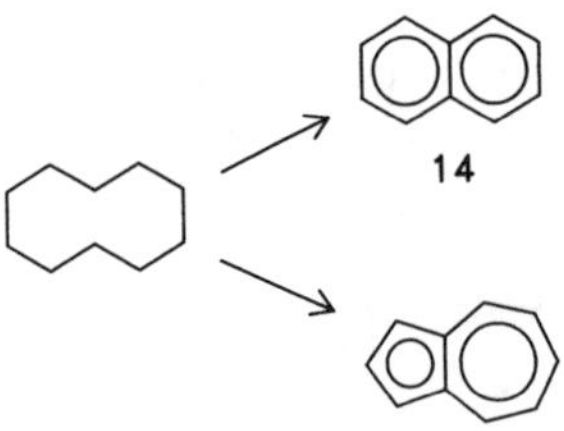

14

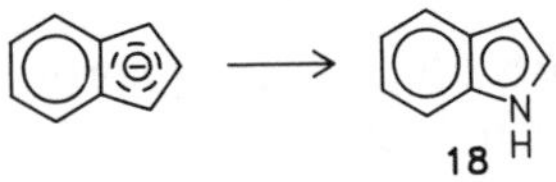

18

15

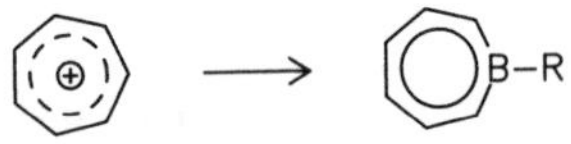

19

Examples of a 1-electron heteroatom replacement are the replacements of =CH— by $=O^{\oplus}$—, $=S^{\oplus}$—, $=NH^{\oplus}$—, =N— or a similar group with one π electron in its p_z AO. (Cf. **20** and **21**.)

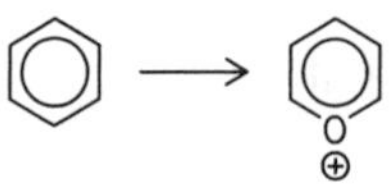

20

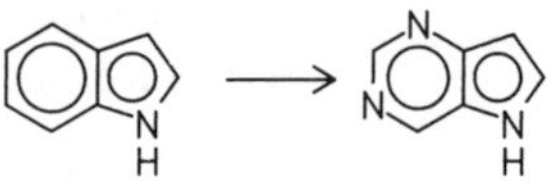

21

Substitution finally refers to the replacement of an H atom by an element from the second or lower row of the periodic table. (Cf. **22** and **23**.) Since all such atoms possess an AO of p_z symmetry, substitution causes an increase in size of the π system, either by conjugation or by hyperconjugation. For some substituents this π effect may be negligible. (Cf. Section 2.4.1.)

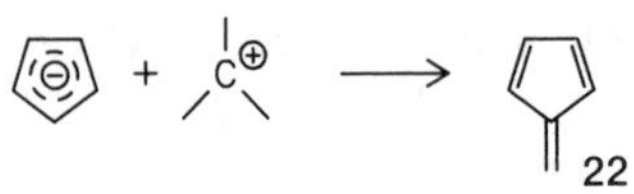

22

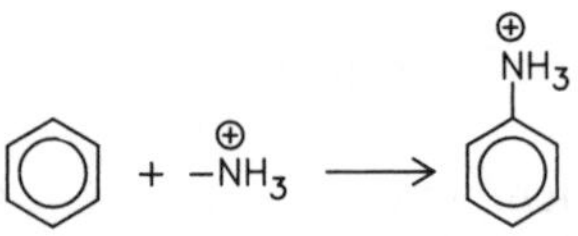

23

All these structural changes may be discussed by applying the concepts of PMO theory (Dewar and Dougherty, 1975), that is, by means of perturbation treatments based on the HMO approximation. From first-order perturbation theory it is seen that the introduction of a resonance integral $\beta_{\varrho\sigma}$ between the perimeter atoms ϱ and σ (cross-linking) or varying the Coulomb integral $\delta\alpha_\varrho$ of the atom ϱ (heteroatom replacement) modifies the energy ε_i of the i-th perimeter orbital by

$$\delta\varepsilon_i = 2\sum_{\varrho-\sigma} c_{\varrho i}c_{\sigma i}\beta_{\varrho\sigma} \tag{2.13}$$

$$\delta\varepsilon_i = \sum_{\varrho} c_{\varrho i}^2\delta\alpha_\varrho$$

The change $\delta\varepsilon_i$ in the orbital energy is proportional to the product $c_{\varrho i}c_{\sigma i}$ or to the square $c_{\varrho i}^2$ of the LCAO coefficients, respectively. The effect of introducing a bond between different fragments R and S (bridging or substitution) on the energy ε_i of the i-th perimeter orbital is approximated by the second-order expression

$$\delta\varepsilon_i = \sum_k \left[\sum_{\varrho-\sigma} \frac{c_{\varrho i}^R c_{\sigma k}^S \beta_{\varrho\sigma}}{(\varepsilon_i^R - \varepsilon_k^S)}\right] \tag{2.14}$$

as long as the orbitals ϕ_i^R and ϕ_k^S are not degenerate.

For applications of the PMO method it is convenient to use real combinations $(1/2)\sqrt{2}(\phi_k + \phi_{-k})$ and $(i/2)\sqrt{2}(\phi_k - \phi_{-k})$ instead of the complex perimeter MOs given in Equation (2.2). For $(4N+2)$-electron perimeters the real MOs of the degenerate pair of HOMOs are

$$\phi_s = \sqrt{2/n}\sum_\mu \left(\cos\frac{2\pi N\mu}{n}\right)\chi_\mu \tag{2.15}$$

$$\phi_a = \sqrt{2/n}\sum_\mu \left(\sin\frac{2\pi N\mu}{n}\right)\chi_\mu$$

where the sum over μ runs from 0 to $n-1$; that is, the perimeter atoms μ and the AOs χ_μ are numbered $0, 1, \ldots, n-1$. The MO ϕ_a possesses a nodal plane perpendicular to the molecular plane and passing through the center $\mu = 0$; it is antisymmetric with respect to this nodal plane. The MO ϕ_s is orthogonal to ϕ_a and is symmetric with respect to the plane of symmetry. The corresponding MOs of the degenerate LUMO pair are*

$$\phi_{s'} = \sqrt{2/n}\sum_\mu \left(\cos\frac{2\pi(N+1)\mu}{n}\right)\chi_\mu \tag{2.16}$$

$$\phi_{a'} = \sqrt{2/n}\sum_\mu \left(\sin\frac{2\pi(N+1)\mu}{n}\right)\chi_\mu$$

---

* In order to be consistent with the nomenclature introduced in Section 1.2.4 for the MOs of alternant hydrocarbons and in order to avoid confusion with the angular momentum quantum numbers of the perimeter MOs ϕ_k and ϕ_{-k}, the original notation ϕ_{-s} and ϕ_{-a} for the LUMOs (Michl, 1978, 1984) has been changed.

Example 2.7:
The location of the nodal planes and hence the signs of the LCAO coefficients of real perimeter MOs obtained from the complex MOs ϕ_k and ϕ_{-k} may be determined quite easily by the following procedure. A polygon with $2k$ vertices is incribed into the perimeter in such a way that it sits on a vertex in the case of the MOs ϕ_a and $\phi_{a'}$ and that it lies on its side in the case of the MOs ϕ_s and $\phi_{s'}$. Lines joining opposite vertices of the polygon identify the location of nodal planes and indicate the regions of the perimeter where the LCAO coefficients have the same sign and the locations where sign changes occur. The absolute magnitude of the coefficient $c_{\mu i}$ is derived by observing how close the vertex μ that represents the atom in question lies to the nearest nodal plane of the MO in question. The smaller the distance, the smaller the coefficient. This is illustrated in Figure 2.14, where the frontier orbitals of benzene and of the [8]annulene dianion are shown together with the HMO coefficients. As usual, the diameter of the circles is proportional to the magnitude of the coefficients and different colors indicate different signs.

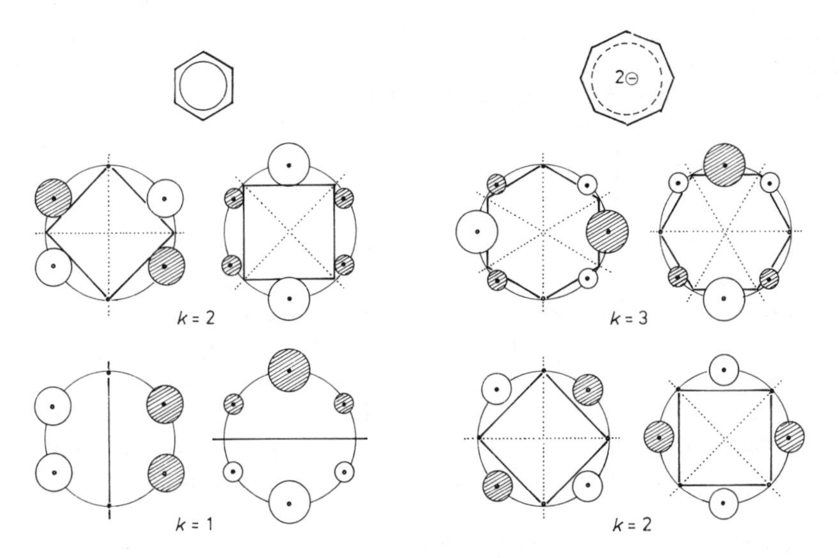

Figure 2.14. Frontier orbitals of benzene and [8]annulene dianion. The nodal planes (···) are obtained from the polygons with $2k$ vertices inscribed into the perimeter, and the magnitude of the LCAO coefficients (indicated by the size of the circles) may be estimated from the location of the nodal planes.

Uncharged perimeters with $4N+2$ π electrons are alternant hydrocarbons and the pairing theorem (cf. Section 1.2.4) applies; ϕ_s and $\phi_{s'}$ are paired, as are ϕ_a and $\phi_{a'}$: their coefficients are equal on all odd atoms μ, $c_{\mu s'} = c_{\mu s}$, and $c_{\mu a'} = c_{\mu a}$, respectively, and they are equal in magnitude but opposite in sign on all even atoms μ, $c_{\mu s'} = -c_{\mu s}$, and $c_{\mu a'} = -c_{\mu a}$. Due to this relation, those perturbations that are proportional to the product $c_{\varrho i}c_{\sigma i}$ of LCAO coefficients according to Equation (2.13) can be divided in two classes: *even perturbations,* for which the sum $\varrho + \sigma$ is even (this includes the case $\varrho = \sigma$), and *odd perturbations,* for which $\varrho + \sigma$ is odd (Moffitt, 1954a).

The following regularities are useful in applying PMO arguments:

When a perturbation is even, as for heteroatom replacements or cross-links that introduce odd-membered rings, $c_{\varrho s}c_{\sigma s} = c_{\varrho s'}c_{\sigma s'}$ and $c_{\varrho a}c_{\sigma a} = c_{\varrho a'}c_{\sigma a'}$. Hence, ϕ_s and $\phi_{s'}$ are either both stabilized or both destabilized by the same amount, and the same is true for ϕ_a and $\phi_{a'}$. It follows that for an even perturbation ΔHOMO = ΔLUMO to the first order and that the orbital order must be either ϕ_s, ϕ_a, $\phi_{s'}$, $\phi_{a'}$ or ϕ_a, ϕ_s, $\phi_{a'}$, $\phi_{s'}$. The relative phase angle $\varphi = \alpha - \beta$ has the value $\varphi = \pi$. (Cf. Example 2.5.)

However, when the perturbation is odd, as for cross-links that introduce even-membered rings, then $c_{\varrho s}c_{\sigma s} = -c_{\varrho s'}c_{\sigma s'}$ and $c_{\varrho a}c_{\sigma a} = -c_{\varrho a'}c_{\sigma a'}$. Thus, if ϕ_s is stabilized, $\phi_{s'}$ is destabilized by the same amount, and if ϕ_s is destabilized, $\phi_{s'}$ is stabilized by the same amount. A similar relation holds between ϕ_a and $\phi_{a'}$. It follows that for an odd perturbation ΔHOMO = ΔLUMO, too. This is fulfilled exactly within the HMO and PPP models since an odd perturbation does not remove the property of alternancy. The orbital order is either ϕ_s, ϕ_a, $\phi_{a'}$, $\phi_{s'}$ or ϕ_a, ϕ_s, $\phi_{s'}$, $\phi_{a'}$, and the relative phase angle can only have the value $\varphi = 0$.

An example is shown in Figure 2.15 where an even perturbation (β_{28}) produces the orbitals of azulene and an odd perturbation (β_{05}) those of naphthalene from the perimeter orbitals of cyclodecapentaene. (Cf. Example 2.5.) The energies of ϕ_a and $\phi_{a'}$ are not affected to the first order by the odd perturbation producing naphthalene, whereas ϕ_s is stabilized and $\phi_{s'}$ destabilized. Therefore, ϕ_a becomes the HOMO and $\phi_{a'}$ the LUMO, and the HOMO-LUMO splitting $\Delta E_{\text{HOMO-LUMO}}$ is the same as for cyclodecapentaene. The HOMO→LUMO transition is referred to as ^1L_a according to Platt. In azulene, on the other hand, ϕ_a is destabilized and becomes the HOMO, whereas $\phi_{s'}$ is stabilized and becomes the LUMO. The HOMO-LUMO splitting is markedly smaller than for cyclodecapentaene, and the HOMO→LUMO transition is of the ^1L_b type.

For a mixed even–odd perturbation one has in general ΔHOMO $\neq$ ΔLUMO, with either ΔHOMO > ΔLUMO or ΔHOMO < ΔLUMO. The effects of π substituents and charged bridges with an odd number of centers (e.g., $C^{\ominus}$ or $C^{\oplus}$) can be viewed as a superposition of an even and an odd perturbation, so ΔHOMO $\neq$ ΔLUMO in these cases.

The fact that due to symmetry there is no mixing between the L states in charged perimeters with $4N+2$ electrons also has consequences for systems derived from these charged perimeters. Even after introducing the perturbation the polarization directions of the $^1L_1$ and $^1L_2$ transitions as well as those of the $^1B_1$ and $^1B_2$ transitions are mutually perpendicular, and the splitting of the ^1L bands is only small. Important examples are indole derivatives

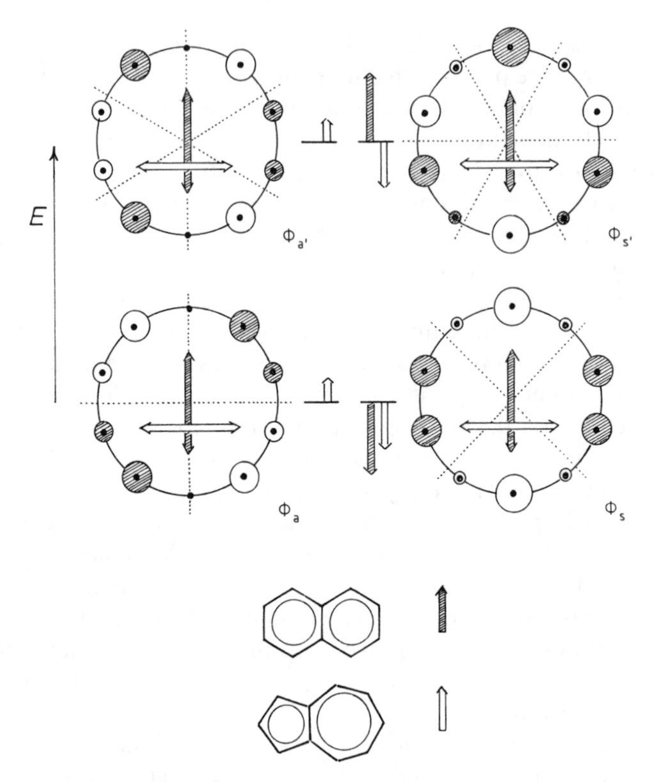

Figure 2.15. The real form of the four frontier MOs of [10]annulene. The double-headed arrows indicate the cross-links that produce naphthalene (black arrows) and azulene (white arrows), respectively. First-order shifts are shown as vertical arrows on the orbital energy levels in the center (by permission from Michl, 1984).

(cf. **18**), which can be produced formally from the [9]annulene anion by one cross-link and one 2-electron heteroatom replacement.

2.2.6 Polyacenes

The polyacenes form an important class of aromatic compounds, which can be derived from uncharged perimeters with $4N+2$ π electrons by introducing cross-links that produce only linearly annelated six-membered rings. Thus, the polyacenes are obtained by odd perturbations.

The interactions between the different configurations of Equation (2.11) vary depending on the nature of the perturbation. The CI problem is simplified considerably if some of the interactions of the CI matrix Equation (2.12) vanish. If the perturbation is odd, the relative phase angle φ vanishes, and only X and U can mix, whereas $Y = (\Phi_1 - \Phi_2)/(i\sqrt{2})$ and $V = (\Phi_3 - \Phi_4)/(i\sqrt{2})$ already correspond to the ^1B_b and ^1L_b states. If the perturbation is even, φ equals π, so X and V can mix, whereas $Y = -(\Phi_1 + \Phi_2)/\sqrt{2}$ and $U = (\Phi_3 + \Phi_4)/\sqrt{2}$ describe the ^1B_a and ^1L_a states (Figure 2.16). In Figure 2.17 the results for the homologous series from benzene to pentacene are presented. The excitation energies shown in Figure 2.17a are calculated without CI. If the interaction is taken into account, the splitting of the states described by X and U increases with increasing size of the conjugated system. As a consequence the ^1L_a state (U) of anthracene and the higher acenes is lower in energy than the ^1L_b state (V), in agreement with experimental observation (Figure 2.17b).

The same considerations may also be applied to the isomers of the polyacenes. If only isomers that are built up from fused benzene rings are considered, all cross-links are odd perturbations. The perturbation affects the energies of all configurations in the same way, and since according to the arguments given above V does not mix with any of the other configurations, the energy of the ^1L_b state is expected, to a first approximation, to be the same for all isomers. But the extent to which U is stabilized through interaction with X depends on the values of the LCAO coefficients at those positions where cross-linking takes place, and is therefore different for different isomers. The critical matrix element $H_{13} = (|a| + |b|)\cos(\varphi/2)$ can be evaluated easily as shown in Example 2.5. Using the numbering scheme in formulae **24** and **25** it is seen that H_{13} is smaller for phenanthrene than

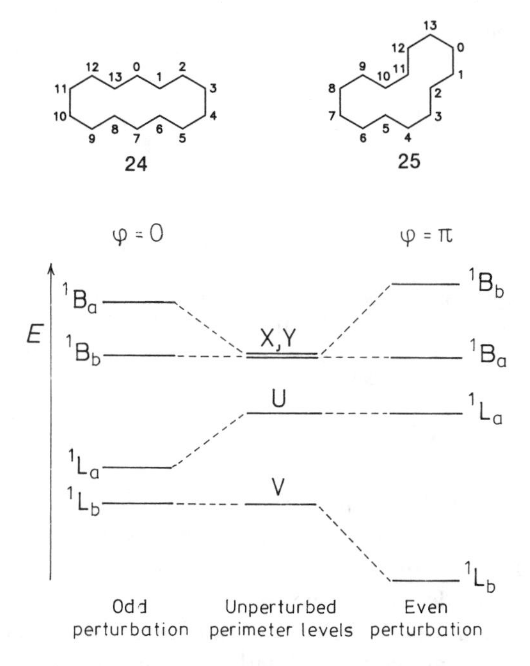

Figure 2.16. The behavior of the lower excited states of cyclic polyenes under odd perturbations (on the left) and under even perturbations (on the right) (adapted from Moffitt, 1954a).

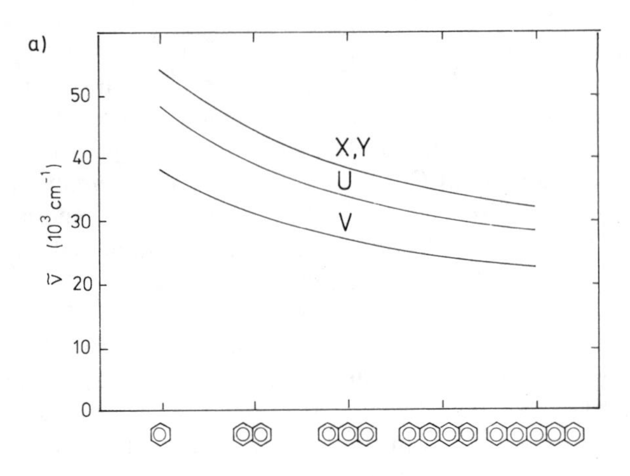

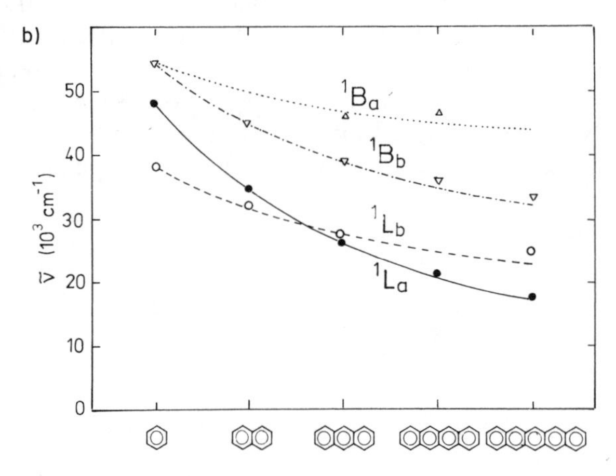

Figure 2.17. Energies of excited states of polyacenes as predicted by the perimeter model; a) without CI, and b) including first-order CI, which results in a splitting of the states described by U and X, together with experimental data (○, ●, ▽, and △) (by permission from Moffitt, 1954a).

for anthracene by a factor $[e^{6\pi i/14} + e^{-6\pi i/14}] - [e^{8\pi i/14} + e^{-8\pi i/14}] = 2\cos(3\pi/7)$. This is in good agreement with the experimental observation that the excitation energy of the ^1L_a state (U) is 3.2 eV for anthracene but 4.0 eV for phenanthrene, while the energies of the next two singlet states (V and Y) are almost the same for both molecules. In Figure 2.18 similar results for the isomers of pentacene are compared with experimental data.

The location of the absorption bands as well as their intensities can be discussed on the basis of the perimeter model and perturbation theory. In Example 2.6 it was shown that the transition dipole moments of the ^1B_a and

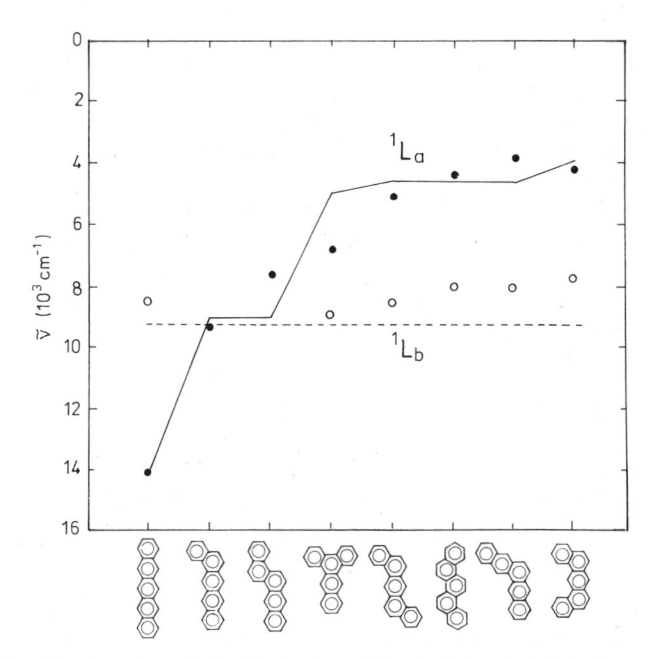

Figure 2.18. The location of the ^1L_a (—) and ^1L_b (---) bands of the isomers of pentacene relative to the ^1B_b band, and comparison with experimental data (● and ○, respectively) (by permission from Moffitt, 1954a).

^1B_b bands of a regular perimeter are of equal magnitude and mutually perpendicular whereas those of the ^1L_a and ^1L_b bands vanish. If the perimeter is distorted so as to have the same shape as a particular cata-condensed hydrocarbon, it is found that the transition moment $\boldsymbol{M}^0_{0Y}$ between the ground configuration and the configuration Y, which cuts across the centers of opposite bonds, increases almost linearly on increasing the number of fused rings. The transition moment $\boldsymbol{M}^0_{0X}$ is perpendicular to $\boldsymbol{M}^0_{0Y}$; its magnitude for benzene is $m = -|e|l_0$ and for a very long polyacene, it approaches $(3/\pi)m$ monotonically. $\boldsymbol{M}^0_{0U}$ is again bounded, remaining small and never exceeding $m/2$ (Moffitt, 1954a). Taking into account the cross-links as odd perturbations leaves the configurations Y and V unchanged. In this approximation, $\boldsymbol{M}^0_{0Y}$ and $\boldsymbol{M}^0_{0V}$ therefore are the transition dipole moments of the intense ^1B_b and the weak ^1L_b bands, respectively. As $\boldsymbol{M}^0_{0U}$ and $\boldsymbol{M}^0_{0X}$ are bounded, the transition moment of the ^1L_a band calculated after configuration mixing cannot exceed $(1/2 + 3/\pi)m/\sqrt{2}$ (Moffitt, 1954a). The intensity of this band in the benzene spectrum, however, is markedly affected through vibronic coupling. (See Example 2.8.)

Example 2.8:
From Figure 2.16 it is seen that the ^1L_a ($^1B_{1u}$) state of benzene can interact with the degenerate ^1B ($^1E_{1u}$) state only via an odd perturbation, and the ^1L_b ($^1B_{2u}$)

state only via an even perturbation. A variation of $\beta_{\varrho\sigma}$ for the bonds between neighboring atoms ϱ and σ is an odd perturbation. The CC bond-length changes associated with the appropriate stretching vibrations of the benzene ring can be described by perturbations of this kind. An even perturbation, on the other hand, would require nonzero $\beta_{\varrho\sigma}$ between nonneighboring atoms ϱ and σ and could be induced only to a small extent by bending vibrations. This implies that vibronic coupling is important for the mixing of the $^1E_{1u}$ with the $^1B_{1u}$ rather than the $^1B_{2u}$ state. This explains why the $^1B_{1u}$ band borrows much more intensity from the $^1E_{1u}$ band than the $^1B_{2u}$ band does, more so than would be expected from the closer proximity of the $^1B_{1u}$ state to the $^1E_{1u}$ state according to second-order perturbation theory, which suggests that the interaction is proportional to the inverse of the energy difference between the two states. (Cf. Moffitt, 1954a.)

2.2.7 Systems with a 4*N* π-Electron Perimeter

If a perimeter has $4N$ rather than $4N+2$ electrons, the situation becomes more complicated. In this case the degenerate MOs ϕ_N and ϕ_{-N} are occupied with a total of only two electrons (SOs). Apart from the reference configuration Φ_0 shown in Figure 2.19 and the two degenerate configurations $\Phi_{N\rightarrow-N}$ and $\Phi_{-N\rightarrow N}$,* there are eight configurations that are singly excited with respect to Φ_0. Instead of a 1×1 and a 4×4 CI problem for the ground state and the four excited states of Equation (2.5), a 3×3 and an 8×8 CI problem are required for the description of the ground and excited states of a $4N$-electron system.

For uncharged perimeters, $\Phi_{N\rightarrow-N}$ and $\Phi_{-N\rightarrow N}$ interact via an electron repulsion term. (Cf. Section 2.2.2.) Within the ZDO approximation, $\Phi_+ = (\Phi_{N\rightarrow-N} + \Phi_{-N\rightarrow N})/\sqrt{2}$ and Φ_0 are degenerate, and $\Phi_- = (\Phi_{N\rightarrow-N} - \Phi_{-N\rightarrow N})/(i\sqrt{2})$ becomes the ground state. Thus, there are five doubly degenerate excited states above the ground state. For charged perimeters the interaction vanishes and the ground state is degenerate. A nondegenerate state and four pairs of doubly degenerate states follow in the order of increasing energy.

The spectroscopic properties that derive from these arguments for perfectly symmetrical antiaromatic systems are of limited practical interest since [4*N*]annulenes are subject to Jahn-Teller and pseudo–Jahn-Teller distortions and do not have C_n symmetry in reality. What makes these results useful is that they provide a starting point for the introduction of perturbations that convert the parent perimeters into π systems of real interest.

---

* Within the HMO model, all three configurations Φ_0, $\Phi_{N\rightarrow-N}$, and $\Phi_{-N\rightarrow N}$ are degenerate.

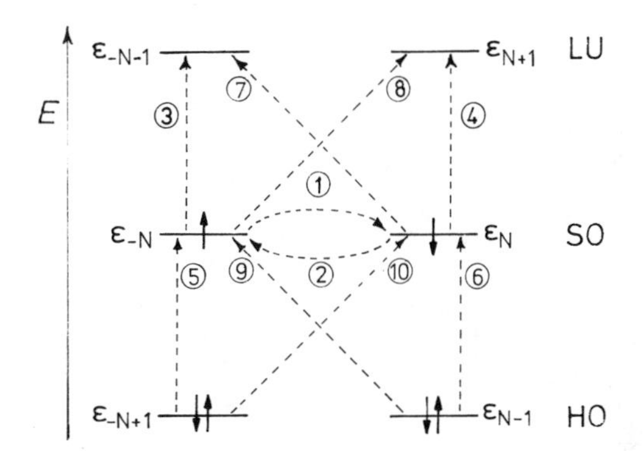

Figure 2.19. Orbital energy scheme of a 4N-electron perimeter. The reference configuration Φ_0 is shown together with the single excitations ①–⑩ between the levels of the frontier orbitals doubly occupied (HO), singly occupied (SO), and unoccupied (LU) in Φ_0.

If a transformation to real orbitals defined by

$$\phi_+^k = (e^{i\varphi/2}\phi_k + e^{-i\varphi/2}\phi_{-k})/\sqrt{2} \tag{2.17}$$

$$\phi_-^k = (e^{i\varphi/2}\phi_k - e^{-i\varphi/2}\phi_{-k})/(i\sqrt{2})$$

is applied to Φ_0, $\Phi_{N\to-N}$, and $\Phi_{-N\to N}$, the configurations $\Phi_1 = |\phi_-^N\bar{\phi}_-^N|$, $\Phi_2 = (|\phi_-^N\bar{\phi}_+^N| + |\phi_+^N\bar{\phi}_-^N|)/\sqrt{2}$, and $\Phi_3 = |\phi_+^N\bar{\phi}_+^N|$ are obtained, where all doubly occupied MOs have been ignored in order to simplify the notation. The interaction matrix for these is given by

$$\begin{array}{ccc} \Phi_1 & \Phi_2 & \Phi_3 \end{array}$$

$$\begin{pmatrix} \frac{1}{2}J[1 + \cos(2\sigma)] - 2|s| & -\frac{1}{2}J\sin(2\sigma) & \frac{1}{2}J[1 - \cos(2\sigma)] \\ -\frac{1}{2}J\sin(2\sigma) & -J\cos(2\sigma) & J\sin(2\sigma) \\ \frac{1}{2}J[1 - \cos(2\sigma)] & J\sin(2\sigma) & -\frac{1}{2}J[1 + \cos(2\sigma)] + 2|s| \end{pmatrix} \tag{2.18}$$

where J is a Coulomb integral, $|s| = |\langle\phi_N|\hat{h}'|\phi_{-N}\rangle|$ describes the interaction between the SOs, and σ is a phase factor. The magnitude of the interaction and hence the nature of the ground-state wave function is determined significantly by the kind of perturbation involved.

Since uncharged 4N-electron perimeters are alternant hydrocarbons, perturbations can be classified as even or odd in the same way as in Section 2.2.5. For even perturbations $\sigma = 0$ [or $2n(\pi/2)$] and the interaction matrix is already diagonal. Figure 2.20a shows the energies of the ground state G, and of the singly and doubly excited states S and D, relative to that of the triplet state as a function of the perturbation strength $|s|$. For odd perturbations $\sigma = \pi/2$ [or $(2n+1)(\pi/2)$] and Φ_1 and Φ_3 will mix. In Figure 2.20b it is

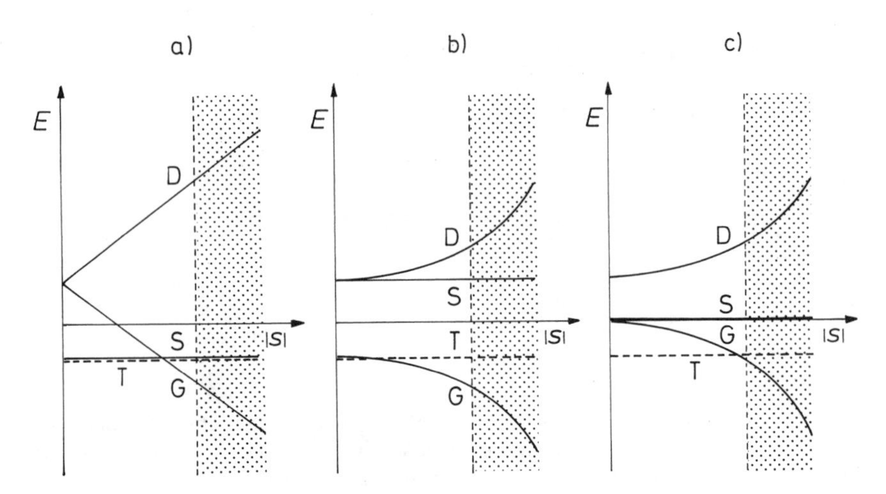

Figure 2.20. Energies of the biradicaloid singlet states G, S, and D relative to the triplet state T as a function of the perturbation $|s|$; a) of uncharged $4N$-electron perimeters for even, and b) for odd perturbations, and c) of charged $4N$-electron perimeters. The discussion of spectra applies only to the shaded region of strong perturbations (adapted from Höweler et al., 1989).

shown how in this case the relative energies of the lowest three singlet states depend on the perturbation strength. (See also Figure 4.20.) As all three singlet states are composed of perimeter configurations that differ only in the occupation of the orbitals ϕ_N and ϕ_{-N}, whose quantum numbers differ by more than a unity, the transition moments vanish. (See Example 2.6.) In Section 4.3.2 it will be shown that the energy ordering given in Figure 2.20a and for $|s| = 0$ for the unperturbed uncharged $4N$-electron perimeter is characteristic for so-called perfect pair biradicals.

In discussing the spectroscopic properties of nonaromatic compounds derived from a $4N$-electron perimeter it is assumed that Φ_1 is a good approximation to the ground-state wave function. From Figure 2.20 it is clear that such is the case when $|s|$ is sufficiently large so that the G state is energetically below the S state. The following arguments apply therefore only if this condition is fulfilled. This is virtually guaranteed for all systems of real interest even if no structural perturbations are present, due to Jahn-Teller and pseudo–Jahn-Teller distortions.

From Figure 2.20a it is easily seen that the dipole-forbidden first absorption of uncharged perimeters, which corresponds to the G→S transition, is expected to occur at longer wavelengths for systems with even perturbations than for systems with odd perturbations. This is confirmed by the absorption spectra of a substituted pentalene (even perturbation) and of biphenylene (odd perturbation) depicted in Figure 2.21. The spectrum of the pentalene also exhibits the slow rise of the absorption profile of the first band starting

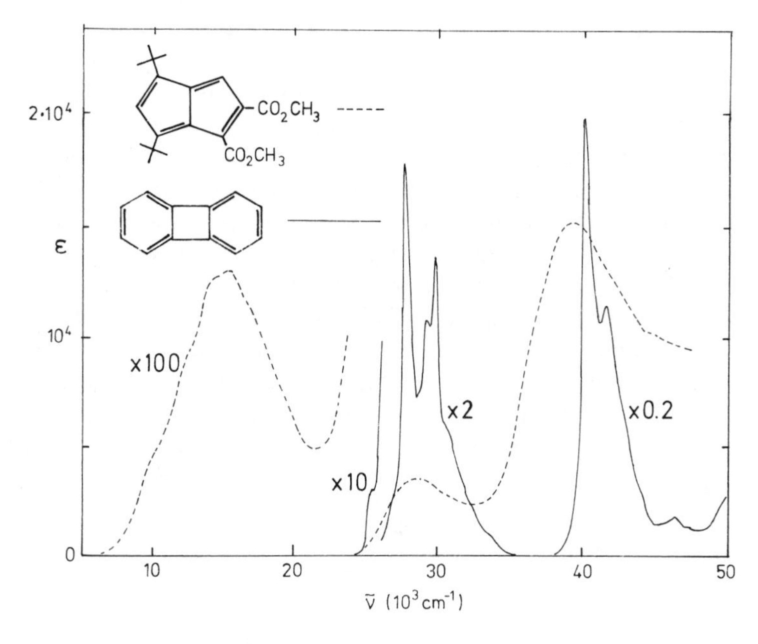

Figure 2.21. Absorption spectra a) of 1,3-di-*t*-butylpentalene-4,5-dicarboxylic ester (adapted from Höweler et al., 1989) and b) of biphenylene (adapted from Jørgensen et al., 1978). Due to the large variation in intensity the various bands are scaled as indicated.

at 10,000 cm^{-1}, which is known as "tailing" and is very typical for systems with even perturbations.

In order to discuss higher states, the orbitals $\phi_{\pm(N-1)}$ and $\phi_{\pm(N+1)}$ or their real combinations obtained by the transformation Equation (2.17) have to be also considered. Four of the configurations that result in this way are doubly excited with respect to Φ_1, and since in general they mix only insignificantly or not at all with singly excited configurations they give rise to states that cannot be observed by one-photon spectroscopy. These high-energy configurations are therefore disregarded in the following discussion. The remaining four configurations are singly excited and depend in a complicated manner on the configurations of the unperturbed system. If the molecule still contains a mirror plane perpendicular to the molecular plane after the perturbation, these configurations interact only in pairs and yield two low-lying minus combinations N_1 and N_2 and two higher plus combinations P_1 and P_2, similar to the L and B states of aromatic compounds. Intensities of G→N transitions are low, due to mutual cancellation of contributions from the two configurations contributing to the excited state. Intensities of the G→P transitions are high, due to their mutual reinforcement. Since transition moments of the interacting configurations are parallel, the corresponding N and P transitions have the same polarization direction. Both N and P transitions are polarized along lines that both pass through either opposite atoms or bonds. A distinction between a and b transitions as in the case of $(4N+2)$-electron perimeters is therefore not possible.

For charged $4N$-electron perimeters the degeneracy of the ground state is lifted by perturbations. The phase angles are not multiples of $\pi/2$; hence a classification of perturbations as even or odd is not possible. Figure 2.20c shows how the energies of the states G, S, and D depend on the perturbation strength $|s|$. The energy order for unperturbed charged perimeters at $|s| = 0$ is typical for perfect axial biradicals. (See Section 4.3.2.)

The classification into N and P as well as the ordering and the polarization of the excited states is analogous to that of systems derived from uncharged $4N$-electron perimeters. These results are important because quinones (**26**), for instance, can be viewed as derivatives of doubly positively charged $4N$-electron perimeters.

26

The results for all cyclic nonaromatic systems are summarized again in Figure 2.22. Perturbations that affect the diagonal elements (e.g., heteroatom replacements) shift the HO, SO, and LU levels that are shown for the

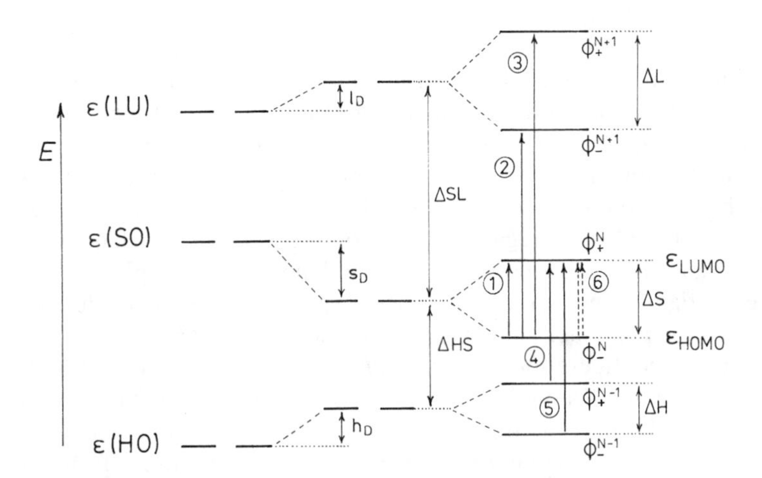

Figure 2.22. Schematic representation of the perturbation-induced shifts and splittings of the degenerate orbitals of $4N$-electron perimeters. The indicated transitions correspond to the main contributions to the observed absorption bands; the broken arrows symbolize the double excitation ⑥ (adapted from Höweler et al., 1989).

general case. Uncharged $4N$-electron perimeters are alternant hydrocarbons, and their HO and LU levels are arranged symmetrically with respect to ε(SO). Off-diagonal perturbations (e.g., bond-length alternation) split the orbital levels by ΔH, ΔS, and ΔL, respectively. Cross-links can affect both diagonal and off-diagonal elements. The transitions denoted by ① and ⑥ correspond essentially to the G→S transition and to the G→D transition, respectively. The latter is a two-electron transition and is therefore expected to be of relatively high energy and very low intensity. The transitions denoted by ②–⑤ correspond to the N and P bands.

2.3 Radicals and Radical Ions of Alternant Hydrocarbons

Radicals with an odd number of electrons can be either uncharged odd hydrocarbons or radical ions of even hydrocarbons or systems derived from such hydrocarbons. Of special interest are the relationships that exist for radicals and radical ions of alternant hydrocarbons.

Figure 2.23a gives a schematic representation of the frontier orbital energy levels of an uncharged odd alternant hydrocarbon. It is seen that

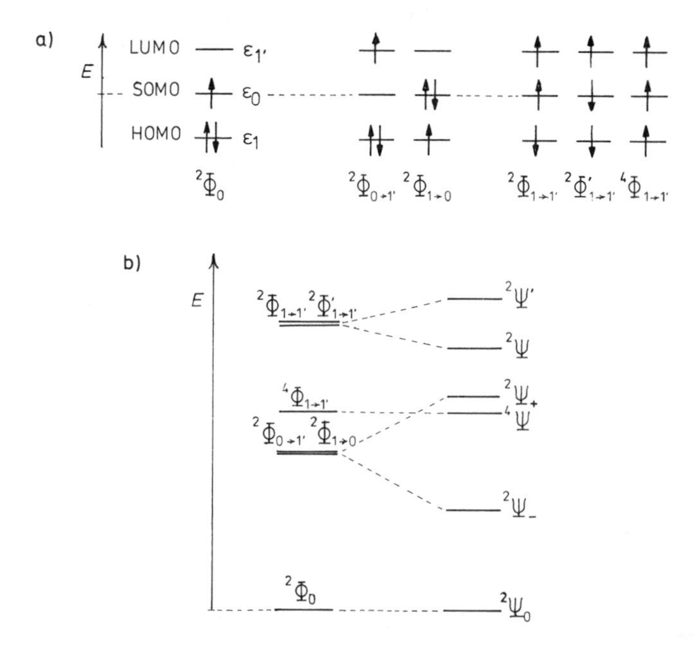

Figure 2.23. Odd alternant hydrocarbon radicals: a) Schematic representation of the frontier orbital energy levels and of the various configurations that are obtained by single excitations from the ground configuration Φ_0. (It should be remembered that spin eigenstates cannot be represented correctly in these diagrams.) b) Energies of these configurations and effect of first-order configuration interaction.

the two transitions that may be referred to as HOMO→SOMO and SOMO→LUMO transitions and that produce the doublet configurations $^2\Phi_{1\to 0}$ and $^2\Phi_{0\to 1'}$, respectively, are degenerate for alternant hydrocarbons. In contrast to closed-shell systems, there are no higher multiplicity transitions that correspond to these lowest-energy excitations. The HOMO→LUMO transition is the first one for which a quartet configuration is possible, in addition to two doublet configurations. One of the doublet configurations, $^2\Phi'_{1\to 1'}$, is obtained from the ground configuration only by flipping the spin of one of the electrons during the excitation. The transition moment between the ground configuration and this configuration therefore vanishes by the spin selection rule. (Cf. Section 1.3.2.)

Figure 2.23b shows the expected energy diagram, including electron repulsion effects. The configurations that correspond to the HOMO→SOMO and the SOMO→LUMO excitation are split into $^2\Psi_-$ and $^2\Psi_+$ states by first-order configuration interaction, familiar from even alternant systems. The components of the transition moment from the configurations $^2\Phi_{1\to 0}$ and $^2\Phi_{0\to 1'}$ either mutually cancel or reinforce, so the longer wavelength transition is forbidden whereas the other one is allowed. In agreement with these expectations the absorption spectrum of the benzyl radical shows a weak band near 450 nm and a strong band near 300 nm (Porter and Strachan, 1958). Calculations for the allyl radical suggest that the quartet state is located between the $^2\Psi_-$ and $^2\Psi_+$ states as indicated in Figure 2.23. At which

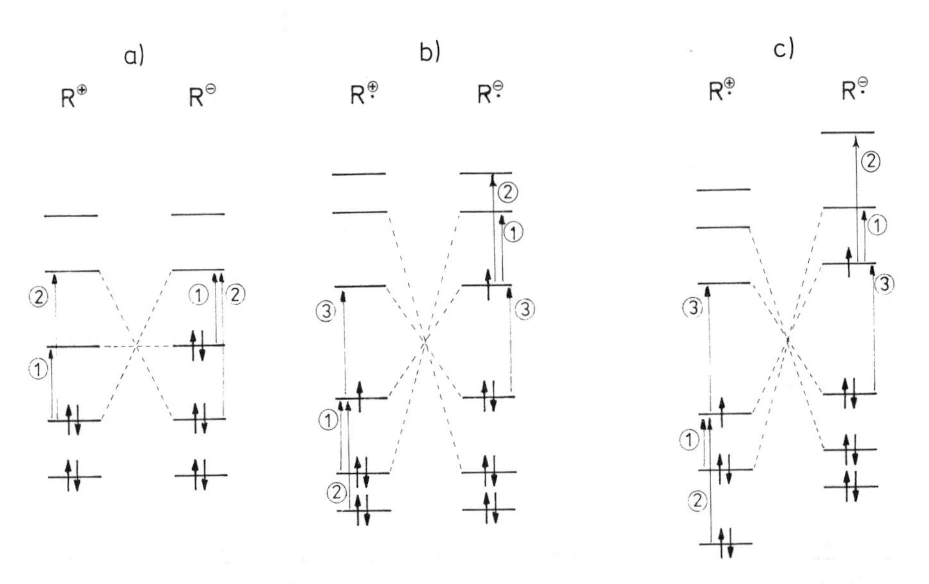

Figure 2.24. Orbital energy levels of alternant hydrocarbon ions: a) anions and cations of odd-alternant systems, and b) radical anions and radical cations of even-alternant systems in the HMO approximation and c) in the PPP approximation.

energies the states arising from $\Phi_{2\to 0}$ and $\Phi_{0\to 2'}$ are to be found in larger systems is not yet entirely settled.

The most characteristic feature in the spectra of ions from odd alternant hydrocarbons as well as from radical ions of even alternant hydrocarbons is the fact that the spectra of the anions and the cations should be identical as long as the pairing theorem is valid. (Cf. Koutecký, 1966.) At the level of the HMO model this is immediately evident from the orbital energy scheme in Figure 2.24. McLachlan (1959) has shown that at the PPP level of approximation the pairing theorem for positive and negative ions has to be modified. The bonding and antibonding MOs of any one radical ion are not paired; rather the bonding MOs of the radical cation are paired with the antibonding MOs of the radical anion and vice versa, as depicted in Figure 2.24c. Such

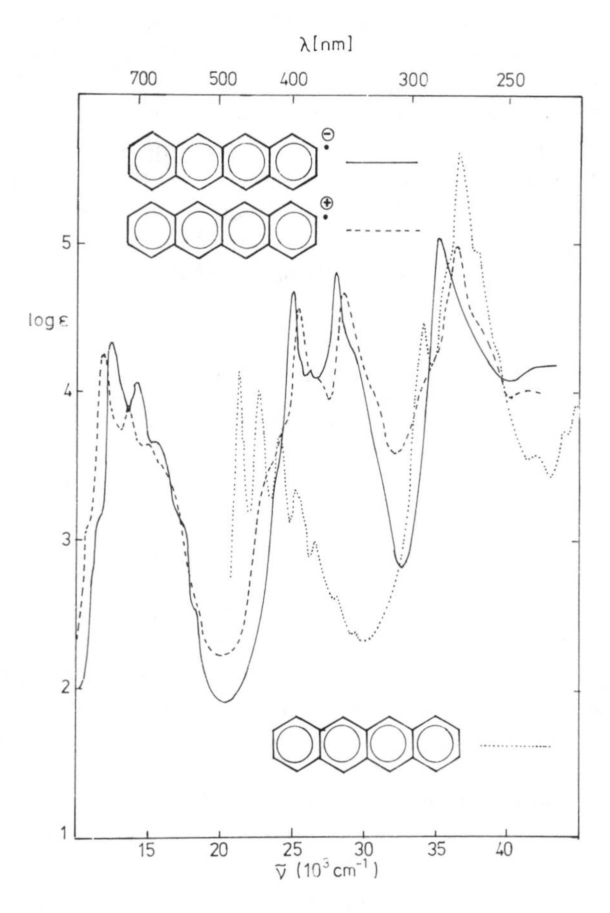

Figure 2.25. Absorption spectra of the radical anion (—) and the radical cation (---) of tetracene in comparison with the spectrum of tetracene (···) (by permission from DMS-UV-Atlas 1966–71).

systems are called *mutually paired.* As a consequence the identity of the absorption spectra of an anion and a cation of an alternant hydrocarbon is retained when electron interaction terms are included in the model. It can be seen from the spectra of the tetracene radical anion and cation shown in Figure 2.25 that this result is in remarkably good agreement with experiment.

2.4 Substituent Effects

Most organic compounds that show absorption in the visible or in the near-UV region have a linear or cyclic π system as the chromophoric system. Therefore, the results of the previous sections may be used and extended to discuss light absorption of all those compounds that can be derived from linear and cyclic hydrocarbons by including the influence of substituents in an appropriate way. (Cf. Michl, 1984.) A complete theory of substituent effects comprises all areas of organic chemistry. Here, only the fundamental concepts of the influence of inductive and mesomeric substituents will be considered. In order to simplify the discussion, substituent effects will be called inductive if in the HMO model they can be represented by a variation of the Coulomb integral α_μ of the substituted π center μ. If they are due to an extension of the π system they will be called mesomeric.

2.4.1 Inductive Substituents and Heteroatoms

The replacement of a C atom in a conjugated system by a heteroatom such as N may be considered the simplest example of a purely inductive substituent effect. The influence on the orbital energies may then be estimated by first-order perturbation theory using the relation

$$\delta\varepsilon_i = c_{\mu i}^2 \delta\alpha_\mu \tag{2.19}$$

For alternant hydrocarbons $c_{\mu 1}^2 = c_{\mu 1'}^2$ for all μ, and there is no first-order energy change for the HOMO→LUMO transition, as is apparent from Figure 2.26. This result is remarkably well confirmed by the absorption spectra of naphthalene, quinoline, and isoquinoline shown in Figure 2.27.

For nonalternant hydrocarbons, however, a bathochromic or a hypsochromic shift may result, depending on the absolute magnitude of the LCAO coefficients of the HOMO and LUMO. This is also shown in Figure 2.26 and is clearly illustrated by the absorption spectra of azulene, 4-azaazulene, and 5-azaazulene, given in Figure 2.28.

If a C atom is replaced by groups such as =SiH—, =P—, =As—, etc., the relation

$$\delta\varepsilon_i = c_{\mu i}^2 \delta\alpha_\mu + 2 \sum_{\mu - \nu} c_{\mu i} c_{\nu i} \delta\beta_{\mu\nu} \tag{2.20}$$

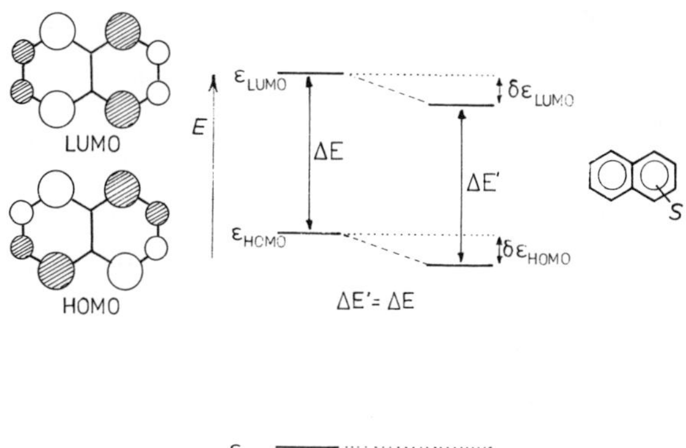

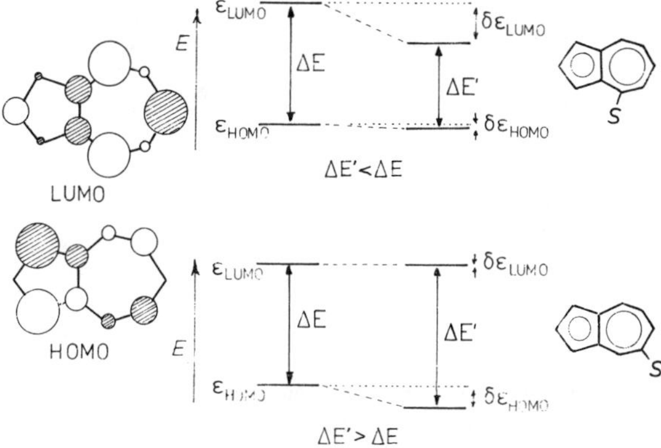

Figure 2.26. HOMO and LUMO of naphthalene and azulene together with the orbital energy changes $\delta\varepsilon_{HOMO}$ and $\delta\varepsilon_{LUMO}$, calculated by first-order perturbation theory for a purely inductive substituent S.

has to be used instead of Equation (2.19). The sum runs over all bonds μ—ν connected to the replacement center μ and takes into account the reduction of the resonance integrals $\beta_{\mu\nu}$ that is to be expected for bonds between carbon and elements of the third or lower rows of the periodic table. The effect of both terms in Equation (2.20) is illustrated in Figure 2.29 for the replacement of =CH— by =SiH—. It is seen that the variation $\delta\beta$ decreases the HOMO-LUMO separation $\Delta\varepsilon_{HL} = \varepsilon_{LUMO} - \varepsilon_{HOMO}$. This results in an appreciable bathochromic shift of the ^1L_b band in going from benzene to silabenzene and 1,4-disilabenzene, as can be seen from the absorption spectra shown in Figure 2.30.

While first-order effects of purely inductive substituents on excitation energies of alternant hydrocarbons vanish, higher-order perturbation theory gives nonzero contributions. Thus, Murrell (1963), using second-order perturbation theory, derived the relation

$$E_i^{(2)}(\varrho) = \sum_{k(\neq i)} B_{ki}(\varrho)\frac{H'^2_{ik}}{E_i - E_k} \tag{2.21}$$

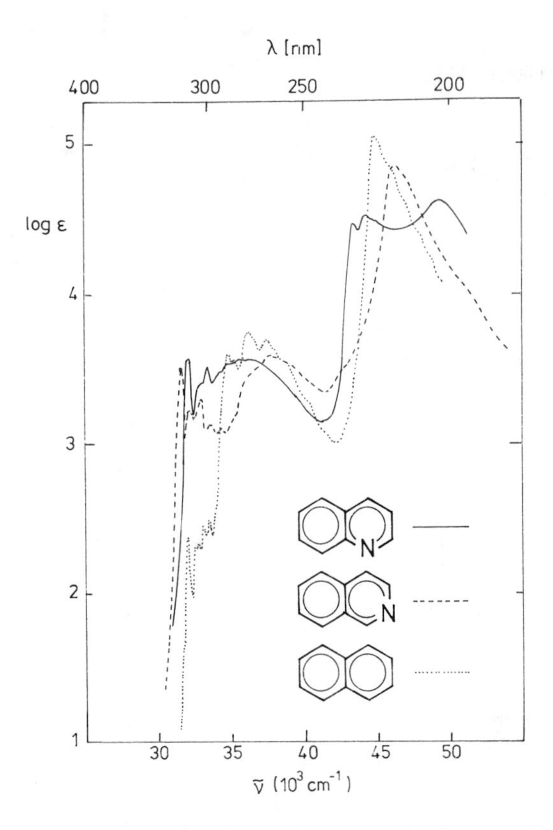

Figure 2.27. Absorption spectra of naphthalene, quinoline, and isoquinoline (by permission from DMS-UV-Atlas 1966–71).

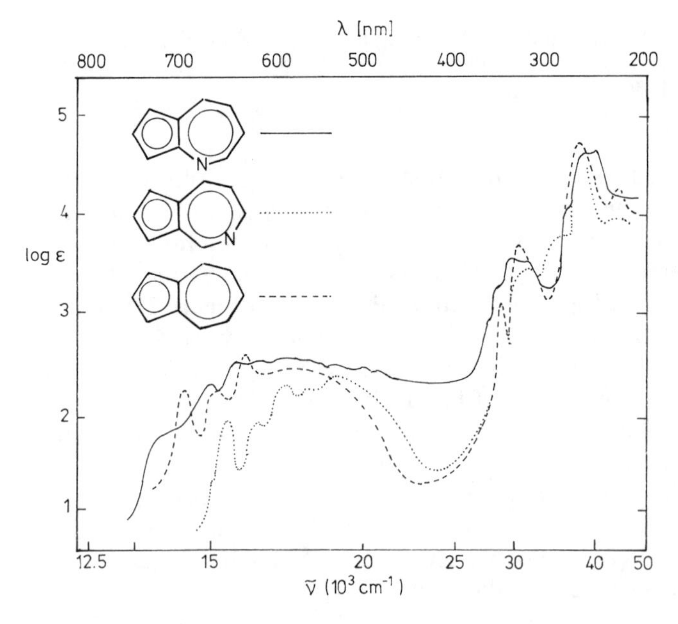

Figure 2.28. Absorption spectra of azulene, 4-azaazulene, and 5-azaazulene (by permission from Meth-Cohn et al., 1985).

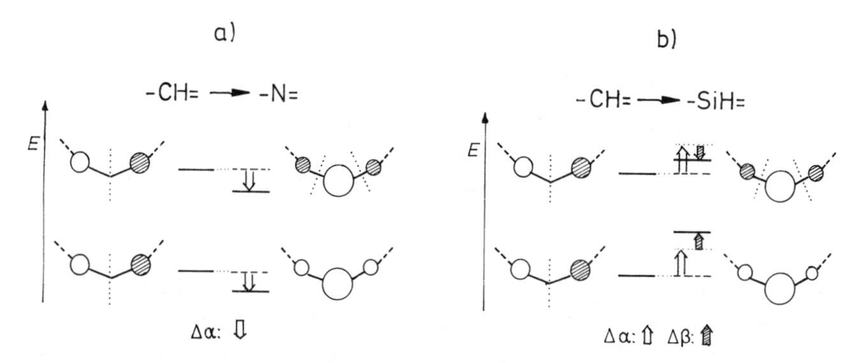

Figure 2.29. First-order effects of one-electron heteroatom replacement on frontier orbital energies of a $(4N+2)$-electron annulene; a) for N as an example of an electronegative second-row element ($\delta\alpha_{\mu} < 0$, $\delta\beta_{\mu\nu} = 0$) and b) for Si as an example of an electropositive third-row element ($\delta\alpha_{\mu} > 0$, $\delta\beta_{\mu\nu} > 0$). The effects of change in electronegativity $\delta\alpha$ are shown as white vertical arrows (by permission from Michl, 1984).

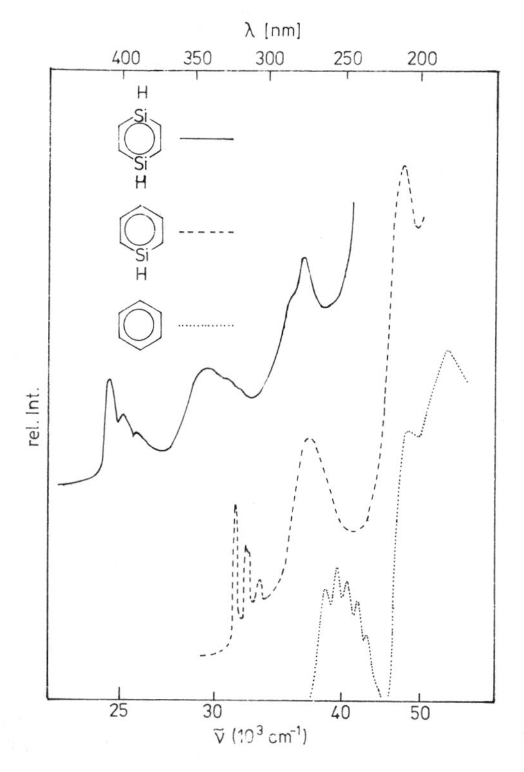

Figure 2.30. Absorption spectra of benzene (···), silabenzene (---), and 1,4-disilabenzene (—) (by permission from Maier, 1986).

for the substituent effect on the energy of the excited state. Here, $H'_{ik} = \langle\Psi_i|\hat{h}'|\Psi_k\rangle$ is the matrix element of the perturbation between the state wave functions Ψ_i and Ψ_k in the case of a monosubstituted benzene. The $B_{ki}(\varrho)$ are constants depending on the number and relative positions of substituents ϱ and are derived from group theoretical arguments.

It follows that the shift of the benzene ^1L_b band can be written as

$$\Delta\tilde{\nu} = -nA + I_\varrho\delta^2 \tag{2.22}$$

where nA represents a perturbation that is proportional to the number n of substituents and the constants I_ϱ and δ depend on the substitution pattern and on the nature of the substituents, respectively. A relation of this type derived from experimental data was first proposed by Sklar (1942).

The intensity of the ^1L_b band arises essentially from a substituent-induced interaction with the ^1B_b state. This can be envisaged by means of the perimeter model as follows. Purely inductive substituents give rise to even perturbations; from Figure 2.16 it is seen that in this case the only interaction is between configurations X and V [see Equation (2.12)] and is given by $(|a| + |b|) = (\Delta\text{HOMO} + \Delta\text{LUMO})/2$. To first order in perturbation theory, this is equal to ΔHOMO. To a first approximation, the amount of the configuration X that is mixed into the ^1L_b state is therefore proportional to ΔHOMO. This contribution causes the intensity increase of the ^1L_b band due to purely inductive substituents. Since the intensity is given by the square

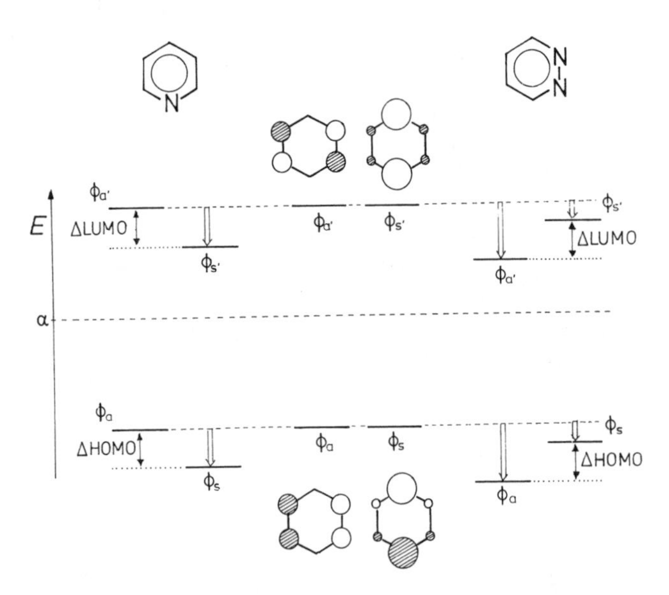

Figure 2.31. The effect of purely inductive substituents in position 1 (on the left) and in positions 1 and 2 (on the right) on ΔHOMO and ΔLUMO of benzene (adapted from Castellan and Michl, 1978).

Table 2.4 Substituent Effects on the Intensity of the ^1L_b Band of Benzene

Substitution Pattern	Intensity Enhancement[a]	Substitution Pattern	Intensity Enhancement[a]
1-	1	1,4-	4
1,2-	1	1,2,3-	0
1,3-	1	1,3,5-	0

[a] Multiples of the value $(\Delta HOMO)^2 = (1/3\delta\alpha_X)^2$ for monosubstitution by a group X.

of the transition moment, it should be proportional to $(\Delta HOMO)^2$. ΔHOMO can be determined quite easily from Equation (2.13) by means of the LCAO coefficients of the real MOs ϕ_s and ϕ_a of Equation (2.15). This is illustrated in Figure 2.31 for substitution in position 1 as well as in positions 1 and 2 of benzene. Depending on the substitution pattern, the perturbation produces the MO order ϕ_s, ϕ_a, $\phi_{s'}$, $\phi_{a'}$ or ϕ_a, ϕ_s, $\phi_{a'}$, $\phi_{s'}$. The intensity enhancements obtained from these arguments for various substitution patterns are collected in Table 2.4.

The polarization direction of a transition depends on the location of the nodal planes of the orbitals involved in the excitation. As the introduction of perturbations may influence the location of the nodal planes, the effect of several simultaneous perturbations may change the polarization directions accordingly. The cross-linking that produces naphthalene from the perimeter causes the ^1L_b and ^1L_a bands to be polarized along the long axis and the short axis of the molecule, respectively. Heteroatom replacement in a $(4N+2)$-electron perimeter results in every MO having either a node or an antinode at the position of the heteroatom. (See Figure 2.29.) Since both the HOMO and LUMO of naphthalene have an antinode in position 1 (Figure 2.26) whereas the adjacent MOs below the HOMO and above the LUMO have a node in this position, there is no change in the polarization direction going from naphthalene to quinoline. The polarization directions of the $^1L_1$ and $^1L_2$ band remain mutually perpendicular. However, the situation is different for isoquinoline, where the location of the nodal planes required by cross-linking and by heteroatom replacement do not coincide. In all four naphthalene MOs of interest, the nodes and antinodes are equally far removed from position 2. In this particular case the effect of the aza group is sufficiently strong to dominate the location of the nodal plane. Hence, the polarization directions of the $^1L_1$ and $^1L_2$ transitions are nearly parallel (Friedrich et al., 1974).

2.4.2 Mesomeric Substituents

Mesomeric substituents possess orbitals of π symmetry and can therefore extend the conjugated system. Benzene substituted by a donor with a doubly occupied p AO or substituted by an acceptor with an empty p AO is isoelectronic with the benzyl anion or benzyl cation, respectively, for which the nonbonding MO is either the HOMO or the LUMO, respectively. (See Figure 2.24a.) A substituent may also possess several orbitals which, depending on their occupancy, may have different donor and acceptor properties.

The substituent effect on the hydrocarbon MOs may again be estimated by perturbation theory. However, it has to be remembered that first-order contributions vanish according to

$$\delta\varepsilon_i^R = -\delta\varepsilon_k^S = c_{\varrho i}^R c_{\sigma k}^S \beta_{\varrho\sigma} \tag{2.23}$$

unless the orbitals ϕ_i^R of the hydrocarbon R and ϕ_k^S of the substituent S are nearly degenerate. In the general case second-order perturbation theory has to be applied and the relation

$$\delta\varepsilon_i^R = \sum_k \frac{(c_{\varrho i}^R)^2 (c_{\sigma k}^S)^2 \beta_{\varrho\sigma}^2}{\varepsilon_i^R - \varepsilon_k^S} \tag{2.24}$$

has to be used, which requires special precautions if ϕ_i^R and ϕ_k^S are nearly degenerate.

$\delta\varepsilon_i^R$ in Equation (2.24) depends not only on $c_{\varrho i}^R$ but also on the quantity $|c_{\sigma k}^S \beta_{\varrho\sigma}|$ as well as on $(\varepsilon_i^R - \varepsilon_k^S)$. It is therefore not possible to characterize the mesomeric effect by a single substituent strength. In fact, depending on ε_i^R the same substituent can act as a donor with respect to one hydrocarbon and yet as an acceptor with respect to another one if it possesses occupied as well as unoccupied orbitals. An example is provided by the vinyl group with its π and π^* orbitals, which acts as an acceptor when attached to the $C_5H_5^{\ominus}$ anion and as a donor when attached to the $C_7H_7^{\oplus}$ cation.

For spectroscopic applications it has proven very useful to characterize the mesomeric effect of particular substituents with respect to an alternant hydrocarbon by the way in which they affect the energy differences ΔHOMO and ΔLUMO between those orbitals, which in the unperturbed $(4N+2)$-electron perimeter are the degenerate highest occupied and lowest unoccupied MOs, respectively. A substituent is referred to as a π donor or $+$M substituent, if ΔHOMO > ΔLUMO, and as a π acceptor or $-$M substituent, if ΔHOMO < ΔLUMO.* If ΔHOMO = ΔLUMO, for instance when a vinyl substituent is attached to benzene, the π electron-donating ability and the π electron-withdrawing ability exactly compensate each other in their net effect. In most instances this definition of donors and acceptors agrees with those based on the measurements of other properties, such as the amount of charge transfer between the substituent and the ring in the

---

* According to Dewar (1969) a π donor is referred to as $-$E substituent and a π acceptor as $+$E substituent. This is the sign convention originally introduced by Lapworth and Robinson. Here, we use the opposite sign convention, which was proposed later by Ingold.

ground state as measured by the π-electron contribution to the dipole moment or by NMR chemical shifts (Michl, 1984).

If a substituent possesses only one p orbital which in most cases is doubly occupied as in $—NH_2$ or —OH, Equation (2.24) is simplified and becomes

$$\delta\varepsilon_i^R = \frac{(c_{\varrho i}^R)^2\beta_{\varrho\sigma}^2}{\varepsilon_i^R - \varepsilon_k^S} \tag{2.25}$$

since $c_{\sigma k}^S = 1$. The interaction of the substituent with the π system will then be characterized by two substituent parameters: the energy of the p_z orbital ε_k^S and the resonance integral $\beta_{\varrho\sigma}$. The relative importance of the two factors will depend on the choice of the hydrocarbon to which the substituent is attached; while the effect of a change in $\beta_{\varrho\sigma}$ is simply multiplicative and is the same for all perimeter MOs, the effect of the variation in the energy difference $(\varepsilon_i^R - \varepsilon_k^S)$ in the denominator is different for each perimeter MO depending on its energy ε_i^R. The distinct effect of nominator and denominator in Equation (2.25) on the frontier orbitals ϕ_a, ϕ_s, $\phi_{a'}$, and $\phi_{s'}$ of a $(4N+2)$-electron perimeter is illustrated in Figure 2.32. Halving the resonance integral $\beta_{\varrho\sigma}$ halves both ΔHOMO and ΔLUMO and the substituent becomes a weaker donor (Figure 2.32b). Subsequent halving of the separation $\Delta\varepsilon = \varepsilon_i^R - \varepsilon_k^S$ between the energies of the donor orbital ϕ_k^S and the HOMO $\phi_i^R = \phi_s$ of the perimeter restores the magnitude of ΔHOMO but hardly affects the more distant $\phi_i^R = \phi_{s'}$ and ΔLUMO (Figure 2.32c). Thus, in the first and the last cases [a) and c) in Figure 2.32] the difference ΔHOMO − ΔLUMO is of comparable magnitude and according to the definition given above the sub-

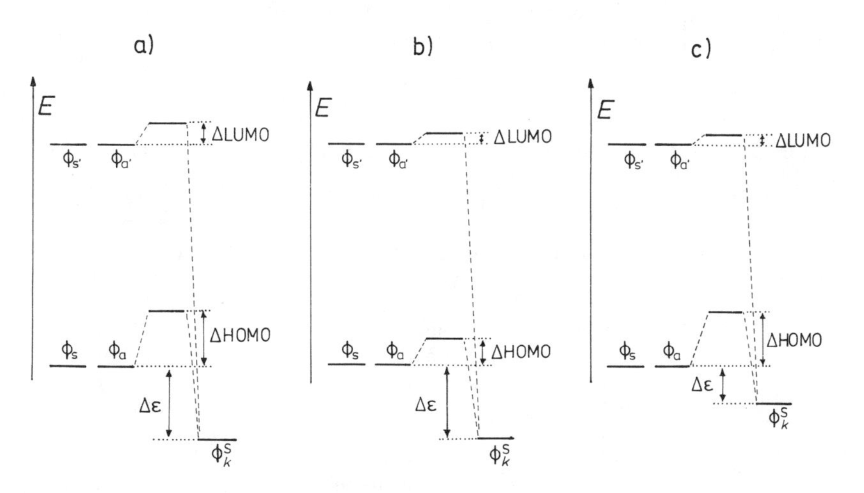

Figure 2.32. Effect of a +M substituent S with one π orbital ϕ_k^S only on the energies of the frontier orbitals of a $(4N+2)$-electron perimeter for different values of the parameters β and $\Delta\varepsilon$; a) β, $\Delta\varepsilon$, b) $\beta/2$, $\Delta\varepsilon$, and c) $\beta/2$, $\Delta\varepsilon/2$, (adapted from Michl, 1984).

stituents possess a similar +M effect. On the other hand, in the second and the third cases [b) and c) in Figure 2.32] ΔLUMO is comparable and the substituents have equal donor strength if measured by the amount of charge transferred into the ring.

For alternant hydrocarbons $c_{\varrho m}^2 = c_{\varrho m'}^2$ and $\varepsilon_m = \alpha + x_m\beta$ with $x_{m'} = -x_m$, so from Equation (2.25) the change $\delta\Delta E$ of the HOMO→LUMO excitation energy is seen to be

$$\begin{aligned}\delta\Delta E &= \frac{c_{\varrho m}^2\beta_{\varrho\sigma}}{-x_m - x_S} - \frac{c_{\varrho m}^2\beta_{\varrho\sigma}}{x_m - x_S} \\ &= \frac{c_{\varrho m}^2[(x_m - x_S) - (-x_m - x_S)]\beta_{\varrho\sigma}}{x_S^2 - x_m^2} = \frac{2x_m}{x_S^2 - x_m^2}c_{\varrho m}\beta_{\varrho\sigma}\end{aligned} \tag{2.26}$$

if the substituent orbital is lower in energy than the HOMO, that is, if $x_S > x_m$; if $x_S < x_m$, x_m in the nominator has to be replaced by $-x_S$. In any case, the mesomeric effect always produces a bathochromic shift and is additive for individual substituents, as long as no new rings are being formed and the change of the coefficients $c_{\varrho m}$ due to substituents can be neglected.

Example 2.9:
The discussion given earlier for weak mesomeric effects offers an explanation for the empirical increment rules proposed by Woodward (1942) for estimating the absorption maxima of dienes and trienes. The distinct values for the homo- and heteroannular arrangement of the double bonds illustrate the configurational influence on the polyene spectra that has been discussed in Example 2.2. From the data collected in Table 2.5 the absorption of methylenecyclohexene (**27**) with an exocyclic double bond and two alkyl substituents is calcu-

Table 2.5 Increment Rules for Diene Absorption (Adapted from Scott, 1964)

Acyclic 217 nm	Homoannular 253 nm		Heteroannular 214 nm
	Increments for:		
	Double bonds extending conjugation		+30 nm
	Exocyclic double bond		+5 nm
	Alkyl group, ring residue		+5 nm
	Polar groups:	O-alkyl	+6 nm
		O-acyl	±0
		S-alkyl	+30 nm
		$N(alkyl)_2$	+60 nm
		Cl	+5 nm
		Br	+5 nm

lated to be at 214 + (2 × 5) + 5 = 229 nm, as compared to the experimental value λ_{max} = 231 nm. Similarly, the values calculated for the steroids **28** and **29** are λ_{max} = 253 + (4 × 5) + (2 × 5) = 283 nm and 214 + (3 × 5) + 5 = 234 nm, respectively, again in excellent agreement with the experimental data (282 nm and 234 nm, respectively). Similar rules for estimating the absorption maxima of unsaturated carbonyl compounds have been proposed by Fieser et al. (1948); see Table 2.6.

27 28 29

Table 2.6 Increment Rules for Enone Absorption (Adapted from Scott, 1964)

O=C(H)–C=C	O=C(R)–C=C	O=C(OR)–C=C
207 nm	215 nm	193 nm

Increments for:

Double bonds extending conjugation					+30 nm
Exocyclic double bond					+5 nm
Homodiene component					+39 nm
Substituents in	α	β	γ	δ	position
R	10	12	18	18	
Cl	15	12			
Br	25	30			
OH	35	30		50	
OR	35	30	17	31	
OAc	6	6	6	6	
NR_2		95			

In general, substituents are neither purely inductive nor purely mesomeric; rather they possess some inductive as well as some mesomeric character. From Equations (2.19) and (2.24) it is seen that inductive as well as mesomeric effects on the HOMO and LUMO are proportional to the square $c_{\varrho i}^2$ of the LCAO coefficient at the substituted center ϱ. It is therefore difficult to differentiate between these two effects from the absorption spectra. This is possible, however, by means of the MCD spectra that depend essentially on the difference ΔHOMO − ΔLUMO. (Cf. Section 3.3.) In the case of uncharged $(4N+2)$-electron perimeters and especially benzene, perturbations due to purely inductive substituents always yield ΔHOMO = ΔLUMO to first order in perturbation theory. Due to the energy difference in the denominator of Equation (2.24), however, ΔHOMO and ΔLUMO can be quite different in the case of mesomeric substituents, depending on the energy of the substituent orbital, as has been demonstrated in Figure 2.32. By correlating the MCD spectra with the difference ΔHOMO − ΔLUMO, it could be shown that hyperconjugation is most important in describing the substituent effects of methyl groups, so that for spectroscopic purposes the methyl group is best considered as a mesomeric substituent. (See Example 3.10.)

Example 2.10:
In nonalternant systems, $(c^2_{\varrho,\text{LU}} - c^2_{\varrho,\text{HO}})$ can be positive or negative, depending on the position ϱ. Purely inductive substituents can therefore produce a shift of the HOMO→LUMO transition to higher or lower energies. In this way, the spectra of methyl-substituted azulenes, for instance, have been analyzed assuming an essentially inductive substituent effect of the methyl group (Heilbronner, 1963). Displacements of the longest-wavelengths band calculated from Equation (2.19) are compared with experimental values in Table 2.7 and give a satisfactory agreement. If, on the other hand, the substituent effect of the methyl group is assumed to be essentially hyperconjugative, the displacements have to be estimated from Equation (2.25), using appropriate values for the energy difference in the denominator. Assuming that $(\varepsilon_{\text{LU}} - \varepsilon_S)$ is approximately twice the size of $(\varepsilon_{\text{HO}} - \varepsilon_S)$, one obtains the values given in the last column of Table 2.7. The agreement with experimental data is even better, especially for 6-methylazulene.

Table 2.7 Observed[a] and Calculated[b] Substituent Effect in Methyl-Substituted Azulenes

		$\Delta\tilde{\nu}_{\text{max}}$		
Substitution	$\tilde{\nu}_{\text{max}}$	Experiment	+ I Effect	+ M Effect
—	17,240	0	0	0
1-Methyl-	16,450	−790	−944	−958
2-Methyl-	17,670	+430	+106	+163
4-Methyl	17,610	+370	+154	+278
5-Methyl	16,890	−350	(−350)	(−350)
6-Methyl	17,700	+460	+288	+428

[a] After Heilbronner, 1963.
[b] Calculated from Equations (2.19) and (2.25), respectively, using $\Delta\tilde{\nu} = -350\ \text{cm}^{-1}$ for 5-methylazulene as reference value.

The more pronounced the π donor strength of a substituent, the greater will be its contribution to the HOMO of the substituted molecule. Excitation from such an orbital into an MO essentially localized in the perimeter will therefore be associated with a charge shift from the substituent into the ring and will thus possess partly some $^1\text{L}_\text{a}$ and partly some CT (charge transfer) character. (Cf. Section 2.6.) For a high-lying substituent orbital ϕ^S_k, the mixing with the perimeter MOs ϕ_s and $\phi_{s'}$ can be of comparable importance.

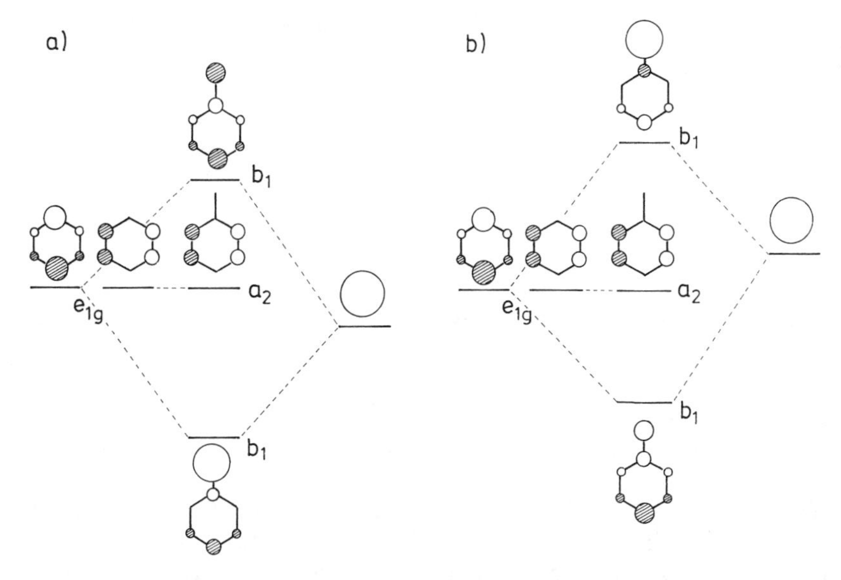

Figure 2.33. Schematic representation of the highest occupied π MOs a) of aniline and b) of dimethylaniline (adapted from Kobayashi et al., 1972).

A very strong mesomeric effect is present in aniline (**30**). From the photoelectron (PE) spectra it is known that the highest occupied MO is largely localized in the benzene ring, whereas in dimethylaniline (**31**) it is more concentrated on the nitrogen. Thus, it follows that the p_π orbital energy of the amino group is very close to the energy of the degenerate π MOs of benzene (see Figure 2.33) and that the longest-wavelength transition of dimethylaniline should have a strong intramolecular CT (charge transfer) contribution.

Perturbation theory is also applicable to substituents with a strong mesomeric effect, but it is to be expected that results gradually become less reliable as the substituent strength increases. Two main models have been proposed for the analysis of strong substituent effects in molecules such as aniline (Godfrey and Murrell, 1964). The localized orbital model makes use of the π MOs of benzene and the p_π orbital of the amino group. (Cf. Figure 2.33.) The energy of the intramolecular CT transition may be estimated from the IP of the donor (10.15 eV for aniline), the EA of the acceptor orbital (-1.6 eV for benzene), and the Coulomb interaction J between the positive charge at the substituent and the charge distribution on the acceptor. For the degenerate MOs ϕ_4 and ϕ_5 of benzene and a C-N distance of 146 pm one obtains the following values for J:

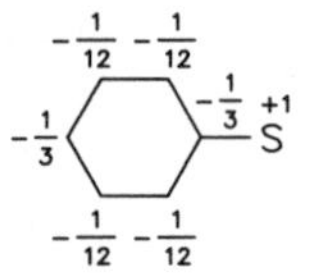

J = 6 eV 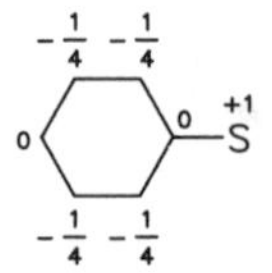   J = 4.8 eV

The CT transition is therefore expected to have a transition energy of

$$\Delta E = 10.15 + 1.6 - 6.0 = 5.75 \text{ eV}$$

and should be found at shorter wavelengths than the ^1L_b band of benzene (4.75 eV). The ^1L_b band is polarized perpendicular to the long axis of the molecule, the CT band parallel to this axis.

The second model makes use of the fact that aniline is isoelectronic with the benzyl anion. From the HMO coefficients it can be seen that the longest-wavelength transition from the doubly occupied nonbonding MO into the lowest antibonding MO is polarized perpendicular to the long axis of the molecule. The transition into the following antibonding MO is polarized parallel to this axis. While both transitions are forbidden in the localized-orbital model, oscillator strengths of $f = 0.15$ and $f = 1.05$, which are much larger than the experimental values, are calculated from the isoconjugate hydrocarbon model. The data collected in Table 2.8 show that the observed oscillator strengths lie between those estimated from the two models. The same is true for the excitation energies and for the ground- and excited-state dipole moments.

Both models can be extended. The localized orbital model is improved by allowing for the interaction between the locally excited benzene states and the CT state, and the isoconjugate hydrocarbon model is improved by taking into account the electronegativity change upon the replacement of the C atom in the benzyl anion by an N atom in aniline. Good agreement with experiment is obtained in both cases.

Both models are useful, but the localized orbital model has the advantage that it is more easily extended to more complicated molecules. The CI

Table 2.8 Comparison of Different Models for Estimating the Substituent Effect in Aniline (Adapted from Murrell, 1963)

Model	Ground state: μ(D)	1st Excited State ΔE(eV)	f	μ(D)	2nd Excited State ΔE(eV)	f	μ(D)
Benzyl anion	5.76	2.18	0.15	13.44	3.87	1.05	9.74
Localized orbital	0	4.71	0	0	5.75	0	—
Experiment	1.02	4.35	0.03	6.0	5.30	0.17	4.44

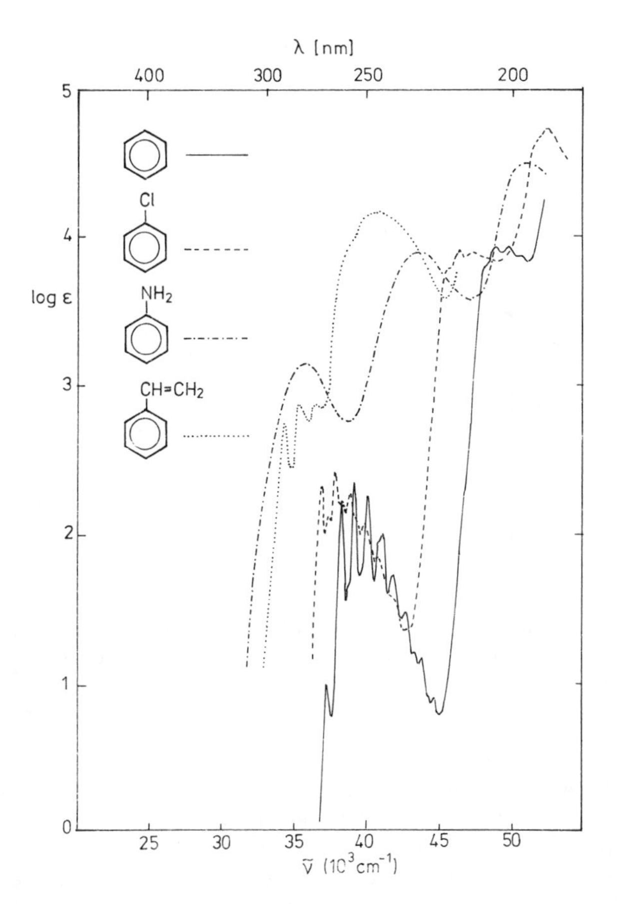

Figure 2.34. Comparison of the absorption spectra of various monosubstituted benzenes (by permission from DMS-UV-Atlas, 1966–71).

method based on this model is also known as the MIM (molecules-in-molecules) method of Longuet-Higgins and Murrell (1955). Using energies of locally excited states of fragments and of charge-transfer states, as well as the interactions between these states, the spectra of complex molecules can be calculated by means of the MIM method as shown for aniline composed of the fragments benzene and ammonia.

In Figure 2.34 the spectra of some monosubstituted benzenes are shown. From these spectra it is evident that there is a continuous change from a weakly perturbed benzene spectrum to a strongly perturbed one, such as that of aniline.

Example 2.11:

It is possible to draw Kekulé structures as well as polar quinoid resonance structures for isomeric disubstituted benzenes with a donor and an acceptor

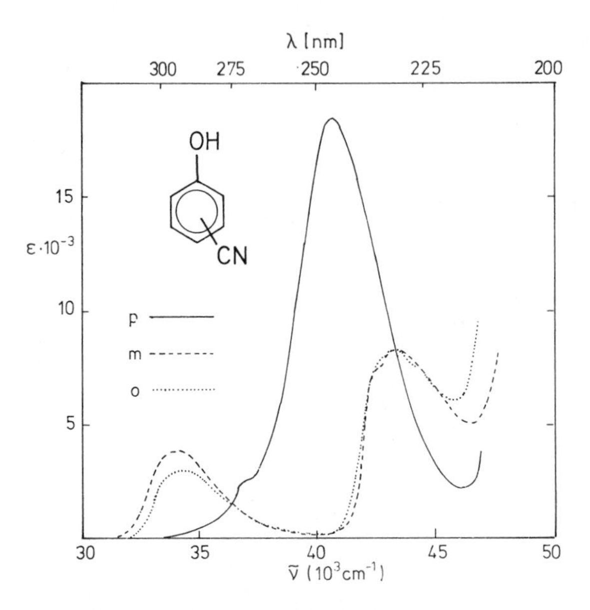

Figure 2.35. Comparison of the absorption spectra of isomeric benzenes with one donor and one acceptor substituent (by permission from Grinter and Heilbronner, 1962).

group in the *o*- or *p*-position to each other, whereas for the *m*-derivative only Kekulé structures are possible. Thus, one could expect that the spectra of the *o*- and *p*-derivatives would be very similar, and that the *m*-derivative should absorb at shorter wavelengths. From Figure 2.35 it is seen, however, that this expectation is not confirmed by experiment. Quantum chemical calculations also give a higher transition energy for the *p*-derivative and similar energies for the *o*- and *m*-derivatives, in agreement with experimental observation (Grinter and Heilbronner, 1962). This example emphasizes that caution is necessary in discussing excited states on the basis of just a few resonance structures.

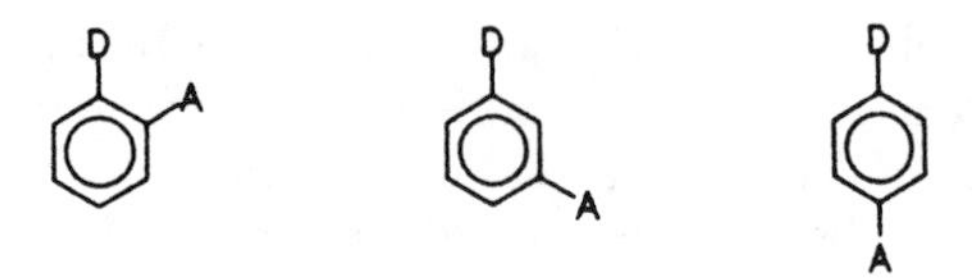

2.5 Molecules with n→π* Transitions

Many heteroatoms carry lone pairs of electrons. Sometimes these are of π symmetry and form a part of the conjugated system, as in aniline (Kasha's l electrons), but in other cases their symmetry is σ (Kasha's n electrons). In addition to $\pi \rightarrow \pi^*$ transitions, molecules with the latter type of heteroatoms may therefore exhibit also n→π^* transitions from a doubly occupied n orbital into a π^* orbital. The most important classes of compounds with n→π^* transitions are carbonyl compounds, azo compounds, and nitrogen heterocycles.

2.5.1 Carbonyl Compounds

The simplest example of a carbonyl compound is formaldehyde. Like the spectrum of ethylene, its spectrum is not yet completely understood, in spite of many elaborate investigations. Only one of the transitions that are to be expected from the orbital energy level diagram shown in Figure 2.36 has been observed in the accessible region of the spectrum, namely, the transition from the ground state into the $^1A_2(n, \pi^*)$ state. The $^3A_2(n, \pi^*)$ and

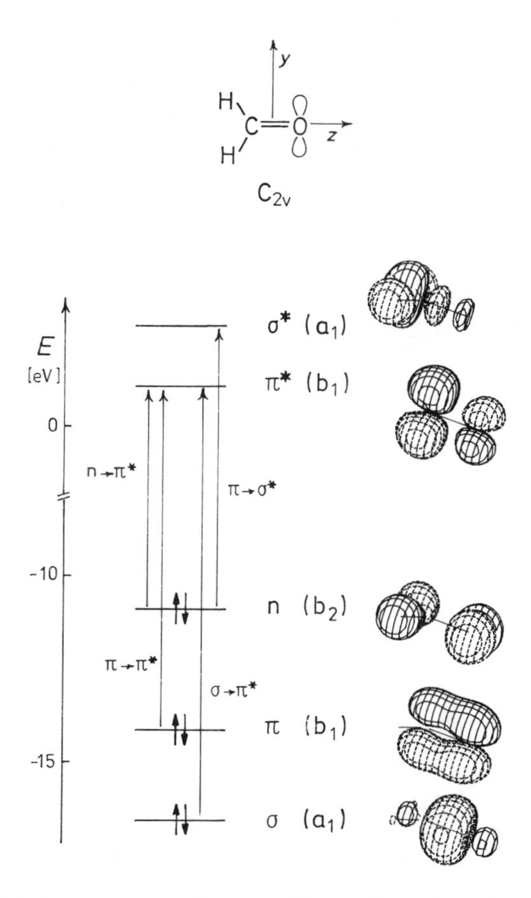

Figure 2.36. Orbital energy level diagram of formaldehyde and qualitative contour diagrams of some MOs.

$^3A_1(\pi, \pi^*)$ states can be detected under special conditions. According to calculations of Colle et al. (1978) the $^1A_1(\pi, \pi^*)$ state is dissociative, and this may be the reason why it has not yet been observed.

The $n \to \pi^*$ transition in the 230–253-nm region has been studied in detail. From vibrational analysis it follows that the CO bond is lengthened from 120.3 pm in the ground state to 132 pm in the $^1A_2(n, \pi^*)$ state, while the CH_2 group is bent out of the molecular plane by 25–30°. (Cf. Table 1.4, Section 1.4.1.) The barrier to inversion is small and gives rise to a prominent inversion doubling with $\tilde{\nu} = 356\ cm^{-1}$ (Moule and Walsh, 1975).

The extinction coefficient, $\varepsilon = 20$, is very small. The low intensity is typical for most $n \to \pi^*$ transitions of carbonyl compounds and can be explained on the basis of the local symmetry. In the case of formaldehyde, which has C_{2v} symmetry, an electronic transition from the n orbital (b_2) into the π^* MO (b_1) is dipole forbidden. This is no longer true for carbonyl compounds of lower symmetry such as acetaldehyde (C_s). Here the n orbital is still essentially of p character, so that the overlap density is approximately $p_x^* p_y$ and still has practically no dipole moment.

An intensity enhancement has been observed for β,γ-unsaturated carbonyl compounds such as **32**, which is due to π–π interactions in the nonplanar system of reduced symmetry (Labhart and Wagniere, 1959). In the series of polyene aldehydes, λ_{max} of the $\pi \to \pi^*$ transition increases with in-

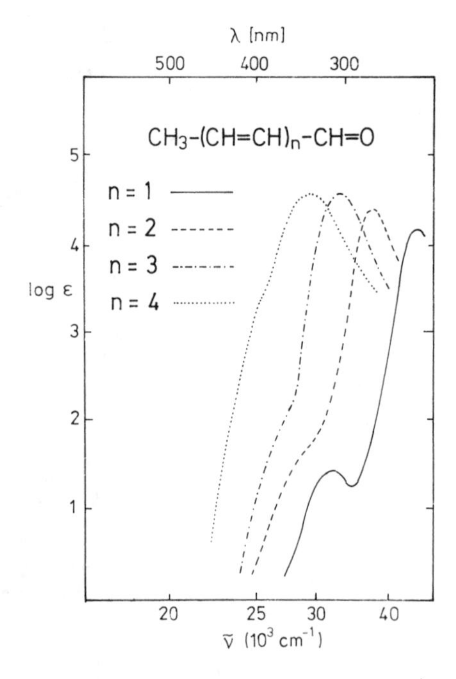

Figure 2.37. Absorption spectra of polyene aldehydes (by permission from Blout and Fields, 1948).

creasing number of double bonds faster than that of the n→π* transition (see Figure 2.37), so that the n→π* band becomes swamped by the more intense π→π* band if the conjugated chain gets long enough. (Cf. Das and Becker, 1978.)

32

Example 2.12:
In glyoxal and other dicarbonyl compounds the lone-pair orbitals n_1 and n_2 at the two carbonyl groups can interact. This results in orbital splitting that can be observed in PE spectra (Cowan et al., 1971). Either the combination $n_a = (n_1 - n_2)$ or $n_s = (n_1 + n_2)$ is higher in energy, depending on whether through-space or through-bond interactions dominate. The data collected in Table 2.9 show the close relationship between the energies of the orbitals n_s and n_a, which are given by Koopmans' theorem as $\varepsilon_i = -IP_i$, and the wave number of the longest-wavelength transition in the absorption spectrum.

Azo compounds also have two lone pairs of electrons that can interact through space or through bond. Azomethane **33** (R = CH_3) has been studied in detail. In the cis form **33a** the n_s orbital is below the n_a orbital and the $n_a \rightarrow \pi^*$ transition at 353 nm is allowed, $\varepsilon = 240$. In contrast, the energy order of the n orbitals is reversed in the trans compound **33b**, as is to be expected from the negative overlap of the sp^2-hybridized n orbitals. The $n_s \rightarrow \pi^*$ transition at 343 nm is therefore symmetry forbidden, as is seen from the value $\varepsilon = 25$ (Haselbach and Heilbronner, 1970). For azobenzene **33** (R = Ph) the corresponding

Table 2.9 PE and UV/VIS Data of Dicarbonyl Compounds

Compound	$IP(n_s)$	$IP(n_a)$	λ_{max}	$\tilde{\nu}_{max}$
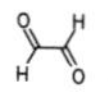 (glyoxal)	10.59	12.19[a]	450[b]	22,500
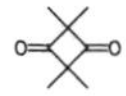 (tetramethylcyclobutanedione)	8.80	9.53[c]		
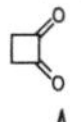 (cyclobutanedione)	9.6	11.7[d]	473	21,100
 (bicyclic dione)	8.85	10.65[e]	526[e]	19,000
 (bicyclic ene-dione)	8.65	9.75[e]	546[e]	18,300
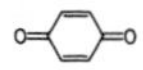 (p-benzoquinone)	10.41	10.11[f]	499[g]	20,000

[a] Turner et al., 1970.
[b] Arnett et al., 1974.
[c] Cowan et al., 1971.
[d] Gleiter et al., 1985.
[e] Martin et al., 1978.
[f] Lauer et al., 1975.
[g] Martin and Wadt, 1982.

data are $\lambda = 430$ nm ($\varepsilon = 1{,}500$) for the cis compound and $\lambda = 440$ nm ($\varepsilon = 500$) for the trans compound (Griffiths, 1972).

R–N=N–R (a) 33 (b)

2.5.2 Nitrogen Heterocycles

Whereas for the aza analogues of benzene and naphthalene the n→π* transitions can be observed as separate bands or as shoulders on the π→π* transitions, they are usually hidden by π→π* transitions in higher aromatics. They are about ten times as intense as the n→π* bands of carbonyl compounds since the n orbital is sp^2 hybridized, and the intraatomic transition moment, which depends on the s character of the nonbonding orbital, no longer vanishes.

The n→π* band is shifted to longer wavelengths as a larger number of nitrogen atoms are introduced into the molecule (Table 2.10). This is due to two factors: the π* orbital is lowered in energy by the nitrogen replacement, and the n orbitals can interact through space and through bonds. From first-order perturbation theory, the stabilization of the π* MO is seen to be proportional to the sum of the squares $\Sigma c_{\mu i}^2$ of the LCAO coefficients of all nitrogen positions μ. If allowance is made for an additional bathochromic shift

Table 2.10 n→π* Absorptions of Some Aza Derivatives of Benzene and Sum of the Squared HMO Coefficients of the HOMO at Positions μ of the Nitrogens; as Most n→π* Absorptions Show Up in the Spectra Only as Shoulders of the π→π* Bands, Their Location has been Characterized by the Wave Number $\nu(\varepsilon = 20)$ for Which the Extinction Coefficient is $\varepsilon = 20$ (Adapted from Mason, 1962)

Compound	Position μ	$\tilde{\nu}(\varepsilon = 20)$ (cm^{-1})	$\sum_{\mu} c_{\mu HOMO}^2$
Pyridine (**34**)	1	34,500	0.333
Pyridazine (**35**)	2, 3	26,600	0.500
Pyrimidine (**36**)	2, 6	30,600	0.500
Pyrazine (**37**)	1, 4	29,850	0.667
sym-Triazine (**38**)	1, 3, 5	31,500	0.500
sym-Tetrazine (**39**)	2, 3, 5, 6	17,100	1.000

of approximately 6,000 cm^{-1} for two nitrogen atoms adjacent to one another, allowing their n orbitals can interact strongly as described in Example 2.12, a good proportionality between theoretical expectations and experimentally observed bathochromic shifts is obtained.

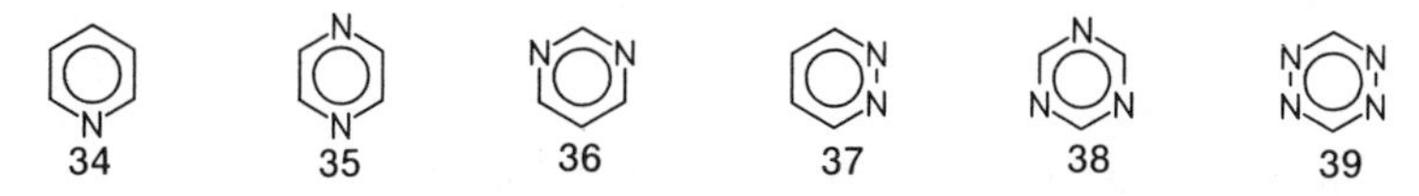

The extent of the n orbital interactions can again be determined experimentally using PE spectroscopy. It has been shown that in most cases the IPs corresponding to the n orbitals are higher than those of the highest occupied π MO (Brogli et al., 1972). The fact that the n→π^* transitions are nevertheless observed at longer wavelengths than the $\pi \rightarrow \pi^*$ transition may be due to the exchange integrals that co-determine the singlet–triplet separation. Like the transition intensity, these exchange integrals depend on the overlap density $\phi_n\phi_{\pi^*}$, and $K_{n\pi^*} < K_{\pi\pi^*}$.

2.6 Systems with CT Transitions

Solutions of a mixture of an electron donor D, such as hexamethylbenzene (**40**), and an electron acceptor A, such as chloranil (**41**), often exhibit an absorption band that is not present in solutions of the pure components. This band has been assigned to an electron transfer from the donor to the acceptor and is therefore referred to as a donor-acceptor or a CT (charge transfer) transition. Mulliken (1952) has developed a theory of CT transitions which can be summarized as follows: If the ground configuration and the polar configuration produced by electron transfer are described by wave functions Φ_{DA} and $\Phi_{D^{\oplus}A^{\ominus}}$, respectively, the interaction of these two configurations yields a weakly stabilized ground state

$$\Psi_0 = \Phi_{DA} + \lambda\Phi_{D^{\oplus}A^{\ominus}} \tag{2.28}$$

and an excited state

$$\Psi_1 = \Phi_{D^{\oplus}A^{\ominus}} + \mu\Phi_{DA} \tag{2.29}$$

where λ and μ are small compared with 1. The CT absorption is then associated with a transition from the ground state Ψ_0 into the excited state Ψ_1.

40

41

Under the assumption that the ZDO approximation is valid for an MO description of the components, the excitation energy of the CT band is easy to estimate. For the transition of an electron from the HOMO of the donor into the LUMO of the acceptor

$$\Delta E = E(\Psi_1) - E(\Psi_0) = \varepsilon^{A}_{LUMO} - \varepsilon^{D}_{HOMO} - C \qquad (2.30)$$
$$= IP_D - EA_A - J + 2K$$

where IP_D is the ionization potential of the donor, EA_A the electron affinity of the acceptor and $C \approx J$ the electrostatic interaction between the ions formed by electron transfer. The exchange term K is nearly equal to zero due to the small value of the overlap density.

Experimentally, CT transitions may be identified by their broad band shape devoid of vibrational fine structure, and by the sensitivity of their wavelength to the polarity of the solvent, which is to be expected from the highly polar nature of the excited state. (See Section 2.7.) The band shape may be explained by the fact that the binding energies are usually rather small and the minima on the ground-state potential energy surface correspondingly shallow, so in solution many different configurations of the complex exist in equilibrium with one another. Also, the equilibrium intermolecular separations in the excited state undoubtedly are smaller than in the ground state, and the vibrational frequency for this motion is very low.

Example 2.13:
Paracyclophanes provide the opportunity to fix the donor and acceptor molecules in a definite geometric structure in order to study the dependence of

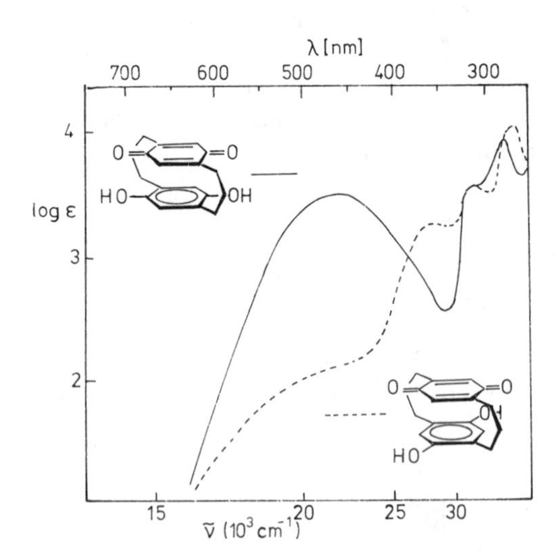

Figure 2.38. Absorption spectra of isomeric [3.3]paracyclophane-quinhydrones in dioxane (by permission from Staab et al., 1983).

donor–acceptor interactions on geometric factors. This concept has been pursued intensely by Staab et al. (1983). Figure 2.38 shows the spectra of the isomeric quinhydrones of the [3.3]paracyclophane series, in which the transannular distance of 310–330 pm is comparable to intermolecular donor–acceptor distances. For the *pseudo-geminal* compound **42** the overlap density is large since the location of nodal planes is the same for the hydroquinone HOMO and the quinone LUMO, which are even identical to a first approximation. This compound exhibits a strong CT band near 500 nm. For the *pseudo-ortho* derivative **43**, however, the overlap density is small due to dissimilar orientation of the nodal planes and the large distance of the oxygen atoms, where the LCAO coefficients of this MO are largest. Only a low-intensity shoulder for **43** is observed in this region, but an additional strong band appears near 360 nm, and has been assigned to a second CT transition.

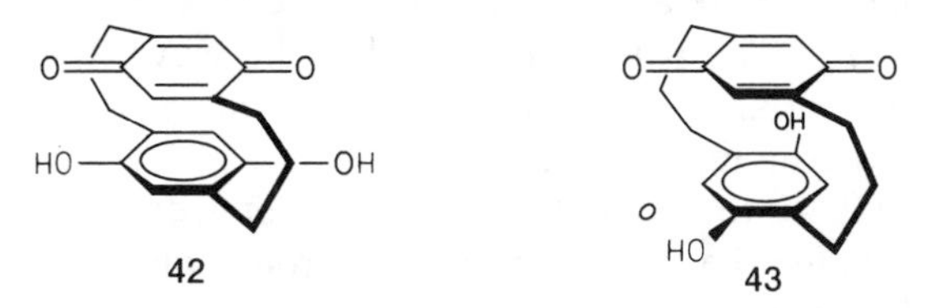

For complexes combining different donors with the same acceptor a linear relation between the ionization potential IP_D of the donor and the wave number of the CT band has been observed. (See Figure 2.39.) This correlation is much better than expected; it requires that the Coulomb term C is either constant or proportional to the donor ionization potential IP_D.

In many series the intensity of the CT band increases as the complex becomes more stable. This can be explained by using the wave functions

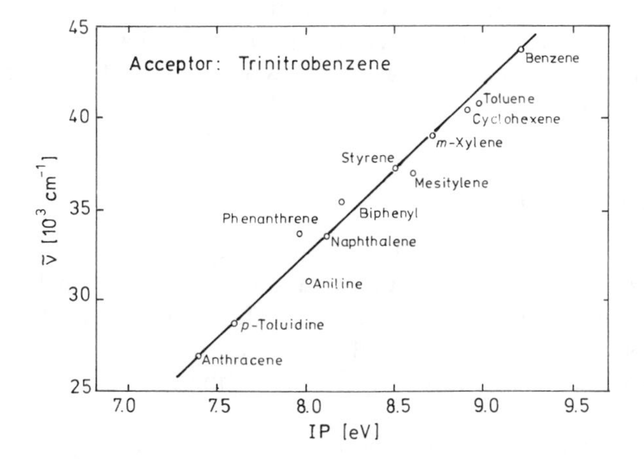

Figure 2.39. Relation between the wave number of the CT band and the ionization potential IP_D of the donor for DA complexes with trinitrobenzene (adapted from Briegleib and Czekalla, 1960; IPs from more recent PE data).

given in Equation (2.28) and Equation (2.29) to calculate the transition moment which then becomes

$$\begin{aligned}<\Psi_0|\hat{M}|\Psi_1> = {} & (1 + \lambda\mu)<\Phi_{DA}|\hat{M}|\Phi_{D^{\oplus}A^{\ominus}}> \\ & + \lambda<\Phi_{D^{\oplus}A^{\ominus}}|\hat{M}|\Phi_{D^{\oplus}A^{\ominus}}> \\ & + \mu<\Phi_{DA}|\hat{M}|\Phi_{DA}>\end{aligned} \tag{2.31}$$

Since the first integral vanishes if use is made of the ZDO approximation, the intensity is seen to be essentially due to the dipole moment of the polar CT configuration, and the transition moment is proportional to the coefficient λ. But in reality the situation is much more complicated, as can be seen from the fact that the intensity can be markedly different from zero even in cases where the ground-state stabilization is very weak and λ is therefore approximately zero. The reason may be that the wave functions Equation (2.28) and Equation (2.29) neglect locally excited states (Dewar and Thompson, 1966). Furthermore, it might be important, at least in cases of weak interaction, to take into account the overlap between donor and acceptor orbitals in calculating the transition moment from Equation (2.31).

If a molecule consists of several weakly coupled chromophores it may be advantageous to speak of intramolecular CT transitions. The MIM method for calculating the absorption spectra of complex molecules, which has been mentioned in the discussion of substituent effects in Section 2.4.2, is based on this idea.

2.7 Steric Effects and Solvent Effects

So far, the discussion has been confined to isolated planar π systems. However, intramolecular and intermolecular interactions, such as steric effects and solvent effects, may influence the spectra considerably. In this section, some examples are discussed as a review of some of the more important aspects of such effects.

2.7.1 Steric Effects

Steric interactions between bulky groups may cause deformations of the geometric structure of a molecule. Of these, twisting of single bonds is by far the most important structural change. This is so because torsion around a single bond by 5° requires less than 0.2 kcal/mol of energy, compared to the 4–8 kcal/mol that is necessary for stretching a bond by 1 pm. (Cf. Rademacher, 1988.)

Thus, in general steric strain in π systems is relieved by rotation around formal single bonds. This stabilizes the ground state with respect to the strained, untwisted system. Since bonds that have a low π bond order in the

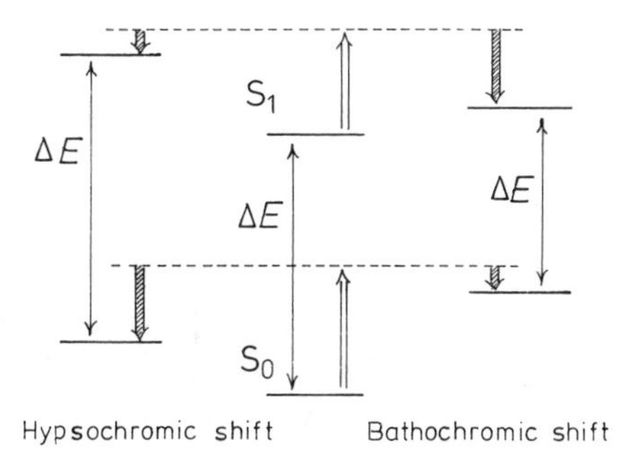

Figure 2.40. Hypsochromic and bathochromic shift due to steric hindrance; the white arrows indicate the steric strain and the black arrows the strain release by molecular deformations (twisting of bonds).

ground state often have a high π bond order in the excited state, that is, a high double-bond character, the twisting is energetically less favorable in the excited state. Steric hindrance is therefore most often expected to result in a hypsochromic shift of the absorption band. (See Figure 2.40.)

However, there are also molecules in which the steric strain can be relieved only by twisting formal double bonds. In this case then, the destabilization of the ground state is larger than that of the excited state, where the double-bond character of the bond to be twisted usually is weaker. This effect, which can be used to explain the longer-wavelength absorption of N,N-dimethylindigo (**44,** λ_{max} = 670 nm) compared to indigo (λ_{max} = 610 nm) (Klessinger, 1978), is sometimes referred to as the Brunings-Corwin effect.

O CH3 N N CH3 O **44** **45** **46**

The effect of bond twisting on the orbital energy ε_i may be estimated by first-order perturbation theory from the relation

$$\delta\varepsilon_i = 2\sum_{\mu}\sum_{\nu} c_{\mu i} c_{\nu i} \delta\beta_{\mu\nu} \qquad (2.32)$$

by allowing $\beta_{\mu\nu}$ of the twisted bond to decrease in absolute magnitude or even to vanish (90° twist). ε_i becomes more positive if $c_{\mu i}$ and $c_{\nu i}$ are of equal sign, and more negative if they are of opposite sign. For alternant hydrocarbons one of the paired MOs is always stabilized and the other one destabilized, and a hypsochromic shift of the HOMO→LUMO transition results if the HOMO is antibonding in the twisted bond. If it is bonding in this bond, a bathochromic shift is produced. Both possibilities are illustrated in Figure

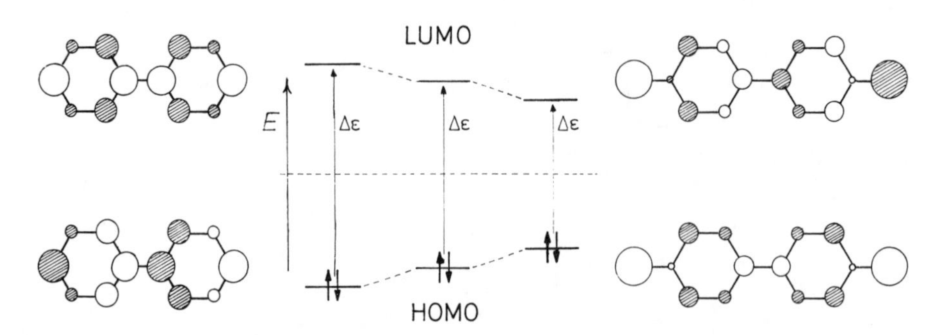

Figure 2.41. Effect of bond twisting on the HOMO and LUMO energies of biphenyl (twisting of a formal single bond) and biphenylquinodimethane (twisting of a formal double bond).

2.41, where a schematic representation of the HOMO and LUMO of biphenyl (**45**) and of biphenylquinodimethane (**46**) is given together with the change in orbital energies due to twisting of the central bond.

The hypsochromic and bathochromic shifts that are to be expected from these arguments are in very good agreement with experimentally observed changes in the spectra. This is demonstrated by the spectra of biphenyl and

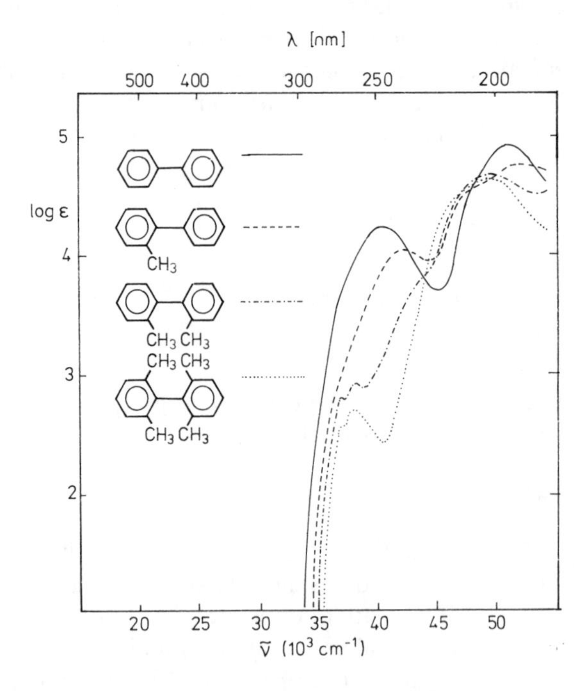

Figure 2.42. The effect of steric hindrance by methyl groups on the absorption spectrum of biphenyl (by permission from DMS-UV-Atlas 1966–71).

some of the methyl derivatives substituted in positions 2, 6, 2′, and 6′ shown in Figure 2.42. The 250 nm band of biphenyl provides an example of a hypsochromic shift with increasing steric hindrance due to a twisting of the central bond for which the HOMO is antibonding.

Example 2.14:

The symmetric cyanine dyes are isoelectronic with anions of odd alternant hydrocarbons. Because of bond equalization in these systems it is a reasonable assumption that steric hindrance will cause all bonds to rotate by roughly the same amount. To a first approximation, the energy of the HOMO will not be affected by such a twist as it is nonbonding in all bonds. Since the LUMO is predominantly antibonding, a uniform rotation of all bonds will lower its energy. Thus, steric hindrance in symmetric cyanine dyes is expected to cause a bathochromic shift, as has in fact been observed for the dyes **47** and **48** (Griffiths, 1976).

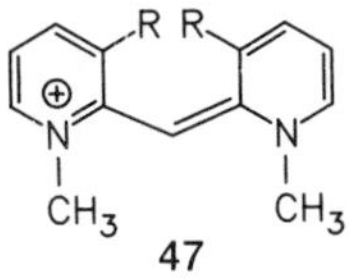

47

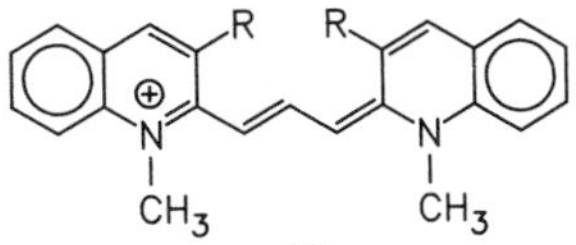

R = H λ_{max} = 520 nm
R = CH_3 λ_{max} = 527 nm

48

R = H λ_{max} = 604 nm
R = CH_3 λ_{max} = 640 nm

Merocyanine dyes and unsymmetrical cyanine dyes, however, have alternating bonds and exhibit in general a hypsochromic shift on steric hindrance; **49** is an interesting dye that, depending on the solvent, shows for R = CH_3 either a bathochromic or a hypsochromic shift with respect to the parent compound with R = H. This effect will be explained in the following section. (Kiprianov and Mikhailenko, 1991.)

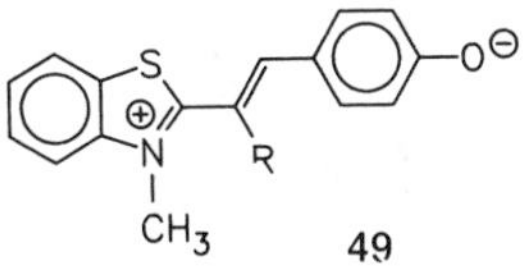

49

2.7.2 Solvent Effects

Light absorption by organic molecules may be influenced by the solvent through either specific or nonspecific interactions between the solute and solvent molecules. In trying to account for the effect of the solvent on the spectrum of a given molecule, one has to evaluate the change in the solute–solvent interaction upon excitation. This change may involve many factors such as dispersion, polarization, and electrostatic forces, and charge-transfer interactions. Dispersion forces are the weak London-van der Waals forces that arise between instantaneous dipoles on the two molecules, solute and solvent. The polarization forces result from Coulombic interactions between a permanent moment on a polar molecule and the induced moments in the partner, whether polar or nonpolar. Finally, electrostatic forces are classic Coulomb forces between permanent moments (usually dipoles) on polar molecules. Figure 2.43 illustrates some of these factors: the dipole moment $\boldsymbol{\mu}$ of the isolated molecule gives rise to a dipole field (Figure 2.43a) that tends to orient the solvent molecules with their permanent and/or induced dipole moments parallel to the field lines (Figure 2.43b). If the electronic and nuclear configurations of all solvent molecules were frozen and the solute molecule were removed, a cavity surrounded by polarized molecules would result. The dipole moments of these solvent molecules generate a field in the cavity that has the opposite direction from the dipole moment of the original solute (Figure 2.43c) and is called the reaction field.

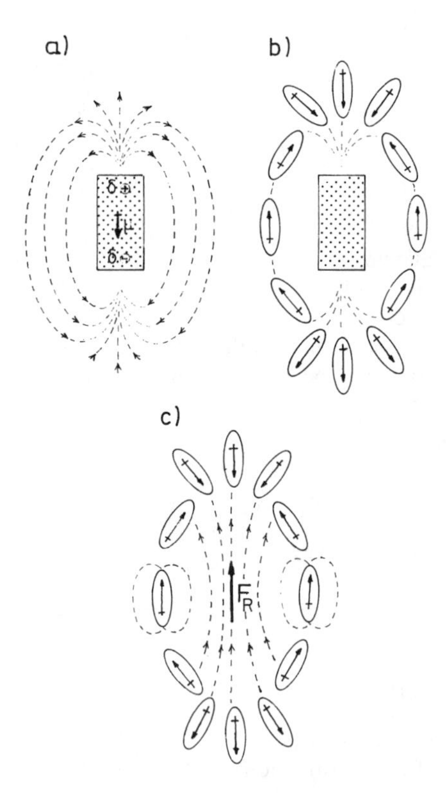

Figure 2.43. Reaction field of a polar solute. a) Dipole field of the isolated molecule, b) orientation of polar solvent molecules parallel to the dipole field, and c) reaction field from the solvent molecules in the cavity if the solute molecule has been removed after freezing all electronic and nuclear coordinates (adapted from Liptay, 1969).

A number of solvent parameters such as the Z values (Kosower, 1958) or E_T values (Reichardt and Dimroth, 1968; Reichardt and Harbusch-Görnert, 1983; Buncel and Rajagopal, 1990) have been proposed. These characterize solvent polarity inclusive of all kinds of solvent–solute interactions.

Three general approaches can be used to evaluate the effects of solvent on absorption spectra: Continuum theories describe the solute as lying in a cavity in contact with a polarizable continuum. In semicontinuum theories the first few shells of solvent molecules around the solute are treated explicitly, and the remaining solvent molecules are treated as a continuum. Fully discrete theories treat as many solvent molecules as possible in full quantum mechanical detail, on the same footing as the solute molecule. The main problem with this approach is the enormous number of degrees of freedom associated with the multidimensional solute–solvent system. Instead of a fully quantum mechanical treatment, it is imperative to use an approximation that separates the system into classical and quantum mechanical parts.

In the classical reaction-field model, the solute molecule is considered embedded in a cavity inside a homogeneous dielectric medium. From the Onsager theory (Onsager, 1936) the "electronic" reaction field at the center of the solute molecule is given by

$$\boldsymbol{F}_{\mathrm{Re}} = -\frac{2}{a^3}\left(\frac{n^2 - 1}{2n^2 + 1}\right)\boldsymbol{\mu}_{\mathrm{s}}$$

where n is the refractive index of the nonpolar medium, and a is the radius of the cavity. In a polar solvent with a macroscopic dielectric constant ε, the reaction field is given by

$$\boldsymbol{F}_{\mathrm{R}} = -\frac{2}{a^3}\left(\frac{\varepsilon - 1}{2\varepsilon + 1}\right)\boldsymbol{\mu}_{\mathrm{s}}$$

and the "orientation" field may be defined as the difference

$$\boldsymbol{F}_{\mathrm{Ro}} = \boldsymbol{F}_{\mathrm{R}} - \boldsymbol{F}_{\mathrm{Re}} = -\frac{2}{a^3}\left(\frac{\varepsilon - 1}{2\varepsilon + 1} - \frac{n^2 - 1}{2n^2 + 1}\right)\boldsymbol{\mu}_{\mathrm{s}}$$

In calculating solvent effects on excitation energies one has to take into account that normally the excitation process is sufficiently rapid for the Franck-Condon principle to hold, which means that while the induced dipole of the solvent molecule can change in response to the change of solute dipole on excitation, the permanent dipoles cannot—that is, there is negligible dipole reorientation. Such reorientation can, however, occur in the time period between absorption and emission and lead to a large Stokes shift in the emission. (Cf. Section 5.3.1.)

Self-consistent reaction-field (SCRF) theories are obtained by introducing solute–solvent interactions into the Hamiltonian. (Cf. Tapia, 1982.) Based on CI wave functions and on reasonable approximation for the cavity radius

a the following useful equation for the solvent shift $\Delta\tilde{\nu}_j$ of the j-th transition was obtained (Karelson and Zerner, 1992):

$$\Delta\tilde{\nu}_j = 22{,}679\,\frac{d}{MM}\left[\frac{\varepsilon - 1}{2\varepsilon + 1}(\boldsymbol{\mu}_0 \cdot \boldsymbol{\mu}_0 - \boldsymbol{\mu}_0 \cdot \boldsymbol{\mu}_j) + \frac{n^2 - 1}{2n^2 + 1}(\boldsymbol{\mu}_0 \cdot \boldsymbol{\mu}_j - \boldsymbol{\mu}_j \cdot \boldsymbol{\mu}_j)\right] \; [\text{cm}^{-1}]$$

MM is the molar mass in amu, d is the mass density of solute in g/cm^3, and $\boldsymbol{\mu}_0$ and $\boldsymbol{\mu}_j$ are the dipole moments of the ground state and the excited state, respectively.

The SCRF model reproduces very well the shifts observed in aprotic solvents. In order to simulate an aqueous solution, calculations on supermolecules containing the chromophore and specifically bound water molecules, all embedded in the dielectric continuum, are required even for merely qualitative agreement. Supermolecule calculations with several water molecules treated as a gas-phase "cluster" often give qualitatively incorrect results (Karelson and Zerner, 1992).

In microscopic approaches the solvent molecules are described as true discrete entities but in some simplified form, generally based on force-field methods (Allinger, 1977). These theories may be of the semicontinuum type if the distant bulk solvent is accounted for, or of the fully discrete type if the solvent description includes a large number of molecules. As an example, the spectrum of formaldehyde in water has been examined using a combination of classical molecular dynamics and ab initio quantum chemical methods and sampling the calculated spectrum at different classical conformations (Blair et al., 1989; Levy et al., 1990). These calculations predict most of the solvent shift as well as the line broadening.

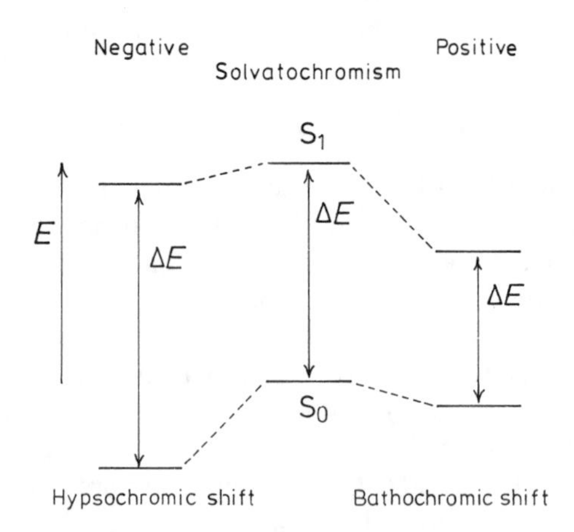

Figure 2.44. Bathochromic and hypsochromic shifts due to differential stabilization of ground and excited states by the solvent.

Other recent microscopic approaches are based on the Langevin dipoles solvent model or on the all-atom solvent model, using a standard force field with van der Waals and electrostatic terms as well as intramolecular terms, and molecular dynamics simulations of the fluctuation of the solvent and the solute, incorporating the potential from the permanent and induced solvent dipoles in the solute Hamiltonian in a self-consistent way (Luzhkov and Warshel, 1991).

The solute–solvent interaction increases with increasing solvent polarity; depending on whether the ground state or the excited state is more strongly stabilized, an increasing bathochromic or hypsochromic shift results. This is referred to as *positive* or *negative solvatochromism*. (See Figure 2.44.) Since the excited state is frequently more polar than the ground state, positive solvatochromism is commonly observed for $\pi \rightarrow \pi^*$ transitions, for example, for the azo dye **50**.

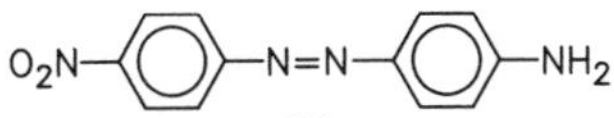

$O_2N-C_6H_4-N=N-C_6H_4-NH_2$

50

Cyclohexane λ_{max} = 470 nm
Ethanol λ_{max} = 510 nm

The situation is quite different for $n \rightarrow \pi^*$ transitions. The lone electron pair is particularly well stabilized by polar and particularly by protic solvents so it becomes energetically more difficult to excite. Figure 2.45 shows the spectrum of N-nitrosodimethylamine in different solvents. Results of calculations indicate that the negative solvatochromism of carbonyl compounds can be explained on the basis of the structural changes due to the formation of hydrogen bonds (Taylor, 1982). Molecular dynamics simulations, however, indicate that the net blue shift is primarily due to electrostatic interactions (Blair et al., 1989). A large number of water molecules around the entire formaldehyde are responsible for the total blue shift; the first solvation shell only accounts for one-third of the full shift.

In the case of $n \rightarrow \pi^*$ transitions in N-heterocyclic compounds the large blue shifts that occur upon solvation can easily be explained as due to the large change of dipole moment associated with the excitation of an electron from a lone-pair orbital to the delocalized π^* orbital (Karelson and Zerner, 1990). This contrasts directly with the idea that the lone-pair orbital is lowered in energy (more than the π^* orbital) due to direct bonding between this orbital and the solvent, or through inductive shifts. In the same study it was found that hydrogen bonding with water, for example, as often as not led to red shifts; the blue shift was reproduced when the entire supermolecule was embedded in a dielectric.

Transition moments and intensities of electronic transitions may also be solvent dependent due to solvent effects on the wave functions of the various states. Large effects are to be expected especially for weak transitions

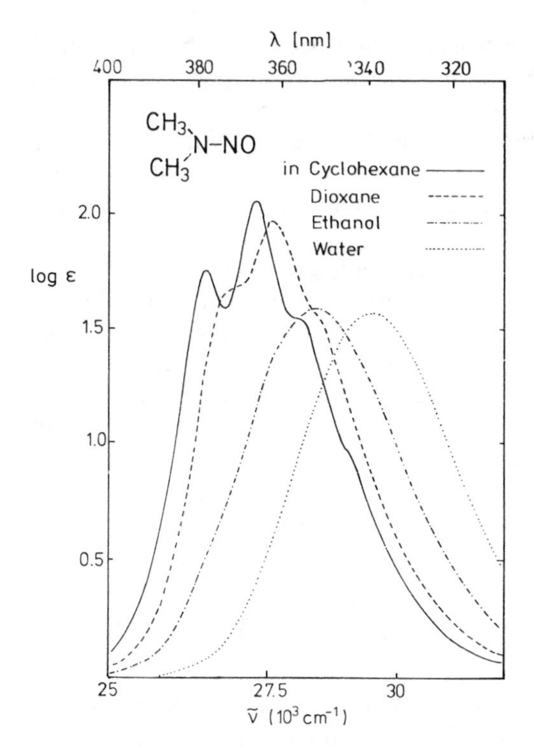

Figure 2.45. Absorption spectrum of N-nitrosodimethylaniline in different solvents (by permission from Jaffé and Orchin, 1962).

adjacent to very intense ones, where small perturbations can produce appreciable changes.

Symmetry-lowering effects of the solvent are referred to as the Ham effect (Ham, 1953; Platt, 1962). Thus, in rigid or fluid solutions, symmetry-forbidden vibrational components of the $^1B_{2u}$ (^1L_b) absorption band of benzene appear with increasing intensity as the polarizability and the polarity of the solvent increase. The fact that for pyrene and 2-methylpyrene the direction of the ^1L_b transition moment is inclined on the average 40° and 20° away from the *y* axis, respectively, has also been ascribed to a symmetry-lowering perturbation by the environment, related to the Ham effect (Langkilde et al., 1983). The associated intensification of otherwise weak vibronic peaks in the L_b band can be used for an investigation of microenvironments such as micelles.

Example 2.15:

Many colored molecules can be described as donor–acceptor chromophores. In such cases, the longest-wavelength absorption depends on the extension of the conjugated system and on the amount of bond length equalization which is related to the donor and acceptor strengths. (Cf. the absorptions of polyenes and cyanines in Figure 2.5.) A simple model to account for solvent effects is based on the notion that direct solvent interactions may influence the donor and acceptor properties and hence affect the absorption through changes in the degree of bond length equalization. Thus the positive solvatochromism of the merocyanine dye **51** can be explained by the assumption that the acceptor strength of the carbonyl group increases through solvation of the $CO^{\ominus}$ group, which makes the π system more similar to that of a symmetric cyanine; that is, bond lengths become more alike.

$$R_2N{-}(CH{=}CH)_{n-1}{-}CH{=}O \underset{\mathbf{51}}{\longleftrightarrow} R_2{}^{\oplus}N{=}CH{-}(CH{=}CH)_{n-1}{-}O^{\ominus}$$

The solvent-dependent steric effect of dye **49** mentioned in Example 2.14 may be understood if the polar solvent stabilizes the polar form (**49b**) to such an extent that the position of the double bonds is reversed compared to the form **49a** that exists in nonpolar solvents.

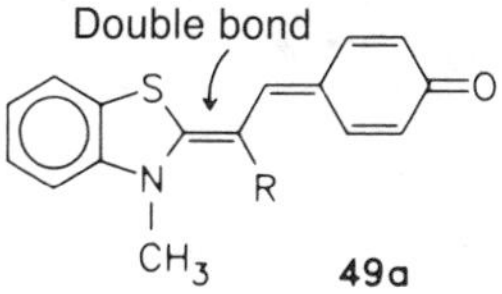

Double bond

49a

in CH_3Cl

Single bond

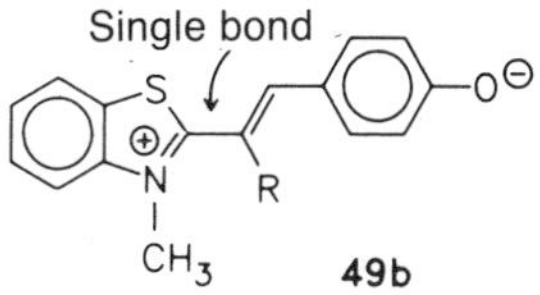

49b

in H_2O

Of particular interest are the so-called *Brooker dyes,* which are also merocyanines of the general formula **49**, but without the steric hindrance by the group R (Brooker et al., 1951). A change from positive to negative solvatochromism is observed for these compounds as the water content of the solvent is increased. This has been interpreted by the assumption that in nonpolar solvents the excited state of the solute is more polar than the ground state. With increasing water content of the solvent, however, the ground state becomes increasingly more polar due to stabilizing interactions. Finally, when the ground state is more polar than the excited state a negative solvatochromism results. However, it could be shown (Liptay, 1966) that the maximum at longer wavelengths is due to a new species F···HS which is formed in an equilibrium reaction by direct interaction of the solvent (HS) with the dye (F), and which disappears again in an additional equilibrium if the water content increases:

$$F + HS \rightleftharpoons F\cdots H{-}S \rightleftharpoons FH^{\oplus} + S^{\ominus}$$

This example demonstrates that one has to be careful in interpreting solvent effects if possibly present special interactions are to be assessed correctly.

Supplemental Reading

UV Spectra of Organic Molecules

Fabian, J., Hartmann, H. (1980), *Light Absorption of Organic Colorants*; Springer-Verlag: Berlin, Heidelberg, New York.

Griffiths, J. (1976), *Colour and Constitution of Organic Molecules*; Academic Press: London, New York.

Jaffé, H.H., Orchin, M. (1962), *Theory and Applications of Ultraviolet Spectroscopy*; Wiley: New York, London.

Fabian, J., Nakazumi, H., Matsuoka, M. (1992), "Near-Infrared Absorbing Dyes," *Chem. Rev.* **92**, 1197.

Hudson, B.S., Kohler, B.E., Schulten, K. (1982), "Linear Polyene Electronic Structure and Potential Surfaces," in *Excited States* **6**; Lim, E.C., Ed.; Academic Press: New York.

Sandorfy, C. (1964), *Electronic Spectra and Quantum Chemistry*; Prentice-Hall: Englewood Cliffs.

Two-Photon Spectroscopy

Friedrich, D.M. (1982), "Two-Photon Molecular Spectroscopy," *J. Chem. Educ.* **59**, 472.

Goodman, L., Rava, R.P. (1984), "Two Photon Spectra of Aromatic Molecules," *Acc. Chem. Res.* **17**, 250.

Hohlneicher, G., Dick, B. (1983), "New Experimental Information from Two-Photon Spectroscopy and Comparison with Theory," *Pure Appl. Chem.* **55**, 261.

Lin, S.H., Fujimura, Y., Neusser, H.J., Schlag, E.W. (1984), *Multiphoton Spectroscopy of Molecules*; Academic Press: New York.

McClain, W.M., Harris, R.A. (1978), "Two-Photon Molecular Spectroscopy in Liquids and Gases," in *Excited States* **3**; Lim, E.C., Ed.; Academic Press: New York.

Neusser, H.J., Schlag, E.W. (1992), "High Resolution Spectroscopy below the Doppler Width," *Angew. Chem. Int. Ed. Engl.* **31**, 263.

PMO Theory

Dewar, M.J.S., Dougherty, R.C. (1975), *The PMO Theory of Organic Chemistry*; Plenum Press: New York, London.

Polyenes

Hudson, B.S., Kohler, B.E., Schulten, K. (1982), "Linear Polyene Structure and Potential Surfaces," in *Excited States* **6**, Lim, E.C., Ed.; Academic Press: New York.

Aromatic Molecules

Clar, E. (1964), *Polycyclic Hydrocarbons*; Academic Press: London.

Michl, J. (1984), "Magnetic Circular Dichroism of Aromatic Molecules," *Tetrahedron* **40**, 3845.

Moffitt, W. (1954), "The Electronic Spectra of Cata-Condensed Hydrocarbons," *J. Chem. Phys.* **22**, 320.

Platt, J.R. (1949), "Classification of Spectra of Cata-Condensed Hydrocarbons," *J. Chem. Phys.* **17**, 484.

Radicals and Radical Ions

Zahradník, R., Čársky, P. (1973), "Physical Properties and Reactivity of Radicals," *Progr. Phys. Org. Chem.* **10**, 327.

Substituent Effects

Petruska, J. (1961), "Changes in the Electronic Transitions of Aromatic Hydrocarbons on Chemical Substitution. II. Application of Perturbation Theory to Substituted-Benzene Spectra," *J. Chem. Phys.* **34**, 1120.

Weeks, G.H., Adcock, W., Klingensmith, K.A., Waluk, J.W., West, R., Vasak, M., Downing, J., Michl, J. (1986), "A Probe for Substituent Hyperconjugative Power: MCD of the Benzene L_b Band," *Pure Appl. Chem.* **85**, 39.

CT Spectra

Foster, R. (1969), *Organic Charge-Transfer Complexes*; Academic Press: London, New York.

Murrell, J.N. (1961), "The Theory of Charge-Transfer Spectra," *Quart. Rev.* **15**, 191.

Steric Effects

Suzuki, H. (1967), *Electronic Absorption Spectra and Geometry of Organic Molecules*; Academic Press: New York.

Solvent Effects

Klopman, G. (1967), "Solvations: A Semi-Empirical Procedure for Including Solvation in Quantum Mechanical Calculations of Large Molecules," *Chem. Phys. Lett.* **1**, 200.

Liptay, W. (1969), "Electrochromism-Solvatochromism," *Angew. Chem. Int. Ed. Engl.* **8**, 177.

Reichardt, Ch. (1988), *Solvent Effects in Organic Chemistry,* 2nd Ed., VCH: Weinheim.

Tapia, O. (1982), "Quantum Theories of Solvent-effect Representation: An Overview of Methods and Results," in *Molecular Interactions* **3**, Orville-Thomas, W.J., Ed.; Wiley: New York.

CHAPTER

3

Optical Activity

3.1 Fundamentals

A compound is considered optically active if it shows the phenomena of *circular birefringence* and *circular dichroism*. These are the manifestations of different refractive indices, n_L and n_R, and different extinction coefficients, ε_L and ε_R, respectively, for left-handed and for right-handed circularly polarized light. Optical activity is therefore closely related to the existence of elliptically and circularly polarized light.

3.1.1 Circularly and Elliptically Polarized Light

For linearly polarized light the direction of the electric field vector $\boldsymbol{E}$ is constant at a given point $x = x_0$, whereas its magnitude changes periodically between $+|\boldsymbol{E}_0|$ and $-|\boldsymbol{E}_0|$. (Cf. Figure 1.2b.) For circularly polarized light, the magnitude is constant and the direction changes continuously. When viewed against the direction of propagation (in the $-x$ direction) the endpoint of the electric field vector moves on a circular path, clockwise for right-handed circularly polarized light and counterclockwise for left-handed circularly polarized light.

When viewed at a given time $t = t_0$ as a function of distance along x, the position of the endpoint of the electric vector of linearly polarized light oscillates in a plane (cf. Figure 1.2a), whereas for circularly polarized light it forms a helix of constant pitch. The helix is right-handed if the light polarization is right-handed and left-handed if the light polarization is left-handed, as shown in Figure 3.1.

The endpoint of the electric field vector of elliptically polarized light describes a distorted helix, and the change in time at a given point corresponds to an ellipse. The ratio of the minor and major axes is given by tan θ where θ is the ellipticity of the light.

According to Equation (1.5) circularly polarized light may be described by the superposition of two linearly polarized waves that are mutually orthogonal and show a phase shift of $\pm\pi/2$ (Figure 3.1). Conversely, linearly polarized light can be viewed as a superposition of right-handed and left-handed circularly polarized light of identical frequency and identical amplitude.

If linearly polarized light of a wavelength that is not being absorbed passes through an optically active sample, the polarization plane is rotated by an angle α due to the different refractive indices n_L and n_R and their effect on the velocity of light $c = c_0/n$. At wavelengths that are being absorbed by the sample, the light becomes elliptically polarized and the major axis of the

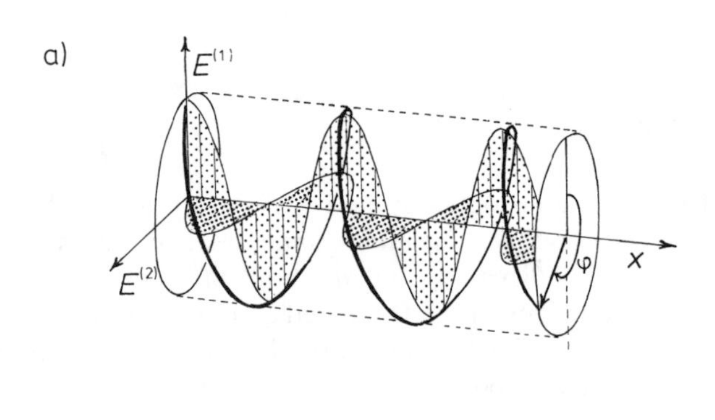

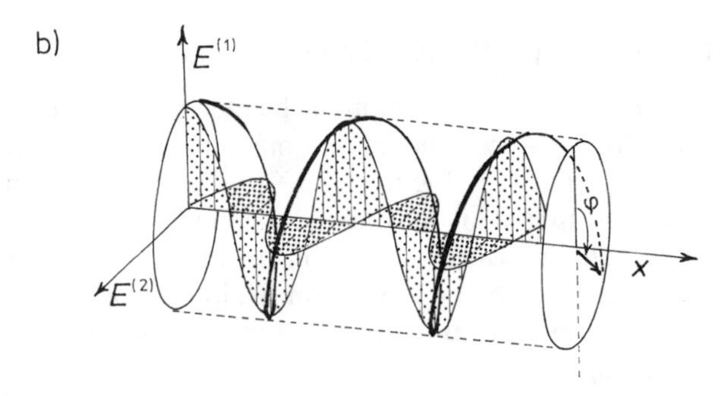

Figure 3.1. a) Right-handed and b) left-handed circularly polarized light as a superposition of two light waves with linear polarizations that are mutually orthogonal and have a phase difference of $\pm\pi/2$. The path of the endpoint of the electric field vector of the circularly polarized light is drawn as a heavy line. The path of the vector endpoint of the linearly polarized light is drawn as a thin line.

ellipse is rotated. The ellipticity θ and the rotation angle α are measured in degrees.

In the following it will be assumed that the sample has the same properties in all directions, that is, that it is isotropic. For anisotropic samples the description of optical activity is much more complicated.*

3.1.2 Chiroptical Measurements

Circular birefringence $n_L - n_R$ and circular dichroism $\varepsilon_L - \varepsilon_R$ can both be measured as a function of the wavelength λ or the wave number $\tilde{\nu}$. The two quantities are not mutually independent. If $n_L(\lambda) - n_R(\lambda)$ is known for $\lambda = 0$ up to $\lambda = \infty$, values of $\varepsilon_L(\lambda) - \varepsilon_R(\lambda)$ for $\lambda = 0$ up to $\lambda = \infty$ can be derived through a mathematical transformation, and vice versa. In spite of the fact that in practice it is not necessary to have values for the complete infinite interval of λ, measurements over a sufficiently large range of values are in general quite difficult, so the transformation is not always possible.

Circular birefringence is most easily determined by measuring the angle by which the polarization direction of linearly polarized light has been rotated upon passing through the sample. It is common to express this angle in degrees and the path length in dm. In this case the rotation angle α shows the following dependence on the frequency $\tilde{\nu}$ (in cm^{-1}) and the circular birefringence $n_L(\lambda) - n_R(\lambda)$:

$$\alpha = 1{,}800\ \tilde{\nu}[n_L(\lambda) - n_R(\lambda)] \tag{3.1}$$

Example 3.1:
The extreme sensitivity of the determination of $n_L - n_R$ obtained by measuring the rotation of the polarization direction can be illustrated in the following example: At $\tilde{\nu} = 16{,}977\ cm^{-1}$ ($\lambda = 589$ nm, Na D line) the circular birefringence of $n_L - n_R = 10^{-7}$ corresponds to a rotation by $\alpha = 1{,}800 \times 16{,}977 \times 10^{-7}\ dm^{-1}$, which, for the commonly used pathlength of 1 dm, is equal to an easily measurable rotation of the polarization direction by $\alpha = 3°$.

The sense of rotation is defined in the way it appears to a spectator when viewed against the light source. It is taken as positive if the polarization direction is rotated clockwise and negative if it is rotated counterclockwise.

The curve $\alpha(\tilde{\nu})$ or $\alpha(\lambda)$ is referred to as *natural optical rotatory dispersion* (ORD) and the instrument used for its measurement is a *spectropolarimeter.* If the optical activity has been induced by an external magnetic field, the

* Even an achiral anisotropic sample can change the polarization of linearly polarized light. This occurs in general in such a way that in regions where the sample does not absorb, the light becomes only elliptically polarized, whereas in regions of absorption the major axis of the ellipse is rotated as well. This is the opposite of the behavior in the case of optical activity. These phenomena have nothing to do with optical activity.

phenomenon is called *magnetic optical rotatory dispersion,* which, interestingly enough, is abbreviated as MORD (German for "murder").

When used as a means of characterization of a chiral compound, optical rotation is often measured at a single wavelength only (commonly the Na D line at $\lambda = 589$ nm). Specific rotation $[\alpha]$ gives the rotation (in deg dm^{-1}) that is measured for a solution of a concentration of 1 g/100 cm^3. The molar rotation $[\alpha]_M$ is then obtained by multiplying with the relative molecular mass M and dividing by 100; it is commonly expressed in units of deg cm^{-1} L mol^{-1}. Then,

$$[\alpha] = 100\ \alpha/cl \tag{3.2}$$

and

$$[\alpha]_M = M[\alpha]/100$$

where α is the measured angle in degrees, c is the concentration in g/100 cm^3, and l is the pathlength in dm.

Circular dichroism is measured using a *dichrograph,* which modulates a monochromatic beam of light between left-handed and right-handed circular polarization. A difference between ε_L and ε_R produces an intensity modulation that can be amplified and displayed. A separate measurement of ε_L and ε_R would be much more exacting because the differences between $\varepsilon_L cl$ and $\varepsilon_R cl$ are usually very small. (The optical densities differ by 10^{-4}–10^{-7} units.)

CD spectra are often plotted not as $\varepsilon_L(\lambda) - \varepsilon_R(\lambda)$, but rather as a so-called molar ellipticity $[\theta]_M(\lambda)$ which is given by

$$[\theta] = 100\ \theta/cl \tag{3.3}$$

and

$$[\theta]_M = M[\theta]/100 = 3{,}298\ (\varepsilon_L - \varepsilon_R)$$

θ is the ellipticity in degrees, c the concentration in g/100 cm^3, and l the pathlength in dm. A difference of 3×10^{-2} in the extinction coefficients coresponds to 1° in ellipticity.

If the optical activity is induced by an external magnetic field it is proportional to the magnetic field strength, and one measures *magnetic circular dichroism* (MCD), commonly expressed as magnetic molar ellipticity per gauss.

Because measurements of ORD and MORD spectra are easier to perform than those of CD and MCD spectra, they were developed earlier. Nowadays, however, except for compounds that are transparent in the easily accessible spectral region above 200 nm and therefore do not show CD or MCD signals, CD and MCD spectra are preferred as they are easier to interpret.

This is related to the spectral form of the contributions of the various electronic transitions to the total spectrum. In a CD or MCD spectrum, the

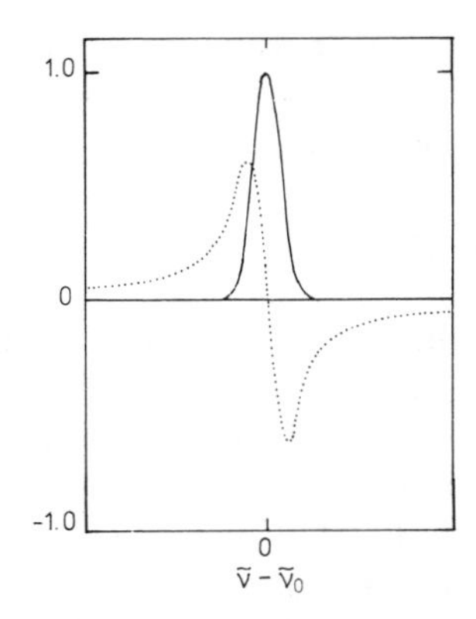

Figure 3.2. Gaussian profile of a CD band (—) and the corresponding dispersion band shape (···) observed in an ORD spectrum.

contribution of a particular electronic transition is restricted to the relatively narrow region of the absorption band where the ordinary extinction coefficient $(\varepsilon_L + \varepsilon_R)/2$ is appreciably different from zero. In contrast, the contribution of the same transition to the ORD or MORD spectrum has a different band shape and extends over a much wider spectral region. This is demonstrated in Figure 3.2, where the Gaussian shape of a CD band is shown together with the dispersion band shape of an ORD spectrum.

As a result, contributions from many transitions often overlap quite strongly in ORD and MORD spectra and are difficult to separate. An additional difficulty is due to the fact that the spectral form of the various contributions to the ORD and MORD spectra is more complicated.

We shall confine ourselves to CD and MCD spectra in the following discussion.

3.2 Natural Circular Dichroism (CD)

3.2.1 General Introduction

If the very minute "parity breaking" effects of "weak interactions" between elementary particles are disregarded, natural optical activity is due entirely to the inherent chirality of the sample. A sample is chiral if it is not superimposable onto its mirror image. Chiral objects are also referred to as *dissymmetric*. Isotropic solutions only are chiral if the individual molecules are chiral.

In reality molecules are often not rigid but exist in different conformations that can rapidly interconvert at room temperature. If none of these conformations is achiral, the molecule is called chiral. As soon as a single achiral conformation is accessible, there will always be the same number of molecules with any given chiral conformation as with the mirror-image conformation. The situation is therefore the same as for a racemate, and the molecule is considered achiral. Thus, whether or not a molecule is chiral may depend on the temperature and on the height of the activation barrier for configurational inversion.

Chiral, that is, dissymmetric, molecules are not necessarily asymmetric, in that they may possess certain symmetry elements. The geometrical prerequisite for chirality is the absence of an improper rotation axis S_n of any order n where S_1 corresponds to a symmetry plane (σ) and S_2 to an inversion center i. This follows immediately from the fact that the rotational strength, which describes the interaction with circularly polarized light, differs from zero only for those transitions for which the electric and the magnetic transition dipole moment have a nonvanishing component in the same direction. (Cf. Section 3.2.2.)

Enantiomers show the same degree of optical activity but with opposite signs. When one of the enantiomers rotates the polarization direction to the left at a given wave number, the other enantiomer rotates it by the same amount to the right. The rotational angles measured at 589 nm (Na D line) are often used to characterize enantiomers as the (−)-isomer (laevororotatory) and as (+)-isomer (dextrorotatory), respectively.

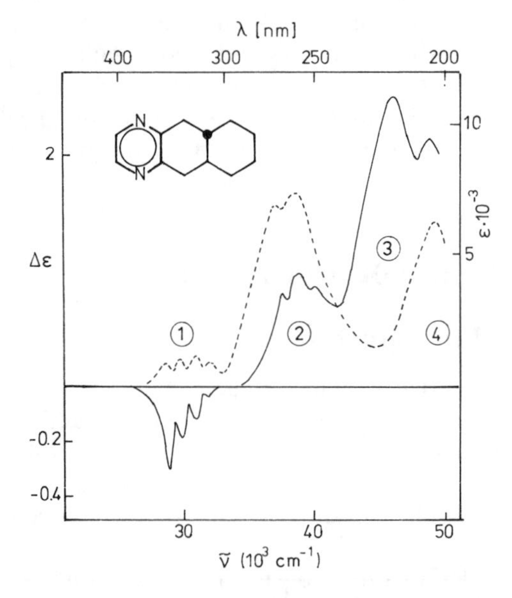

Figure 3.3. UV (---) and CD spectra (—) of octahydrobenzoquinoxaline. Band ③, which stems from a dipole-forbidden transition, is only seen in the CD spectrum (by permission from Snatzke, 1982).

The following holds for the circular dichroism of enantiomers: if at a given wave number one enantiomer has the extinction coefficients $\varepsilon_L^{(+)}$ and $\varepsilon_R^{(+)}$, the other one $\varepsilon_L^{(-)}$ and $\varepsilon_R^{(-)}$, then $\varepsilon_R^{(-)} = \varepsilon_L^{(+)}$ and $\varepsilon_R^{(+)} = \varepsilon_L^{(-)}$.

Even though contributions from the various transitions are important only within a narrow spectral region, band overlap similar to that occurring in ordinary absorption spectra is possible in CD spectra as well. The relative intensities of different transitions in the CD spectrum may, however, be quite distinct from those in the absorption spectrum. Accordingly, low-intensity absorption bands of chiral compounds, which are hidden by other bands, may well be detectable in the CD spectrum. As examples, the UV and CD spectra of an octahydrobenzoquinoxaline are shown in Figure 3.3.

The most important application of CD spectroscopy is in the field of structural elucidation. An empirical utilization of CD spectra is based on the assumption that similar chiral molecules show similar CD spectra, whereas chiral molecules that are approximately mirror images of each other should give CD spectra that are also approximately mirror images. As this general assumption is not always valid, a successful application of CD spectroscopy in structural elucidation requires some understanding of the underlying theory.

3.2.2 Theory

A quantitative measure of the optical activity of an electronic transition is the rotational strength $\mathcal{R}$, which can be determined from the CD spectrum by integration over the corresponding band. In units of Debye × Bohr magneton one has

$$\mathcal{R} = 7.51 \times 10^{-5}\int[\theta]_M d\tilde{\nu}/\tilde{\nu} = 0.248\int(\varepsilon_L - \varepsilon_R)d\tilde{\nu}/\tilde{\nu} \tag{3.4}$$

where the numerical factor is given by universal natural constants as in Equation (1.27). (Cf. Michl and Thulstrup, 1986.)

The theoretical expression for the rotational strength of a transition is given by the Rosenfeld formula

$$\mathcal{R}_{0\to f} = \mathrm{Im}\{<\Psi_0|\hat{M}|\Psi_f> \cdot <\Psi_f|\hat{\boldsymbol{M}}|\Psi_0>\} \tag{3.5}$$

as the imaginary part of the scalar product of the electric and the magnetic dipole moments $<\Psi_f|\hat{M}|\Psi_0>$ and $<\Psi_f|\hat{\boldsymbol{M}}|\Psi_0>$ of the transition from the ground state Ψ_0 into the excited state Ψ_f (Rosenfeld, 1928).

If the wave functions Ψ_0 and Ψ_f are real as usual, the scalar product within the braces is purely imaginary, since the operator $\hat{\boldsymbol{M}}$ is also purely imaginary:

$$\hat{\boldsymbol{M}} = \sum_j \hat{\boldsymbol{m}}_j = \frac{-|e|}{2m_e c}\sum_j(\boldsymbol{r}_j \times \hat{\boldsymbol{p}}_j) \tag{3.6}$$

Here, e, m_e, $\hbar$, and c are well-known natural constants, $\hat{\boldsymbol{p}}_j = -i\hbar\hat{\nabla}_j$ is the linear momentum operator, and $\boldsymbol{r}_j$ is the position operator of electron j. For real wave functions the Rosenfeld formula may be written in a simpler way as

$$\begin{aligned}\mathcal{R}_{0\rightarrow f} &= \langle\Psi_0|\hat{\boldsymbol{M}}|\Psi_f\rangle \cdot \langle\Psi_f|\hat{\mathbf{M}}/\mathrm{i}|\Psi_0\rangle \\ &= |\hat{\boldsymbol{M}}_{0\rightarrow f}|\,|\mathbf{M}_{0\rightarrow f}|\cos(\boldsymbol{M}_{0\rightarrow f}, \mathbf{M}_{0\rightarrow f})\end{aligned} \tag{3.7}$$

where $\boldsymbol{M}_{0\rightarrow f}$ is the transition moment vector defined in Equation (1.35) and the real vector $\mathbf{M}_{0\rightarrow f} = \langle\Psi_f|\hat{\mathbf{M}}/\mathrm{i}|\Psi_0\rangle$is equal to the imaginary part of the magnetic transition moment. The prediction of optical activity for a transition is therefore straightforward if the two transition moments are known. It vanishes if a least one of the transition moments is equal to zero or if both are nonzero but perpendicular to each other. This is always the case if an improper rotation axis S_n is present. As already mentioned, optical activity cannot occur in molecules that belong to one of the corresponding point groups.

Although the Rosenfeld formula looks very simple, it is by no means easy to predict the length and the relative orientation of the two vectors theoretically. The results are very sensitive to the quality of the wave functions Ψ_0 and Ψ_f, and it is the wave function Ψ_f of the excited state that is particularly difficult to calculate. In general, the CD spectrum is measured in the UV and the visible region; Ψ_f is then the wave function of an electronically excited state. Recently, CD measurements in the IR region became feasible as well; Ψ_f is then a vibrationally excited state. (Cf. Mason, 1981.)

According to Moffitt and Moscowitz (1959) two classes of chiral molecules can be distinguished, that is:

- Molecules with an inherently chiral chromophore
- Molecules in which the chromophore is achiral but is located in a chiral environment within the molecule

In both classes there may be molecules that contain two (or more) identical or similar chromophores in a chiral arrangement. For each of these cases a different theoretical method of computation has proven useful.

It is possible to calculate, using, for example, the MO-CI method, Ψ_0 and Ψ_f for the complete molecule. Although this procedure can in principle be applied to every molecule, it is particularly suited for inherently dissymmetric chromophores. The method is very demanding as far as computing is concerned. Its main disadvantage, however, is that every molecule represents a special case, so each additional molecule requires a brand-new calculation. General trends or relationships between molecules in homologous series are then difficult to recognize.

For molecules with an achiral chromophore located in a chiral environment that is transparent in the region of the spectrum characteristic of the chromophore, it is advantageous to consider the effect of the chiral environment on the transition moments $\boldsymbol{M}_{0\rightarrow f}$ and $\mathbf{M}_{0\rightarrow f}$ of the achiral chromophore as a perturbation of the localized wave functions Ψ_0 and Ψ_f of the chromophore. An important example of such an achiral chromophore is the carbonyl group that occurs in a large number of chiral ketones.

The complete mathematical description of the influence of a chiral environment on the optical activity of a chromophore is rather complex and contains, apart from possible inherent contributions, quite a number of contributions that may be divided into various groups that correspond to different "mechanisms" of inducing optical activity.

The "one-electron" mechanism is based on a mixing of various transitions of the chromophore induced by the chiral environment and is somewhat comparable to vibronic coupling (Section 1.3.4). Interactions between electric transition dipoles of groups in a chiral arrangement with a central chromophore may be described as coupled oscillators. The interaction of an electric transition dipole moment with a magnetic transition dipole moment or an electric transition quadrupole moment forms the basis of the "electromagnetic" mechanism of induced optical activity. In general, contributions from all different mechanisms have to be considered. In practice, however, quite often one of the mechanisms predominates to such an extent that all other mechanisms can be neglected. It is therefore possible to speak of a *one-electron model* of optical activity, or of the *exciton-chirality model,* which forms a special case of the model of coupled oscillators, etc.

Finally, the optical activity of polymers that contain a very large number of inherently achiral chromophores in a chiral arrangement may be treated by methods developed in solid-state physics, like the other optical properties of polymers. The resulting theoretical description can be viewed as a generalization of the exciton–chirality model.

3.2.3 CD Spectra of Single-Chromophore Systems

The field of natural products has been particularly fertile for the application of CD spectra to structural elucidation. The method has proven especially useful for carbonyl compounds. These have been investigated very thoroughly and will therefore be discussed here in some detail. Other important naturally occurring chromophores such as alkenes, dienes, disulfides, and aromatics will be mentioned more briefly.

For a carbonyl group in a C_{2v} symmetry environment, such as in formaldehyde, the dipole approximation for the $n \rightarrow \pi^*$ transition yields $\mathbf{M}_{0\rightarrow f} \gg 0$ and $\boldsymbol{M}_{0\rightarrow f} = 0$. Although the transition is magnetically strongly allowed and polarized along the CO axis, it is electrically forbidden. The absorption therefore is of very weak intensity and the rotational strength is equal to zero. Perturbations by vibrations or by an achiral solvent can affect Ψ_0 and Ψ_f to such an extent that a small nonzero electric dipole transition moment results; but again this produces only a very small absorption intensity and

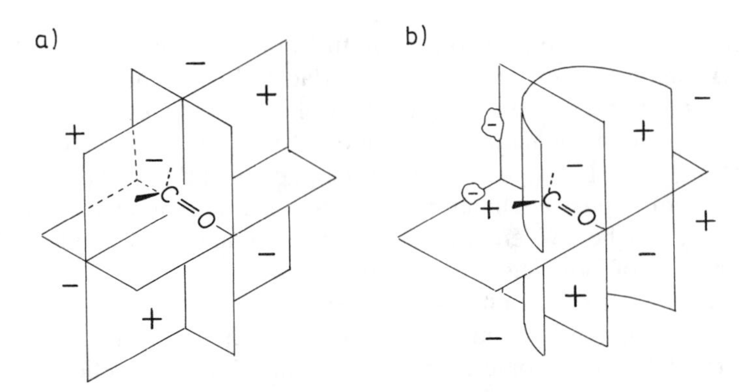

Figure 3.4. The octant rule. a) Subdivision of the space into "octants" and sign of the contribution of a group in each octant to the CD effect of the n→π^* transition of the carbonyl group. b) Computed position of the octant planes (adapted from Bouman and Lightner, 1976).

no optical activity. This requires a chiral perturbation. As the magnetic dipole moment is already inherently large it will be hardly affected by the perturbation. Thus, the optical activity is essentially due to the fact that the chiral perturbation induces a component of the electric transition moment in the direction of the CO axis, either parallel or antiparallel to the magnetic transition moment.

The optical activity of carbonyl compounds has been studied using various methods. The first analyses of the perturbation of the symmetrical carbonyl chromophore by groups in a chiral arrangement led to the so-called *octant rule* (Moffitt et al., 1961; see also Wagnière, 1966). According to this rule the sign of the rotatory strength of the n→π^* transition depends in the following way on the spatial coordinates of the groups that produce the chiral perturbation:

If a right-handed Cartesian coordinate system is placed with its origin in the center of the C═O bond such that the z axis points from the carbonyl carbon toward the oxygen and the y axis is in the plane of the carbonyl chromophore, space is divided into eight octants. A sign is associated with each octant and corresponds to the sign of the product xyz of the coordinates of the points in that octant. (Cf. Figure 3.4a.) The sign of each octant then gives the sign of the contribution to the rotatory strength by a group located in this octant. Groups that are located in the xy, yz, or zx plane do not contribute to the rotatory strength.

Example 3.2:

In Figure 3.5 the application of the octant rule to (+)-3-methylcyclohexanone is demonstrated. In the boat form as well as in the chair form the methyl group is located in a positive octant if it is in an equatorial position and in a negative

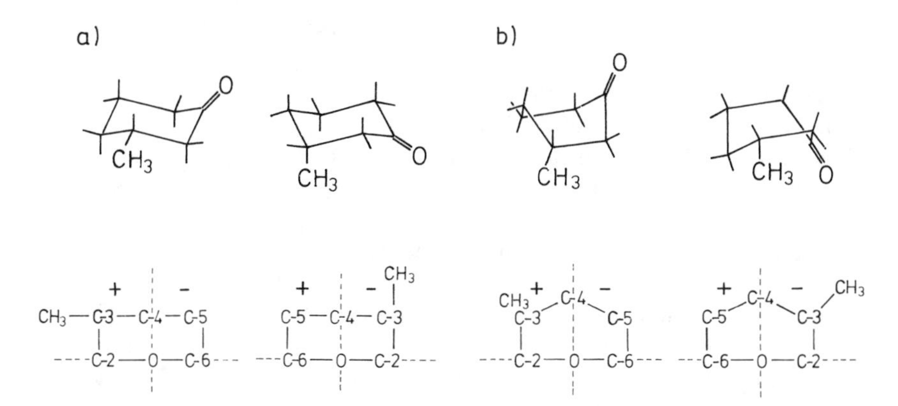

Figure 3.5. Projections of (+)-3-methylcyclohexanone into the octants a) for the chair form and b) for the boat form.

octant if it is in an axial position. Since all other groups are located in either the *yz* or the *xz* plane or pairwise in octants with opposite signs, they do not contribute to the rotatory strength. Decisive therefore is the difference between the methyl group in a positive octant and the hydrogen atom in a negative one. As the effect of the methyl group is stronger, a positive rotatory strength is expected for the more stable conformation with the methyl group in the equatorial position. This in fact is what has been found experimentally. This example illustrates how the octant rule permits one to either establish the absolute configuration of compounds with rigid structure and known conformation or to determine the conformation of compounds of known absolute configuration. However, it is not possible to determine the absolute configuration and the conformation at the same time in this way.

If viewed from the oxygen, most optically active carbonyl compounds have their substituents only in the rear octants. The appearance of the plane that separates the rear octants from the front octants is not determined by the symmetry of the isolated chromophore. Calculations have shown that it has approximately the shape depicted in Figure 3.4b. Many examples have verified the validity of the octant rule, but there are also cases where it is not applicable, at least not in its original, simple form. This is true for ketones with a cyclopropane ring in the α, β-position and for fluorosubstituted ketones, for which the experimentally observed sign can be reproduced only if the perturbation due to the fluorine atom is assumed to be smaller than that due to the hydrogen atom. More recent detailed calculations solved some of these problems. (Cf. Charney, 1979.)

The CD effect of simple alkenes occurs at shorter wavelengths and is therefore more difficult to measure. The spectra are usually interpreted on the basis of a modified octant rule. The interpretation is much less reliable, however, because in the spectral region of interest at least two transitions overlap.

Dienes are inherently chiral if they are twisted around the central CC bond. This produces a rotational strength that is positive if the twisted butadiene chromophore forms a right-handed helix. But since this inherent effect is not always predominant, contributions of other substituents have to be taken into account as well in order to be able to predict the CD spectrum. Substituents in allylic positions are frequently most important.

The first absorption band of disulfides corresponds to an excitation from the energetically higher combination of the two 3p AOs of the sulfur lone pairs of electrons into the antibonding σ^* orbital of the S—S bond. This is again an inherently dissymmetric chromophore. In this case, however, the expected sign of the CD effect depends not only on whether the R_1—S—S—R_2 helix forms a right-handed or a left-handed screw, but also on whether the dihedral angle is smaller or larger than 90°. For dihedral angles of approximately 90° the energies of the plus and minus combinations of the lone pair orbitals on the two sulfur atoms are more or less equal, and as a result the longest-wavelength absorption band near 250 nm is approximately degenerate. The rotatory strengths of the two transitions are of the same magnitude but of opposite sign, so they just cancel each other. For dihedral angles of 0° or 180° the first band is nondegenerate and at long wavelengths ($\lambda \approx 350$ nm) but the inherent contribution to the rotatory strength is again very small. In these cases, the resultant optical activity is frequently determined by chiral perturbations due to other atoms. If the helix forms a right-handed screw the rotatory strength of the first band is large and positive for acute dihedral angles ($0° < \varphi < 90°$) but negative for obtuse angles ($90° < \varphi < 180°$) (Linderberg and Michl, 1970). This has been used to investigate disulfide bridges in proteins (Niephaus et al., 1985).

Example 3.3:

If the absorption responsible for the CD effect can be described with sufficient accuracy as an electronic transition from MO ϕ_i into MO ϕ_k, the orientation of the electric and the magnetic transition moments may be derived from the nodal properties of these MOs, thus giving the sign of the CD effect. This will be demonstrated using an inherently chiral disulfide as an example. In Figure 3.6, the orbitals involved in the excitation as well as the overlap densities are given for the case of a positive torsional angle $\varphi < 90°$. Taking into account the negative charge of the electron, the direction of the transition dipole may be derived from the overlap densities. [Cf. Equation (1.39).] The *right-hand rule* yields the direction of the magnetic transition dipole if the p AO components of the initial orbital ϕ_i are rotated by the smaller one of the two possible angles in such a way that they are converted into the p AO components of the final MO ϕ_k. Using this procedure it is easy to show (Figure 3.6c) that the components of the magnetic transition dipole moment are arranged along the S—R_1 and S—R_2 bonds such that the total magnetic transition dipole, which is given

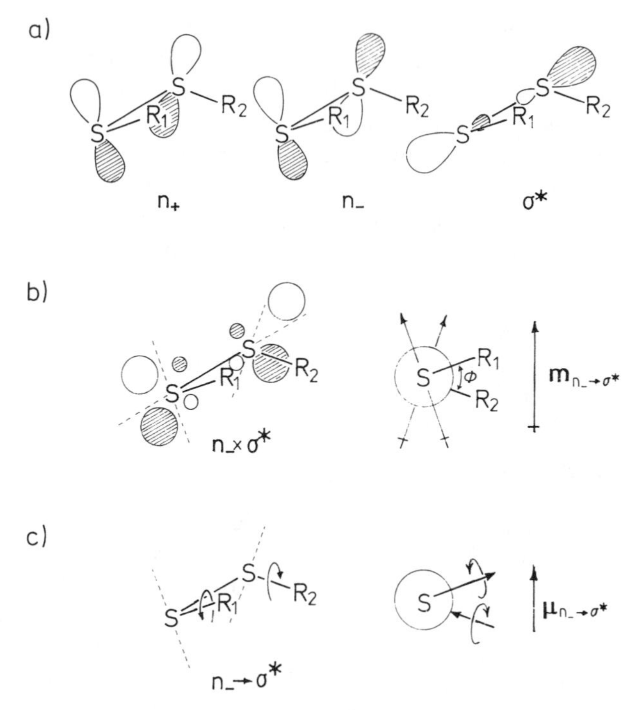

Figure 3.6. The chiral disulfide chromophore with a positive torsional angle $\varphi < 90°$: a) orbitals involved in the n→σ^* transition; b) transition density with positive and negative regions corresponding to overlap of orbitals with equal or opposite signs shown light and shaded, respectively, as well as the transition dipoles resulting from taking into account the negative charge of the electron; c) right-hand rule and the resulting magnetic transition dipole (by permission from Snatzke, 1982).

as their sum, is parallel to the electric transition dipole. The CD effect for this conformation of the disulfide chromophore is therefore positive, in agreement with experimental results for 2,3-dithia-α-steroids (**1**).

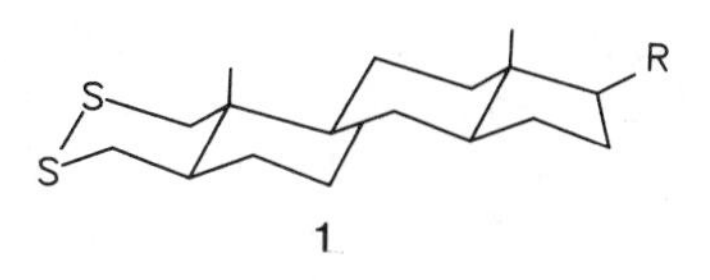

Typical for aromatic hydrocarbons and their derivatives are the ^1L_b, ^1L_a, ^1B_b, and ^1B_a bands. Chirally twisted aromatic compounds, for example, hexahelicene (**2**), are inherently dissymmetric π chromophores. The rotatory strength of the various transitions can be calculated by means of the common π-electron methods.

More numerous, however, are aromatic compounds with an effectively planar π system, the optical activity of which comes from chiral perturbations. Benzene derivatives such as phenylalanine (**3**) are particularly important. The ^1L_b band of the benzene chromophore ($\lambda \approx 260$ nm) is neither magnetically nor electrically allowed. The symmetry selection rule may be broken by vibronic interactions or due to substituents. The interpretation of the observed rotatory strength is not easy, but empirical sector rules have been proposed. (Cf. Charney, 1979; Pickard and Smith, 1990.)

2

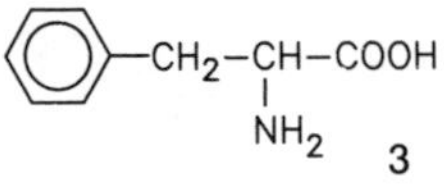

$C_6H_5-CH_2-CH(NH_2)-COOH$

3

3.2.4 Two-Chromophore Systems

For the description of the CD spectra of molecules with two (or more) identical or at least similar chromophores in a chiral arrangement the method of coupled oscillators (exciton chirality method) has proven particularly successful. Such systems can occur either as chiral "dimers," or they can be obtained by introducing suitable chromophores into a chiral molecule. The best-known example is given by the dibenzoates derived from 1,2-cyclohexanediol of the general formula **4**:

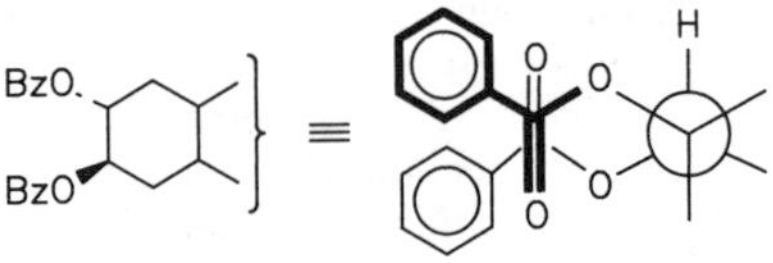

4

In all these cases it is possible to derive from the CD spectrum the absolute spatial orientation of the electric transition dipoles of the two chromophores. If the direction of the transition dipole within the chromophore is known, the absolute stereochemistry of the compound may be determined using these methods.

If Φ_{k0} and Φ_{kf} ($k = 1, 2$) are localized ground- and excited-state wave functions of the chromophores k, the ground state of the two-chromophore system may be described by $\Psi_0 = \Phi_{10}\Phi_{20}$, whereas the excited states $\Psi_f^{\pm} = N_{\pm}(\Phi_{1f}\Phi_{20} \pm \Phi_{10}\Phi_{2f})$ are degenerate in zero-order approximation. The exciton–chirality model only takes into account the interaction between the transition dipole moments $\boldsymbol{M}_1$ and $\boldsymbol{M}_2$ localized in the chromophores. Thus, the interaction gives rise to a Davydov splitting by $2V_{12}$ of the energies of combinations Ψ_f^+ and Ψ_f^- of locally excited states. From the dipole–dipole approximation one obtains

$$V_{12} = (\boldsymbol{M}_1 \cdot \boldsymbol{M}_2)/R_{12}^3 - 3(\boldsymbol{M}_1 \cdot \boldsymbol{R}_{12})(\boldsymbol{M}_2 \cdot \boldsymbol{R}_{12})/R_{12}^5 \quad (3.8)$$

where $\boldsymbol{R}_{12}$ is the vector from the center of one chromophore to the center of the other chromophore and $R_{12} = |\boldsymbol{R}_{12}|$.

The rotational strengths $\mathcal{R}^+$ and $\mathcal{R}^-$ of these two transitions are of the same magnitude but of opposite sign:

$$\mathcal{R}^{\pm} = \pm\tfrac{1}{2}[\pi\tilde{\nu}\boldsymbol{R}_{12} \cdot \boldsymbol{M}_1 \times \boldsymbol{M}_2 + \mathrm{Im}(\boldsymbol{M}_1 \cdot \boldsymbol{M}_2 + \boldsymbol{M}_2 \cdot \boldsymbol{M}_1)] \qquad (3.9)$$

In applications of this relation, the contribution from the coupling with the magnetic transition moments $\boldsymbol{M}_k = \langle\Phi_{kf}|\hat{\boldsymbol{M}}/i|\Phi_{k0}\rangle$ is often neglected. The sign of the rotational strength of the lower-energy transition is then the same as the chiral sign for a two transition dipole moment system. A right-handed screw (clockwise rotation) is here considered as positive and a left-handed screw (counterclockwise rotation) as negative. The CD spectrum then shows

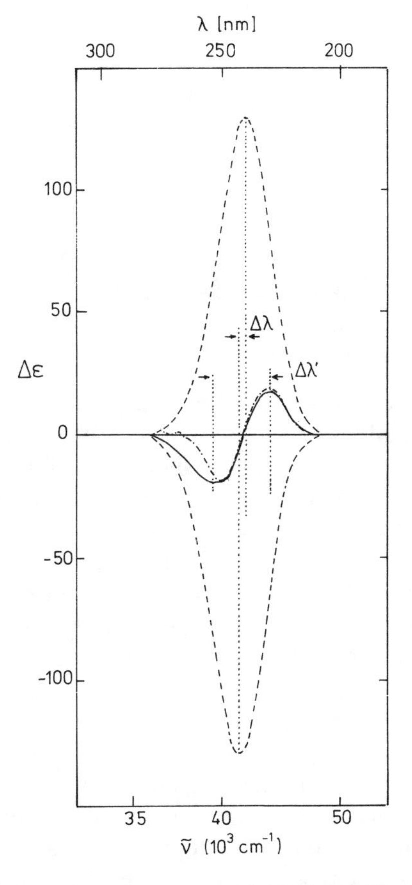

Figure 3.7. CD spectrum of cholest-5-ene-3β,4β-bis(p-chlorobenzoate); contributions of the two transitions with Davydov splitting $\Delta\lambda$ (----), sum curve (—), and observed curve (-·-·-) (by permission from Harada and Nakanishi, 1972).

a combination of a positive and a negative CD band (a "couplet"). The sign of the longer-wavelength band is the same as the sign of the chirality.

Example 3.4:
The transition moment of the strong absorption band at 230 nm of the benzoyloxychromophore is directed roughly along the long axis; in a benzoate it is therefore approximately parallel to the C—O bond of the alcohol, irrespective of the conformation relative to rotation around the C—O bond. (Cf. **4**.) This is also true for *p*-substituted benzoyloxy derivatives with a long-wavelength shift of this band, which then usually overlaps much less with other transitions. Figure 3.7 shows a part of the CD spectrum of the bis(*p*-chlorobenzoate) of the vicinal diol **5**, a steroid with negative chirality of the C—O bonds. As expected, the CD band at longer wavelengths is negative, and the one at shorter wavelengths is positive. This method has been used to determine the absolute configuration of quite a number of natural products.

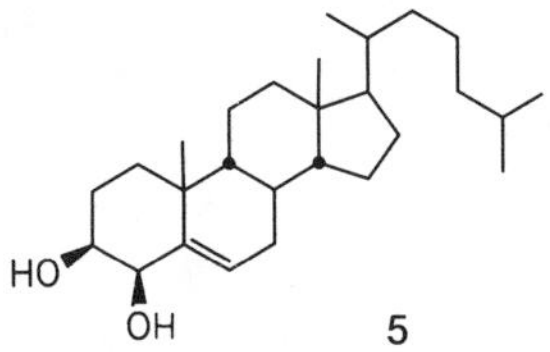

HO
OH
5

3.3 Magnetic Circular Dichroism (MCD)

3.3.1 General Introduction

In an external magnetic field all matter becomes optically active. This observation was first made by Faraday in 1845; the magnetically induced rotation of the plane of polarized light is therefore referred to as the *Faraday effect*. In recent years, the common mode of study of this phenomenon has been the measurement of magnetic circular dichroism (MCD). Similarly to natural circular dichroism, magnetic circular dichroism is defined as the difference $\Delta\varepsilon = \varepsilon_L - \varepsilon_R$ of the extinction coefficients for left-handed and right-handed circularly polarized light as caused by the presence of a magnetic field directed parallel to the light propagation direction. To a very good approximation, the effect is proportional to the strength of the magnetic field.

A common mode of display of MCD spectra is a plot of magnetically induced molar ellipticity $[\theta]_M$, normalized for unit strength [1 Gauss (G) = 10^{-4} Tesla (T)] of the magnetic field, against wavelength λ or wave number $\tilde{\nu}$. This quantity is related to $\Delta\varepsilon$ in the same way as shown in Equation (3.3) for natural CD. It is a good approximation to say that for chiral samples that are naturally optically active, the magnetically induced effect is superimposed in an additive fashion onto their natural CD effect.

An MCD spectrum is measured in the same way as a CD spectrum, except that the sample is placed inside the coil of a superconducting magnet or between the poles of an electromagnet in which holes have been bored to permit the light beam to pass through. Only that component of the magnetic field that is parallel to the beam of light contributes to the effect. Reversal of the field direction changes the sign of the measured ellipticity, providing a simple method to cancel out the natural CD effect.

Like UV absorption and natural CD spectra, the MCD spectrum observed in the UV/VIS region is a superposition of contributions from all electronic transitions in the molecule. Whereas in ordinary isotropic absorption spectra or in natural CD spectra each transition can be characterized by its band shape and a single scalar quantity, the oscillator strength or the rotatory strength, three numbers, A_i, B_i, and C_i, are necessary to describe the contribution of each electronic transition i to the MCD spectrum of an isotropic sample. The three numbers are known as the A term, the B term, and the C term of the i-th transition. The contributions of the B term and of the C term to the MCD curve $[\theta]_M(\tilde{\nu})$ have the shape of an absorption peak, which in the simplest case has a Gaussian profile as shown in Figure 3.8. They differ insofar as the contribution of the B term is temperature independent, whereas that of the C term is inversely proportional to the absolute temperature and can therefore dominate the spectra at very low temperatures. If nonzero B and C terms are both present, a measurement of the temperature dependence of the spectrum is needed to separate them.

An MCD band can be either positive or negative just as in natural CD spectra. According to the nomenclature accepted in organic MCD spectros-

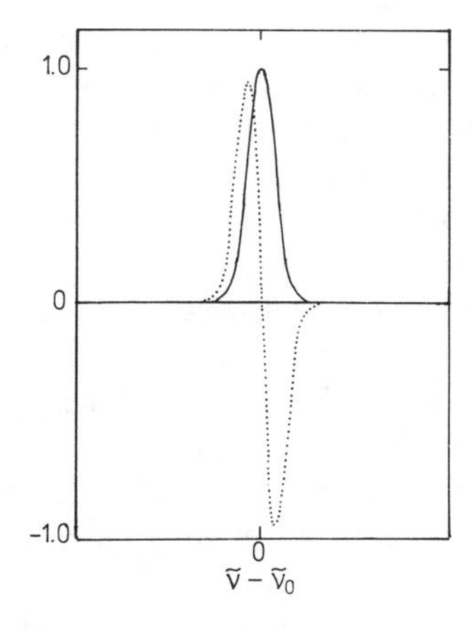

Figure 3.8. Gaussian profile of a MCD band corresponding to a nonzero B or C term (—) and its derivative, which is the band shape of contributions from the A term (···).

copy, a B term or a C term is called *positive if it corresponds to a negative MCD peak* ($[\theta]_M < 0$), *and negative if it corresponds to a positive MCD peak.*

The contributions from the A term do not have the shape of an absorption peak, but rather that of the derivative of an absorption peak, which resembles a letter S on its side; this is shown schematically in Figure 3.8 by the dotted curve. The A term is called positive if $[\theta]_M(\tilde{\nu})$ is negative at lower wave numbers and positive at higher wave numbers; it is called negative if the contributions to the $[\theta]_M(\tilde{\nu})$ curve are positive at lower and negative at higher $\tilde{\nu}$ values. At the absorption maximum the contribution from the A term vanishes. If A and B (or C) terms are both nonzero the resulting band shape is a superposition of the various contributions.

In order to obtain a simple mathematical description of the total contribution of the i-th electronic transition to the MCD spectrum, a normalized line-shape function $g_i(\tilde{\nu})$ and its derivative $f_i(\tilde{\nu})$ are defined. The total spectrum may then be written as a sum over all electronic transitions as

$$[\theta]_M(\tilde{\nu}) = -21.3458 \sum_i [A_i f_i(\tilde{\nu}) + (B_i + C_i/kT) g_i(\tilde{\nu})] \tag{3.10}$$

where k is Boltzmann's constant and T the absolute temperature. Common units are $D^2\beta_e$ for A_i and $D^2\beta_e/cm^{-1}$ for B_i and for C_i/kT, where D stands for Debye and β_e for Bohr magneton.

The value of C_i is zero unless the ground state is degenerate. This is very rare for organic molecules, so the MCD spectra of organic molecules are in general temperature independent except for cases where conformational changes or similar temperature-dependent processes change the molecule itself. In the following it will be assumed that all C terms vanish.

In that case the A term of an electronic transition is nonzero only if the excited state is degenerate.* This can happen in highly symmetrical molecules with an axis of rotation C_n of order $n \geq 3$. Again, such molecules are more numerous in inorganic chemistry than in organic chemistry. If the A_i term is sufficiently large compared to the B_i term, both a positive and a negative MCD peak are observed. In all other cases each absorption peak corresponds exactly to an MCD band.

Since relative intensities of the various bands in the absorption and MCD spectra, just as in the CD spectra, can be very different, it is possible to use MCD spectroscopy in order to detect transitions that are absent or difficult to recognize in the absorption spectrum. MCD spectroscopy and polarization spectroscopy are in many cases complementary in this respect, since transitions with identical polarizations may have B terms of opposite signs

---

* If the ground state is degenerate the A term is nonzero even if the excited state is nondegenerate.

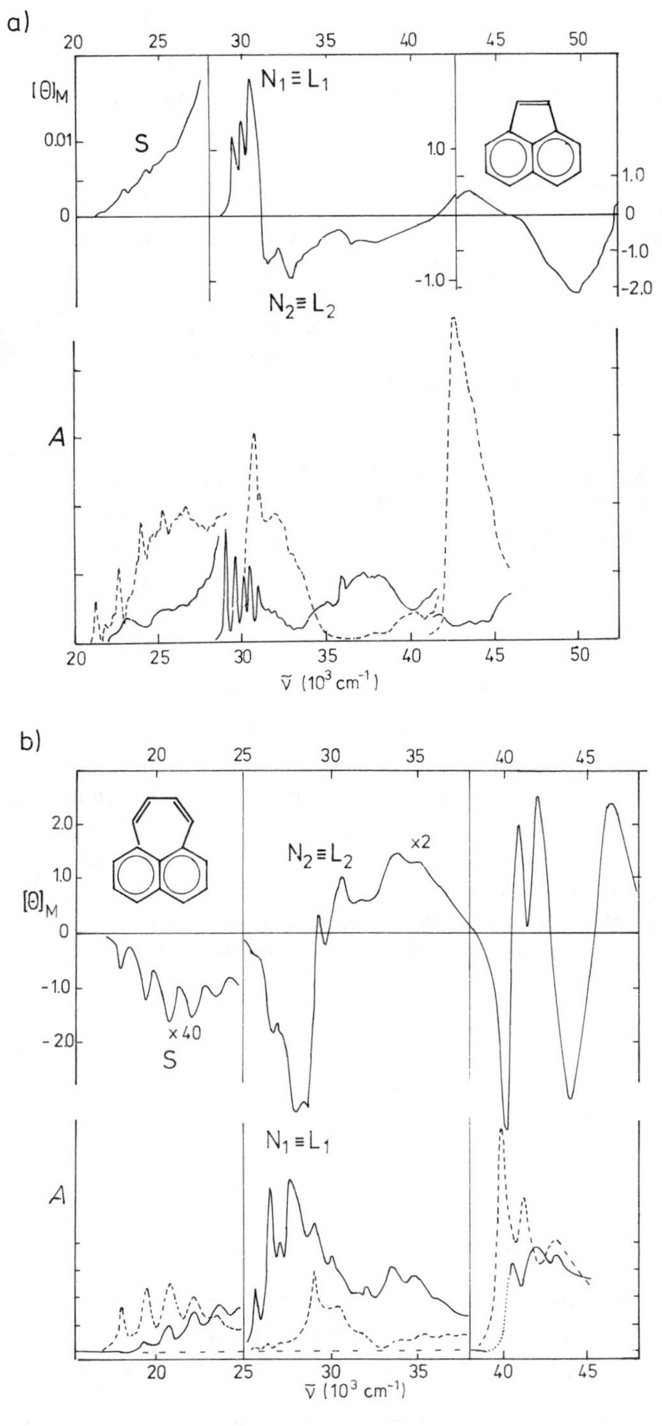

Figure 3.9. Magnetic circular dichroism (top) and absorption polarized along the z axis (—) and along the y axis (---) of a) acenaphthylene and b) pleiadiene (by permission from Kolc and Michl, 1976, and from Thulstrup and Michl, 1976).

and transitions with B terms of equal sign may differ in the polarization direction.

Example 3.5:
Figure 3.9 shows the MCD spectra of acenaphthylene and pleiadiene. The second and the third absorption bands overlap strongly. They are the N_1 or L_1 and N_2 or L_2 bands, which differ in the sign of the B term and are therefore easy to recognize in the MCD spectra. For comparison the linearly polarized spectra, obtained from measurements on stretched sheets, are also shown. They also make it possible to distinguish the two bands, since they have different polarization directions.

Another spectroscopic application of MCD spectroscopy is important for highly symmetrical molecules. The presence of nonzero A terms reveals clearly which transitions are degenerate and which are not. This distinction is in general not possible from ordinary absorption spectra, even if no bands overlap. The magnetic moment of the excited state may be derived from the A_i value by means of the relationship

$$\mu_i = -2A_i/D_i \tag{3.11}$$

where D_i is the dipole strength of the transition i. The sign of the magnetic moment corresponds to that excited state that is being generated by left-handed circularly polarized light. A positive value means that the projection of the magnetic moment onto the direction of the light is positive.

Values of the A_i, B_i, and C_i terms are most easily obtained if the bands belonging to different transitions overlap as little as possible. In this case one can integrate over the transition in question and obtain the following expressions in the usual units given above and in D^2 for D_i:

$$A_i = 33.53^{-1} \int_i d\tilde{\nu}(\tilde{\nu} - \tilde{\nu}_0)[\theta]_M/\tilde{\nu} \tag{3.12}$$

$$B_i + C_i/kT = -33.53^{-1} \int_i d\tilde{\nu}[\theta]_M/\tilde{\nu} \tag{3.13}$$

$$D_i = 9.188 \times 10^{-3} \int_i d\tilde{\nu}\varepsilon/\tilde{\nu} \tag{3.14}$$

Here, $\tilde{\nu}$ is the wave number of the center of the absorption band and $[\theta]_M$ and ε are measured in the usual units (deg L m^{-1} mol^{-1} G^{-1} or L mol^{-1} cm^{-1}, respectively).

Example 3.6:
The MCD spectrum of the croconate dianion $C_5O_5^{2\ominus}$ is shown in Figure 3.10. The molecule possesses a fivefold symmetry axis and belongs to the point

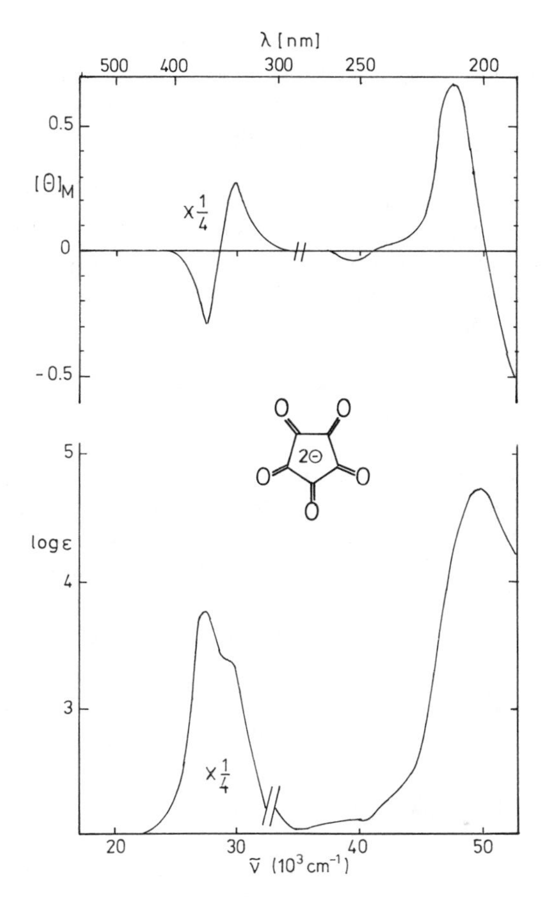

Figure 3.10. MCD spectrum (top) and absorption spectrum (bottom) of the croconate dianion $C_5O_5^{2\ominus}$ (by permission from West et al., 1981).

group D_{5h}; the first excited singlet state is degenerate. This can be verified from the fact that the first band in the MCD spectrum exhibits a positive A term. Numerical integration gives A_i and D_i values, from which the magnetic moment of the excited state can be derived as $\mu = -0.25\ \beta_e$.

If two or more transitions overlap, the quantitative interpretation of the MCD spectrum becomes much more difficult and computational curve fitting that requires simple band shapes is needed. Often, the vibrational structure of the MCD spectrum, which will in general be different from the one in the absorption spectrum, will be so pronounced that this assumption is not fulfilled. Although uncommon, there are some cases known where some sections of the vibrational envelope associated with a single electronic transition are positive and some negative. Thus, the appearance of both positive and negative MCD peaks in a spectral region does not necessarily mean that a degenerate electronic state is involved or that more than one electronic transition is present in the region. Reliable assignment of the number of electronic transitions usually requires a painstaking analysis of a number of different kinds of spectra. The resulting detailed information is extremely useful for the spectroscopist.

For applications in organic chemistry, on the other hand, knowledge of the symmetries of excited states, of the energy ordering of the orbitals, or of the relative magnitude of orbital coefficients is of primary interest. Such applications of the MCD method require at least a superficial understanding of the fundamental theory.

3.3.2 Theory

The quantum mechanical expressions for the A_i and C_i terms of the electronic transition i derived by means of perturbation theory are relatively simple, whereas the expression for the B_i term is much more complicated in that it involves two infinite summations over all electronic states Ψ_j of the molecule. Using Greek indices to distinguish different components of degenerate electronic states, the various terms may be written as follows:

$$A_i = \frac{1}{2g}\sum_{\alpha\gamma} (<\Psi_{i\alpha}|\hat{\mathbf{M}}|\Psi_{i\alpha}> - <\Psi_{0\gamma}|\hat{\mathbf{M}}|\Psi_{0\gamma}>) \cdot \mathrm{Im}(<\Psi_{0\gamma}|\hat{M}|\Psi_{i\alpha}> \times <\Psi_{i\alpha}|\hat{M}|\Psi_{0\gamma}>) \tag{3.15}$$

$$B_i = \frac{1}{g}\sum_{\alpha\gamma} \mathrm{Im}\{\sum_{k\varkappa, k\neq 0} [<\Psi_{k\varkappa}|\hat{\mathbf{M}}|\Psi_{0\gamma}> \cdot <\Psi_{0\gamma}|\hat{M}|\Psi_{i\alpha}> \times <\Psi_{i\alpha}|\hat{M}|\Psi_{k\varkappa}>]/[E_k - E_0] + \sum_{k\varkappa, k\neq i} [<\Psi_{i\alpha}|\hat{\mathbf{M}}|\Psi_{k\varkappa}> \cdot <\Psi_{0\gamma}|\hat{M}|\Psi_{i\alpha}> \times <\Psi_{k\varkappa}|\hat{M}|\Psi_{0\gamma}>]/[E_k - E_i]\} \tag{3.16}$$

$$C_i = \frac{1}{2g}\sum_{\alpha\gamma} <\Psi_{0\gamma}|\hat{\mathbf{M}}|\Psi_{0\gamma}> \cdot \mathrm{Im}(<\Psi_{0\gamma}|\hat{M}|\Psi_{i\alpha}> \times <\Psi_{i\alpha}|\hat{M}|\Psi_{0\gamma}>) \tag{3.17}$$

$$D_i = \sum_{\alpha\gamma} <\Psi_{0\gamma}|\hat{M}|\Psi_{i\alpha}> \cdot <\Psi_{i\alpha}|\hat{M}|\Psi_{0\gamma}> \tag{3.18}$$

Here, g is the degree of degeneracy of the ground state, E_j is the energy of the state Ψ_j, and $\hat{M}$ and $\hat{\mathbf{M}}$ are the operators of the electric and magnetic dipole moments, respectively. The wave functions of the ground state Ψ_0, of the final state Ψ_i, as well as that of the intermediate state Ψ_k, over which the summation runs, are unperturbed wave functions defined in the absence of the magnetic field.

The physical meaning of these terms is easy to comprehend qualitatively. They are based on the Zeeman effect, which results in a magnetic-field-induced splitting of degenerate states. This depends on the magnitude of the magnetic field strength and on the magnetic dipole moment of the molecule.

If the ground state of the free molecule is degenerate, the population of the individual components, which in the presence of the magnetic field are no longer degenerate, depends on temperature and results in temperature-dependent contributions from the C terms. If, on the other hand, the excited state is degenerate, one component of this state will be excited by left-handed and the other one by right-handed circularly polarized light. Depending on which of these components is shifted to lower energies by the Zeeman splitting in the magnetic field, $\Delta\varepsilon$ will become initially either positive or negative as the wave number $\tilde{\nu}$ increases, so a negative or a positive A term results.

Sometimes it is possible to predict the sign by means of group theory and in this way to obtain an assignment of state symmetries. This is particularly true for those molecules to which the perimeter model discussed in Section 2.2.2 is applicable.

Some of the results of the perimeter model, which are essential for the following discussions, are briefly repeated in Figure 3.11, which is basically a portion of Figure 2.11: For a $(4N+2)$-electron perimeter ($N \neq 0$) HOMOs ϕ_N and ϕ_{-N} and LUMOs ϕ_{N+1} and ϕ_{-N-1} occur as degenerate pairs. On the average, an electron in ϕ_k ($k = N, N+1$) is found to move counterclockwise along the perimeter when viewed from the positive end of the z axis and an electron in ϕ_{-k} tends to move clockwise. Since it is negatively charged, an electron in ϕ_k with a positive angular momentum produces a negative z component of the magnetic moment μ_z and an electron in ϕ_{-k} a positive z component of equal magnitude. The transition $\phi_N \rightarrow \phi_{N+1}$ requires a photon of left-handed circularly polarized light that corresponds to a counterclockwise-rotating electromagnetic field. The transition $\phi_{-N} \rightarrow \phi_{-N-1}$ requires a photon of right-handed circularly polarized light.

From the orbital magnetic moments produced by electron circulation in the various orbitals, the z component of the magnetic moment resulting for each electron configuration may be derived. A degenerate state whose components are connected with magnetic moments μ_z and $-\mu_z$, respectively, undergoes a Zeeman splitting in the magnetic field. If B_z is the component of the magnetic field in the direction of the light beam, the interaction is given by $-B_z\mu_z$. Thus, the component with positive μ_z will be stabilized, and that with negative μ_z will be destabilized. Depending on whether the

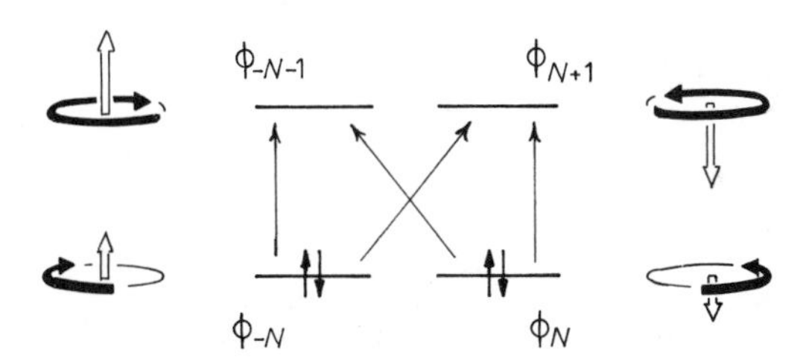

Figure 3.11. HOMO and LUMO of an [n]annulene with $4N+2$ electrons. The sense of electron circulation in these MOs and the resulting orbital magnetic moments are shown schematically (by permission from Souto et al., 1980).

energetically more stable component will be produced by left-handed or by right-handed circularly polarized light, a positive or a negative A term results.

Example 3.7:
The first singlet excitation of the hypothetical unsubstituted cyclobutadiene dication $C_4H_4^{2\oplus}$ (**6a**) corresponds to an electronic transition from the lowest molecular orbital ϕ_0 into the degenerate nonbonding π MOs ϕ_1, ϕ_{-1}. Similarly, the first singlet excitation of the hypothetical unsubstituted cyclobutadiene dianion $C_4H_4^{2\ominus}$ (**6b**) corresponds to an electronic transition from the nonbonding π MOs ϕ_1, ϕ_{-1} into the antibonding π MO ϕ_2 (which is identical with ϕ_{-2}). (Cf. Figure 3.12.) In both molecules this transition is allowed and degenerate and should therefore give rise to an A term in the MCD spectrum. The ground states of $C_4H_4^{2\oplus}$ and $C_4H_4^{2\ominus}$ are nondegenerate and have no magnetic moment, but in the excited states the magnetic moment is no longer zero. In $C_4H_4^{2\oplus}$ one electron remains in ϕ_0 and contributes nothing to the orbital magnetic moment, but the other one will be either in the ϕ_1 or ϕ_{-1}, depending on whether a left-handed or a right-handed circularly polarized photon has been absorbed. It thus produces either a negative (ϕ_1) or a positive (ϕ_{-1}) contribution to the z component of the magnetic moment. In $C_4H_4^{2\ominus}$ the electron that has been pro-

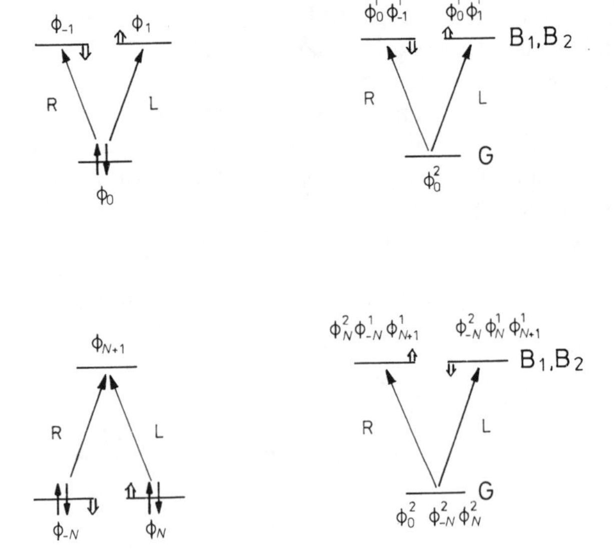

Figure 3.12. MO (left) and state (right) energies for a 2π-electron annulene (top) and for a 2π-hole annulene (bottom). Full arrows indicate the two possible one-electron promotions (left) leading to an excitation from the ground state G to the two components B_1 and B_2 of the degenerate excited state (right). The required photon is right-handed (R) or left-handed (L). Double arrows show the Zeeman shifts (adapted from Michl, 1984).

moted to ϕ_2 ($\equiv \phi_{-2}$) contributes nothing to the magnetic moment. Depending on whether one electron has been promoted by the absorption of a left-handed photon from ϕ_1 to ϕ_2 or by a right-handed photon from ϕ_{-1} to ϕ_2, the positive contribution of one of the two electrons in ϕ_{-1} or the negative contribution of one of the two electrons in ϕ_1 to the z component of the magnetic moment is no longer compensated. Thus, the absorption of a left-handed circularly polarized photon by $C_4H_4^{2\oplus}$ produces an excited state with a negative z component of the magnetic moment, whereas absorption by $C_4H_4^{2\ominus}$ produces one with a positive z component. Signs of the z components are reversed for the absorption of a right-handed circularly polarized photon.

The component with positive μ_z will be stabilized by the Zeeman splitting. In $C_4H_4^{2\oplus}$, that is the component that has been produced by a right-handed photon; in $C_4H_4^{2\ominus}$ it is the one that has been produced by a left-handed photon. Consequently for $C_4H_4^{2\oplus}$, ε_R will differ from zero for lower and ε_L for higher photon energies, whereas for $C_4H_4^{2\ominus}$ the opposite is true. The MCD spectrum is a plot of $\varepsilon_L - \varepsilon_R$, so for $C_4H_4^{2\oplus}$ one anticipates an MCD curve that first dips to negative values and then becomes positive as the photon energy increases. This corresponds to a positive A term. For $C_4H_4^{2\ominus}$ the MCD curve should have the opposite shape, so a negative A term is to be expected.

Similar arguments apply to other highly symmetrical aromatic compounds with only two π electrons or two π "holes." Thus, the positive A term of the croconate dianion $C_5O_5^{2\ominus}$ (**7**) may be rationalized by realizing that it can be formally written as a five times deprotonated pentahydroxy derivative of the trication $C_5H_5^{3\oplus}$. (Cf. Example 3.6.)

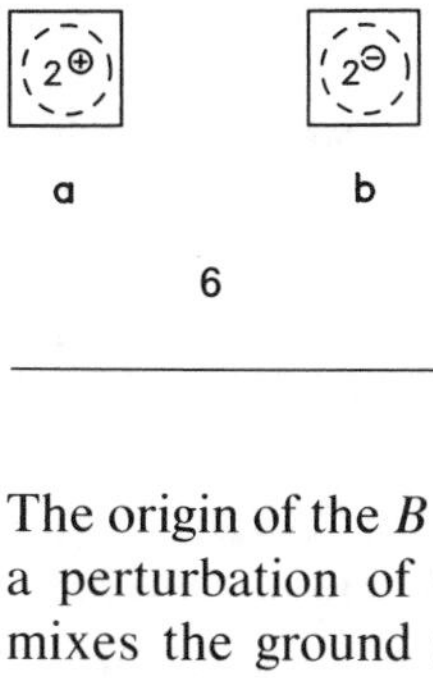

a b

6

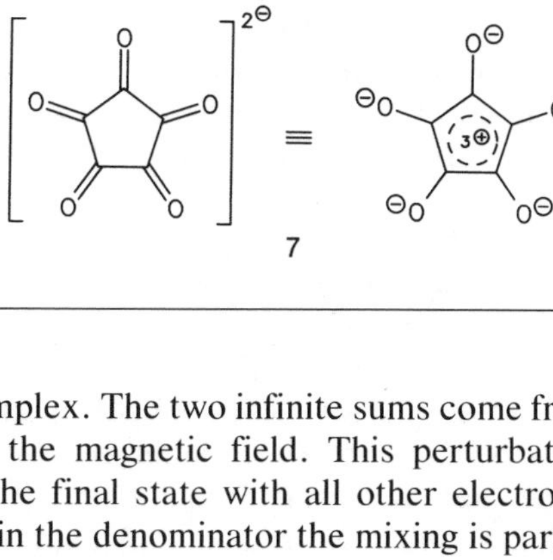

7

The origin of the B terms is more complex. The two infinite sums come from a perturbation of the molecule by the magnetic field. This perturbation mixes the ground state as well as the final state with all other electronic states. Due to the energy difference in the denominator the mixing is particularly effective if both states are close to each other in energy. The second sum, which involves the mutual mixing of excited states, is therefore generally more important than the first one.

The mixing of two states Ψ_i and Ψ_j is dictated by the coupling matrix element $<\Psi_i|\hat{\boldsymbol{M}}|\Psi_j>$. This is small unless the states Ψ_i and Ψ_j can be expressed predominantly by configurations that differ only in promotions of a single electron and that show the appropriate symmetries. Furthermore, it is necessary that the transition moments $<\Psi_0|\hat{\boldsymbol{M}}|\Psi_i>$ and $<\Psi_i|\hat{\boldsymbol{M}}|\Psi_k>$ or $<\Psi_0|\hat{\boldsymbol{M}}|\Psi_i>$ and $<\Psi_k|\hat{\boldsymbol{M}}|\Psi_0>$, respectively, be not parallel but rather as close as possible to mutually perpendicular. This again introduces symmetry restrictions. In practice only a few important terms are left over from the infinite sums, thus making the origin of the B term much easier to understand than might be anticipated.

Finally, it should be noted that for states Ψ_i and Ψ_j that are produced by mutual magnetic mixing, the contributions to the B terms are of equal magnitude but of opposite sign. Whether the contribution of the energetically lower or the higher transition to the B term is positive depends on the orientation of the electric transition moment $<\Psi_0|\hat{M}|\Psi_i>$ relative to the magnetic transition moments $<\Psi_k|\hat{\boldsymbol{M}}|\Psi_0>$ and $<\Psi_i|\hat{\boldsymbol{M}}|\Psi_k>$.

3.3.3 Cyclic π Systems with a $(4N+2)$-Electron Perimeter

The perimeter model for the description of electronic states of aromatic molecules discussed in Section 2.2.2 is also suited for deriving relations between the structure of these compounds and the sign of the energetically lowest MCD bands.

With very few exceptions the HOMO and LUMO of highly symmetrical [n]annulenes with $4N+2$ electrons are degenerate. (Cf. Figure 3.11.) Structural perturbations that produce the actual molecule of interest from the pure perimeter will, in general, lower the symmetry and remove the double degeneracy. The splittings will be denoted by ΔHOMO and ΔLUMO, respectively. (See Section 2.2.5.) According to the perimeter model, the relative magnitude of ΔHOMO and ΔLUMO determines the sign of the B term of the ^{1}L and ^{1}B bands of aromatic molecules.

When ΔHOMO is much larger than ΔLUMO, the signs of the four B terms are $+, -, +, -$ in order of increasing transition energies. When ΔHOMO is much smaller than ΔLUMO, the signs are $-, +, -, +$. These signs are primarily determined by the magnetic mixing of the $^1L_1$ with the $^1L_2$ state and of the $^1B_1$ with the $^1B_2$ state. Such MCD chromophores are referred to as *positive-hard* (ΔHOMO $\gg$ ΔLUMO) and *negative-hard* (ΔHOMO $\ll$ ΔLUMO), since additional perturbations have only a small effect on the relative values of ΔHOMO and ΔLUMO, and thus their MCD signs are hard though not impossible to modify.

A pictorial way of understanding the origin of these so-called μ^+ contributions, which are proportional to the sum $\mu_{N+1} + \mu_N$ of the magnetic moments of the perimeter orbitals involved, is given in Figure 3.13. Since the two perimeter orbitals ϕ_k and ϕ_{-k} ($k = N, N+1$) are exactly degenerate in the absence of a magnetic field, they will enter with exactly equal weights into each of the two MOs that result from their mixing, so the z components of the angular momentum just cancel each other. The perturbation is said to quench the angular momentum. In the magnetic field the degeneracy of ϕ_k and ϕ_{-k} will be removed by Zeeman splitting, and depending on the ratio of the Zeeman splitting to the total splitting, their relative weights in the perturbed MOs will be different, and the angular momentum will be correspond-

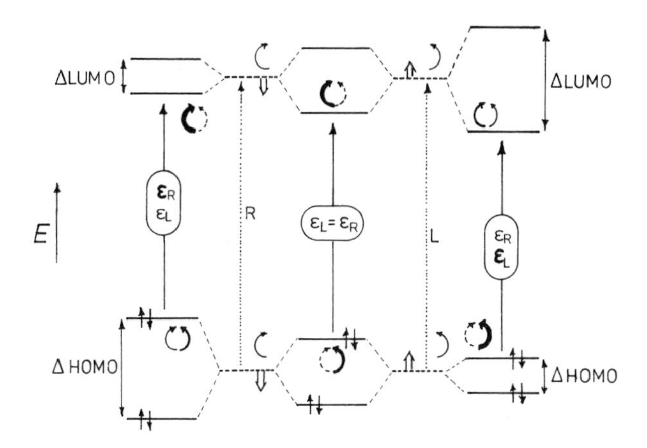

Figure 3.13. Origin of the μ^+ contributions to the B term. The broken horizontal lines indicate the energies of the complex MOs in the absence of the magnetic field, the double arrows give the Zeeman splittings and the round arrows show the sense of electron circulation (viewed from the positive z axis). The canonical MOs of the perturbed annulene, whose energies are given by full lines, result from a pairwise mixing of the perimeter MOs, whose relative weights are indicated by the thickness of the round arrows (by permission from Michl, 1978).

ingly less completely quenched. This is indicated in Figure 3.13 by the varying thickness of the round arrows. The MCD sign can be immediately verified from this diagram.

When ΔHOMO = ΔLUMO, the μ^+ contributions to the B terms vanish. The latter is then determined only by the so-called μ^- contributions, which are proportional to the difference $\mu_{N+1} - \mu_N$ in the magnetic moments of the perimeter orbitals involved in the transition and are therefore primarily determined by the nature of the perimeter and not the nature of the perturbation. For positively charged and uncharged perimeters four MCD bands with signs +, +, +, − in the order of increasing transition energies are to be expected. The μ^- contributions to the first two B terms are very weak and may even vanish depending on the perturbations. MCD chromophores of this type are called *soft* because of the ease with which the MCD signs can be changed by minor further perturbations that remove the equality of ΔHOMO and ΔLUMO.

When ΔHOMO and ΔLUMO differ only slightly, the μ^+ contributions are nonzero, but they may easily be so small that the μ^- contributions are significant as well. This is particularly true for the two B bands whose μ^- contributions are relatively large.

Example 3.8:
Acenaphthylene is a negative-hard chromophore (ΔHOMO < ΔLUMO); pleiadiene is a positive-hard one (ΔHOMO > ΔLUMO). The MCD spectra of these *peri*-condensed aromatics were shown in Figure 3.9. The relative magnitude of

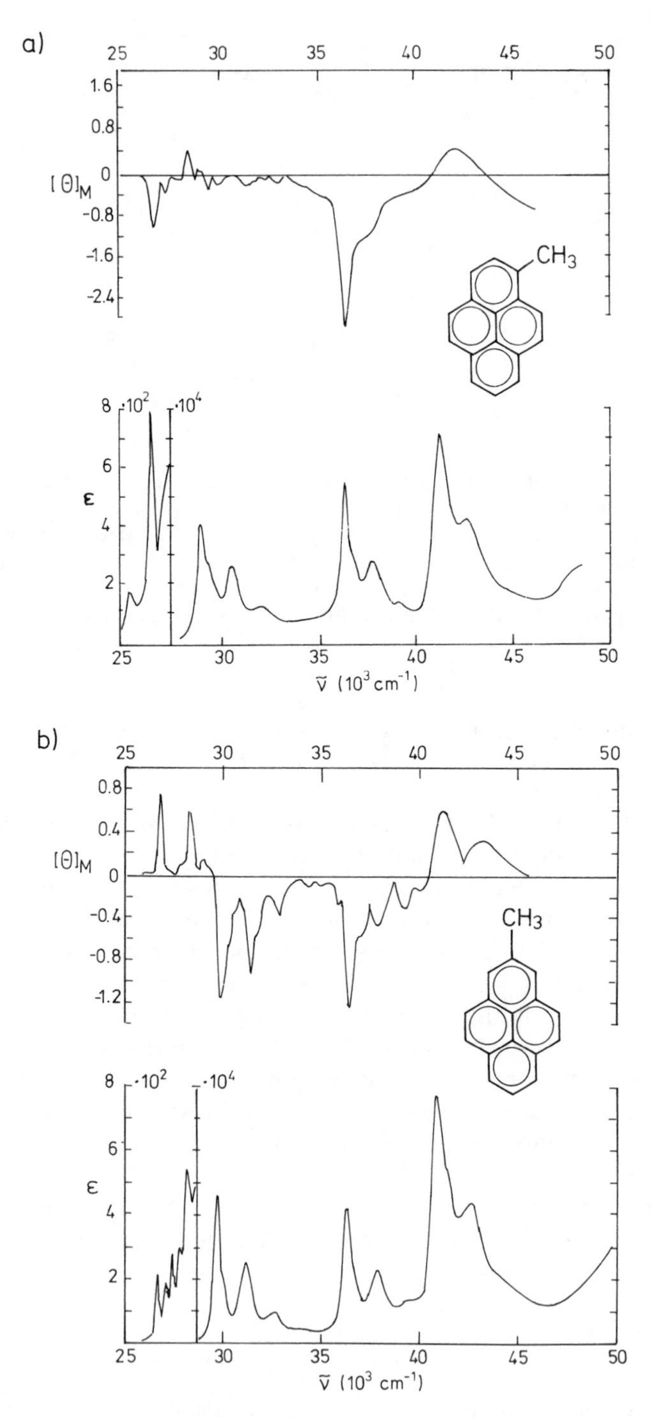

Figure 3.14. MCD (top) and absorption (bottom) spectra of a) 1-methylpyrene and b) 2-methylpyrene (by permission from Michl, 1984).

ΔHOMO and ΔLUMO may be derived from a perturbational treatment of the union of the [11]annulenyl cation and $C^{\ominus}$ or of the [13]annulenide anion and $C^{\oplus}$, respectively. This also shows that these molecules have further excited states in addition to those derived from a $(4N+2)$-electron perimeter. This is true for the longest-wavelength transitions of both molecules, which therefore cannot be labeled within the framework of Platt's nomenclature. Hence, a prediction of their MCD sign on the basis of the perimeter model is also impossible. However, the next two bands correspond to the $^1L_1$ and $^1L_2$ states of a $(4N+2)$-electron perimeter and show the expected behavior in the MCD spectra. In acenaphthylene the order of the B terms is $-$, $+$ and in pleiadiene $+$, $-$. These signs are not changed by perturbing substituents since the difference between ΔHOMO and ΔLUMO is too large. Both molecules represent hard chromophores.

The situation is different for pyrene. This is an alternant hydrocarbon for which ΔHOMO = ΔLUMO must hold to a first approximation (cf. Section 2.2), so the μ^+ contributions, which usually are dominant, vanish. The MCD of the ^1L bands of pyrene is not only very weak but is also affected even by very small perturbations. This is evident from the MCD spectra of 1-methylpyrene and 2-methylpyrene shown in Figure 3.14, from which it is easy to recognize that the signs of the ^1L_b and ^1L_a bands of these two compounds are reversed. The fact that ΔHOMO is somewhat larger than ΔLUMO in 1-methylpyrene, but somewhat smaller than ΔLUMO in 2-methylpyrene, can be understood easily using arguments based on perturbation theory. This problem will be discussed again in a more general way in Example 3.11.

3.3.4 Cyclic π Systems with a 4N-Electron Perimeter

Although the description of electronic states by means of the perimeter model is somewhat less satisfactory for molecules that can be derived formally from an antiaromatic $4N$ perimeter than for aromatic molecules, simple statements about MCD signs are still possible. While nothing can be said about the S and D bands, which according to the perimeter model have zero electric transition moments and which experimentally are found to be very weak (the latter is normally inobservable), predictions are possible for the strong absorptions that are referred to as the N_1, N_2, P_1, and P_2 bands according to the nomenclature given in Section 2.2.7. The parameters that are essential for the MCD spectra of systems derived from a $4N$-electron perimeter are

$$\Delta HL = \Delta H - \Delta L \tag{3.19}$$

and

$$\Delta HSL = (2\varepsilon_N - \varepsilon_{N-1} - \varepsilon_{N+1}) + 2s_D - h_D - l_D \tag{3.20}$$

where according to Figure 2.22, ΔH and ΔL are HOMO and LUMO splittings, respectively, and h_D, s_D, and l_D denote the shifts of average energies of the two components of the HOMO, SOMO, and LUMO, respectively, which were degenerate in the unperturbed perimeter. The values ε_i are the

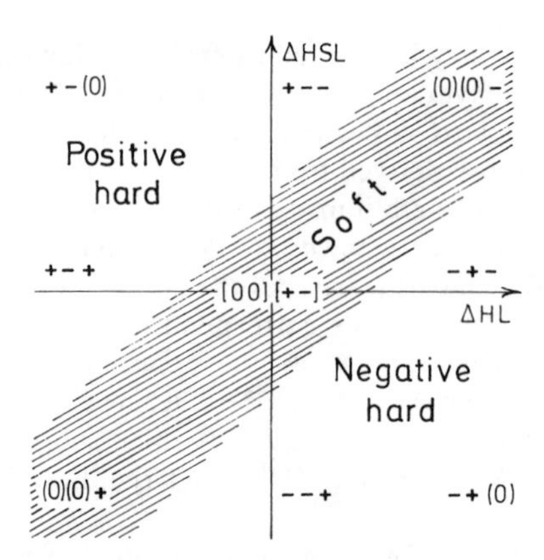

Figure 3.15. *B* term signs of the N and P bands of π systems derived from a $4N$-electron perimeter as a function of ΔHL and ΔHSL. In soft chromophores the intensity of the two longer-wavelength MCD bands is nearly zero. In hard chromophores with $|\Delta HSL| \neq |\Delta HL|$ this is true for the third band. No *B* terms but rather two *A* terms are to be expected for the unperturbed perimeter, of which the first is zero and the second positive, indicated by [00] and [+ −], respectively (by permission from Höweler et al., 1989).

orbital energies of the unperturbed perimeter, so for uncharged $4N$ perimeters the pairing theorem yields $(2\varepsilon_N - \varepsilon_{N-1} - \varepsilon_{N+1}) = 0$. ΔHSL = ΔHS − ΔSL is therefore a measure of the asymmetry between the HOMO-SOMO separation ΔHS and the SOMO-LUMO separation ΔSL that is due to a shift in the average energies of the MO pairs upon perturbation, whereas ΔHL measures the asymmetry of the splitting of these orbital pairs.

In hard chromophores, ΔHL and ΔHSL differ markedly from each other. The cases for which either ΔHL or ΔHSL is nearly zero are particularly important. When ΔHL and ΔHSL are of similar magnitude, the chromophore is soft. For the unperturbed perimeter one has ΔHL = ΔHSL = 0.

The signs of the first three N and P bands expected for different situations can be taken from Figure 3.15. They are independent of the ordering of the transitions, N_1, N_2, P_1, as usual, or N_1, P_1, N_2. The P_2 transition is in general below 200 nm and is therefore difficult to observe.

Example 3.9:

The π systems of acenaphthylene and pleiadiene may not only be formally derived from $(4N+2)$-electron perimeters as in Example 3.7, but just as well from $4N$-electron perimeters. Simple perturbational arguments for the union of the [11]annulenyl anion with $C^{\oplus}$ and of the [13]annulenyl cation with $C^{\ominus}$, respectively, yield relative values for the parameters ΔHL and ΔHSL. For acenaphthylene one finds ΔHL ≈ 0 and ΔHSL < 0; hence it is a negative-hard chromophore; pleiadiene with ΔHL ≈ 0 and ΔHSL > 0, on the other hand, is positive-hard.

The first weak band of both molecules is an S band for which no prediction of the MCD sign can be made. The next two bands, which according to Platt's nomenclature are the $^1L_1$ and $^1L_2$ bands, are now labeled N_1 and N_2. From Figure 3.15 the signs of the *B* terms are −, + for acenaphthylene and +, − for pleiadiene in the order of increasing excitation energies, in agreement with the MCD spectra in Figure 3.9. Thus, the application of the perimeter model to these species, viewed once as derivatives of [4*N*+2]annulenes and once as derivatives of [4*N*]annulenes, results in different nomenclature for their absorption bands but in the same predictions for the signs of their *B* terms.

Not all cyclic conjugated π systems can be formally derived from a (4*N*+2)-electron perimeter as well as from a 4*N*-electron perimeter. Examples are pentalene (**8**) and heptalene (**9**) whose MCD spectra could not yet be measured because of the great instability of these hydrocarbons. Pentalene and heptalene are obtained from the corresponding annulene by symmetrical introduction of a single cross-link, which is an even perturbation and yields $\Delta HL \approx 0$ and $|\Delta HSL| \gg |\Delta HL|$. From the signs of the MO coefficients, which may easily be derived as indicated in Section 2.2.5, it is seen that ΔHSL is negative for pentalene and positive for heptalene. Hence both molecules are hard chromophores and the expected signs of the *B* terms of the N_1, N_2, and P_1 bands are −, +, + for pentalene and +, −, − for heptalene. The MCD spectra of simple derivatives of pentalene and heptalene are shown in Figure 3.16. They both start with an extremely weak low-energy band assigned to the S transi-

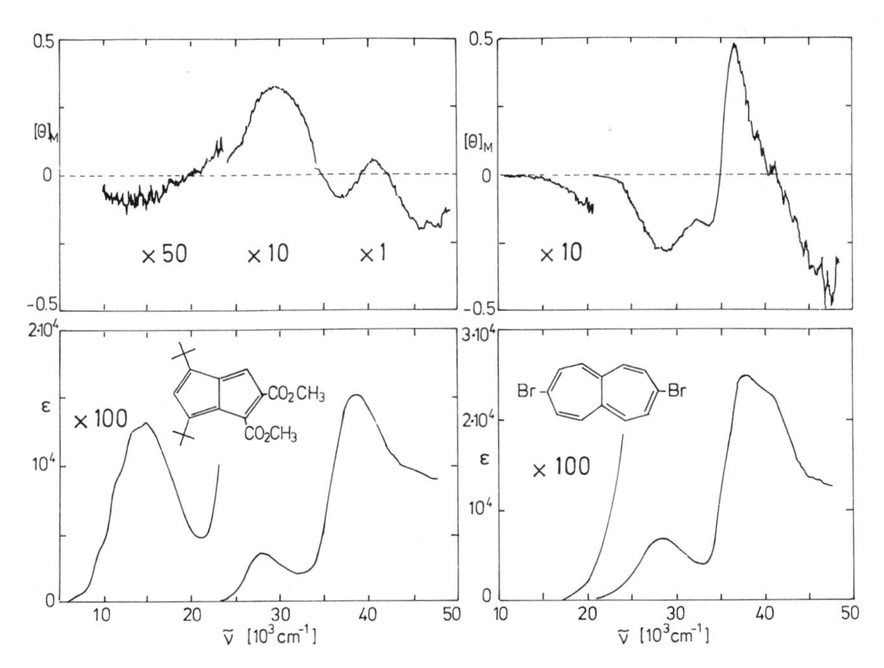

Figure 3.16. MCD and absorption spectra of a) 1,3-di-*t*-butyl-pentalene-4,5-dicarboxylic ester and b) 3,8-dibromoheptalene (by permission from Höweler et al., 1989).

tion, and at higher energies contain the stronger N_1 and N_2 bands, which show the expected signs.

8 9

3.3.5 The Mirror-Image Theorem for Alternant π Systems

Due to the pairing theorem, the absorption spectra and transition polarization directions of two mutually paired alternant systems should be identical, as shown in Figure 2.25 for the radical anion and the radical cation of tetracene. Under the same conditions (i.e., $\beta_{\mu\nu} = 0$ if μ and ν are nonneighbors), it can be shown that the MCD spectra of mutually paired alternant π systems should be mirror images of each other (Michl, 1974c).

From this mirror-image theorem it follows that MCD spectra of uncharged alternant systems that are paired with themselves should be zero. This is only true within the confines of the above approximations, in which the μ^- contributions to the MCD are neglected. The μ^+ contributions are in fact zero for uncharged alternant hydrocarbons since from the pairing theorem it follows that ΔHOMO = ΔLUMO.

The originally only theoretically derived mirror-image theorem for MCD

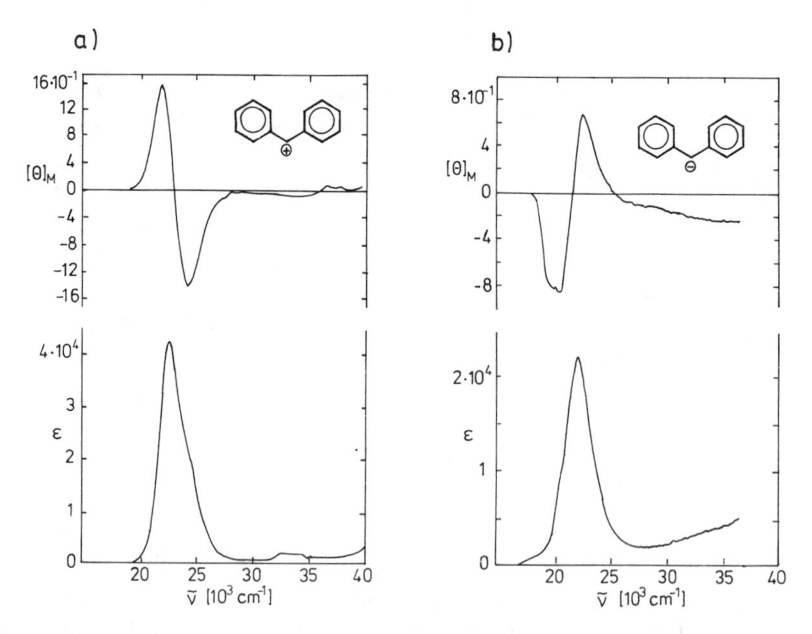

Figure 3.17. MCD and absorption spectra of a) diphenylmethyl cation and b) diphenylmethyl anion (by permission from Tseng and Michl, 1976).

spectra of paired π systems has been confirmed experimentally on several examples. One of these is shown in Figure 3.17.

The fact that the benzyl anion and benzyl cation should have opposite MCD signs according to the mirror-image theorem makes it easy to understand why benzene derivatives with mesomeric donor substituents, which are isoelectronic with the benzyl anion, and benzene derivatives with mesomeric acceptor substituents, which are isoelectronic with the benzyl cation, show opposite MCD signs. This fact can be used for a qualitative and even a quantitative characterization of mesomeric substituent effects. (See also Section 2.4 and Example 3.10.)

3.3.6 Applications

MCD measurements are useful not only in purely spectroscopic investigations, for example, in the detection of hidden absorption bands or for the identification of degenerate absorptions, but also in structural organic chemistry. There is, for instance, a rule that describes the influence of the molecular environment on the MCD effect of the $n \rightarrow \pi^*$ band of a ketone (Seamans et al., 1977; Linder et al., 1977). However, the method is clearly most useful for investigating cyclic conjugated π systems.

All of the examples selected here to illustrate the utility of MCD spectroscopy are based on systems that may be formally derived from a $(4N+2)$-electron perimeter. For such systems the MCD effect is most simply and lucidly connected through the quantities ΔHOMO and ΔLUMO with the form and the ordering of the molecular orbitals. It can therefore be used to draw conclusions regarding the electronic and chemical structure of the system under investigation. A structural class for which the perimeter model analysis has been particularly fruitful is that of the porphyrins (Goldbeck, 1988) and related macrocycles (Waluk et al., 1991; Waluk and Michl, 1991).

Applications become particularly simple if a series of such derivatives is studied that can all be derived from the same π system by introducing structural perturbations. The prediction of changes in the MCD spectrum as a function of such perturbation requires the knowledge of three factors:

1. The nature of the perturbation
2. The location of the perturbation on the molecular framework
3. The energy ordering of the four frontier orbitals derived from the $(4N+2)$-electron perimeter and defined by their nodal properties

A knowledge of all three factors permits a correct prediction of MCD effects, as has been shown in the previous sections. Conversely, if only two of the three factors are known and the MCD spectra are measured, it is possible to draw conclusions concerning the unknown third factor.

For most aromatic systems the appearance of the four frontier orbitals, that is, their nodal properties and the relative magnitude of the LCAO coefficients as well as their energy ordering, is well known. When substituents are introduced into such a system at known locations, MCD spectroscopy can be used to investigate the nature of the electronic perturbation caused by the substituent by examining changes in the relative magnitude of ΔHOMO and ΔLUMO. The effects are particularly pronounced for systems for which ΔHOMO = ΔLUMO, that is, for alternant hydrocarbons. From the data in Example 3.10 it can be seen that even very weak hyperconjugative effects of substituents like SiH_3 and CH_3 may be assessed quantitatively.

Example 3.10:
Since for benzene the MOs are well known and all positions are equivalent the MCD spectra of monosubstituted benzene derivatives are particularly suited for a determination of substituent constants. A more quantitative analysis on the basis of the perimeter model neglecting the μ^- contributions yields the following result for the B term of the ^1L_b transition of substituted benzenes:

$$B = 3.7 \times 10^{-12}(\Delta HOMO + \Delta LUMO)(\Delta HOMO - \Delta LUMO)$$

where B is in units of $D^2\beta_e/cm^{-1}$ and ΔHOMO and ΔLUMO are in cm^{-1} (Weeks

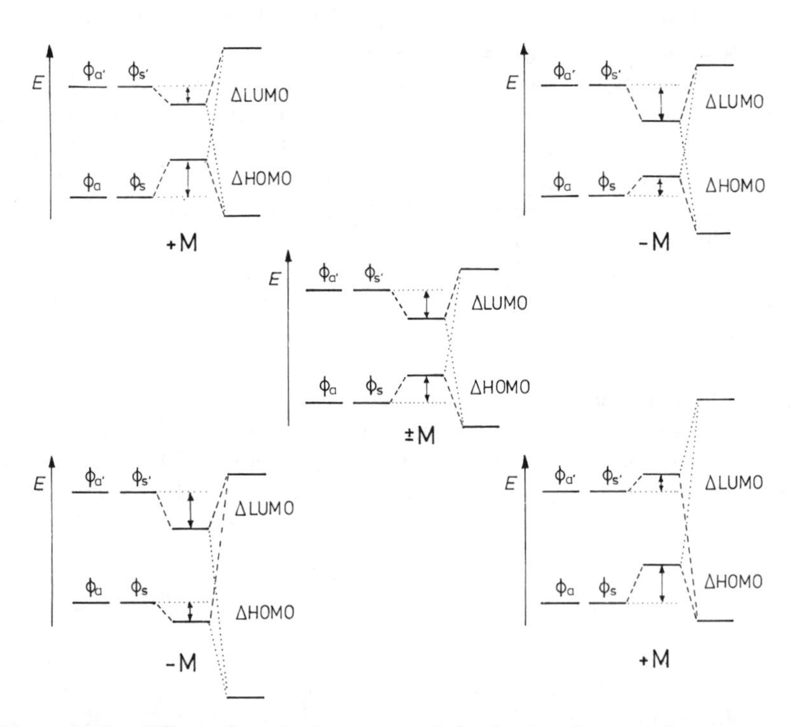

Figure 3.18. Effect of a substituent containing both a donor and an acceptor orbital of π symmetry on the energies of the frontier orbitals of a $(4N+2)$-electron annulene perimeter (schematic). Strong interactions are indicated by dashed lines, weak interactions by dotted lines (by permission from Michl, 1984).

Table 3.1 *B* Terms of the ^1L_b Transition of Substituted Benzenes C_6H_5X (in Units of 10^{-5} $D^2\beta_e/cm^{-1}$)[a]

X	$B - 0.3$	X	$B - 0.3$
H	0	CH_2CMe_3	1.7
CH_3	2.1	CH_2SiMe_3	14.3
CH_2F	−0.9	CH_2GeMe_3	18.7
CHF_2	−2.6	CH_2SnMe_3	27.7
CF_3	−3.1	SiH_3	−5.7
CH_2CN	−0.6	SiH_2Me	−4.9
$CH(CN)_2$	−1.5	$SiHMe_2$	−4.2
$C(CN)_3$	−3.2	$SiMe_3$	−3.6
CH_2SiMe_3	14.3	CMe_3	1.8
$CH(SiMe_3)_2$	13.4	$SiMe_3$	−3.6
$C(SiMe_3)_3$	10.4	$GeMe_3$	−1.8
		$SnMe_3$	−1.5

[a] From Michl (1984); the values shown have been normalized against benzene as the standard by subtracting 3×10^{-6} $D^2\beta_e/cm^{-1}$.

et al., 1986). For a series of weak substituents the percentage changes in (ΔHOMO + ΔLUMO) will be much smaller than those in the much smaller quantity (ΔHOMO − ΔLUMO). To a good approximation the former can therefore be absorbed in the proportionality constant; at least for weak substituents the *B* term will then be proportional to ΔHOMO − ΔLUMO. In a better approximation for strong substituents, variations of the proportionality factor 3.7×10^{-12} (ΔHOMO + ΔLUMO) with substituent strength have to be considered as well. Since to first order in perturbation theory, ΔHOMO − ΔLUMO does not depend on the σ effect of a substituent (cf. Section 2.4), the *B* value may be considered a fair measure of ΔHOMO − ΔLUMO and will therefore be a suitable approximate measure of the net π-electron-donating and -accepting ability of a substituent. Some substituents may act simultaneously as both π donor and π acceptor. When both effects are equal, ΔHOMO = ΔLUMO (Figure 3.18) and the *B* term vanishes as in benzene itself. This is the case for a vinyl substituent on benzene, and the *B* term of the ^1L_b band of styrene is in fact very small. If the π donor effect of the substituent is stronger than the π acceptor effect, ΔHOMO > ΔLUMO is obtained according to Figure 3.18 and the *B* term will be more positive as the donor strength of the substituent becomes more pronounced. If, on the other hand, the substituent is predominantly a π acceptor rather than a π donor, ΔHOMO < ΔLUMO is to be expected from Figure 3.18 and the resultant *B* term will be a measure of the acceptor strength of the substituent. Table 3.1 summarizes the results obtained in this way for a series of hyperconjugative substituents. The nonvanishing *B* term of benzene itself, due to vibronically borrowed intensity, is assumed to be invariant in all of the cases listed and has been subtracted.

If the σ and π effects of the substituents are known, the substitution pattern may be determined in the same way as described in Example 3.11 for 1-methyl and 2-methylpyrene. Since for many heterocyclic compounds

ΔHOMO and ΔLUMO are not very different from each other, the same procedure can be applied to such compounds in order to gain information about the protonation site, for example, in the various aza analogues of indolizine (**10**), or about the tautomerism in heterocycles as exemplified by the lactam–lactim isomerism in 2-pyridone (**11**) and about positional isomerism such as in the tautomeric 3,5- and 3,6-diazaindoles (**12**, **13**). This is possible since the protonation of a nitrogen or the replacement of a carbon by a heteroatom can be viewed as a special case of substitution.

10 11 12 13

Example 3.11:
According to Figure 3.14, the MCD spectra of 1-methyl- and 2-methylpyrene differ in the sign of the first *B* term, which is either positive or negative, depending on whether ΔHOMO − ΔLUMO is larger or smaller than zero. The relative magnitude of the LCAO coefficients of the two highest occupied and the two lowest occupied MOs of pyrene may be estimated from Figure 3.19. For a π donor such as the methyl group, the influence of the unoccupied fron-

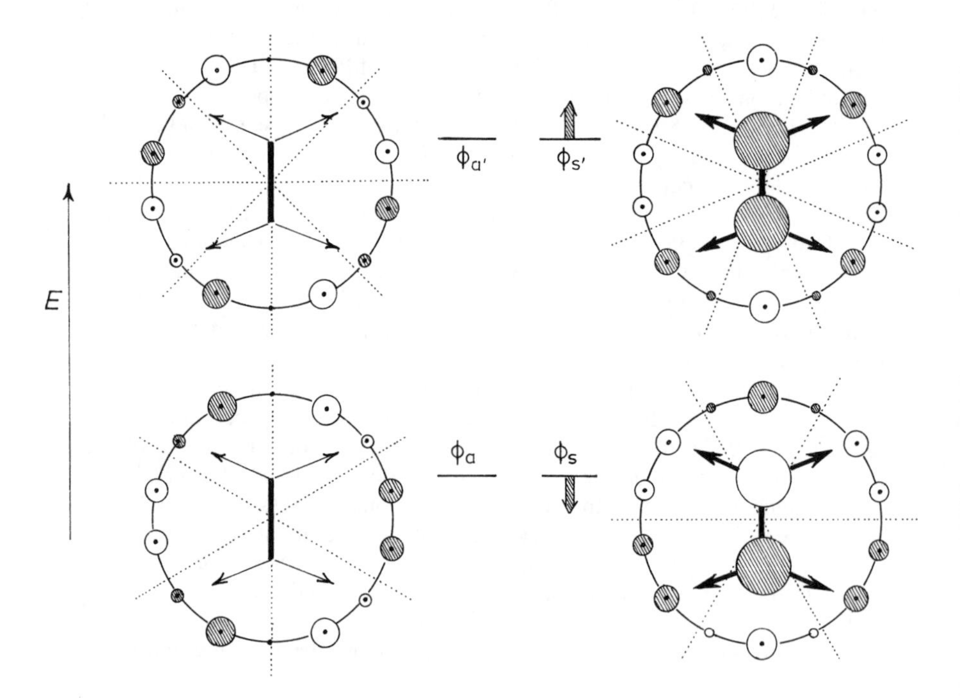

Figure 3.19. The real form of the four frontier orbitals of the 14 π-electron [14]annulene perimeter and the effect of bridging by an ethylene unit (by permission from Michl, 1984).

tier orbitals may be assumed to be negligible compared to the occupied frontier MOs, which are energetically much closer to the donor orbital of the substituent. The interaction with the donor orbital will cause a shift of the occupied orbitals to higher energies that is approximately proportional to the square of the coefficient at the position of the substituent. Hence, it follows that for 1-methylpyrene the antisymmetric MO is more strongly raised than the symmetric one, and ΔHOMO increases. Methyl substitution in the 2-position, however, raises only the symmetric orbital, since the antisymmetric orbital has a node at this position; ΔHOMO then decreases. Therefore, 1-methylpyrene will have a positive first B term because $\Delta HOMO - \Delta LUMO > 0$, whereas for 2-methylpyrene $\Delta HOMO - \Delta LUMO < 0$ and the B term will be negative. If the position of the substituents had not been known in advance, the two isomers could have been easily assigned by means of the MCD spectra.

Finally, if the perturbation due to the substituent as well as its position in the molecular framework is known, MCD spectroscopy can be used to investigate the nature and the ordering of the frontier orbitals. In this way, the relative importance of transannular interactions compared to substituent effects and geometrical distortions for the electronic structure of bridged annulenes such as 1,6-methano[10]annulene (**14**) may be assessed by determining the energy ordering of the frontier orbitals of substituted derivatives of **14** from their MCD spectra. This is shown in detail in Example 3.12.

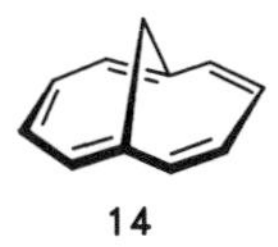

Example 3.12:
From a schematic representation of the frontier orbitals ϕ_s, ϕ_a, $\phi_{a'}$, $\phi_{s'}$ of the [10]annulene perimeter shown in Figure 3.20 it is easy to recognize that trans-

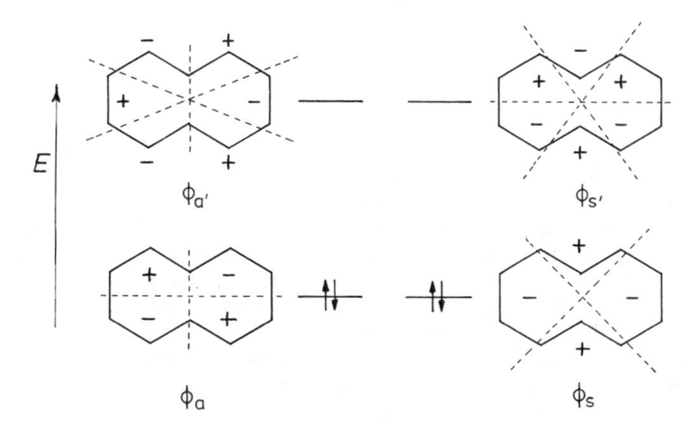

Figure 3.20. Nodal properties of the four frontier orbitals of the [10]annulene perimeter.

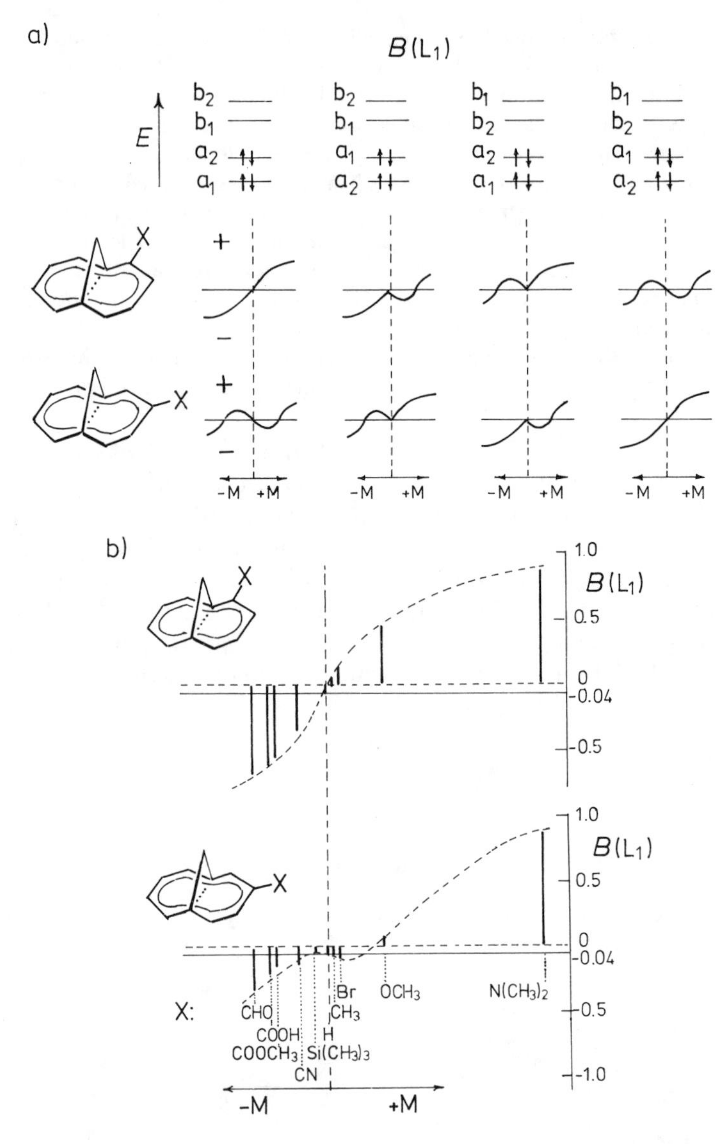

Figure 3.21. Determination of the transannular interaction in 1,6-methano[10]-annulene from an analysis of substituent effects on the MCD spectra: a) theoretically expected dependence of the B term of the ^1L_b transition on substituent strength (+M corresponds to a π donor and −M to a π acceptor) for four possible arrangements of the frontier orbitals that are labeled according to the C_{2v} symmetry of the parent compound, and b) observed B terms of the ^1L_b transition for different substituents in units of 10^{-3} $D^2\beta_e/cm^{-1}$. The solid horizontal lines indicate the B term of the unsubstituted hydrocarbon ($-4 \cdot 10^{-5}$ $D^2\beta_e/cm^{-1}$) (by permission from Klingensmith et al., 1983).

annular interaction, that is, direct through-space interaction between the bridgehead carbons in 1,6-methano[10]annulene, will not affect the energies of the MOs ϕ_a and $\phi_{a'}$ that have a nodal plane through the bridgehead carbons but will stabilize the MO ϕ_s that has no node across the new resonance integral, and destabilize the $\phi_{s'}$ that has a node across the new resonance integral. Thus the anticipated orbital arrangement is ϕ_s below ϕ_a in energy. All other possible perturbations destabilize ϕ_s and leave ϕ_a unchanged, therefore producing the opposite ordering of the occupied orbitals, with ϕ_a below ϕ_s. For instance, dihedral twisting of the partial double bonds in the molecule results in a reduction of the resonance integrals for the four π bonds originating at the bridgehead carbon atoms, and will therefore destabilize the MO ϕ_s, which is binding in these bonds. In order to derive the orbital ordering from the MCD spectrum, use can be made of the fact that according to Figure 3.20 the LCAO coefficients of the MOs ϕ_s and $\phi_{s'}$ are larger in absolute value in position 3 than in position 2, whereas the opposite holds for ϕ_a and $\phi_{a'}$. Therefore, all that needs to be done in order to determine the order of the four frontier orbitals is to measure the B term of the ^1L_b band of some derivatives of 1,6-methano[10]annulene with substituents in positions 2 or 3. The theoretically expected shapes for the dependence of the B term of the ^1L_b band on substituent strength for both positions of substitution are shown in the upper part of Figure 3.21 for four possible orbital orderings. The measured B terms for various substituents are displayed in the lower part of Figure 3.21. The resulting shapes show unambiguously that the orbital ordering is that expected only if transannular interaction is the dominant effect; that is, ϕ_s (labeled a_1 in Figure 3.21) below ϕ_a (a_2) and $\phi_{a'}$ (b_1) below $\phi_{s'}$ (b_2).

Supplemental Reading

General

Caldwell, D.J., Eyring, H. (1971), *The Theory of Optical Activity*; Wiley-Interscience: New York.

Natural Circular Dichroism

Barron, L.D. (1982), *Molecular Light Scattering and Optical Activity*; Cambridge University Press: Cambridge.

Charney, E. (1979), *The Molecular Basis of Optical Activity*; Wiley: New York.

Mason, S.F. (1982), *Molecular Optical Activity and the Chiral Discriminations*; Cambridge University Press: Cambridge.

Nakanishi, K., Berova, N., Woody, R.W. (1994), *Circular Dichroism: Principles and Applications;* VCH Publishers, Inc.: New York.

Snatzke, G. (1979), "Circular Dichroism and Absolute Conformation—Application of Qualitative MO Theory to Chiroptical Phenomena", *Angew. Chem. Int. Ed. Engl.* **18**, 363.

Exciton Chirality Model

Harada, N., Nakanishi, K. (1983), *Circular Dichroic Spectroscopy*; University Science Books: New York.

Magnetic Circular Dichroism

Michl, J. (1984), "Magnetic Circular Dichroism of Aromatic Molecules", *Tetrahedron* **40**, 3845.

Michl, J. (1978), "Magnetic Circular Dichroism of Cyclic π-Electron Systems", *J. Am. Chem. Soc.* **100**, 6801, 6812, 6819.

Piepho, S.B., Schatz, P.N. (1983), *Group Theory in Spectroscopy*; Wiley: New York.

CHAPTER

4

Potential Energy Surfaces: Barriers, Minima, and Funnels

The theoretical treatment of photophysical and photochemical processes within the framework of the Born-Oppenheimer approximation requires knowledge of the potential energy surfaces of the ground state and of one or more excited states, and a qualitative understanding of the way in which these surfaces govern nuclear motion. Some of the fundamental principles necessary for such a discussion are presented in the first section of this chapter. There follows a brief outline of correlation diagrams, which in many cases provide very illuminating qualitative descriptions of the essential characteristics of potential energy surfaces. Detailed quantum chemical calculations of such surfaces for low-symmetry systems of interest to organic chemists have very recently become feasible, but the high dimensionality of the problem makes the searches for the topologically relevant features of the surfaces relatively demanding. Many fundamental problems can be solved using simpler model systems and these have provided a useful and intuitively satisfying basis for a general discussion of complex photoprocesses. This is illustrated in the last two sections of this chapter.

4.1 Potential Energy Surfaces

4.1.1 Potential Energy Surfaces for Ground and Excited States

Within the Born-Oppenheimer approximation—that is, after separating off the nuclear motion—adiabatic potential energy surfaces are obtained by solving the electronic Schrödinger equation for a great number of nuclear geometries. A plot of the lowest energy $E(\boldsymbol{Q})$ as a function of all geometrical variables (combinations of bond lengths, angles, etc.) $Q_1, Q_2, \ldots, Q_F \equiv \boldsymbol{Q}$ provides an F-dimensional surface for the ground state in a space that also contains the energy as an additional dimension. This is sometimes referred to as an F-dimensional hypersurface in an $(F+1)$-dimensional hyperspace. All isomers that can be formed from a given collection of atomic nuclei and electrons have a common ground-state surface and correspond to minima therein. Since the Schrödinger equation has an infinite number of solutions at each nuclear geometry $\boldsymbol{Q}$, an infinite number of Born-Oppenheimer surfaces can be obtained in principle. Among these, however, usually only the ground-state surface and the lowest few excited-state surfaces are of interest.

The potential energy surfaces form the basis for a detailed description of the reaction process. Its quantum mechanical treatment requires a solution of the time-dependent nuclear Schrödinger equation with $E(\boldsymbol{Q})$ as the potential. Classically, nuclear motion is handled by finding the classical trajectory of a point that moves without friction on the surface $E(\boldsymbol{Q})$. The forces that operate on the molecule at a given nuclear arrangement $\boldsymbol{Q}$ are given by the steepest slope on the surface at that particular point (i.e., by minus the gradient of the surface at that point). When there are no interactions with the environment, the total energy is constant, but the kinetic and the potential energy fractions vary continuously as the point moves along the surface. Most often, we display only one vibrational mode at a time in a plot of energy against displacement in nuclear configuration space. In such a drawing, the potential energy is represented by the potential energy curve and the amount of kinetic energy residing in this particular mode is shown by the height of a point above the surface. Although the total energy of the molecule is constant, its distribution among the various modes of nuclear motion varies in time, and so does the height of the point that represents the molecule in this two-dimensional plot. Travel over barriers is possible only occasionally, when the mode happens to have acquired enough kinetic energy and the point is high enough to pass over the barrier.

The following discussion makes frequent use of such one-dimensional cross sections through potential energy surfaces. The reaction coordinate Q used as the abscissa often remains unspecified in schematic representations. Caution is required in interpreting such cross sections. What appears as a minimum, barrier, or saddle point in one cross section may look quite different in another one. A typical example is a maximum of a reaction profile, which appears as a minimum in a cross section perpendicular to the reaction coordinate.

Since spin-orbit coupling is normally not included in the Born-Oppenheimer Hamiltonian, singlet and triplet states can be distinguished. In a discussion of photochemical processes, large areas of the nuclear configuration space are of interest, and it is useful to label the energy surfaces in a way that differs from the one conventionally used by spectroscopists. At any

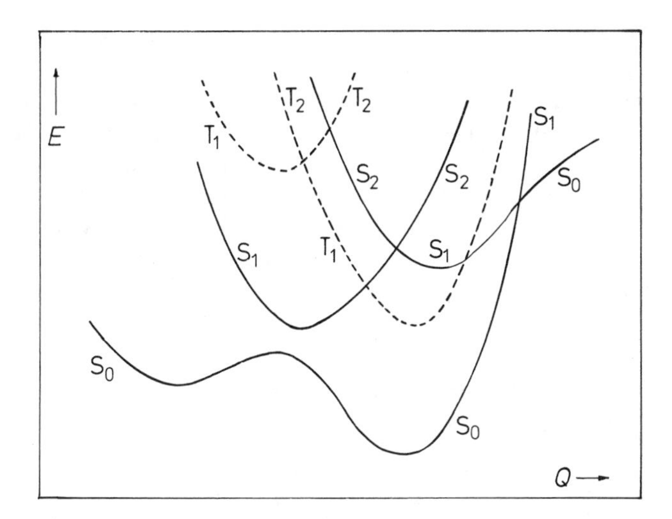

Figure 4.1. Schematic representation of Born-Oppenheimer potential energy surfaces. Using the photochemical nomenclature, the ground-state surface of a closed-shell system, which is the lowest singlet surface, is labeled S_0, followed by S_1, S_2, etc. in order of increasing energies. The triplet surfaces are similarly labeled T_1, T_2, etc.

geometry, states of a given multiplicity are labeled sequentially in the order of increasing energies. Surfaces of different multiplicity can cross freely; such crossings do not disturb the labeling of the surfaces by S_0, S_1, S_2, . . . and T_1, T_2, T_3, . . . On the other hand, if states of the same multiplicity do cross, they exchange their labels as defined by the convention used here. (See Figure 4.1.) Therefore, the surface S_n (T_n) can touch the surfaces S_{n-1} (T_{n-1}) and S_{n+1} (T_{n+1}) but cannot cross them. The situation depicted in Figure 4.1 corresponds to a touching of S_0 and S_1, S_1 and S_2, and T_1 and T_2. In polyatomic molecules, such touching of surfaces of identical spin and spatial symmetry is allowed (v. Neumann and Wigner, 1929; Teller, 1937; Herzberg and Longuet-Higgins, 1963; Salem, 1982), although along many paths "intended" touchings are still avoided. (Cf. Example 4.1.)

The convenience gained by the use of the "photochemical" nomenclature is that all isomers of the same formula have a common ground-state surface S_0. In the spectroscopist's convention, labels of crossing surfaces are kept and S_0 of one isomer could lie above S_1 at the geometry of another isomer or even above itself, at the same geometry, provided two or more surfaces cross in an appropriate manner in several dimensions.*

*Figure 4.2b (Section 4.1.2) demonstrates that in this convention S_0 can lie above itself. Both surfaces would have to be labeled S_0 if one followed a cut from the minimum in S_0 to the intersection and further up, and then sideways and back to the original geometry, avoiding now the intersection.

Using the convention adopted here, it becomes possible to make general statements such as the following: "During a photochemical reaction every molecule, independent of any geometric changes it undergoes, will eventually return to the surface labeled S_0 (or T_1)." The disadvantage of this convention is that the physical nature of any given state can vary enormously and abruptly for different nuclear geometries, particularly in areas with unavoided or weakly avoided surface touchings. Thus the lowest excited-singlet state S_1 can be of (n,π^*) nature for some geometries of a molecule, and (π,π^*) for others.

4.1.2 Funnels: True and Weakly Avoided Conical Intersections

Regions of surface touching or near touching are referred to jointly as funnels (Michl, 1972). As we shall discuss, they play a central role in photochemistry.

For a long time it was believed that most surface touchings are at least weakly avoided and that true touchings are relatively rare except at high-symmetry geometries, where the touching states can be of different electronic symmetry. In diatomic molecules, whose nuclear configuration (geometry) is defined by a single degree of freedom ($F = 1$), it can be shown rigorously that touching (crossing) of surfaces can occur only if the electronic symmetries of the two states are different (*noncrossing rule*).

Example 4.1:
Consider two many-electron basis functions Φ_1 and Φ_2, interacting through the Hamiltonian $\hat{H}$:

$$\begin{vmatrix} H_{11}-E & H_{12} \\ H_{21} & H_{22}-E \end{vmatrix} = 0, \quad H_{ij} = \int \Phi_i \hat{H} \Phi_j \mathrm{d}\tau$$

Since $H_{21} = H_{12}$, the two resulting state energies are

$$E_{1,2} = \frac{1}{2}\{H_{11} + H_{22} \pm [\,(H_{11} - H_{22})^2 + 4H_{12}^2]^{1/2}\}$$

Two conditions need to be fulfilled for the two energies E_1 and E_2 to be equal:

$$H_{11} = H_{22}$$
$$H_{12} = 0$$

A diatomic molecule has only the internuclear distance Q as an internal coordinate ($F = 1$). Unless H_{12} vanishes because Φ_1 and Φ_2 are of different symmetry, it is in general impossible to find a value of Q that would satisfy simultaneously both conditions. The energies E_1 and E_2 therefore are different, and in one dimension, two states of the same electronic symmetry cannot cross. In a system with two independent internal coordinates Q_1 and Q_2 ($F = 2$), for

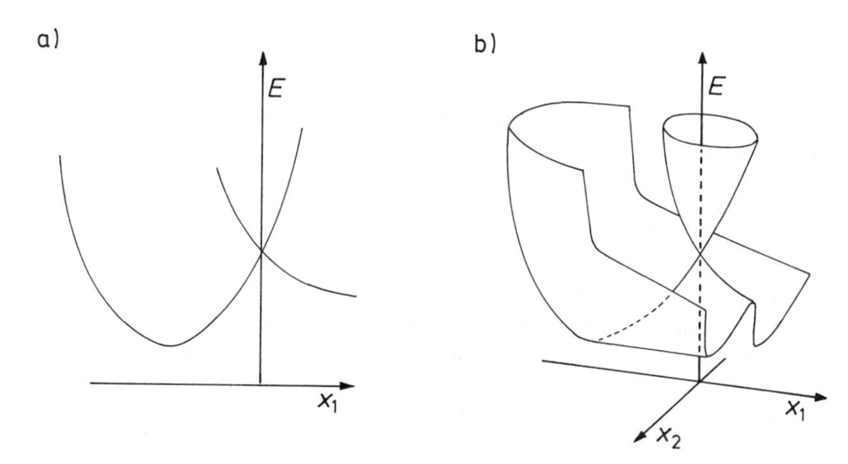

Figure 4.2. Conical intersection of potential energy surfaces with different symmetries with respect to a symmetry element which is present only along the x_1 axis; a) cross section along this x_1 axis; b) three-dimensional representation, with symmetry lowering along the x_2 axis (adapted from Lorquet et al., 1981).

instance, in a triatomic molecule constrained to be linear, there may be one or more zero-dimensional subspaces (points) at which the two conditions given above are fulfilled by a judicious choice of Q_1 and Q_2. (Cf. Figure 4.2.) In polyatomic molecules with a larger number of degrees of freedom, it is generally possible to satisfy the two conditions simultaneously, and surfaces of the same electronic symmetry can cross.

Recent work has shown that in polyatomic molecules ($F > 2$) the unavoided touching of states of equal electronic symmetry is actually quite common (Bernardi et al., 1990b; Xantheas et al., 1991), hence that the earlier belief in its scarcity was wrong. The dimensionality of the subspace in which the energies of the two touching states are equal (the *intersection coordinate subspace*) is $F - 2$. In the remaining two dimensions (x_1, x_2) of the total F-dimensional nuclear configuration space, which define the *branching space,* the touching is at least weakly avoided. In the immediate vicinity of a touching point, a plot of the surface energies has the form of a double cone. At the cone touching point, the two states are degenerate, and as one moves away by an infinitesimal amount in the x_1, x_2 plane, the degeneracy is lifted. Thus, at most points in the total F-dimensional space, the touching is avoided, but in an ($F - 2$)-dimensional "hyperline," it is not (Atchity et al., 1991).

Example 4.2:
In the adiabatic basis obtained by the diagonalization of $\hat{H}$, the vector x_1 is defined as the gradient difference

$$x_1 = \frac{\partial(E_1 - E_2)}{\partial q}$$

and in this direction the difference in the slopes of the upper and the lower surface is the largest. The vector x_2 is given by

$$x_2 = \langle\Psi_1|\frac{\partial\Psi_2}{\partial q}\rangle$$

and this is the direction of nuclear displacement that mixes the two adiabatic electronic wave functions of the cone point the best. If the electronic wave functions Ψ_1 and Ψ_2 are of different symmetries at the cone point, x_2 is the symmetry-lowering nuclear coordinate that permits them to mix. (Cf. Figure 4.2b.) The vectors x_1 and x_2 are in general not collinear and often are close to orthogonal.

Efficient algorithms have been developed for locating conical intersections and for locating the points of minimum energy within the $(F - 2)$-dimensional intersection coordinate subspace: that is, the true bottom of the funnel (Yarkony, 1990; Ragazos et al., 1992; Bearpark et al., 1994), and in recent years, quite a few have been found. (See Chapter 7.) Note that the gradient on neither of the touching potential energy surfaces is zero at the minimum energy point of the intersection, as it would be in a true stationary point. Rather, it is only the projection of the gradient into the $(F - 2)$-dimensional intersection coordinate subspace that is zero. Therefore, the bottom of the funnel is not a minimum in the sense of having a zero gradient but only in the sense of having the lowest energy (except if the funnel is very tilted). Thus, gradient-based search routines will not recognize this minimum in the F-dimensional space as such.

Unlike the general case of conical intersection in polyatomic molecules, which occurs between states of equal electronic symmetry and has not been well documented until relatively recently, the special case in which the condition $H_{12} = 0$ (Example 4.1) is enforced by a difference of symmetries of the two states has been well known and studied for a long time. This class of conical intersections is much easier to access computationally and to predict from correlation diagrams. The reaction path along which high symmetry is preserved, and along which H_{12} vanishes, corresponds to x_1. The symmetry-lowering direction that makes H_{12} nonzero, and thus induces a mixing of the two zero-order states, corresponds to x_2 (Figure 4.2).

The expression "avoided state touching (crossing)" can refer to one of two situations (Salem et al., 1975). First, an avoided touching can be encountered along paths that pass close to a real state touching but miss the tip of the cones and do not quite reach the required values of x_1 and x_2 to enter the $(F - 2)$-dimensional intersection coordinate subspace. An example would be a path that does not quite preserve the high symmetry that is characteristic of some conical intersections. Second, it can be encountered upon improvement of a simpler quantum mechanical approximation in which a crossing was unavoided because H_{12} was approximated by zero or in which H_{11} and H_{22} were set equal.

The wave functions of the states we have dealt with so far diagonalize the complete electronic Hamiltonian and are called *adiabatic*. As we shall discuss in more detail in Section 6.1, they govern the nuclear motion if it is sufficiently slow. For description of fast motion on surfaces, it is sometimes an advantage to work with surfaces that do not diagonalize the electronic Hamiltonian and therefore cross, such as those obtained in the approximate treatments referred to above. Such states are called *diabatic* or *nonadiabatic*. They are described by wave functions that do not greatly change their physical nature at the crossing. Thus, in the region of an avoided crossing, those two parts of the adiabatic potential energy surface that in the nonadiabatic approximation correspond to the same surface (e.g., the parts of the surfaces at the upper left and lower right of Figure 4.3), will be described by very similar wave functions, whereas the character of the wave function changes appreciably along the adiabatic potential surface. If the nuclei are moving so fast that the electronic motion cannot adapt instantaneously to the nuclear configuration as required in the adiabatic limit, the molecule may follow the nonadiabatic surface instead; that is, a jump from one to another adiabatic state may occur. In many dimensions, the situation is more complicated, but in general, arrival of a nuclear wave packet into a region of an unavoided or weakly avoided conical intersection means that the jump to the lower surface will occur with high probability upon first passage. The expression "funnel" is used for regions of a surface in which passage to another surface is so fast that there is no time for vibrational equilibration before the jump, and it can refer both to an unavoided and to a weakly avoided conical intersection (Section 6.1).

In perturbation theory language, one can say that the adiabatic Born-Oppenheimer states are a good zero-order approximation for describing nuclear motion if the non–Born-Oppenheimer corrections are small. This is

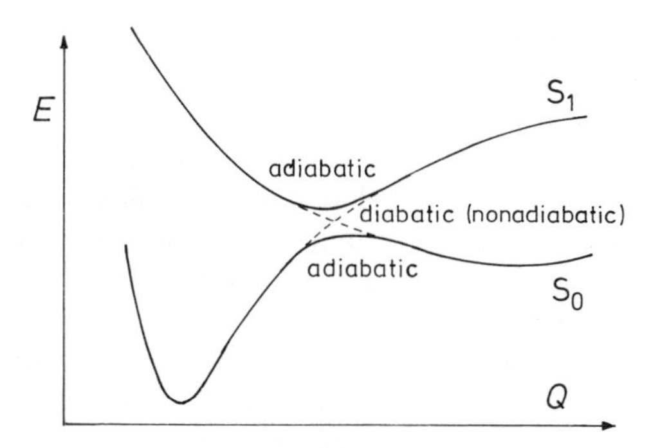

Figure 4.3. Schematic representation of adiabatic and diabatic (nondiabatic) potential energy surfaces S_0 and S_1 (adapted from Michl, 1974a).

because large terms that cause the crossing to be avoided—for example, electron repulsion—are already included in the description of the Born-Oppenheimer states. Diabatic states, on the other hand, are a good zero-order approximation in the opposite case.

4.1.3 Spectroscopic and Reactive Minima in Excited-State Surfaces

A qualitative anticipation of the location of minima and funnels on excited-state potential energy surfaces, in particular the S_1 and T_1 surfaces, is in general more difficult than the estimation of the minima on the ground-state surface S_0 that correspond to stable ground-state species. There are three basic types of geometries where one would intuitively expect minima and/or funnels in S_1 and T_1—namely, near ground-state equilibrium geometries, at exciplex and excimer (complex) geometries, and at biradicaloid geometries.

Minima on excited-state surfaces in the region of ground-state equilibrium geometries may be referred to as *spectroscopic minima*. Spectroscopic transitions from the ground state to such minima or vice versa are in general easy to observe, even if the probability may be somewhat constrained by the Franck-Condon principle. Excitation from the ground state to spectroscopic minima is approximately vertical. Generally, upon return from these minima to the ground state by fluorescence, phosphorescence, or radiationless processes, an excited molecule will end up in the initial minimum of S_0, and the excitation-relaxation sequence corresponds to a *photophysical process* without chemical conversion. In particular in larger molecules such as naphthalene, the promotion of an electron from a bonding into an antibonding MO causes such a small perturbation of the total bonding situation that its effect on the equilibrium geometry of the excited state is relatively small. It can, however, also be more significant. Thus, formaldehyde is pyramidal in the $^1(n,\pi^*)$ as well as in the $^3(n,\pi^*)$ state, and not planar as in the ground state. (Cf. Section 1.4.1.)

Exciplex minima can be viewed as a particular case of spectroscopic minima. Their presence in the excited surfaces reflects the fact that molecules are more polarizable, more prone to charge-transfer processes, and generally "stickier" in the excited state. These minima occur at geometries that correspond to a fairly close approach of two molecules at their usual ground-state geometries, for example, the face-to-face approach of two π systems generally associated with excimers and exciplexes. The ground-state surface normally does not have a significant minimum at the complex geometry when the very shallow van der Waals minimum is disregarded, except in the so-called charge-transfer complexes and similar cases. Radiative or nonradiative return from this type of minimum in an excited state to the ground state leads to the reformation of the two starting molecules in an overall photophysical process.

If a pair of nearly degenerate approximately nonbonding orbitals is occupied with a total of only two electrons in the simple MO picture of the ground state, the molecule is called a *biradical* or a *biradicaloid*. Various types of biradicals and biradicaloids will be discussed in some detail in Section 4.3. A usual prerequisite for the presence of degenerate orbitals is a specific nuclear arrangement that is referred to as biradicaloid geometry. Minima and funnels at such biradicaloid geometries, that is, biradicaloid minima or funnels, normally are *reactive minima* or *funnels*. They are typically characterized by a small or zero (in the case of critically heterosymmetric biradicaloids, see Section 4.3.3) energy gap between the potential energy surfaces; a usually fast radiationless transition from the excited state to the ground state; and normally a shallow minimum in the ground-state surface, if any. A molecule in a biradicaloid minimum generally is a very short-lived species. After return to S_0, deeper minima at geometries other than the initial one will usually be reached, so the net process corresponds to a photochemical conversion. The relevance of biradicaloid minima for photochemical reactivity was first pointed out by Zimmerman (1966, 1969) and van der Lugt and Oosterhoff (1969). They represent funnels or leaks in the excited-state potential energy surface (Michl, 1972, 1974a).

Biradicaloid geometries are in general highly unfavorable in the ground state, since the two electrons in nonbonding orbitals contribute nothing to the number of bonds. As a result, the total bonding is less than that ordinarily possible for the number of electrons available. When the geometry is distorted to one in which the two orbitals are forced to interact, a stabilization will result. This is evident from Figure 4.4, which shows that the two orbitals combine into one bonding and one antibonding orbital, and both

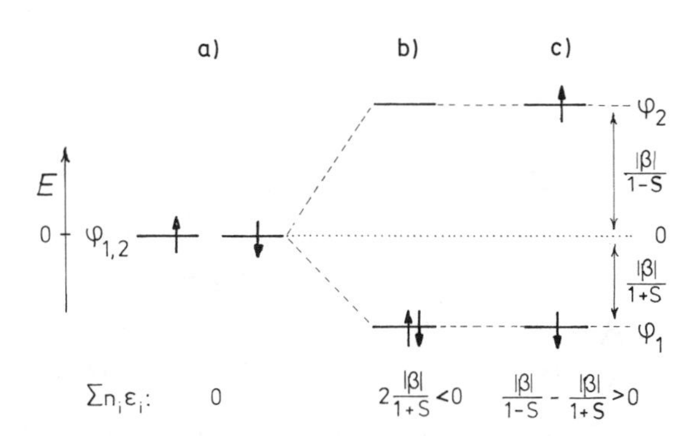

Figure 4.4. Orbital energy scheme for two orbitals, φ_1 and φ_2, which are occupied with a total of two electrons: a) for biradicaloid geometries φ_1 and φ_2 are degenerate; from the three possible singlet configurations only one is shown; b) if the degeneracy is removed the ground state will be stabilized; c) excited configurations will usually be destabilized. The doubly excited configuration and the triplet configuration will be destabilized correspondingly.

electrons can occupy the bonding orbital. In the excited state, however, any distortion from a biradicaloid geometry that causes the two orbitals to interact, and split into a bonding and an antibonding combination, is likely to lead to destabilization, since only one of the two electrons can be kept in the bonding MO, while the other one is kept in the antibonding MO, and the destabilizing effect of the latter predominates.

Here, we consider the two simplest and most fundamental cases of biradicaloid geometries, reached from ordinary geometries by stretching a single bond or by twisting a double bond. These two cases together underlie much of organic photochemistry.

The stretching of an H_2 molecule provides an illustration of the dissociation of a prototypical single bond (see Figure 4.5): The minimum in S_0 occurs at small internuclear distances (74 pm) and is definitely not a biradicaloid minimum since the σ_g MO is clearly bonding and the σ_u^* MO clearly antibonding at this geometry. In the dissociation limit, the S_0 state corresponds to a pair of H atoms coupled into a singlet and has a high energy, the same as the T_1 state. In the latter, the coupling of the electron spins is different, but this has no effect on the energy since the H atoms are far apart. The T_1

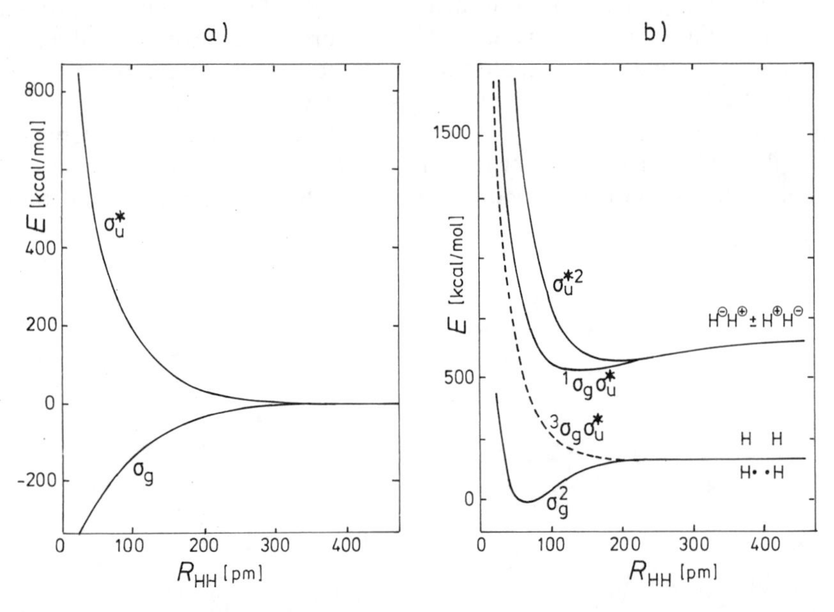

Figure 4.5. Energies a) of the bonding MO σ_g and the antibonding MO σ_u^*, as well as b) of the electronic states of the H_2 molecule as a function of bond length (schematic). On the left, the states are labeled by the MO configuration that is dominant at the equilibrium distance; on the right, they are labeled by the VB structure that is dominant in the dissociation limit. Rydberg states are ignored (by permission from Michl and Bonačić-Koutecký, 1990).

state has its minimum at the infinite separation of the two H atoms where both MOs are nonbonding. The fully dissociated geometry thus clearly corresponds to a biradicaloid minimum. The S_1 state has a minimum at 130 pm, that is, at intermediate nuclear separations. For this geometry the MOs σ_g and σ_u^* are much less bonding and antibonding, respectively, than for the ground-state geometry, so the minimum can also be referred to as biradicaloid. The difference in the minimum geometry of T_1 and S_1 is easily understood when it is realized that S_1 and S_2 are of zwitterionic nature and may be described by the VB structures $H^{\ominus}H^{\oplus}$ and $H^{\oplus}H^{\ominus}$. (See Example 4.3.) The location of the minimum in S_1 represents a compromise between the tendency to minimize the energy difference between σ_g and σ_u^*, favoring a large internuclear distance, and to minimize the electrostatic energy corresponding to the charge separation in the zwitterionic state, favoring a small internuclear distance.

Even though H_2 may not appear to be of much interest to the organic chemist, the importance of Figure 4.5 cannot be overemphasized, and we shall refer to it frequently in the remainder of the text. This is because it represents the dissociation of a single bond in its various states of excitation, and the breaking of a single bond, usually aided by the bond's environment, underlies most chemical reactions. In a sense, Figure 4.5 represents the simplest orbital and state correlation diagrams, to which Section 4.2 is dedicated. Introduction of perturbations by the environment of the bond converts its parts a and b, respectively, into correlation diagrams for the orbitals (e.g., Figures 4.10 and 4.11) or the states (e.g., Figure 4.13) of systems of actual interest for the organic photochemist. In Section 4.3 we shall examine the wave functions of an electron pair in its various states of excitation in more detail, and shall describe the relation between the MO-based description (left-hand side of Figure 4.5b) and the VB-based description (right-hand side of Figure 4.5b). A recent illustration of the same principle in the case of a C—C bond, more directly relevant to organic photochemistry, is provided by the contrast between the Franck-Condon envelopes of the $S_0 \rightarrow S_1$ and the $S_0 \rightarrow T_1$ transitions in [1.1.1]propellane (**1**) (Schafer et al., 1992). These make it clear that the length of the central bond is nearly the same in the ground state and the $^1(\sigma,\sigma^*)$ excited state, which can be thought of as a contact ion pair. In contrast, in spite of the constraint imposed by the tricyclic cage, the central bond is substantially longer in the $^3(\sigma,\sigma^*)$ state, which can be thought of as a repulsive triplet radical pair.

1

However, it is also important to note the essential limitations of the diagrams presented in Figure 4.5: they apply to the dissociation of single bonds that (1) are nonpolar and (2) connect atoms neither of which carries lone-pair AOs or empty AOs nor is involved in multiple bonds to other atoms. Thus, as we shall see later (Section 6.3.3), they apply to what is known as nonpolar bitopic bond dissociation. For example, Figure 4.5 applies to the dissociation of the Si—Si bond in a saturated oligosilane (see Section 7.2.3), but not to the dissociation of the C—$S^{\oplus}$ bond in an alkylsulfonium salt, to the dissociation of the C—O bond in an alcohol (see Section 6.3.3), or to the dissociation of the C—C bond adjacent to a carbonyl group. (See Section 6.3.1.) Introduction of an electronegativity difference between the two bond termini causes changes in the course of the correlation lines in Figure 4.5 and is discussed in Section 4.3. Introduction of lone pairs, empty orbitals, or adjacent multiple bonds introduces new low-energy states and is discussed in Section 6.3.3.

The twisting of the ethylene molecule provides an illustration of a prototypical π bond dissociation (Figure 4.6): The minimum in S_0 occurs at a planar geometry (twist angle of 0°) and is definitely not a biradicaloid minimum since the π_u MO is clearly bonding and the π_g^* MO clearly antibonding at this geometry. In contrast, the T_1, S_1, and S_2 states have a minimum at 90° twist, where both MOs are nonbonding. This obviously is a biradicaloid min-

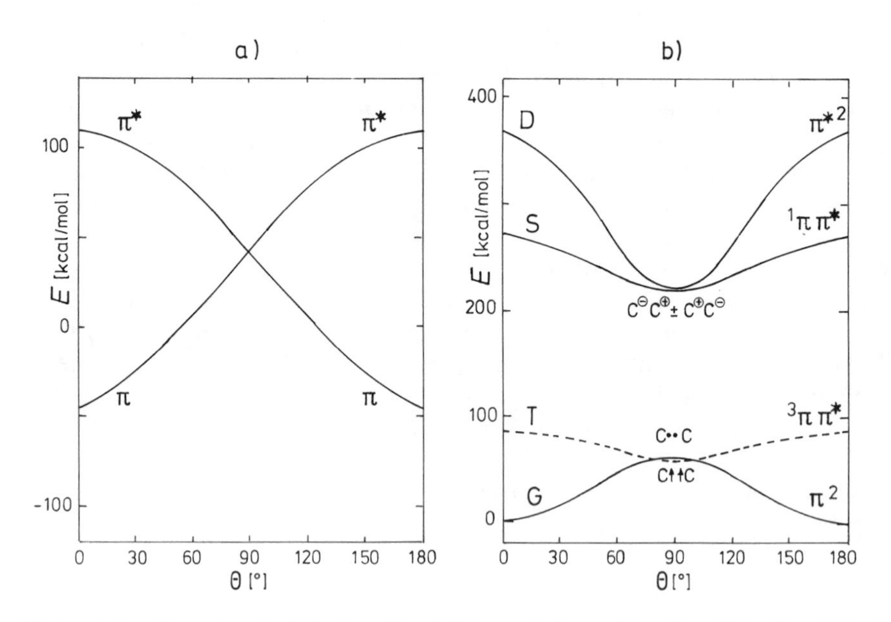

Figure 4.6. Energies a) of the bonding MO π and the antibonding MO π^* and b) of the π-electronic states of ethylene as a function of the twist angle θ. On both sides the states are labeled by the MO configuration dominant at planar geometries; in the middle, they are labeled by the VB structure that is dominant at the orthogonally twisted geometry (by permission from Michl and Bonačić-Koutecký, 1990).

imum, and the wave functions of the molecular states are analogous to those discussed above for the case of H_2. The S_0 and T_1 states are represented by VB structures containing a single electron in each of the carbon 2p orbitals of the original double bond, prevented from interacting by symmetry, and not by distance as in the H_2 case. The S_1 and S_2 states are of zwitterionic nature and are represented by VB structures containing both electrons in one of these orbitals and none in the other.

Unlike the biradicaloid minima located along the dissociation path of a σ bond, one of which occurred at a loose (T_1) and one at a tight (S_1) geometry, those located along the dissociation path of a π bond are at nearly the same geometry. The reason for this difference can be readily understood in terms of the simple model of biradicaloid states outlined in Section 4.3.

Figure 4.6 again represents a very simple correlation diagram, and is of similar fundamental importance for reactions involving the dissociation of a π bond as Figure 4.5 is for reactions in which a σ bond dissociates. It underlies correlation diagrams such as that for the cis-trans isomerization of stilbene (Figure 7.3). Its use is subject to similar limitations as the use of Figure 4.5: (1) The course of the potential energy curves changes when the electronegativity of the two bond termini differs (e.g., Figures 4.22 and 7.6), and such modifications are essential for the understanding of the relation of the potential energy diagrams of protonated Schiff bases and TICT (twisted internal charge transfer) molecules to those of simple olefins. (2) The number of low-energy potential energy surfaces increases when one or both bond termini carry lone pairs or empty orbitals, or if they engage in additional multiple bond formation. Examples of such situations are the cis-trans isomerization of Schiff bases (Figure 7.8) and azo compounds (Figures 7.10 and 7.11).

A more complete discussion of the results embodied in Figures 4.5 and 4.6 than can be given here is available elsewhere (Michl and Bonačić-Koutecký, 1990).

Geometries with large distances between the radical centers such as the T_1 state of H_2 are called *loose biradicaloid geometries*; those with small distances such as the S_1 state of H_2 are called *tight biradicaloid geometries*. Since biradicaloid minima in T_1 occur preferentially for loose geometries, whereas those in S_1 occur for tight geometries, return from the S_1 and T_1 to the ground state S_0 will frequently take place at different geometries. This presumably quite general behavior may be one of the primary reasons for the difference in the photochemistry of molecules in the singlet and triplet states (Michl, 1972; Zimmerman et al., 1981).

Example 4.3:

The simplest model to describe a σ as well as a π bond is the two-electron two-orbital model, which will be used frequently in the remainder of the text. The MO and the VB treatment of the H_2 molecule may serve as an example. In the simple MO picture the singlet states G, S, and D are approximated by the configurations

$$^1\Phi_0 = |\phi_1\bar{\phi}_1|$$
$$^1\Phi_1 = [|\phi_1\bar{\phi}_2| + |\phi_2\bar{\phi}_1|]/\sqrt{2}$$

and

$$^1\Phi_2 = |\phi_2\bar{\phi}_2|$$

respectively. In the first of these, the bonding MO $\phi_1 \equiv \sigma_g$ (HOMO) is doubly occupied. In the second, ϕ_1 and the antibonding MO $\phi_2 \equiv \sigma_u^*$ (LUMO) are both singly occupied, and in the third, ϕ_2 is doubly occupied. The triplet state T may be written as

$$^3\Phi_1 = [|\phi_1\bar{\phi}_2| - |\phi_2\bar{\phi}_1|]/\sqrt{2}$$

In the basis of orthogonalized AOs χ_a and χ_b, the "VB" structures for the singlet states are

$$^1(\chi_a\chi_b) = [|\chi_a\bar{\chi}_b| + |\chi_b\bar{\chi}_a|]/\sqrt{2}$$
$$(\chi_a^2) = |\chi_a\bar{\chi}_a|$$

and

$$(\chi_b^2) = |\chi_b\bar{\chi}_b|$$

and for the triplet state

$$^3(\chi_a\chi_b) = [|\chi_a\bar{\chi}_b| - |\chi_b\bar{\chi}_a|]/\sqrt{2}$$

It can be verified by direct insertion of $\phi_1 = (\chi_a + \chi_b)/\sqrt{2}$ and $\phi_2 = (\chi_a - \chi_b)/\sqrt{2}$ that the relation between the MO configurations and the "VB" structures is

$$^1\Phi_0 = [(\chi_a^2 + \chi_b^2)/\sqrt{2} + {}^1(\chi_a\chi_b)]/\sqrt{2}$$
$$^1\Phi_1 = (\chi_a^2 - \chi_b^2)/\sqrt{2}$$
$$^1\Phi_2 = [(\chi_a^2 + \chi_b^2)/\sqrt{2} - {}^1(\chi_a\chi_b)]/\sqrt{2}$$

and

$$^3\Phi_1 = {}^3(\chi_a\chi_b)$$

Note that the triplet MO configuration is equal to the triplet "VB" structure and that in the case of nonpolar bonds the singly excited MO configuration is equal to the zwitterionic "VB" structure $(\chi_a^2 - \chi_b^2)/\sqrt{2}$. Configuration mixing provides a better description of the states G $= {}^1\Phi_0 - \lambda{}^1\Phi_2$ and D $= {}^1\Phi_2 + \lambda{}^1\Phi_0$, whose ionic character can vary from purely covalent to purely ionic as the value of λ changes. A transformation to standard VB theory based on non-orthogonal AOs is also possible but is more complicated (Michl and Bonačić-Koutecký, 1990: Section 4.1).

Whereas simple MO theory does not provide any clue as to why the S state and even the D state of a σ bond with the antibonding MO ϕ_2 singly and doubly occupied, respectively, are bound and not dissociative like the T state, these minima are readily understood on the basis of VB structures, which closely approximate the actual state wave functions: The high energy of the zwitterionic structures (χ_a^2) and (χ_b^2) and their increase with increasing internuclear distance are plausible as a result of charge separation, and the rise of all curves at short internuclear distances is clearly due to poorly screened internuclear repulsion. While the bond in the S_0 state is covalent, the bond in the S_1 (and S_2) state is purely electrostatic, as in a contact ion pair. At large internuclear separations, the singlet ground configuration ${}^1\Phi_0$ is a very poor approximation for the true ground-state function, and its energy is much higher than that of the triplet state. In summary, the simple MO description is useful at short distances, where individual configurations represent reasonable approximations to electronic states. However, at intermediate and larger internuclear separations (biradicaloid geometries), introduction of configuration interaction into the MO picture is absolutely necessary. In this region, the VB description provides a more intuitive rationalization of the shapes of the energy curves.

In the case of twisted ethylene the wave functions of the molecular states are quite analogous to those of H_2, with the carbon 2p AOs of the original double bond at 90° twist being prevented from interacting by symmetry, and not by distance as in the H_2 case.

4.2 Correlation Diagrams

Valuable information concerning the excited-state potential energy surface areas where minima or barriers are to be expected can be obtained from state correlation diagrams. Such diagrams are a very useful tool in discussing photochemical reaction mechanisms, in spite of their shortcomings. In particular, they are frequently the easiest to construct along reaction paths that preserve some symmetry elements, and yet, minima in S_1 and S_0–S_1 touchings are most likely to occur at geometries devoid of all symmetry. Correlation diagrams usually provide only an approximate guide to the location of these important topological features.

4.2.1 Orbital Symmetry Conservation

An orbital correlation diagram for a reaction path is obtained if the (qualitative) orbital energy schemes at both ends of the path are known and if energy levels of corresponding orbitals are connected or correlated. The already familiar Figure 4.5a, which shows the energies of the bonding σ and the antibonding σ^* orbitals for H_2 at the equilibrium geometry on the left and for two infinitely separated hydrogen atoms on the right, may serve as an example.

Each of the various assignments of electrons to orbitals corresponds to a different electron configuration. From the orbital correlation diagram one can thus deduce which reactant and product configurations correlate with each other and obtain a configuration correlation diagram. If correlation lines of equal symmetry cross in this diagram, the noncrossing rule is violated. These crossings will usually be avoided when configuration interaction is taken into account, and a state correlation diagram is obtained. Such a state correlation diagram can be interpreted as a cross section through the various potential energy surfaces of the system, and gives at least a qualitative description of the location of energy maxima and minima along the reaction path. Figures 4.5b and 4.6b are examples of state correlation diagrams.

In Figure 4.7 the orbital correlation diagram for the concerted disrotatory ring opening of cyclobutene to butadiene is given together with the resulting configuration and state correlation diagrams as an example of a ground-state–forbidden reaction. According to the principle of conservation of orbital symmetry, those reactant and product orbitals associated with the reaction were correlated that show the same symmetry behavior (S = symmetric, A = antisymmetric) with respect to a symmetry element that is preserved along the reaction path. In the case in question, this is the symmetry plane bisecting the butadiene C2—C3 bond and perpendicular to the molecular plane. The HOMO-LUMO crossing in the orbital correlation diagram induces a crossing in the configuration correlation diagram, which is avoided in the state correlation diagram since both configurations that cross contain only doubly occupied orbitals, and are therefore totally symmetric. It should be noted, however, that the real photochemical reaction is not likely to follow a path that preserves any symmetry. Removal of the exact degeneracy

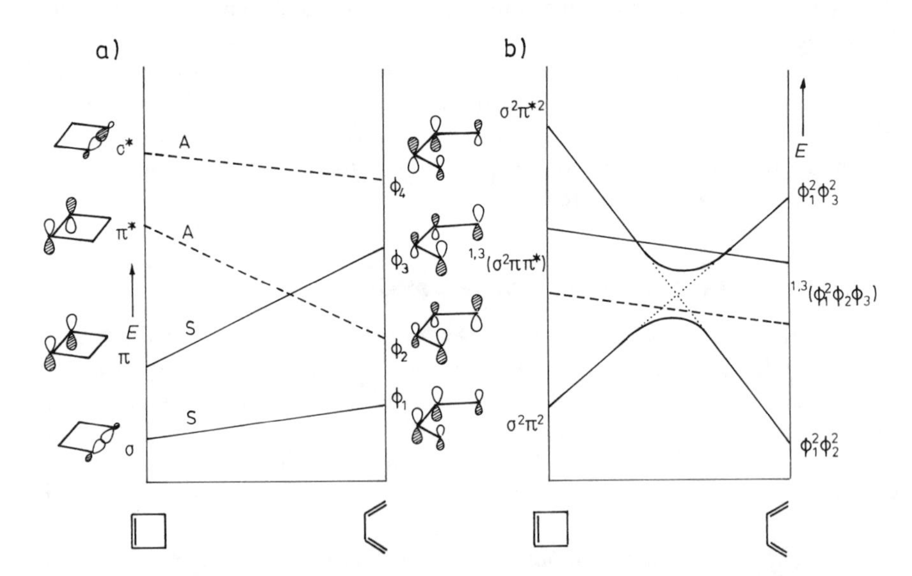

Figure 4.7. Disrotatory ring opening of cyclobutene to butadiene; a) orbital correlation diagram, b) configuration correlation diagram (dotted lines) and effect of configuration interaction, which converts the diagram into a state correlation diagram (solid lines). The triplet state is indicated by a broken line.

of the nonbonding orbitals at the biradicaloid geometry by a suitable symmetry-lowering distortion has the potential for reducing the S_0–S_1 gap to zero. (Cf. the discussion of critically heterosymmetric biradicaloids in Sections 4.3 and 4.4.) The resulting conical intersection represents a very efficient funnel for returning the excited molecules back to the ground-state surface.

The appearance of this correlation diagram is very typical for a symmetrical ground-state–forbidden concerted reaction path. At the midway point, the crossing of a doubly occupied and an unoccupied MO results in two orbitals of equal energy that are occupied by a total of two electrons, thus producing a biradical at half-reaction. As mentioned earlier, reduction of symmetry has the potential to reduce the S_0–S_1 gap to zero, but it is not clear whether the energy of the point of intersection lies below or above that of the lowest S_1 energy at the symmetrical geometries shown in Figures 4.7 and 4.8.

Except in the case of very exothermic reactions (cf. Example 6.7), molecules that have traveled from the reactant geometry to the biradicaloid minimum will have little chance of reaching the product geometry on the same surface, because the return to S_0 from the pericyclic minimum or a nearby conical intersection will be very efficient.

As shown in the correlation diagram in Figure 4.7b, the T_1 state is not affected by first-order configuration interaction. It is therefore very likely that along the reaction path there are no other minima than the spectroscopic

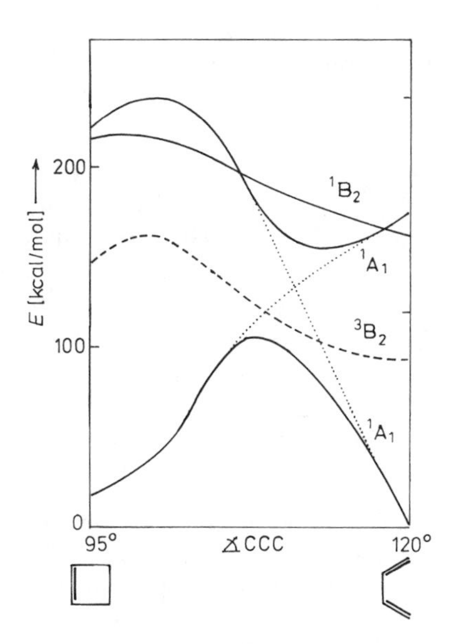

Figure 4.8. Calculated state correlation diagram for the disrotatory ring opening of cyclobutene to butadiene (by permission from Grimbert et al., 1975).

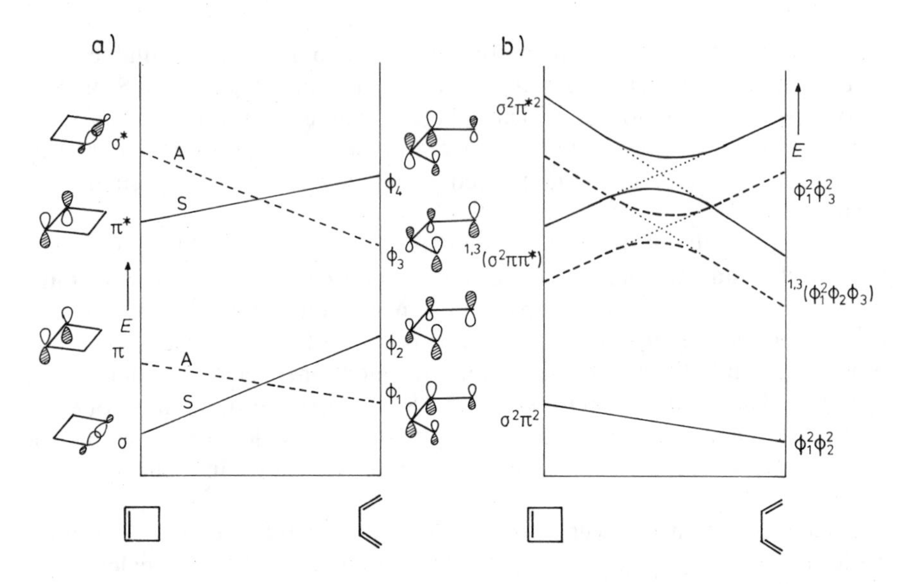

Figure 4.9. a) Orbital correlation diagram, b) configuration correlation diagram (dotted lines), and state correlation diagram (solid lines) for the conrotatory ring opening of cyclobutene as an example for a pericyclic reaction that is allowed in the ground state. Triplet states are indicated by broken correlation lines.

minima near the reactant and product geometries. In particular, a pericyclic minimum is not to be expected on the T_1 surface in this case (uncharged perimeter). If no other reaction path is available and if the reactant geometry is energetically more favorable than the product geometry, the molecule will stay in this geometry after excitation into the triplet state and will only undergo photophysical processes; that is, it will return to the initial minimum in the ground-state surface, either by emission or via radiationless processes. If, however, the triplet minimum at the product geometry is lower, triplet excited product may be formed. This can be detected by triplet–triplet absorption spectroscopy. If phosphorescence can compete with radiationless deactivation, product emission can be also observed. Systems derived from a charged perimeter are different in that a special stabilization of T_1 is expected at the pericyclic geometry, and a minimum is probably present. (See Michl and Bonačić-Koutecký, 1990.)

Example 4.4:

Calculated potential energy curves for the symmetrical disrotatory ring-opening path from cyclobutene to butadiene are shown in Figure 4.8. It is seen that the biradicaloid minimum for this reaction occurs on a surface S_1 whose wave function is similar to that of the S_2 surface at the reactant geometry. This is due to a touching of the potential energy surfaces of the first two excited singlet states. This touching is avoided in the absence of full symmetry. Since there are no other important correlation-imposed barriers between the planar reactant geometry and the pericyclic minimum, and because the return to S_0 is remarkably efficient, neither reactant fluorescence nor the formation of a fluorescent product can be observed. Finally, it is seen from the diagram that a concerted ring opening of cyclobutene in its triplet state T_1 is not to be expected (Grimbert et al., 1975).

In Figure 4.9 the conrotatory ring opening reaction of cyclobutene is shown as a typical example of an orbital correlation diagram for a ground-state–allowed pericyclic reaction and of the resulting configuration and state correlation diagrams. In this case the spectroscopic product minimum is not separated by a funnel from the initial geometry, so at a first glance it should be easily accessible. However, the concerted reaction path in S_1 shows a correlation-imposed barrier that inhibits the formation of the product.

4.2.2 Intended and Natural Orbital Correlations

Difficulties are often encountered in constructing orbital correlation diagrams because no symmetry element is conserved during the reaction, or because the relative energies of reactant and product orbitals are not known. In these cases a two-step procedure has proven very helpful (Michl, 1972; 1973; 1974b). This is based on the decomposition of a relatively complicated system into simpler subunits. First, a correlation diagram is constructed for the noninteracting subunits, and in a second step, interaction between the subunits is added. This will be illustrated using the disrotatory and the conrotatory ring opening of cyclobutene, shown in Figures 4.10 and 4.11 as an example.

First, the breaking of the σ bond is considered with the assumption that an imaginary wall prevents interaction of this subunit with the remainder of the system. Then ring opening leads to two nonbonding hybrid AOs from the σ and σ^* MOs, whereas the π orbitals (and all other orbitals not explicitly treated here) remain unaffected. In the second step the imaginary wall is removed so that orbitals with comparable nodal properties can interact with each other. In this way the orbitals of the biradical formed during the first step are converted into the orbitals of the more stable butadiene. In estimating the effects of orbital interaction it has to be remembered that they will become larger as the energy difference between the orbitals of interest becomes smaller and as the LCAO coefficients at the interacting centers become larger.

From a comparison of Figures 4.10a and 4.11a it is evident that the nodal properties of the newly generated π components of the original σ and σ^* combinations are opposite to each other for the disrotatory and the conrotatory ring opening. The π^* MO of the ethylenic bond has a node across the bond and is suitable for interaction with the π component of the original σ^* combination if the opening is disrotatory (Figure 4.10a) and with the π com-

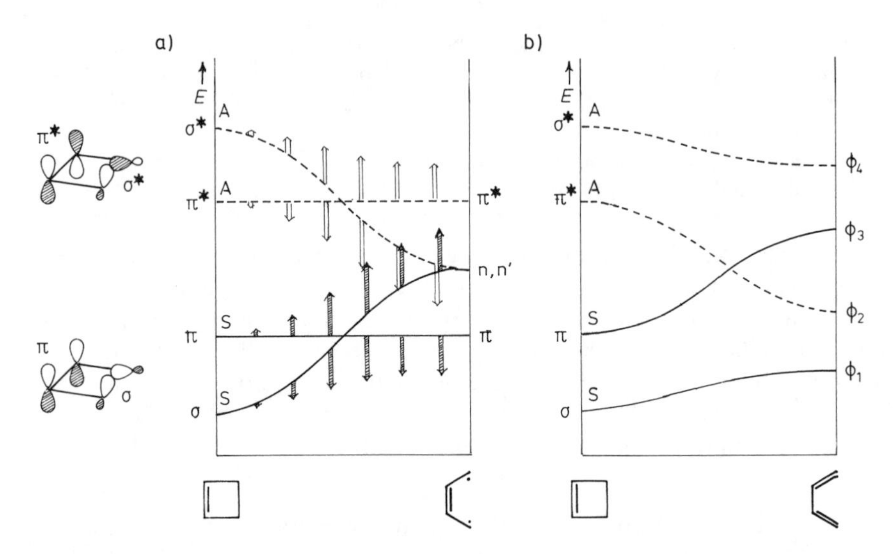

Figure 4.10. Two-step procedure for the derivation of the orbital correlation diagram of the disrotatory ring opening of cyclobutene; a) "intended" orbital correlation that results in the absence of interaction between orbitals of the σ bond that is being broken and orbitals of the π bond. The schematic representation of the orbitals shows that due to different nodal properties only σ–π and σ^*–π^* MO interactions are possible. Orbital symmetry labels π and σ apply strictly only at the planar cyclobutene and butadiene geometries. Dark and light arrows indicate the magnitude of these interactions, which are switched on in the second step and produce the orbital correlation diagram shown in b); labels S and A or solid and broken lines, respectively, indicate the different symmetry properties with respect to the symmetry plane that is preserved along the reaction path (by permission from Michl, 1974b).

ponent of the original σ combination if the opening is conrotatory (Figure 4.11a). The π orbital has no node across the bond, so here the situation is reversed. In Figures 4.10a and 4.11a the resulting interactions are indicated by arrows, and in Figures 4.10b and 4.11b the final orbital correlation diagrams are shown. It can be seen that the stabilization of the biradical by orbital interaction is much more efficient for the conrotatory opening of cyclobutene than for the disrotatory reaction. As a consequence of the HOMO-LUMO crossing, the latter has to pass through a biradicaloid geometry in spite of the potentially stabilizing interaction.

Since the two-step derivation of the orbital correlation diagram with the help of the imaginary wall makes use only of orbital nodal properties, it is generally applicable, even in those cases where no symmetry element is preserved during the reaction. Relations obtained in the first step can be referred to as "intended" correlations; they demonstrate how the original orbitals would change during the reaction if no other relations could be established through additional interactions. Thus, in the case of the disrotatory ring opening shown in Figure 4.10a, the π and σ as well as the π^* and

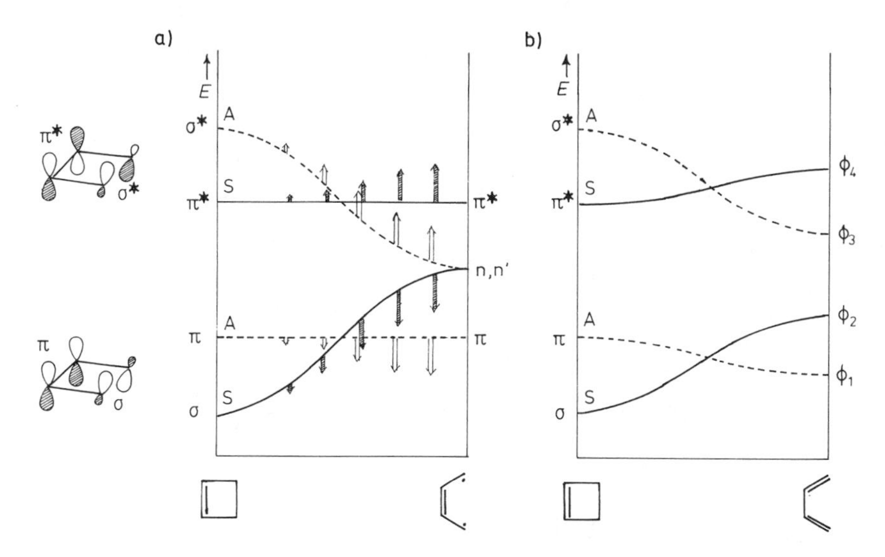

Figure 4.11. Derivation of the orbital correlation diagram for the conrotatory ring opening of cyclobutene; a) intended correlation, b) correlation including interaction between σ and π^* and between π and σ^* MO's, respectively. Orbital symmetry labels π and σ apply strictly only at the planar cyclobutene and butadiene geometries. Labels S and A, or solid and broken correlation lines, respectively, indicate the symmetry behavior with respect to the twofold-symmetry axis (by permission from Michl, 1974b).

σ^* correlation lines cross in the first step, whereas the crossing is avoided if interactions are taken into account. From the correlation lines shown in Figure 4.10b this avoided crossing can still be visualized; and since orbital interactions can be assessed as described earlier, the shape of the correlation lines is not arbitrary. In the following section it will be shown that such avoided crossings in the orbital correlation diagram may in fact show up on potential energy surfaces.

If a correlation diagram is constructed by connecting reactant and product orbitals of equal symmetry that are localized in the same spatial region of the molecule and that show the same sign relations for the LCAO coefficients, crossings of correlation lines of equal symmetry may occur. The *natural* orbital correlations (Devaquet et al., 1978) obtained in this way are very similar to the intended correlations described earlier. The main difference is that in the first step the orbitals of the product are now not taken to be the AOs of the bond that is being broken, but rather MOs constructed from them.

In Figure 4.12 the natural orbital correlation is shown once again for the disrotatory ring opening of cyclobutene. The π and π^* MOs correlate with the butadiene MOs ϕ_1 and ϕ_4, for which the LCAO coefficients are largest for the two inner carbon atoms and have either equal or opposite signs, re-

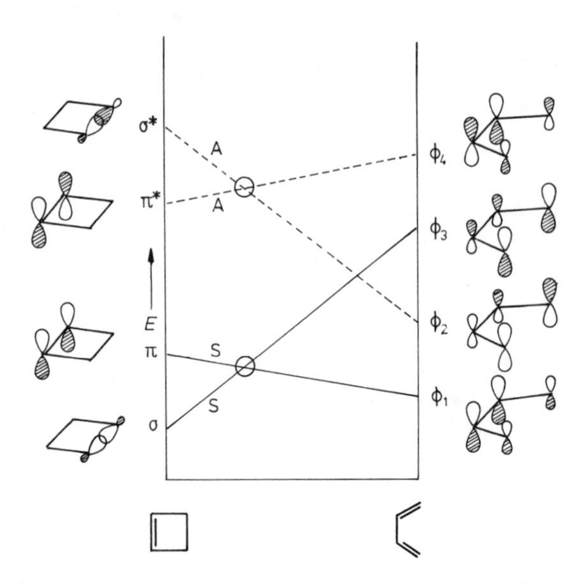

Figure 4.12. Natural orbital correlation for the disrotatory ring opening of cyclobutene to butadiene (by permission from Bigot, 1980).

spectively. Similarly, the σ and σ^* MOs correlate with ϕ_2 and ϕ_3 for which the contributions from the end atoms dominate.

4.2.3 State Correlation Diagrams

If intended or natural correlations are used, one has the choice either to switch on the interaction between subunits first and then to construct the configuration and state correlation diagrams, or to construct the configuration correlation diagram first and then to add the interaction. The first alternative is identical with the procedure described in Section 4.2.2; the second one is illustrated in Figure 4.13, once again using the disrotatory ring opening of cyclobutene as an example. It is seen that due to the natural correlation of the σ and π^* MOs of cyclobutene with the antibonding butadiene MOs ϕ_3 and ϕ_4 and of the π and σ^* MOs with the bonding MOs ϕ_1 and ϕ_2, the correlations starting at the ground configuration and at the $\pi \rightarrow \pi^*$ excited configuration of cyclobutene go uphill, whereas those starting at the $\sigma \rightarrow \sigma^*$ excited configurations go downhill. Hence, the avoided crossings produce barriers in the lowest triplet state as well as in the two lowest singlet states that are not obtained if ordinary symmetry correlations are used. The calculated potential energy surfaces show similar barriers, which have their origin in the fact that the MOs "remember" the intended or natural correlation. This becomes evident when Figure 4.13 and Figure 4.8 are compared.

The situation is complicated by the fact that sometimes it proves useful to derive the orbital correlations not on the basis of MOs but rather on the

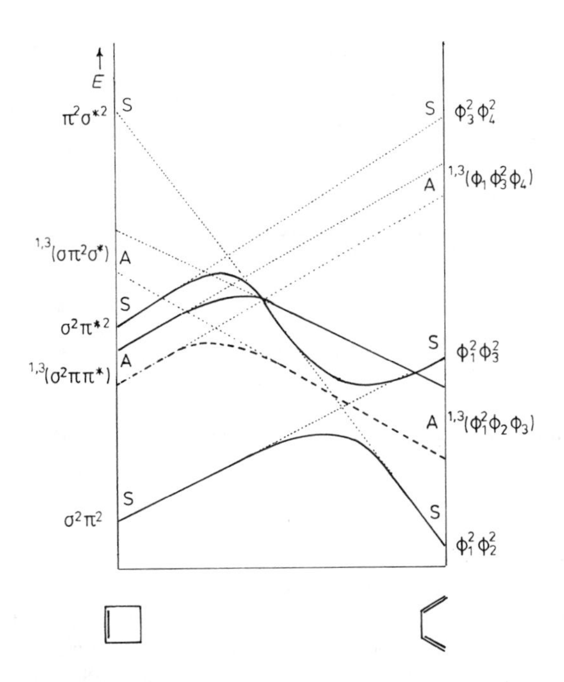

Figure 4.13. Configuration correlation diagram (dotted lines) and the resulting state correlation diagram (solid and broken lines, respectively) for the disrotatory ring opening of cyclobutene, based on natural orbital correlations (by permission from Bigot, 1980).

basis of AOs, within the framework of a VB description of the wave function.

Altogether there are the following possibilities, which are collected in Figure 4.14: On the one hand, with the help of the imaginary wall which prevents orbital interaction, configurations of a system can be constructed either from the interaction-free AOs or from the MOs obtained from these AOs. The configurations and states of interest are then obtained by turning on the interaction, that is, by removing the imaginary wall. This corresponds to the use of interaction-free one-electron functions for constructing many-electron wave functions in the framework of the VB or MO theory and taking into account the interactions at the level of the many-electron wave functions, that is, either at the MO configuration or VB structure correlation level or even at the state correlation diagram level, which is often very useful. On the other hand, the interaction may be switched on already at the level of one-electron functions; this leads to MOs with interactions included, which are then used to construct the many-electron wave functions of configurations and states. In principle, all these procedures will give the same results, but depending on the problem in question, one or another procedure will prove advantageous.

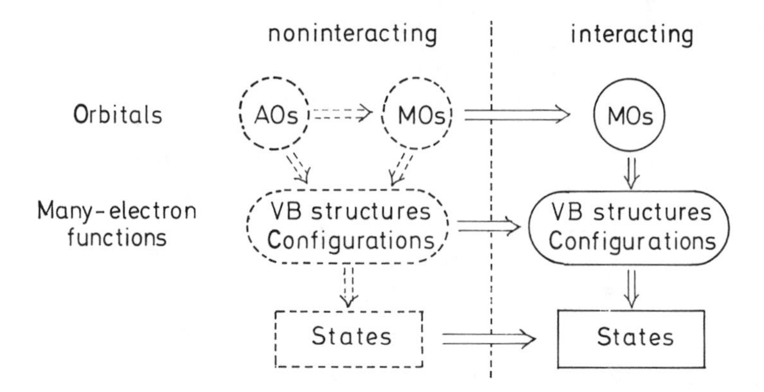

Figure 4.14. Schematic representation of the different possibilities for constructing state correlation diagrams by switching on the orbital interaction at various levels.

Example 4.5:

The orbital correlation diagram for the concerted dimerization of ethylene to form cyclobutane or for the reverse reaction, the fragmentation of cyclobutane into two ethylenes, may be obtained most easily by applying the principle of conservation of orbital symmetry. A mirror plane perpendicular to the molec-

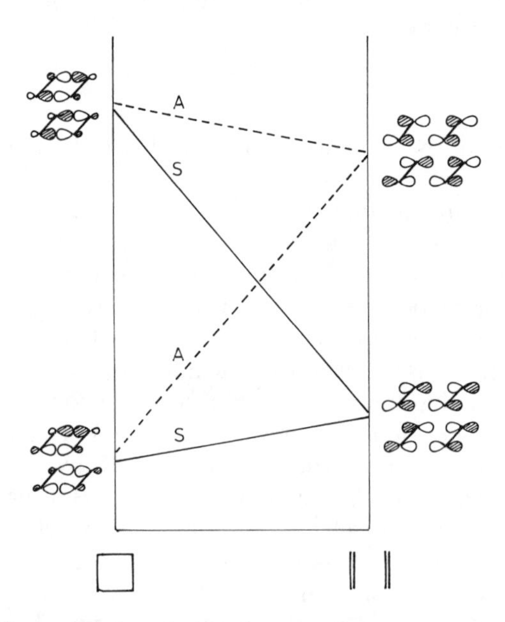

Figure 4.15. Symmetry-based orbital correlation of the orbitals involved in the fragmentation of cyclobutane into two ethylene molecules. Indicated is the symmetry behavior (S or A) with respect to a symmetry plane perpendicular to the molecular plane and cutting the ethylene bonds into halves.

ular plane and either across the σ bonds that will break, or across the π bonds that will be formed, is used as the symmetry element. It is easily seen that both choices give the same result. Next, we notice that the σ and σ^* MOs of cyclobutane are stronger bonding and antibonding, respectively, than the π and π^* MOs of ethylene, and obtain the orbital correlation diagram of Figure 4.15.

The corresponding MO correlation diagram for the fragmentation of cyclobutenophenanthrene into phenanthrene and acetylene is most easily obtained if in a first step the orbital correlation diagram of Figure 4.15 is superimposed on the π MO energy-level diagram of biphenyl and in a second step interactions between orbitals of equal symmetry (S and A with respect to the mirror plane perpendicular to the molecular plane of phenanthrene) are introduced, as shown in Figure 4.16.

In hydrogen atom abstraction by a ketone the following orbitals are important: for the reactant $R_2'CO$ + RH, the orbitals σ_{CH} and σ_{CH}^* of the R—H bond to be broken and the orbitals n_O, π_{CO}, and π_{CO}^* of the ketone chromophore. For the product $R_2'\dot{C}OH$ + ·R, the orbitals σ_{OH} and σ_{OH}^* of the newly formed OH bond, plus the orbitals π_O and π_C of the protonated ketyl radical and the orbital p_C of the odd electron on the residue R. Their relative energies can be easily estimated qualitatively. Since both n_O and σ_{OH} are essentially localized

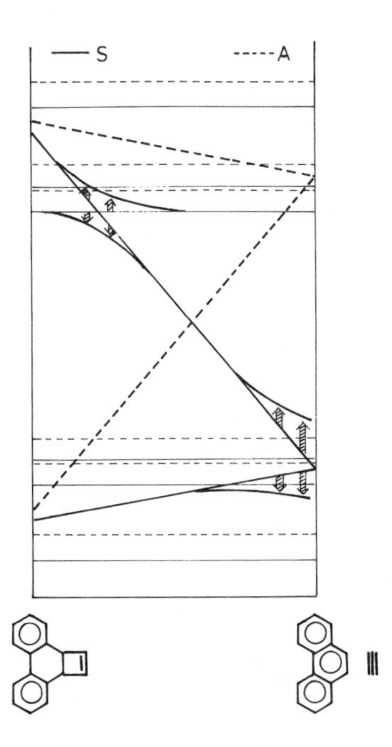

Figure 4.16. Derivation of the correlation diagram for the concerted fragmentation of cyclobutenophenanthrene from the orbital energy-level scheme of biphenyl and the orbital correlation diagram for the fragmentation of cyclobutane into two ethylenes. (The additional double bond has been neglected in the simplified treatment.) The arrows indicate the magnitude of orbital interactions between the two superimposed systems (by permission from Michl, 1974b).

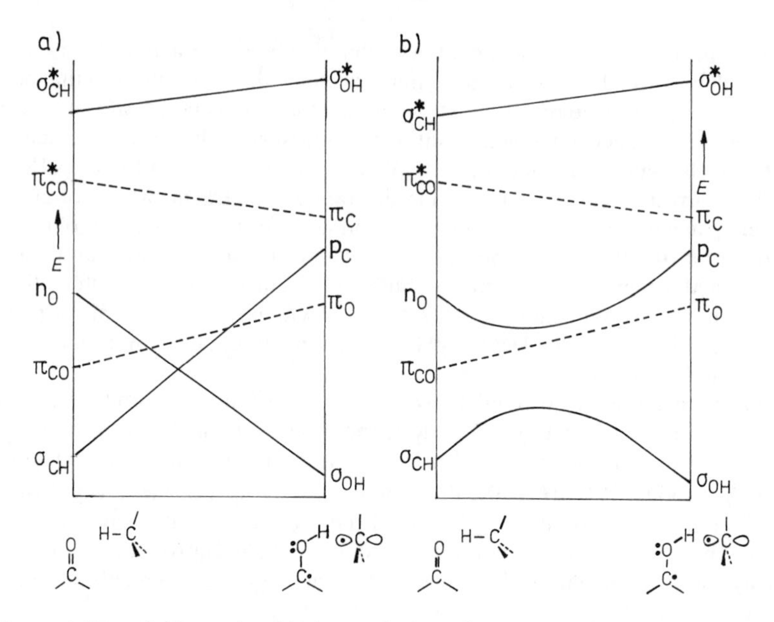

Figure 4.17. a) Natural orbital correlation diagram for hydrogen atom abstraction by a carbonyl compound and b) orbital correlations after introducing interactions between orbitals of equal symmetry.

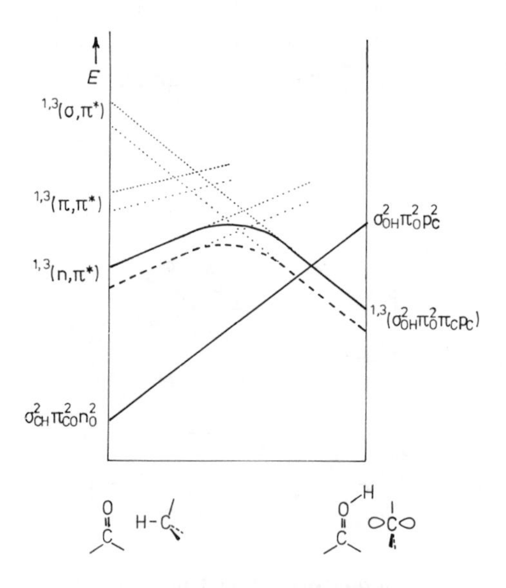

Figure 4.18. State correlation diagram for hydrogen atom abstraction by a carbonyl compound derived from the natural orbital correlations. The configuration correlations are shown by dotted lines, taking into account the interaction yields the correlations shown by solid and broken lines for singlet and triplet states, respectively.

on the oxygen atom, and σ_{CH} and p_C are on the carbon atom of the residue R, the natural orbital correlation shown in Figure 4.17a is obtained. This leads to the configuration correlation diagram shown in Figure 4.18 by dotted lines. From this the state correlation diagram shown in solid and broken lines is obtained by taking into account the avoided crossings, which produce barriers in the S_1 and T_1 correlation lines, and which correspond closely to those obtained through detailed calculations of the relevant potential curves (Bigot, 1980). If, however, the interaction is already introduced at the orbital correlation level, so that σ_{CH} correlates with σ_{OH} and n_O with p_C (Figure 4.17b), a state correlation diagram without barriers and minima resulting from the avoided crossing is obtained. This is in agreement with the Salem diagram derived on the basis of VB structures, which will be discussed in Section 6.3.2.

4.3 Biradicals and Biradicaloids

In the beginning of this chapter it was mentioned that biradicaloid minima are of great importance in photochemical reactions. Due to their very short lifetime, molecules in biradicaloid minima are usually not as easy to observe as molecules in spectroscopic minima with ordinary geometries. This is particularly true in the singlet state, whose decay is not slowed down by the need for electron spin inversion. Much of the knowledge of reactive minima is therefore based primarily on theoretical arguments, some of which will be considered in this section. (Cf. Bonačić-Koutecký et al., 1987; Michl 1991.)

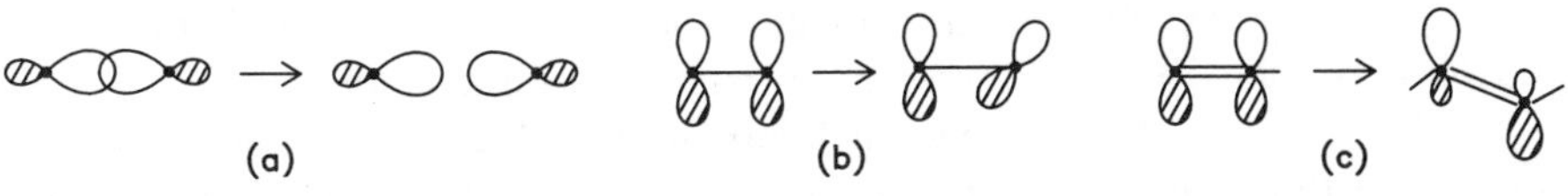

(a) (b) (c)

4.3.1 A Simple Model for the Description of Biradicals

For the discussion of potential energy surfaces of excited states it is particularly important that at appropriate geometries closed-shell molecules can turn into biradicaloids. We have already seen briefly in Section 4.1.2 that such biradicaloid geometries can be derived from equilibrium geometries by suitable distortions such as the stretching of a single bond (**a**), twisting of a double bond (**b**), or bending of a triple bond (**c**).

The "antiaromatic" geometry found along the concerted path of ground-state–forbidden pericyclic reactions, which is topologically equivalent to an antiaromatic Hückel [4*n*]annulene or Möbius [4*n* + 2]annulene, is a particularly interesting type of biradicaloid geometry. (Cf. Section 4.4.) Other biradicaloid geometries and combinations of those mentioned are equally possible.

Whereas at equilibrium geometries the electronic structures of even-electron molecules are in general quite well described by a single closed-shell configuration, this is not so at biradicaloid geometries. At the latter, only two electrons occupy two nearly degenerate approximately nonbonding orbitals in all low-energy states, and as a minimum, one needs to consider all configurations that result from the various possible occupancies of these two orbitals. If the active space is restricted to these two nonbonding orbitals, chosen to be orthogonal and labeled φ and φ', these are the following configurations:

$$\begin{aligned} {}^1\mathrm{B} &= {}^1(\varphi\varphi') = (1/\sqrt{2})\{|\varphi\bar{\varphi}'| + |\varphi'\bar{\varphi}|\} \\ {}^3\mathrm{B} &= {}^3(\varphi\varphi') = (1/\sqrt{2})\{|\varphi\bar{\varphi}'| - |\varphi'\bar{\varphi}|\} \end{aligned} \tag{4.1}$$

with both orbitals singly occupied and

$$\begin{aligned} \mathrm{Z}_1 &= {}^1(\varphi'^2) = |\varphi'\bar{\varphi}'| \\ \mathrm{Z}_2 &= {}^1(\varphi^2) = |\varphi\bar{\varphi}| \end{aligned} \tag{4.2}$$

with one of them doubly occupied. (Cf. Example 4.3.) In the simplest model for the electronic structure of biradicaloids, attention is limited to these three singlet configurations and one triplet configuration, so only three singlet states and one triplet state result (Salem and Rowland, 1972; Michl, 1972; Bonačić-Koutecký et al., 1987; Michl, 1988). Before comparing with experiment, it has to be remembered that upon going to a more accurate description by taking into account all interactions with other configurations, the four states of this model can be influenced differentially. Such interactions, including effects known as "dynamic spin polarization" (Kollmar and Staemmler, 1978), may thus change the energetic order of closely spaced states. Nevertheless, those aspects of the model that do not depend on small energy differences are generally correct, and it provides a very pictorial description of many of the interesting properties of biradicaloids.

If, for instance, the orthogonal nonbonding orbitals are AOs χ_a and χ_b located at the atoms A and B and prevented from interacting either by excessive separation, as in a dissociated single bond, or by symmetry, as in a twisted double bond—that is, if $\varphi = \chi_a$ and $\varphi' = \chi_b$—the configurations given in Equations (4.1) and (4.2) are the VB structures which may be represented by the formulae **2–7**.

$A-B$ or $\cdot AB\cdot$	$A-B$:	$:A-B$
(^{1,3}B)	(Z$_1$)	(Z$_2$)
 **2**	 **3**	 **4**
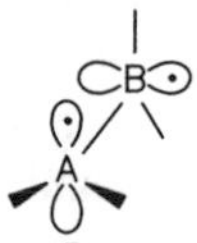 5	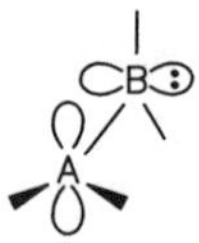 6	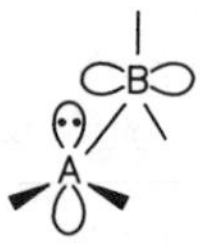 7

For systems such as a dissociated H_2 molecule (**8**) or twisted ethylene (**9**) the structures ^{1}B and ^{3}B are nonpolar and can be represented as A—B and

$\cdot AB\cdot$, respectively. Because of their "double radical" appearance, they will be referred to as "dot–dot" structures. Sometimes they are also called biradical structures, or covalent structures. The label covalent is derived from the Heitler–London description of a covalent bond, and is therefore somewhat inappropriate in the case of the triplet, which does not correspond to a bond at all, but to an antibond. The polarized structures Z_1 and Z_2 may be represented by the formulae $A^{\oplus}B^{\ominus}$ and $A^{\ominus}B^{\oplus}$, respectively, and will be referred to as hole-pair structures. In the case of dissociated H_2 and twisted ethylene, they are zwitterionic. Because of the charge separation, their energy is high.

In general, however, there is no simple relation between the dot–dot or hole-pair nature of these VB structures and their zwitterionic nature. For instance, in the twisted aminoborane (**10**) the dot–dot structure ${}^{\ominus}\cdot BH_2—NH_2{}^{\oplus}\cdot$ is zwitterionic and high in energy, whereas the hole-pair structure Z_1 corresponds to $BH_2—\ddot{N}H_2$, carries no formal charges, and is of low energy. In singly charged systems such as **11–13**, neither the dot–dot structure ${}^1B$ nor the hole-pair structure Z_1 is zwitterionic, and their energies can be quite comparable.

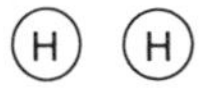

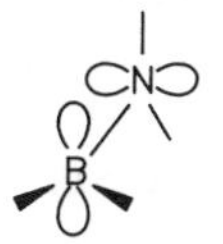

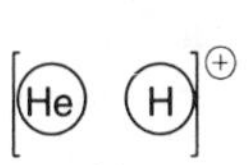

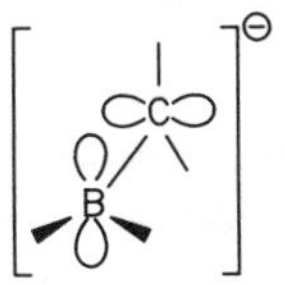

It is not necessary to choose the orthogonal orbitals φ and φ' to be the most localized ones, as we have done in our discussion so far. Since in the simple model we consider all configurations that can be constructed from the two nonbonding orbitals by any permissible electron occupancy, the energies and wave functions of the resulting states are invariant to any mixing of these orbitals. Thus, we may equally well choose them to be the most delocalized orthogonal molecular orbitals, related to the most localized ones by

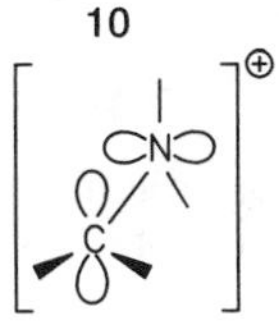

$$\phi_1 = (\chi_a + \chi_b)/\sqrt{2} \tag{4.3}$$

and

$$\phi_2 = (\chi_a - \chi_b)/\sqrt{2}$$

or to be the complex orthogonal orbitals defined in Figure 7.53. In addition to these three canonical choices of orthogonal orbital pairs, an infinite number of others are available (Bonačić-Koutecký et al., 1987; Michl, 1991,

1992). The appearance of the final singlet-state wave functions is different ("open-shell," "closed-shell") when expressed through one or another choice of orthogonal orbitals, even though they remain the same wave functions. By a suitable choice of one of the canonical orbital sets, any one of the three resulting singlet states can be made to have an "open-shell" ("dot–dot") appearance, with two singly occupied orbitals. For instance, the "open-shell" configuration $^1B = {}^1(\phi_1\phi_2)$ in the most delocalized orbital (MO) basis is identical with the combination of two "closed-shell" ("hole-pair") structures in the localized ("VB") basis, $\chi_a^2 - \chi_b^2$, whereas $Z_1 - Z_2 = \phi_1^2 - \phi_2^2$ in the delocalized basis is identical with the "open-shell" structure $^1(\chi_a\chi_b)$ in the localized basis. (Cf. Example 4.3.) The triplet-state wave function has the same open-shell appearance in all possible orbital bases.

Although the various choices of orthogonal orbital basis are all equivalent in principle, often one or another is more convenient to use. In general, the most delocalized choice provides more physical insight for molecules at ordinary geometries, and the most localized choice for molecules at biradicaloid geometries. It was for this reason that we used MO-based labels for the states of the ordinary H_2 and ethylene molecules on the left-hand sides of Figures 4.5 and 4.6, respectively, and VB-based labels for the states of the dissociated H_2 and twisted ethylene molecules in the same figures.

At times, it is best to work in the basis of nonorthogonal orbitals (GVB calculations), but we shall not use it here.

4.3.2 Perfect Biradicals

In a perfect biradical, the orthogonal nonbonding orbitals φ and φ' have the same energy and do not interact, no matter how they are chosen. Examples are provided by the earlier cases of fully dissociated H_2 and 90°-twisted ethylene. The energy-level diagram for a perfect biradical is characterized by only two parameters, the exchange integrals K and K', and is shown in Figure 4.19.

The energy diagram may be derived equally well starting with any orbital choice, but the three canonical ones are the simplest to use. We shall outline the derivation starting with the most localized orbitals. If for the moment spin is ignored, the four configurations $\chi_a(1)\chi_a(2)$, $\chi_b(1)\chi_b(2)$, $\chi_a(1)\chi_b(2)$, and $\chi_b(1)\chi_a(2)$ would have the same energy in the absence of electron repulsion (HMO approximation). Electron repulsion is much higher for the charge-separated functions of the hole-pair type than for those of the dot–dot type and causes a symmetric splitting of their energies by $2K'_{ab} = (J_{aa} + J_{bb})/2 - J_{ab}$ with respect to the average energy E_0, where $J_{ab} = (aa|bb)$ is the usual Coulomb integral. Finally, when the properly symmetry-adapted combinations $[\chi_a(1)\chi_a(2) \pm \chi_b(1)\chi_b(2)]/\sqrt{2}$ and $[\chi_a(1)\chi_b(2) \pm \chi_b(1)\chi_a(2)]/\sqrt{2}$ are used, the in-phase (plus) combinations are destabilized and the out-of-phase (minus) ones are stabilized by K_{ab}. Here, $K_{ab} = (ab|ab)$ is the usual exchange integral. Thus, the resulting four energy levels are $E_0 \pm K'_{ab} \pm K_{ab}$.

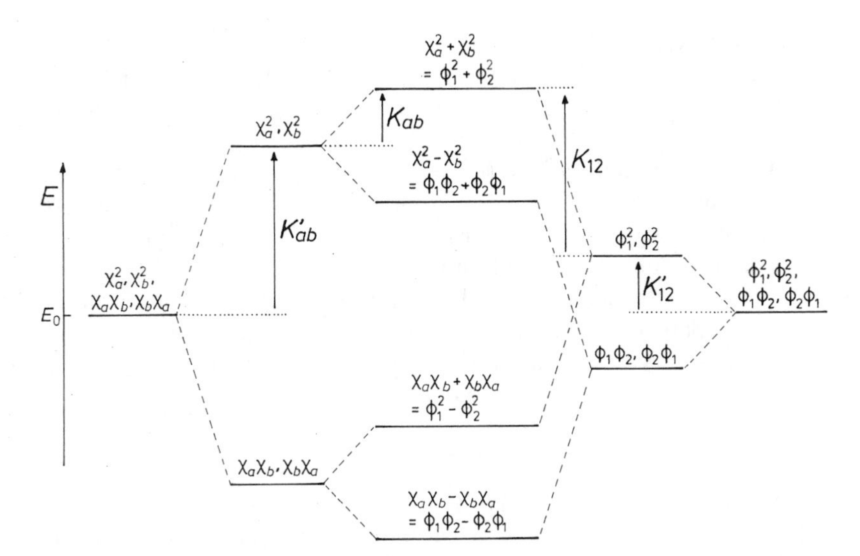

Figure 4.19. Wave functions and energy levels of a perfect biradical (center), constructed from the most localized orbitals χ_a and χ_b (left) and from the most delocalized orbitals ϕ_1 and ϕ_2 (right) (adapted from Bonačić-Koutecký et al., 1987).

Starting from delocalized orbitals ϕ_1 and ϕ_2, and introducing K_{12} and $K'_{12} = (J_{11} + J_{22})/2 - J_{12}$ as the corresponding electron repulsion integrals,

$$K_{12} = K'_{ab} \tag{4.4}$$

and

$$K'_{12} = K_{ab}$$

(cf. Equation 4.3), one obtains similar results (Figure 4.19).

The overlap density $\varphi\varphi'$ is smaller for the localized orbitals χ_a and χ_b, which occupy as distinct parts of space as at all possible, than for the delocalized orbitals ϕ_1 and ϕ_2, so

$$K'_{ab} \geq K_{ab} \tag{4.5}$$

and

$$K'_{12} \leq K_{12}$$

The wave function $[\chi_a(1)\chi_b(2) - \chi_b(1)\chi_a(2)]/\sqrt{2}$ is antisymmetric with respect to an interchange of electrons 1 and 2 and needs to be multiplied with one of the three symmetric two-electron spin functions to describe a triplet state T. The other three functions are symmetric, need to be multiplied with the antisymmetric two-electron spin function, and describe three singlet states, S_0, S_1 and S_2. The amusing isomorphism of the two-electron ordinary and spin functional space has been discussed elsewhere (Michl, 1991, 1992).

Since both K_{ab} and K'_{ab} are non-negative, in this model the triplet state T is the most stable of all four states. From Figure 4.19 it is also clear that (1) the wave function $(\chi_a^2 + \chi_b^2) = (\phi_1^2 + \phi_2^2)$ represents the least stable of the four states, S_2; (2) the wave function $(\chi_a^2 - \chi_b^2)$, or $^1(\phi_1\phi_2)$, represents S_1; and (3) the wave function $^1(\chi_a\chi_b)$, or $(\phi_1^2 - \phi_2^2)$, represents S_0. In the general case K'_{ab} and K_{ab} are different from each other, so the energies of the four states, T, S_0, S_1, and S_2 are all different. Biradicals with $K_{ab} = 0$ are referred to as *pair biradicals,* since the condition can be strictly satisfied only if the separation between the centers *A* and *B* is infinite so that the biradical in fact consists of a pair of distant radical centers. In practice, even twisted ethylene already is very nearly an ideal pair biradical. In pair biradicals, S_0 and T as well as S_1 and S_2 are pairwise degenerate.

Example 4.6:
On the basis of the two-electron two-orbital model it is easy to understand the differences in the potential energy curves describing the dissociation of a σ bond and a π bond. The H_2 molecule at infinite internuclear separation (right-hand side of Figure 4.5) and the ethylene molecule at 90° twist (center of Figure 4.6) are perfect biradicals. The former is a perfect pair biradical ($K_{ab} = 0$); the latter is a nearly perfect one (small K_{ab}). This difference is reflected in perfect T–S_0 and S_1–S_2 degeneracies in the former and only near degeneracies in the latter within the framework of the simple model. However, upon going to a more accurate calculation, the state order within both nearly degenerate state pairs of the twisted ethylene changes, for reasons that are now well understood (Mulder, 1980; Buenker et al., 1980).

Biradicals for which $K'_{ab} = K_{ab}$ are referred to as *axial biradicals* and in these, S_0 and S_1 are degenerate. This condition is normally enforced by the presence of a threefold or higher axis of symmetry (e.g., in O_2 and the pentagonal cyclopentadienyl cation), but alone, this is not sufficient (e.g., square cyclobutadiene is almost a pair biradical).

4.3.3 Biradicaloids

Imperfect biradicals, for which at least one of the necessary conditions for perfect biradicals is not fulfilled, are called biradicaloids. In the general case they are *nonsymmetric biradicaloids*. In these, the localized orbitals χ_a and χ_b have different energies (electronegativities) and also interact. An example is propene partially twisted about its double bond. In *homosymmetric biradicaloids* the localized orbitals χ_a and χ_b have equal energies, but interact. An example is partially twisted ethylene. In *heterosymmetric biradicaloids* the localized orbitals χ_a and χ_b have different energies, but do not interact. An example is 90°-twisted propene. In the simple model, the triplet state of a biradicaloid is still described by the wave function $^3(\varphi\varphi')$, but the singlet wave functions no longer are those given in Figure 4.19. The mixing (CI) between the eigenfunctions of the perfect biradical is due to interaction elements introduced by the perturbation that converted it into a biradicaloid.

In homosymmetric biradicaloids, a bonding interaction between the most localized orbitals is the only perturbation ("covalent perturbation"). Its magnitude is described by γ, approximately equal to twice the resonance integral of simple theories. Its presence causes a mixing of the state functions $^1(\chi_a\chi_b)$ and $(\chi_a^2 + \chi_b^2)$ of the perfect biradical, which stabilizes its ground state S_0 and destabilizes its S_2 state relative to the triplet state, whose energy we take as a reference (Figure 4.20a). When γ is large enough, an ordinary nonpolar covalent bond results, with a large percentage of dot–dot character, $^1(\chi_a\chi_b)$, and a small amount of symmetrized hole-pair character, $(\chi_a^2 + \chi_b^2)$, so that an ordinary molecule is formed (Figure 4.20a). The S_1 state is not affected, and the energy gap ΔE between the S_0 and the S_1 state increases with increasing interaction.

Stretched H_2 and twisted ethylene again may serve as examples. As the H—H separation decreases, and as the twist angle in ethylene decreases, the covalent perturbation γ increases in absolute magnitude. In both cases, this results in a stabilization of the S_0 state (bond formation) relative to the antibonding T state. It also results in the stabilization of the S_1 and S_2 states of H_2, although in the usual description the first of these states has as many electrons in the antibonding as in the bonding orbital, and the second only

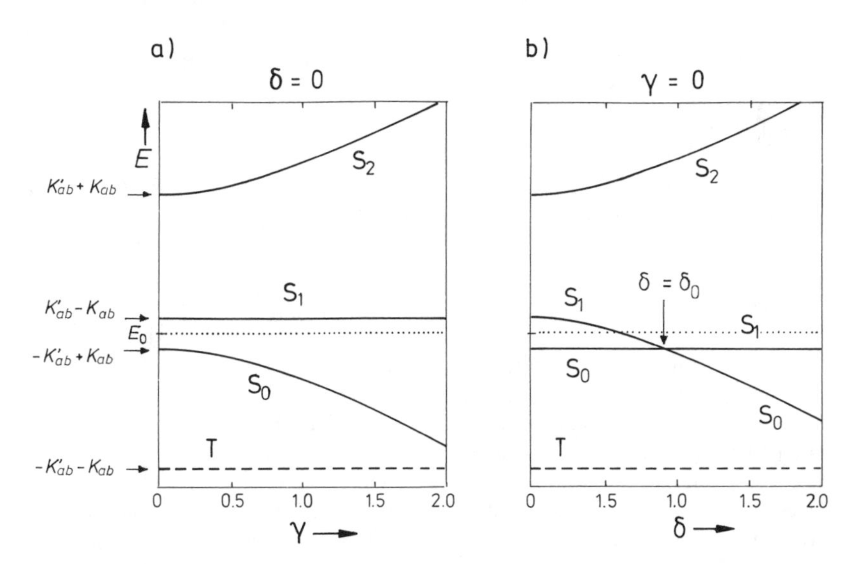

Figure 4.20. The two-electron two-orbital model a) for a homosymmetric biradicaloid and b) for a heterosymmetric biradicaloid: schematic representation of state energies, relative to the T state, as a function of the interaction integral γ or as a function of the electronegativity difference δ of the orbitals χ_a and χ_b (by permission from Michl and Bonačić-Koutecký, 1990).

has electrons in the antibonding orbital, and thus neither might be expected to be bound. In this regard, the effect of untwisting the ethylene molecule, which results in an energy increase for both the S_1 and S_2 states, is comforting. The difference is clearly due to the fact that along the ethylene twisting path, K'_{ab} hardly changes as $|\gamma|$ grows, so the magnitude of γ alone dominates the state energies, while along the H_2 association path, K'_{ab} drops dramatically as the charge separation energy decreases at shorter distances, so it counteracts the effect of the growing $|\gamma|$. The bonding in S_1 and S_2 is not covalent; rather, it is similar to bonding in an ion pair.

In order to understand the concurrent changes in the energy of the reference state T within the simple model, it is necessary to consider overlap explicitly (Bonačić-Koutecký et al., 1987). As the overlap between the nonorthogonal AOs grows, the energy of the orthogonalized localized orbitals χ_a and χ_b increases, and so does the energy of the T state.

In heterosymmetric biradicaloids, the only perturbation is an electronegativity difference between the two most localized orbitals ("polarizing perturbation"). This is described by their energy difference δ, equal to half the energy difference between the structures χ_a^2 and χ_b^2. Its presence causes the in-phase and out-of-phase combinations of the hole-pair structures, $(\chi_a^2 + \chi_b^2)$ and $(\chi_a^2 - \chi_b^2)$, to mix. This results in an increase in the energy of the former (S_2), dominated by the less stable of the two hole-pair structures, and a decrease in the energy of the latter (S_1), dominated by the more stable hole-pair structure, relative to the energy of the triplet state. The dot–dot structure $^1(\chi_a\chi_b)$ remains an eigenstate of the system, independent of δ. In perfect biradicals with a small K_{ab}, close to the pair biradical limit (e.g., twisted ethylene), the S_1 and S_2 states are nearly degenerate, and even a weak polarizing perturbation δ causes an essentially complete uncoupling of the hole-pair configurations χ_b^2 and χ_a^2. This imparts a highly polar character to the S_1 and S_2 states and underlies an effect that is known in the literature as *sudden polarization* (Bonačić-Koutecký et al., 1975).

Example 4.7:

The sudden polarization of the zwitterionic state of a nonsymmetric biradicaloid, starting with a perfect biradical, requires the presence of a small polarizing perturbation δ and the presence of a covalent interaction γ whose magnitude can be varied from well above δ to less than δ. As long as γ is much larger than δ, the effect of the latter on the composition of the eigenstates is essentially nil, and the S_1 and S_2 states are represented by nearly exactly balanced mixtures of the hole-pair structures. As γ becomes smaller than δ, the situation changes abruptly, and now the effect of δ dominates the composition of the wave functions: the S_1 and S_2 states are each nearly exactly represented by a single hole-pair structure.

The results of ab initio calculations for the charge distribution in the S_1 state of *s-cis-s-trans*-1,3,5-hexatriene plotted in Figure 4.21 as a function of the twist

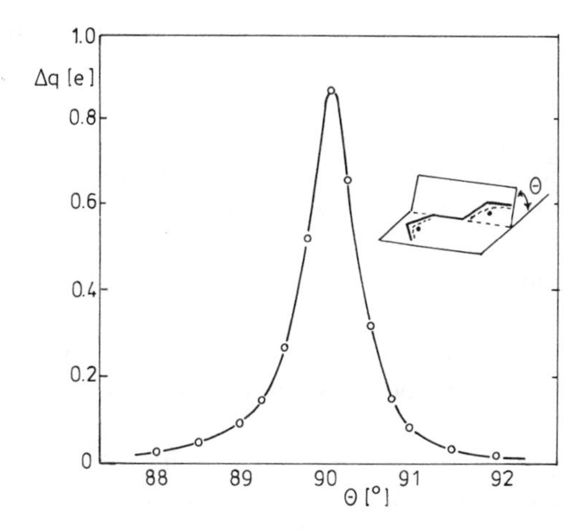

Figure 4.21. Charge separation Δq in the excited singlet state S_1 of twisted *s-cis-s-trans*-1,3,5-hexatriene as a function of the twist angle θ (by permission from Bonačić-Koutecký et al., 1975).

angle θ, which controls the magnitude of the covalent perturbation γ, make this evident. For nearly all values of θ the charge is distributed evenly among the two allylic units according to the ionic structures **14a** ↔ **14b**. If the allylic groups are nearly at right angles to each other ($\theta = 90°$), the covalent perturbation vanishes and sudden polarization occurs as a consequence of the small asymmetry δ between the cis- and trans-connected allyl groups (Bonačić-Koutecký et al., 1975).

a **14** **b**

The fact that the S_1 state rather than the S_0 state of a perfect biradical is stabilized upon polarizing perturbation has an important consequence. When the perturbation becomes strong enough, it causes a surface crossing as indicated in Figure 4.20b. The degeneracy of the S_0 and S_1 states has important consequences for photochemistry, since it provides a facile point of return for singlet excited molecules to the ground state. (See Chapter 6.)

The point at which the states S_0 and S_1 are degenerate is given by the relation

$$\delta = \delta_0 = 2\sqrt{K'_{ab}(K'_{ab} - K_{ab})} \tag{4.6}$$

Three important situations are of interest. For $\delta < \delta_0$ the biradicaloid is *weakly heterosymmetric*. The lowest singlet, S_0, similar to that of the perfect biradical, is represented by $^1(\chi_a\chi_b)$, and its separation from the T state is unaffected by δ. The S_1 state is represented by a mixture of hole-pair configurations, (χ_a^2) and (χ_b^2), with (χ_b^2) dominant. This situation, which corresponds to a large energy gap ΔE, is usually encountered if the dot–dot structure involves no formal charges. Examples are unsymmetrical 90°-twisted double bonds, such as twisted propene, or twisted ethylene pyramidalized at one carbon.

For $\delta > \delta_0$ the biradicaloid is referred to as *strongly heterosymmetric*. In the lowest singlet S_0, represented nearly exclusively by the more stable of the hole-pair configurations, both electrons are kept virtually exclusively in the more stable orbital χ_b localized on the center B, often as a "lone pair," while the less electronegative orbital χ_a is empty. If (χ_b^2) involves formal separation of charge, such species are usually referred to as zwitterions or ion pairs, with a positive charge on center A and a negative charge on B. However, really large δ values are normally encountered in systems in which the dot–dot structure $^1(\chi_a\chi_b)$ carries separated formal charges, while the hole-pair configurations (χ_b^2) does not. Examples of strongly heterosymmetric biradicaloids are molecules containing a noninteracting donor–acceptor pair, such as the 90°-twisted aminoborane (**10**); the TICT (twisted internal charge transfer) states represented by the zwitterionic excited state S_1 of these biradicaloids lead to a very pronounced solvent dependence of the emission wavelength (Grabowski and Dobkowski, 1983; Lippert et al., 1987).

Finally, a species with $\delta \approx \delta_0$ is a *critically heterosymmetric* biradicaloid. In this case, the dot–dot structure has the same energy as the more stable of the hole-pair structures, and the simple model predicts S_0 and S_1 to be degenerate. This situation is most readily obtained if neither $^1(\chi_a\chi_b)$ nor (χ_b^2) involves formal charge separation, that is, in charged biradicaloids where excitation results only in a translocation of formal charge. These arguments are in good agreement with detailed calculations for the ethyleneiminium ion $CH_2{=}NH_2^{\oplus}$, for which the energy gap $\Delta E(S_1 - S_0)$ is just equal to zero in the 90°-twisted geometry. (See Section 7.1.5.)

These results are equally applicable to σ and π bond dissociations, and suggest the general rule that the S_0 and S_1 hypersurfaces of a biradicaloid are expected to closely approach each other and to touch if the energies of the dot–dot (A—B) and hole-pair VB structures (AB:) are equal and if the two structures cannot interact. The energy equality can be reached by an appropriate choice of atoms or groups A and B, including the choice of substituents and their orientation (e.g., by choosing the particular double bond in a conjugated sequence around which the molecule will be twisted), the choice of transannular interactions in pericyclic biradicaloids, and the choice of the solvent (Bonačić-Koutecký et al., 1984; Bonačić-Koutecký and Michl, 1985b).

We can now return to the bond dissociation diagrams in Figures 4.5 and 4.6 and generalize them from the nonpolar case ($\delta = 0$) to a polar case. A major change will occur in the dissociated (biradicaloid) limit, since the state energies plotted there apply for $\delta = 0$ and will change in the manner shown in Figure 4.20b as δ increases. The results are shown in Figure 4.22 for the dissociation of a σ bond and Figure 4.23 for the dissociation of a π bond.

In the case of a σ bond, the bond length is plotted from left to right so that the dissociated limit ($R = \infty$) is on the right, with the energy of the dot–dot singlet and triplet states independent of δ and the energy of the hole-pair state dropping as δ increases toward the viewer, as expected from Figure 4.20b. The back side of the plot shows the nonpolar dissociation curves already familiar from Figure 4.5, and cuts at constant δ taken closer to the viewer provide dissociation curves at increasing values of δ. Clearly, the major change is that the S_1 state becomes dissociative, and at $\delta = \delta_0$ goes to the same dissociative limit as the triplet. This situation is nearly reached for the charged bonds such as C—N$^{\oplus}$, for which the dot–dot and the hole-pair dissociation limits differ by charge translocation rather than charge separation, and have comparable energies.

For even larger values of δ, the nature of the ground state changes. It is increasingly well described by a single hole-pair structure, and the entity in question can be called a dative bond, or a charge-transfer complex. The

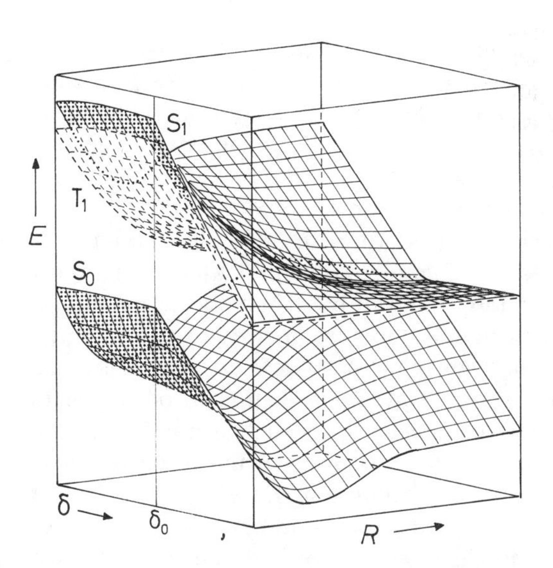

Figure 4.22. Schematic representation of the state energies of a dissociating σ bond. Shown are the S_0, S_1, and T_1 energies as a function of the bond length R and the electronegativity difference δ of orbitals χ_a and χ_b (by permission from Michl and Bonačić-Koutecký, 1990).

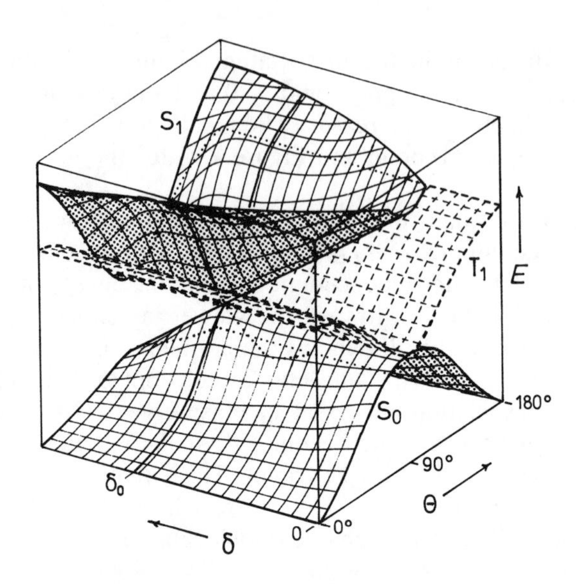

Figure 4.23. Schematic representation of the state energies of a dissociating π bond as an example of a heterosymmetric biradicaloid. Shown are the S_0, S_1, and T_1 energies as a function of the twist angle θ and the electronegativity difference δ of orbitals χ_a and χ_b (by permission from Bonačić-Koutecký and Michl, 1985b).

singlet excited state is zwitterionic, and the S_1–T singlet–triplet splitting is small. A simple example is $^{\oplus}H_3N—BH_3^{\ominus}$. In organic chemistry, this situation is frequently encountered in donor–acceptor pairs. In most cases, the donor, the acceptor, or both have a complicated internal structure and possess low-lying locally excited states that need to be added for a complete description, using the techniques of Section 4.2.

While the nonpolar case ($\delta = 0$) of S_1 dissociation clearly applies to an isolated H_2 molecule, Figure 4.22 suggests that such a symmetrical dissociation path is probably followed by no others. In polyatomic molecules, the two bond termini always have an opportunity to acquire different electronegativities (a nonvanishing δ value), by intramolecular means such as changes in the degree of pyramidalization or of conjugation with adjacent groups, or by intermolecular means such as unsymmetrical solvation. Since such a reduction of symmetry will permit an uncoupling of the two hole-pair configurations (cf. Example 4.7), permitting a decrease in the energy of the S_1 state, it is hard to imagine circumstances under which it would fail to occur. One can therefore expect single bond dissociations in S_1 to proceed along unsymmetrical pathways and to produce contact ion pairs, even in the nominally nonpolar cases such as Si—Si bond dissociation. Internal conversion to S_0 corresponds to back electron transfer and may be slow if the excitation energy is high. Fluorescence, intersystem crossing, and attack by external or intramolecularly present nucleophiles or electrophiles on the ion pair may then compete successfully, and it is indeed possible to account for most if not all of the singlet photochemistry of organo-oligosilanes on this basis. It is not known to what degree this is correct, and to what degree their reactions involve motion into those parts of the potential energy surfaces where the S_1 state is of a dot–dot character. (See Section 4.4.)

Figure 4.23 is exactly analogous but deals with the dissociation of a π bond. The energies of the S_0, S_1, and T states are shown as a function of the twist angle θ and the electronegativity difference δ. The nonpolar case, $\delta = 0$, is shown in front and corresponds to the potential energy curves of Figure 4.6. A cut through the surfaces at $\theta = 90°$ yields the curves familiar from Figure 4.20b: the singlet and triplet dot–dot structures are nearly degenerate since K_{ab} is small, and do not change their energy as δ is increased, while one of the hole-pair structures is stabilized until it meets the dot–dot singlet at $\delta = \delta_0$, and thereafter its energy descends further below that of the dot–dot structures.

Cuts through the surfaces of Figure 4.23 at various constant values of δ give the π-bond dissociation curves of polar bonds. The different values of δ could be achieved intramolecularly, by a suitable variation of the molecular geometry, or intermolecularly, by changes in the molecular environment, or even by such modifications in the molecular composition itself as changes in the substituents on the double bond or in the nature of the doubly bonded atoms.

The cut at δ_0 contains a critically heterosymmetric biradicaloid and a S_1—S_0 touching. This situation is normally encountered with charged π bonds, for example, protonated Schiff bases and many cationic dyes. (See Section 7.1.5 and Figure 7.6.) Inasmuch as a variation of δ can be accomplished by geometrical distortions of a molecule, Figure 4.23 can be viewed as depicting a funnel in its S_1 potential energy surface. This provides a nice illustration of the physical significance of the vectors $\boldsymbol{x}_1$ and $\boldsymbol{x}_2$ (Section 4.1.2) in the nuclear configuration space that define the branching space of a conical intersection corresponding to a critically heterosymmetric biradicaloid: $\boldsymbol{x}_1$ is the δ coordinate (i.e., the direction of the fastest change in the energy difference between the two "nonbonding" orbitals) and $\boldsymbol{x}_2$ is the γ coordinate (i.e., the direction of the fastest change in the degree of interaction between these two orbitals). Recall that the electronic energy that is converted into nuclear kinetic energy by downhill motion on the S_1 surface toward the tip of the cone usually tends to accelerate the nuclei in the direction of $\boldsymbol{x}_1$ (since that is usually the steepest slope on the upper cone); the electronic energy that is converted into nuclear kinetic energy by a jump to the lower surface accelerates the nuclei in the direction $\boldsymbol{x}_2$. The paths followed after return to S_0 through a "critically heterosymmetric" funnel are therefore likely to follow some linear combination of the directions of the fastest change in γ and δ.

Cuts at larger values of δ describe the behavior of π-donor–π-acceptor systems connected by a single bond (TICT molecules, cyanine dyes, etc.) and make their otherwise perhaps somewhat mysterious propensity toward twisting understandable. A simple example is $H_2N—BH_2$ (**10**). More common are systems with more complicated donor and acceptor structures and low-lying locally excited states, which need to be considered using the methods of Section 4.2.

The nonpolar case ($\delta = 0$) is not likely to be followed in the singlet reaction of any real organic molecule, for reasons already given in the treatment of σ dissociation (Section 4.1.3). A real twisting double bond, even in ethylene, let alone in stilbene, is likely to find a way to make the electronegativities of the orbitals at the two bond termini different in one of the ways mentioned in Section 4.1.3. This will permit the two hole-pair configurations to be uncoupled and the energy of the S_1 state to be lowered, for instance, by pyramidalization or a more severe distortion of the positions of the substituents. (Cf. the discussion of sudden polarization above and Section 7.1.) If it is possible for the magnitude of the polarizing perturbation δ to be brought close to δ_0, the S_1–S_0 gap will be particularly small and the return to S_0 particularly facile.

These considerations illustrate the general tendency of S_1 surfaces to have minima at unsymmetrical geometries. The T surface has a similar tendency, for different reasons, and it is probably fair to say that in photochemical reactions, the unsymmetrical reaction paths are the ones normally followed. In spite of this, most illustrations of potential energy curves in the literature and in this book are for symmetrical paths, since these are much easier to calculate or guess. It is up to the reader of any of the theoretical photochemical literature to keep this in mind and to correct for it the best he or she can, using the principles outlined in this chapter.

Example 4.8:

The case of a critically heterosymmetric biradicaloid with $\delta = \delta_0$ is nearly reached for charged bonds such as $C—N^{\oplus}$. Figure 4.24a shows the results of a CI calculation for protonated methylamine $CH_3NH_3^{\oplus}$, with Rydberg configurations ignored. Although the detailed results depend on the geometries assumed, the S_0, S_1, and T_1 curves clearly come close together at the dissociation limit, as is to be expected from the simple model in Figure 4.22. The case $\delta > \delta_0$ may be exemplified by a dative bond such as that in the ammonia–borane adduct $H_3N^{\oplus}—BH_3^{\ominus}$. The bond in the S_0 state is weaker than a charged or uncharged covalent bond. In the dissociation limit the energy gap between the S_0 state (χ_b^2) and the charge-transfer excited states $^{1,3}(\chi_a\chi_b)$ is large, as is seen from Figure 4.24b. In most cases of interest, the donor, the acceptor, or both have a complicated internal structure and possess low-lying locally

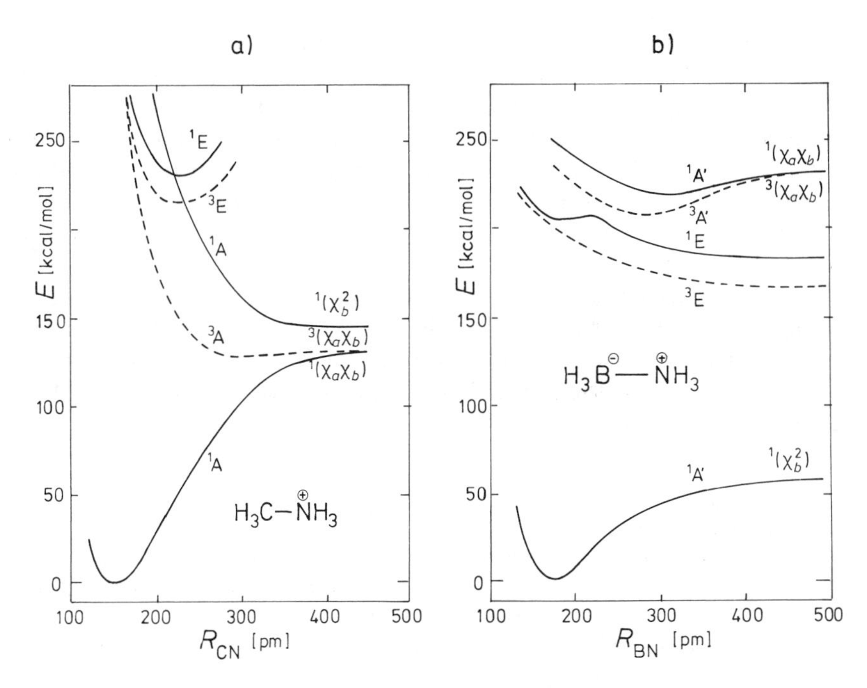

Figure 4.24. Potential energy curves for the dissociation a) of the C—N bond in protonated methylamine and b) of the B—N bond in the ammonia–borane adduct. Results of large-scale CI calculations with all geometric parameters except the bond distance R held fixed (by permission from Michl and Bonačić-Koutecký, 1990).

excited states that need to be added to the simple model for a complete description, using the techniques of Section 4.2.

4.3.4 Intersystem Crossing in Biradicals and Biradicaloids

Triplet biradicals and biradicaloids are frequent intermediates in photochemical reactions. (See Section 6.3.) Typically, the biradical is initially generated from a photochemical triplet precursor, and its initial electron spin function is an essentially pure triplet. At one or another time in the process of forming the singlet final products, the reacting molecule needs to cross to the S_0 surface, and the knowledge of geometries at which such crossing is probable is important for the understanding of the nature of the products.

In addition to the solvent-induced electron spin relaxation by independent spin flips at two well-separated radical centers (spin-lattice relaxation), which is quite slow in the absence of paramagnetic impurities (ordinarily, on the order of 10^5–10^6 s^{-1} in monoradicals), there are two important mechanisms of spin-flipping in triplet biradicals and biradicaloids. The first of these is due to the difference of the g values of the unpaired electron spins in the presence of a magnetic field, and to their hyperfine interaction with the spin of magnetic nuclei located in the vicinity (Closs et al., 1992; for earlier views see Doubleday et al., 1989). It is well known from the theory of CIDNP (chemically induced dynamic nuclear polarization, Salikhov et al., 1984). The second important mechanism for intersystem crossing is spin-orbit coupling. (Cf. Section 1.3.2.)

The "g-value difference and hyperfine coupling mechanism" is important in biradicals in which the two radical centers are relatively far apart (1,6-biradicals and longer), and the spin-orbit coupling is weak at almost all biradical geometries, and the T_1 and S_0 levels are almost exactly degenerate (at most a few cm^{-1} apart).

Example 4.9:

In long biradicals the singlet–triplet separation falls off roughly exponentially with distance (De Kanter et al., 1977; for an updated view see Forbes and Schulz, 1994) and with the number of intervening bonds (Closs et al., 1992). In a strong magnetic field, e.g., in EPR experiments, only the T_0 component of the triplet state is usually sufficiently close to S_0 for fast crossing.

The coupling of each unpaired electron to the magnetic nuclei in its vicinity is fundamentally a local effect, and is nearly independent of the distance between the two radical centers. Each nuclear spin substate in each of the three low-energy triplet levels of the biradical behaves essentially independently, except that ordinary spin relaxation processes may cause transitions among them and more slowly, into the singlet level, as mentioned earlier. Since the singlet–triplet splitting is so small at nearly all conformational geometries of the long-chain biradical, even the very small g-value and hyperfine terms (typically 0.01 cm^{-1}) cause spin function mixing, and each sublevel becomes an "impure" triplet, with a small weight of singlet character. The weight oscillates very rapidly in time, but only its average value is important. When the conformational motion of the chain that proceeds over small barriers on the triplet surface brings the two radical centers into close vicinity such that the two singly occupied orbitals interact, the covalent perturbation γ suddenly increases greatly. This causes the singlet–triplet splitting to increase in magnitude well above the mixing terms, whose effect thus becomes negligible, and the new spin eigenfunctions of the system once again become an essentially pure triplet or an essentially pure singlet, with probabilities given by the respective weights of these two spin functions in the "impure" triplet at the time immediately before the encounter of the radical ends. Most of the time, the decision falls in favor of the triplet, and the nuclei then feel forces dictated by the T potential energy surface. This is typically repulsive and causes the radical centers to separate. No reaction occurs; the "impure" triplet biradical is reformed, and continues to live to try to recombine another time. After a few hundred or thousand excursions into geometries characterized by a large value of γ, the decision falls in favor of the singlet; that is, intersystem crossing occurs. In this event, the spin-orbit coupling mechanism of intersystem crossing, which tends to be important at just these geometries, as discussed later, provides additional singlet–triplet mixing and may play a significant role, but it does not discriminate between the various nuclear substates. From now on, the nuclear motion is governed by the shape of the S_0 surface, which typically slopes steeply downhill to a ring-closure or a disproportionation product without a barrier. Singlet product formation from all molecules that have undergone intersystem crossing is thus virtually guaranteed. This is not to say that the S_0 surface does not have small conformational barriers at various geometries, just like the triplet—only that these are not likely to lie between the geometry at which intersystem crossing was forced by a large increase in the S_0–T_1 splitting, and the geometry of a singlet product.

Note that the region of geometries where the covalent interaction just sets in is particularly favorable for a jump between surfaces: it contains the turning point for vibrations on the T surface, which just begins to rise, as well as on the S_0 surface, which just begins to drop, and the Franck-Condon overlap of the vibrational wave functions is sizable. At the same time, the amount of energy that needs to be transferred from electronic to vibrational motion in the non–Born-Oppenheimer jump from one to the other surface is still relatively small. A motion along a direction that increases the magnitude of the covalent perturbation is therefore well suited for promoting intersystem crossing.

The rate of intersystem crossing is then proportional to the frequency of end–end encounters and to the weight of the singlet spin function in the "impure" triplet, dictated by a combination of spin-orbit and hyperfine coupling terms. Because of the latter, it may be different for each nuclear spin sublevel. At times, this leads to spectroscopically detectable nuclear spin polarization in the products, and to large isotope effects.

The second mechanism of intersystem crossing in biradicals and biradicaloids, spin-orbit coupling, requires a significant degree of covalent interaction between the two radical centers, through space or through bond. It needs to be considered in biradicals of any length at geometries in which the two radical centers are close, becomes important at all geometries in radicals of intermediate length, and is dominant in short biradicals, which occur as intermediates in numerous photochemical reactions (Chapter 7). In 1,2-, 1-3-, and 1,4-biradicals the S_0–T splitting tends to be relatively large at most geometries, from a few to several thousand cm^{-1} (Doubleday et al., 1982, 1985; Caldwell et al., 1988) and the g-value and hyperfine effects are then negligible.

In the absence of magnetic field, the three triplet levels are separated in energy by the zero field splitting (roughly 10^{-2}–10^{-1} cm^{-1} in short biradicals) and are quantized with respect to the molecular axes dictated by the principal directions of the electron spin–spin dipolar coupling tensor primarily responsible for the zero-field splitting. (Spin-orbit coupling affects the zero-field splitting, too.) Their spin functions are labeled Θ_x, Θ_y, and Θ_z and are given by

$$\begin{aligned}\Theta_x &= -[\alpha(1)\alpha(2) - \beta(1)\beta(2)]/\sqrt{2} \\ \Theta_y &= \mathrm{i}[\alpha(1)\alpha(2) + \beta(1)\beta(2)]/\sqrt{2} \\ \Theta_z &= [\alpha(1)\beta(2) + \beta(1)\alpha(2)]/\sqrt{2}\end{aligned} \tag{4.7}$$

while the singlet spin function Σ is given by

$$\Sigma = [\alpha(1)\beta(2) - \beta(1)\alpha(2)]/\sqrt{2} \tag{4.8}$$

The three sublevels of the triplet couple to the S_0 singlet to different degrees, dictated by the singlet–triplet separation and by the matrix elements H_x^{SO}, H_y^{SO}, and H_z^{SO}, respectively. The nuclear spin states are immaterial. Often, it is assumed that the relaxation between the three levels is rapid, or that all three are initially populated more or less equally, and the "total spin-orbit coupling strength" of the S_0 state is then defined as $(H_x^{SO2} + H_y^{SO2} + H_z^{SO2})^{1/2}$. This can be thought of as the length of the "spin-orbit coupling vector" $\boldsymbol{H}^{SO}$ with components H_u^{SO}, $u = x, y, z$. In short organic biradicals, values above about 0.1 cm^{-1} are considered large, and typical magnitudes do not exceed a few cm^{-1}.

The original qualitative conclusions of Salem and Rowland (1972) concerning spin-orbit coupling in biradicals were based on the two-electron–two-orbital model of biradicals. They have been supported more recently to a remarkable degree by the ab initio calculations mentioned earlier, as long as one permits the radical centers to delocalize somewhat into the nearby bonds and does not insist on the original simplest interpretation in which they were strictly viewed as carbon 2p orbitals, thus precluding through-bond contributions. In the two-electron–two-orbital model, the description of the presumably dominant one-electron part of spin-orbit coupling is relatively simple (Salem and Rowland, 1972; Michl and Bonačić-Koutecký, 1990; Michl, 1991). If the singlet state functions S_i and the triplet state functions T_u ($u = x, y, z$) are written as

$$\begin{aligned}S_i = \sqrt{1/2}\ \{&C_{i,-}[\varphi(1)\varphi(2) - \varphi'(1)\varphi'(2)] \\ &+ C_{i,+}[\varphi(1)\varphi)(2) + \varphi'(1)\varphi'(2)] \\ &+ C_{i,0}[\varphi(1)\varphi'(2) + \varphi'(1)\varphi(2)]\}\Sigma\end{aligned} \tag{4.9}$$

$$T_u = \sqrt{1/2}\ [\varphi(1)\varphi'(2) - \varphi'(1)\varphi(2)]\Theta_u \tag{4.10}$$

with the spin functions Σ and Θ_u given by Equations (4.7) and (4.8), and if the spin-orbit coupling operator of Equation (1.42) is used, one obtains, after performing the spin integration (Example 1.8),

$$<T_u|\hat{H}_{SO}|S_i> = C_{i,+}\frac{e^2\hbar}{2m_e^2c^2}<\varphi|\sum_{\mu}\frac{Z_\mu}{|\boldsymbol{r}^\mu|^3}\hat{\boldsymbol{l}}_u^\mu|\varphi'> \tag{4.11}$$

where the sum is over all atoms μ. Because of the presence of the factor $|\boldsymbol{r}^\mu|^{-3}$, contributions from pairs of atomic orbitals that are both located on atoms other than μ can be ignored when evaluating the μ-th term in the sum. The magnitude of the coefficient $C_{i,+}$ does not depend on the choice of the nonbonding orbitals φ and φ' as localized, delocalized, or even intermediate, as long as they are real and orthogonal. In the following, we adopt the most convenient choice, the most localized orbitals χ_a and χ_b. For these, the hole-pair functions can be identified with the zwitterionic functions if the biradical carries no formal charges at the radical centers. The form of Equation (4.11) most useful for our purposes (Michl, 1991) is finally arrived at by the standard procedure of correcting for the effects of the electrons of the fixed core, including inner-shell electrons, and for the effect of the neglected two-electron part of the spin-orbit operator, by introducing an atomic spin-orbit coupling parameter, ζ_μ, whose magnitude increases roughly with the fourth power of the atomic number Z_μ. For the spin-orbit coupling vector $\boldsymbol{H}^{SO}$ we obtain

$$\boldsymbol{H}^{SO} = C_{i,+} \sum_{\mu} \zeta_\mu <\chi_a|\hat{\nabla}^\mu|\chi_b> \tag{4.12}$$

The sum runs over all atoms μ. Each term includes only contributions from those pairs of atomic orbitals in which at least one of the partners is located at the atom in question. The x, y, and z components of the vector operator $\hat{\nabla}^\mu$, which differs from the angular momentum operator $\hat{l}^\mu$ only by the factor $\hbar/i$, are $\partial/\partial\xi$, $\partial/\partial\eta$, and $\partial/\partial\zeta$, respectively, where ξ, η, and ζ are the angles of rotation around the x, y, and z axes passing through the μ-th nucleus. The action of this operator on atomic orbitals located on atom μ is as follows: The s functions are annihilated, and for p functions,

$$\begin{array}{lll} \hat{\nabla}_z p_x = -p_y & \hat{\nabla}_z p_y = p_x & \hat{\nabla}_z p_z = 0 \\ \hat{\nabla}_y p_z = -p_x & \hat{\nabla}_y p_x = p_z & \hat{\nabla}_y p_y = 0 \\ \hat{\nabla}_x p_y = -p_z & \hat{\nabla}_x p_z = p_y & \hat{\nabla}_x p_x = 0 \end{array} \tag{4.13}$$

(Cf. Example 1.8.) This result shows that in the 3×3 model, the spin-orbit coupling vector depends on three factors: the coefficient $C_{i,+}$ of the in-phase $^1(\chi_a^2 + \chi_b^2)$ character of the singlet state, the spin-orbit coupling parameter ζ_μ (heavy atom effect), and the spatial disposition of the orbitals χ_a and χ_b. The actual intersystem crossing rate will also depend on the Franck-Condon–weighted density of states. (Cf. the Fermi golden rule, Section 5.2.3.)

First of all, we consider the significance of the presence of $C_{i,+}$, the coefficient of the in-phase combination of the two hole-pair functions in the S_0 wave function, in Equation 4.12. In the simple model for a perfect biradical, this in-phase combination is exactly equal to the wave function of the S_2 state, and it does not enter into those of the S_0 and S_1 states at all. Thus, in this approximation, S_0 does not spin-orbit couple to the triplet. The same is true in weakly heterosymmetric biradicaloids ($0 < \delta < \delta_0$), in which the in-phase hole-pair character is shared by S_1 and S_2, but not S_0, and the former two spin-orbit couple to T, but S_0 does not. In strongly heterosymmetric biradicaloids ($\delta > \delta_0$), the in-phase hole-pair character is shared by S_2 and S_0. (Cf. Figure 4.20b.) Now, S_0 spin-orbit couples to T, but in such donor–acceptor pairs the two states are ordinarily far removed from each other in energy and the S_0–T intersystem crossing is relatively slow. The S_1 state of such species is nearly degenerate with T, but does not spin-orbit couple with it in this approximation.

The situation changes dramatically in the presence of a covalent perturbation, $\gamma \neq 0$, which causes the in-phase hole-pair function to mix into the S_0 state. In the resulting homosymmetric or nonsymmetric biradicaloid, T may spin-orbit couple to S_0, and the total spin-orbit coupling strength will be proportional to the coefficient $C_{i,+}$ of the in-phase hole-pair spin function in the wave function of S_0. This is sometimes expressed in a simplified way (Salem and Rowland, 1972) such as "T–S_0 spin-orbit coupling in a biradical requires ionic character in the S_0 state," but it may be more accurate to state that it requires covalent interaction (nonzero resonance integral) between the orthogonal localized orbitals at the radical termini, since it is only an admixture of the in-phase and not the out-of-phase combination of the two hole-pair ("ionic") configurations that matters. The simultaneous presence of a polarizing perturbation in nonsymmetrical biradicaloids does not facilitate spin-orbit coupling, but might have a positive effect on the intersystem crossing rate by affecting the S_0–T gap.

Note that the geometries needed for the required covalent perturbation are not likely to be located at energy minima in the triplet surface, since after all, the triplet state is antibonding with respect to the two radical centers. At least a small thermal activation barrier to reaching the optimal spin-orbit coupling geometries thus is to be expected. This could well be different in different conformers, possibly leading to temperature-dependent stereoselectivity and isotopic selectivity in product formation even in those cases in which the conformers as such had identical free energies.

The tendency of the spin-orbit coupling matrix element to be largest just at the covalently perturbed biradicaloid geometries, which lie on the path to singlet products, is undoubtedly at least partly responsible for the perception that intersystem crossing in triplet biradicals ordinarily gives closed-shell singlet products rather than floppy singlet biradicals. (Cf. Example 4.11.)

Next, we consider the significance of the sum over atoms on the right-hand side of Equation 4.12. Each atom contributes a vector $<\chi_a|\hat{\nabla}^\mu|\chi_b>$, weighted by its spin-orbit coupling parameter ζ_μ, and this clearly reflects the "heavy atom effect." However, it indicates equally clearly that a contribution from an atom does not need to be large just because the atom is heavy: the vector contribution from that atom might have a small length or an unfortunate direction that cancels contributions from other atoms. (Note that "inverse" heavy atom effects are therefore possible; cf. Turro et al., 1972b.)

This brings us to the last factor to consider—the size and direction of the atomic vector contributions $<\chi_a|\hat{\nabla}^\mu|\chi_b>$. The size can only be significant if the coefficients of the 2p orbitals located on atom μ in at least one of the most localized orbitals χ_a and χ_b are large (and indeed, in the first approximation only the two atoms carrying the radical centers were considered, Salem and Rowland, 1972), but this condition is not sufficient. Specifically, in view of Equation (4.13), the z component of the vector contribution from atom μ consists of two parts that are added algebraically. The first will be large when the coefficient on the $p_{\mu y}$ orbital in χ_b is large, and when one or more of the atomic orbitals that have a large overlap with $p_{\mu x}$ (including $p_{\mu x}$ itself) have large coefficients in χ_a. The second will be large when the coefficient on the $p_{\mu x}$ orbital in χ_b is large, and when one or more of the atomic orbitals (including $p_{\mu y}$) that have a large overlap with $p_{\mu y}$ have large coefficients in χ_a. The signs of the two contributions are dictated by the signs of the orbital coefficients, by the signs in Equation (4.13), and by the signs of the overlaps. Similar results hold for the x and y contributions to the atomic vector provided by atom μ.

Example 4.10:
The atomic vectors $<\chi_a|\hat{\nabla}^\mu|\chi_b>$ contain through-space and through-bond contributions. Their origin is most easily visualized for conformations in which the axes of the singly occupied orbitals at the two radical centers are mutually perpendicular. It is somewhat unfortunate that these are often just the conformations at which the covalent perturbation γ of a perfect biradical is zero, making $C_{i,+}$ vanish, and causing the S_0–T_1 spin-orbit mixing to be negligible even if the resultant of the atomic vectors is large (it is then the experimentally uninteresting S_2–T_1 coupling that is large). This is indeed the case in both of the following simple examples.

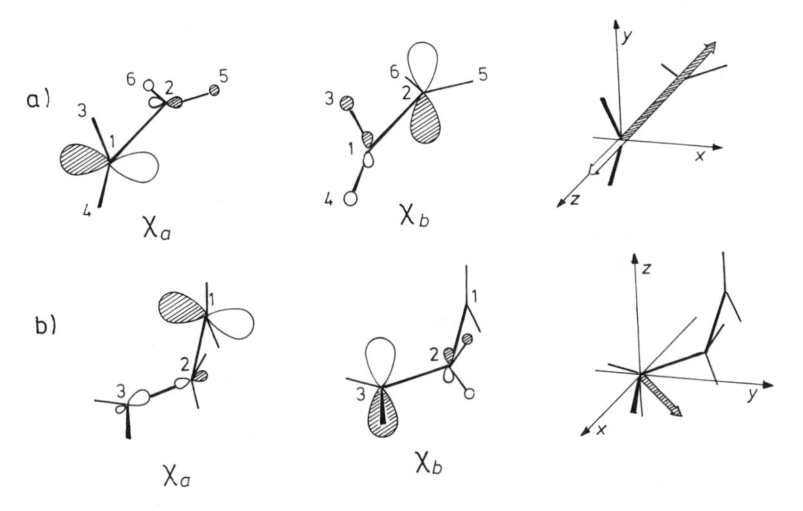

Figure 4.25. Sum-over-atoms factor in the spin-orbit coupling vector $\boldsymbol{H}^{SO}$ a) in orthogonally twisted ethylene and b) in (0, 90°) twisted trimethylene biradical, using Equation (4.12) and (4.13); most localized orbitals χ_a, χ_b and non-vanishing atomic vectorial contributions from χ_b (white: through-space, black: through-bond).

In orthogonally twisted ethylene, the localized orbitals χ_a and χ_b are related by symmetry. Each is primarily located on a free p orbital of one CH_2 group and is hyperconjugatively delocalized into the CH bonds of the other CH_2 group (Fig. 4.25a). The principal axes of the zero-field-splitting tensor are dictated by symmetry. Only the z component of the atomic vectors $<\chi_a|\hat{\nabla}^\mu|\chi_b>$, directed along the CC axis, is different from zero, and spin-orbit coupling will mix S_0 with only one of the three triplet components, T_{1z}. We can therefore locate the x and y axes arbitrarily as long as all three axes are mutually orthogonal, and we choose them to lie in the planes of the two CH_2 groups.

The nonbonding orbitals then have the form

$$\chi_a = c_1 p_{1x} - c_2 p_{2x} - c_3(s_5 - s_6)$$
$$\chi_b = c_1 p_{2y} - c_2 p_{1y} - c_3(s_3 - s_4)$$

where $c_1 \gg c_2, c_3 > 0$.

The resulting atomic vectors for the hydrogen atoms are negligible; those for the two carbon atoms ($\mu = 1,2$) are large and equal. We illustrate the evaluation $<\chi_a|\hat{\nabla}_z^1|\chi_b>$:

$$<\chi_a|\hat{\nabla}_z^1|\chi_b> = c_1^2<p_{1x}|\hat{\nabla}_z^1|p_{2y}> - c_1c_2<p_{1x}|\hat{\nabla}_z^1|p_{1y}> - c_1c_3<p_{1x}|\hat{\nabla}_z^1|s_3 - s_4> + \textit{small terms}$$

where the small terms either contain a product of two small coefficients, c_2c_3, and/or contain no atomic orbitals located on C(1) (e.g., $c_1c_2<p_{2x}|\hat{\nabla}_z^1|p_{2y}>$).

The first term on the right-hand side is of the through-space type. It would be present even if there were no hyperconjugation, that is, even if the nonbonding orbitals were strictly localized on C(1) and C(2), respectively ($c_1 = 1$, $c_2 = c_3 = 0$). The second and third terms would vanish in this limit and are of the through-bond type.

The action of the operator $\hat{\nabla}_z^1$ on a π-symmetry orbital on its right is to rotate it by 90° around the z axis in the clockwise sense when viewed against the direction of z (cf. Equation 4.13); for instance,

$$<p_{1x}|\hat{\nabla}_z^1|p_{2y}> = <p_{1x}|p_{2x}> = S_{12} \approx 0.25$$
$$<p_{1x}|\hat{\nabla}_z^1|p_{1y}> = <p_{1x}|p_{1x}> = 1$$

Thus, the through-space term makes a positive contribution to the z component of the atomic vector $<\chi_a|\hat{\nabla}^1|\chi_b>$, and the two through-bond terms make negative contributions. It is impossible to evaluate the sign of the resultant without at least a rough numerical evaluation of the opposing contributions: the through-space term contains the overlap integral S_{12} as a relatively small multiplier while the through-bond terms contain the small coefficient c_2 or c_3 as a multiplier, but there are more of them. In a minimum basis set approximation, a computation shows that the through-bond terms dominate. Inclusion of the two-electron part of the spin-orbit operator in the calculation reduces the magnitude of the result by a factor of about two, but does not change its sign.

Note that the opposed signs of the through-space and through-bond contributions are a result of the nodal properties of the orbitals χ_a and χ_b. As is seen in Figure 4.25a, they have a node separating the main part of the orbital from the minor conjugatively delocalized part. However, the presence of the node is sensitive to the details of the structure, and so is the relative sign of the through-space and through-bond contributions. For instance, in orthogonally twisted H_2N—BH_2, the empty orbital located on boron does not have such a node, and the dominant through-bond contribution, due to the one-center term on the N atom, now has the same sign as the through-space contribution.

The through-bond terms will not always dominate, but are likely to be quite generally important. In our second example, a singly twisted 1,3-trimethylene biradical (Figure 4.25b), the leading term ($\mu = 3$) arises from the hyperconjugation of the singly occupied orbital of the CH_2 group on C(1), whose hydrogens are twisted out of the CCC plane, with the C(2)C(3) bond. In the coordinate system of Figure 4.25b, chosen so as to diagonalize the zero-field splitting tensor, χ_a then contains small contributions from the p_{3x} and p_{3y} orbitals on C(3), an atom on which χ_b has its dominant component, p_{3z}. The action of either the $\hat{\nabla}_x^3$ or the $\hat{\nabla}_y^3$ operator rotates p_{3z} into overlap with χ_a. As a result, both the x and y components of the atomic vector $<\chi_a|\hat{\nabla}^3|\chi_b>$ are nonzero, and the through-bond mechanism of spin-orbit coupling will be the primary cause of the mixing of both T_{1x} and T_{1y} with S_0.

Example 4.11:
The magnitude of the total spin-orbit coupling strength can change dramatically as a function of biradical conformation (Carlacci et al., 1987). The through-space part can be roughly approximated by

$$|\boldsymbol{H}^{SO}| = 15 \times |S| \sin \omega \text{ cm}^{-1}$$

where ω is the acute angle between the axes of the p orbitals containing the unpaired electrons and S is their overlap integral. The overlap S is related to the resonance integral between these two orbitals and thus to the coefficient of the in-phase combination of the two hole-pair functions in the S_0 wave function, $C_{0,+}$. The factor sin ω originates from the operator $\hat{\nabla}^{\mu}$ in the matrix element $<\chi_a|\hat{\nabla}^{\mu}|\chi_b>$, which rotates the orbital χ_b according to Equation (4.13) before its overlap with χ_a is taken.

No such simple generally valid approximation is currently available for the through-bond part of the spin-orbit coupling strength, which results from the delocalization of even the most localized form of the nonbonding orbitals into the σ skeleton, onto nearby carbon atoms μ located between the radical centers. This produces nonzero coefficients on the p orbitals on these atoms and permits them to contribute to the sum in Equation (4.12). A similar mechanism operates when atoms carrying lone pairs, such as oxygen, are located between the radical centers. The low-lying electronic states of biradicals of this type are more numerous and the spin-orbit coupling is less likely to be properly described by the simple model that led to this equation. The availability of additional states involving promotion from the lone pairs appears to make spin-orbit coupling particularly effective. Note that the placement of lone-pair carrying or heavy atoms into positions that do not lie between the two radical centers cannot be expected to have much effect, since then only one of the most localized nonbonding orbitals has significant coefficients on their orbitals.

As discussed in Example 4.9, the probability of the intersystem crossing to the singlet will be determined by the weight of the singlet spin function in the "impure" triplet at a time when the molecular motion brings its geometry to a region in which the singlet–triplet splitting increases significantly, so that a decision between an essentially pure triplet and an essentially pure singlet state must be made. If it occurs, further nuclear motions will be dictated by the S_0 surface, and if it does not, the biradical will continue its conformational motion in the impure triplet state. The geometries at which the singlet–triplet splitting becomes large are those at which the covalent interaction between the two localized singly occupied orbitals, described in the simple model by the quantity γ, is large. At many of these, the spin-orbit coupling is also large. At these geometries, the S_0 surface will typically slope steeply toward a product minimum, with no barrier in the way, and the probability that a newly formed singlet molecule might escape the likely fate of falling into this abyss is undoubtedly minimal. The somewhat startling conclusion, that intersystem crossing in triplet biradicals can and normally does produce closed-shell singlet products and not floppy singlet biradicals, was initially suggested by Closs. It has been gradually gaining recognition (De Kanter and Kaptein, 1982; Wagner, 1989; Wagner et al., 1991). Indeed, intersystem crossing caused by random spin relaxation at the two independently acting radical centers, which can be accelerated by the addition of paramagnetic impurities, and which would be expected to generate floppy singlet biradicals at a variety of geometries, yields different product ratios (Scaiano, 1982).

Geometries expected to be favorable for intersystem crossing are those at which both the covalent interaction γ as well as the spin-orbit coupling strength become large. This requires that the two most localized singly occupied orbitals overlap enough to establish a significant covalent perturbation of the biradical, and that they are mutually oriented in such a manner that they still overlap after the action of the angular momentum operator $\hat{l}^{\mu}$ on one of them, that is, after a 90° rotation of the p orbital around an axis that passes through its center (particularly if it is located on an atom of a high atomic number). If the orbitals overlap through space, the likely singlet reaction is covalent bond formation between the two radical centers, as exemplified in Figure 4.26a for π bond formation in 1,2-biradicals by planarization, and in Figures 4.26b and c for disrotatory and conrotatory σ bond formation in 1,n-biradicals by radical recombination. It also occurs when they interact in a through-bond fashion, for example, in 1,4-biradicals, in which case the likely singlet reactions are disproportionation or fragmentation, depending on the details of the geometry as indicated in Figures 4.26d and e. In order to understand the rate of intersystem crossing and the nature of the products, it is therefore essential to consider not only the shape of the T potential energy surface, which determines the occupancy of the various conformer minima and the frequency with which the various molecular ge-

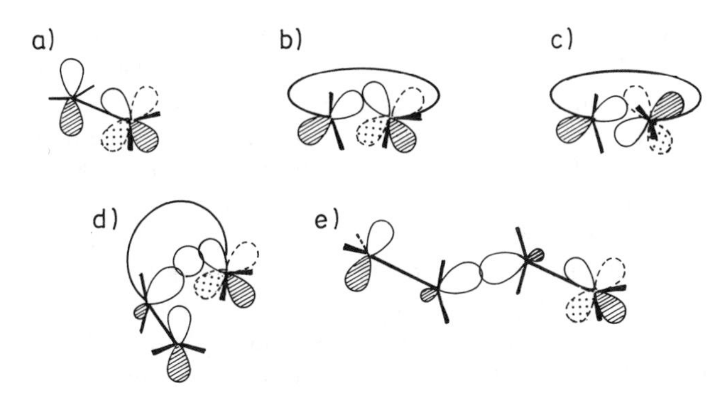

Figure 4.26. Examples of biradicaloid geometries expected to be favorable for spin-orbit coupling: overlap leading to covalent interaction γ between the localized nonbonding orbitals (full lines) and nonzero overlap after the action of the angular momentum operator $\hat{l}^{\mu}$ on one of them (dotted lines); a) partial double-bond twist, b) disrotatory, and c) conrotatory ring closure, d) disproportionation, and e) fragmentation.

ometries are visited, but also the magnitude of the spin-orbit coupling element and of the singlet–triplet splitting as a function of geometry.

To summarize the result of the simple theory, an admixture of S_0 character into the triplet wave function of a biradical is expected to be large at those geometries at which covalent interaction and spin-orbit coupling are both simultaneously large, that is, at which the most localized singly occupied orbitals overlap sufficiently before as well as after a 90° rotation of one of the important p orbitals. A jump to an essentially pure singlet, followed by motion on the S_0 surface, occurs upon excursion of the biradical geometry into an area of strong covalent perturbation (large singlet–triplet splitting, singlet stabilization), with a probability dictated by the degree of admixture of singlet character into the triplet wave function.

The simple theory of spin-orbit coupling in biradicals has been found useful for the interpretation of the lifetimes of triplet biradicals as a function of their structure and conformation (Johnston and Scaiano, 1989; Adam et al., 1990) and of the stereochemistry of their reaction products (Chapter 7).

4.4 Pericyclic Funnels (Minima)

The global term "pericyclic funnel" will be used to refer to the funnel or funnels in the S_1 surface that occur at the critically heterosymmetric biradicaloid geometries reached near the halfway point along the path of a thermally forbidden pericyclic reaction, and the minima in S_1 that are encountered along one-dimensional cuts along reaction paths that miss the conical intersections (in particular, those along high-symmetry paths, which pass through the geometry of a perfect biradical, and in which the S_1–S_0 touching is therefore avoided). This region of the S_1 surface was usually called "pericyclic minimum" because it was thought that the touching is most likely weakly avoided everywhere. Thanks to the extensive recent work of the groups of Bernardi, Olivucci, and Robb (Bernardi et al., 1990a–c, 1992a–c; Olivucci et al., 1993, 1994a,b) this is now known not to be so (cf. Section 4.1.3), and for quite a few reactions the exact location of the bottom of this funnel has been determined in considerable detail. An understanding of the electronic states in the region of the pericyclic funnel is of fundamental significance in many photoreactions such as ground-state–forbidden cycloadditions, electrocyclic reactions, or sigmatropic rearrangements, and it appears essential to discuss these states in some detail.

We shall base the discussion on an analysis of a simple four-electron–four-orbital model, exemplified by H_4 (the "20 × 20 CI model"), which contains all the essential ingredients (Gerhartz et al., 1976, 1977), as shown by the more recent calculations on actual molecules (two of these are discussed in more detail in Section 6.2.1). Although it might thus appear that an understanding of the electronic states of a tetraradical is necessary, we shall see that already an understanding of the states of biradicals and biradicaloids at the level of the two-electron–two-orbital ("3 × 3 CI") model (Section 4.3) is immensely helpful, although not quite a substitute for the full 20 × 20 CI description. For those familiar with valence-bond theory, an even simpler "2 × 2" VB model (Bernardi et al., 1988, 1990b) will be described, and the strengths and weaknesses of the 3 × 3 and 2 × 2 models will be compared.

4.4.1 The Potential Energy Surfaces of Photochemical $[2_s + 2_s]$ and $x[2_s + 2_s]$ Processes

Pericyclic processes involving interactions of four electrons in four overlapping orbitals arranged in a cyclic array (*A*, *C*, *D*, *B*, labeled clockwise along the perimeter) correspond in general to the switching of two bonds originally connecting atoms *A* with *B* and *C* with *D*, in the reactants to connect atoms *A* with *C* and *B* with *D* in the final product (Scheme 1). Examples are the conversion of butadiene into cyclobutene (cf. Section 4.2.2) or norbornadiene (**15**) into quadricyclane (**16**):

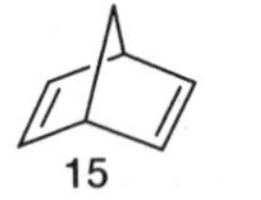
15

16

17

If the new bonds are formed in a suprafacial way, the process is referred to as $[2_s + 2_s]$. The singlet states of the hypothetical H_4 molecule at square geometries calculated by Gerhartz et al. (1976) can be thought of as a simple model for the electronic states involved.

A C B D ←x[$2_s + 2_s$]— A–B C–D —[$2_s + 2_s$]→ A–C B–D

Scheme 1

Analogous processes in which the final product has atom *A* attached to *D* and atom *B* attached to *C* (cross-links in the perimeter) are also known (Scheme 1) and could be called cross-pericyclic (the term "cross-cycloaddition" is relatively common in the literature). An example is the conversion of 1,3-butadiene to bicyclobutane (**17**). If the new bonds are formed in a suprafacial way, the process is referred to as $x[2_s + 2_s]$. The singlet states of the hypothetical H_4 molecule at triply right tetrahedral geometries calculated by Gerhartz et al. (1977) can be thought of as a simple model for the electronic states involved.

The results of model 20 × 20 CI calculations on H_4 led to the proposal (Gerhartz et al., 1977) that the $[2_s + 2_s]$ and $x[2_s + 2_s]$ processes proceed through the same pericyclic funnel in the S_1 state. This has been supported by recent more realistic calculations (Olivucci et al., 1993, 1994b), and the two processes are therefore best discussed together.

To simplify a relatively complex situation, we shall however first pretend that only a relatively high-symmetry version of the $[2_s + 2_s]$ path is accessible, in which the four orbitals in question (modeled by the four H atoms) are located at the corners of a rectangle. Halfway along this reaction path, they are located at the corners of a square. At this point we shall find a fairly strongly avoided touching of the S_1 and S_0 states and a "pericyclic minimum" in the S_1 state. Of course, in H_4, this is not at all a minimum, since several types of distortion, such as an increase in the size of the square, lower the energy. A realistic example of such a path is the one used in the correlation diagrams in Figures 4.7 and 4.8.

Subsequently, we shall relax the condition that all four H atoms have to be in the same plane, and we shall find a downhill path that leads from this pericyclic "minimum" to an S_1–S_0 conical intersection as the square is puckered and the two diagonals, *AD* and *BC*, shortened. A real molecule is likely to follow a lower-symmetry path straight to this "cross-bonded" pericyclic conical intersection. Once on the S_0 surface, it will end up as a $[2_s + 2_s]$, as an $x[2_s + 2_s]$ product, or as the starting material.

Figure 4.27 is based on the results of a full 20 × 20 CI calculation for H_4 and shows the energies calculated for the lowest singlet states of the reaction $H_2 + H_2' \rightleftharpoons 2HH'$ with geometries corresponding to the hypothetical best ground-state rectangular path. The nomenclature used to label the states refers to the whole H_4 system or "supermolecule." The singly excited S state correlates smoothly from left to right without a barrier; the calculation produces, however, a broad minimum at the biradicaloid geometry. We shall see

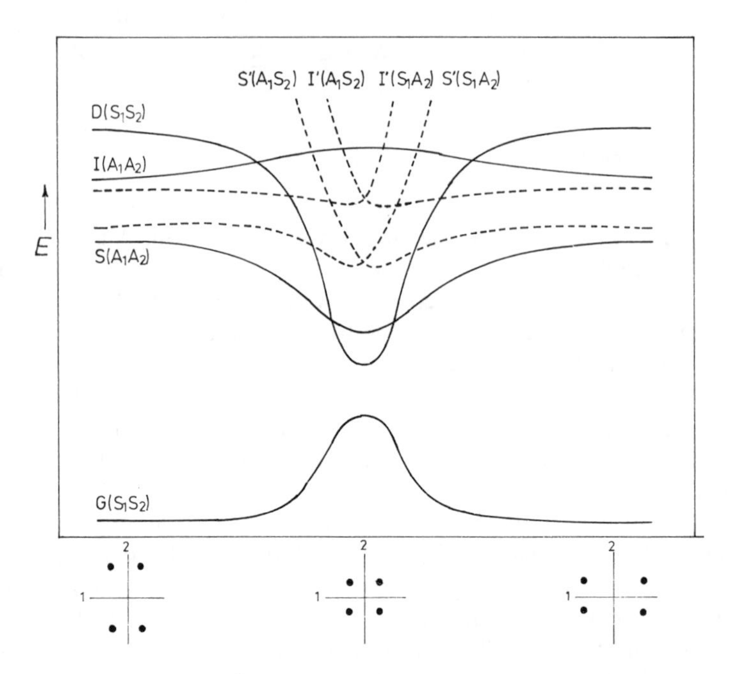

Figure 4.27. Schematic state correlation diagram for the $H_2 + H_2' \rightleftharpoons 2HH'$ reaction along a rectangular path. Solid and broken correlation lines refer to states of different symmetry (by permission from Michl, 1977).

(Section 5.4.2) that this corresponds to the excimer minimum. The avoided crossing between the ground state G and the doubly excited state D produces the "pericyclic minimum" in the excited-state and the barrier in the ground-state surface. In the nomenclature of the simpler 3 × 3 CI model (Bonačić-Koutecký et al., 1987), square H_4 corresponds to a perfect biradical close to the "pair" biradical limit (uncharged 4*N*-electron perimeter), with near degeneracy between the T and S_0 states and between the S_1 and S_2 states, and a fairly large gap between the two pairs of states. This agrees well with the relatively large degree to which the intended S_0–S_1 touching in Figure 4.27 is avoided. However, in another respect the simplified 3 × 3 CI description fails, as it renders incorrectly the order of the states within the nearly degenerate pairs (it is inherently incapable of placing the D state below the S state, as indicated below: Example 4.12).

The ordering of states at the geometry of the pericyclic minimum provided by the full four-electron–four-orbital treatment (20 × 20 CI) may be rationalized only when we combine the two simpler and intuitively more easily grasped models, 3 × 3 CI (MO) and 2 × 2 VB. The comparison will clearly illustrate the strengths and the shortcomings of the two simplified models. Unlike the full 20 × 20 CI model, which contains all the possible MO configurations and all the possible VB structures, the 3 × 3 model only contains one dot–dot (covalent) structure and two hole-pair (zwitterionic) structures based on the most localized nonbonding MOs. It confines two of the four electrons into a nonpolarizable core (analogous to the most stable MO of square cyclobutadiene), and ignores the availability of the antibonding MO (analogous to the least stable MO of square cyclobutadiene). It will be particularly poor at geometries at which one or both of the two MOs that have been ignored are energetically close to the two nonbonding orbitals that are being considered (e.g., at tetrahedral geometries) or those of large squares. The 2×2 VB model considers all four electrons explicitly but permits only the two possible covalent structures. By ignoring all zwitterionic structures, it gives up all hope of describing the excited states allowed for absorption from the ground state, the excimer state, and any states with separated charges. In H_4, the energies of structures with separated charges are quite unfavorable, but in real organic molecules they will be stabilized by dynamic correlation with the electrons of the other bonds present, and they can be quite competitive with the excited covalent structures, particularly when the opposite charges are not far from each other.

Figure 4.28 shows the energies E_G, E_D, and E_S of the G, D, and S states of H_4 at the square geometries obtained from the full 20×20 calculation as a function of the square size. In the limit of a very small square, the ordering of $E_G < E_S < E_D$, the same as obtained from the simple 3×3 CI description. (Cf. Example 4.12.) In the limit of an infinitely large square, the ordering is $E_G = E_D < E_S$, the same as in the simple 2×2 VB description, with four H atoms coupled into a covalent overall singlet in two different ways. According to Figure 4.28, neither simple description is adequate for square sizes of practical interest. It is perhaps not necessary to go to a full 20×20 CI, but

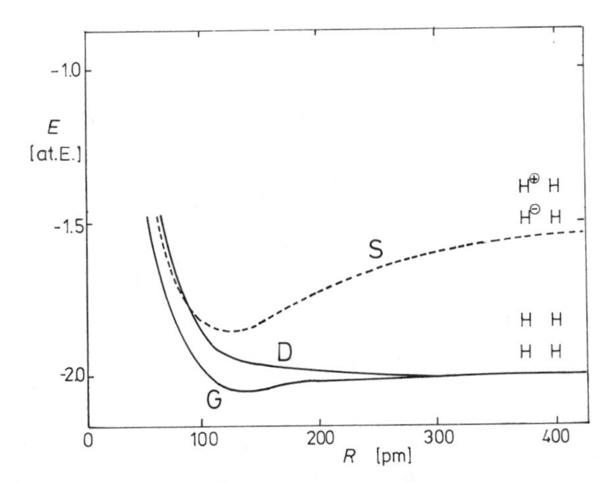

Figure 4.28. Calculated state correlation diagram for H_4 at square geometries as a function of the size of the square. Solid and broken correlation lines, respectively, refer to states of different symmetry (by permission from Michl, 1977).

certainly more than three configurations are needed when the MO approach is used, and the VB approach definitely requires zwitterionic structures. Hence, arguments based on the simplified models of bonding have to be interpreted with due care.

Example 4.12:
A 3 × 3 CI calculation for H_4 takes into consideration the configurations

$$\Phi_1 = |\phi_2\bar{\phi}_2|$$
$$\Phi_2 = 1/\sqrt{2}\{|\phi_2\bar{\phi}_3| + |\phi_3\bar{\phi}_2|\}$$
$$\Phi_3 = |\phi_3\bar{\phi}_3|$$

which can be constructed from the MOs $\phi_2 = \frac{1}{2}(\chi_1 + \chi_2 - \chi_3 - \chi_4)$ and $\phi_3 = \frac{1}{2}(\chi_1 - \chi_2 - \chi_3 + \chi_4)$. The most stable MO $\phi_1 = \frac{1}{2}(\chi_1 + \chi_2 + \chi_3 + \chi_4)$ is neglected in this example; it is always doubly occupied and forms a nonpolarizable core. At square or nearly square geometries the system is a perfect biradical or a homosymmetric biradicaloid, respectively, and the energy ordering of the three singlet states may be obtained from Figure 4.19 or 4.20:

$$E_G < E_S < E_D$$

For an explicit calculation of the energy order, Slater rules yield the CI matrix elements

$$<\Phi_1|\hat{H}|\Phi_1> = <|\phi_2\bar{\phi}_2|\hat{H}|\phi_2\bar{\phi}_2|> = 2I_2 + J_{22}$$
$$\begin{aligned}<\Phi_2|\hat{H}|\Phi_2> &= \tfrac{1}{2}\{<|\phi_2\bar{\phi}_3|\hat{H}|\phi_2\bar{\phi}_3|> + <|\phi_3\bar{\phi}_2|\hat{H}|\phi_3\bar{\phi}_2|> \\ &\quad + <|\phi_2\bar{\phi}_3|\hat{H}|\phi_3\bar{\phi}_2|> + <|\phi_3\bar{\phi}_2|\hat{H}|\phi_2\bar{\phi}_3|>\} \\ &= I_2 + I_3 + J_{23} + K_{23}\end{aligned}$$
$$<\Phi_3|\hat{H}|\Phi_3> = <|\phi_3\bar{\phi}_3|\hat{H}|\phi_3\bar{\phi}_3|> = 2I_3 + J_{33}$$
$$<\Phi_1|\hat{H}|\Phi_3> = <|\phi_2\bar{\phi}_2|\hat{H}|\phi_3\bar{\phi}_3|> = K_{23}$$
$$<\Phi_1|\hat{H}|\Phi_2> = <|\Phi_3|\hat{H}|\Phi_2|> = 0$$

where the last line follows from symmetry arguments. Thus the CI problem factorizes into a 2 × 2 problem and a 1 × 1 problem. Using the ZDO approximation, one has $I_2 = I_3$ and $J_{22} = J_{33} = J_{23}$. Thus $K'_{23} = [(J_{22} + J_{33})/2 - J_{23}]/2 = 0$ and the state energies can be written down immediately:

$$E_1 = E_G = <\Phi_1|\hat{H}|\Phi_1> - <\Phi_1|\hat{H}|\Phi_3> = 2I_2 + J_{22} - K_{23}$$
$$E_2 = E_D = <\Phi_3|\hat{H}|\Phi_3> + <\Phi_1|\hat{H}|\Phi_3> = 2I_2 + J_{22} + K_{23}$$
$$E_3 = E_S = <\Phi_2|\hat{H}|\Phi_2> = 2I_2 + J_{22} + K_{23}$$

Since all electron repulsion terms are positive, it follows that

$$E_G < E_S = E_D$$

Since in the ZDO approximation K'_{23} vanishes, the S and D states are degenerate. Interactions with additional configurations are needed in order to stabilize D to such an extent that $E_D < E_S$.

We are now ready to give up the fiction that a high-symmetry path along rectangular geometries will be followed in the excited-state pericyclic process and to explore downhill paths from the square version of the pericyclic minimum in H_4 thus far considered. It was recognized early on (Gerhartz et al., 1977) that the 20 × 20 CI wave function of the G state of a square array of four orbitals exhibits predominantly singlet local coupling along the perimeter and predominantly triplet local coupling across the diagonals, and that the opposite is true of the covalent part of the wave function of the D state. Naively, in the 2 × 2 VB approximation, one could say that the G state is reached by bringing two ground-state H_2 molecules together side by side, and the D state by placing them across each other. These are indeed the state correlations found in the full calculation. Figure 4.29 shows cuts through the potential energy surfaces of H_4 at the 20 × 20 CI level along the already discussed path of rectangles and along the path of triply right tetrahedra, which result when the two diagonals of a square are pulled apart in opposite directions perpendicular to the plane of the square but their lengths are kept unchanged. Along the latter path, the G state increases and the D state decreases in energy, and the two become degenerate when the geometry of a regular tetrahedron is reached. Afterward, the D state represents the S_0 surface. Ultimately, if the lengths of the diagonals were permitted to adjust to the ground-state equilibrium values in H_2, the energy would drop much more, and the ordinary ground state of two H_2 molecules would result. As this situation is approached, the zwitterionic contributions to the D state are reduced in size, and the 2 × 2 VB description, which ignores them completely, would be quite adequate. Already at the geometry of the G–D state crossing (S_0–S_1 state touching), the 2 × 2 VB model is quite good in a qual-

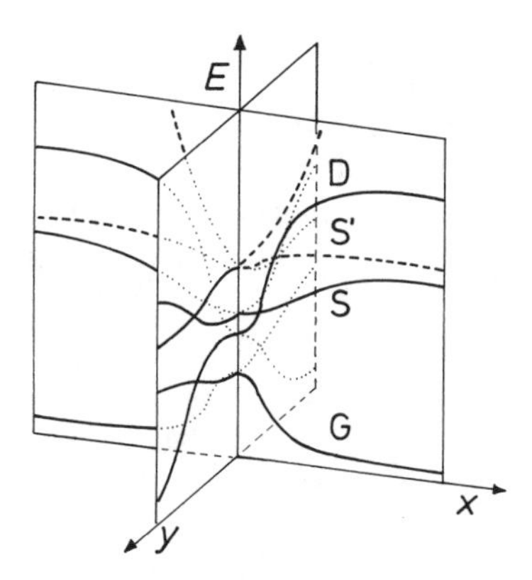

Figure 4.29. Schematic state correlation diagram for the $H_2 + H_2' \rightleftarrows 2HH'$ reaction along the rectangular path (x) and along the tetrahedral path for $R_1 = R_2$ (y).

itative sense. In contrast, the 3 × 3 CI model, incapable of describing more than a single covalent function, is unsuitable.

One can expect other downhill paths from the square "pericyclic minimum" in the D state, in which advantage will be taken of the diagonal singlet coupling, but the two diagonals will not remain of equal length. For instance, a distortion of the square into a rhombus will shorten one and extend the other diagonal. A continuation of this path will not produce two H_2 molecules but only one, plus two hydrogen atoms, in the right electronic state for subsequent barrierless recombination into a second H_2 molecule. In an organic molecule, this path would correspond to the formation of a 1,4-biradical ready to collapse with the formation of an additional bond. The effects of this type of distortion are also well described qualitatively by the 2 × 2 VB model. (See Example 4.13.)

Now, however, even the 3 × 3 CI model predicts an S_1–S_0 touching, since unequal diagonal interactions in a square represent a heterosymmetric perturbation δ. (Cf. Section 4.3.) When δ is allowed to reach a critical value, a heterosymmetric biradicaloid with an S_0–S_1 degeneracy will result. The perturbation due to the introduction of resonance integrals across a conjugated perimeter was discussed in some detail in Sections 2.2.3–2.2.7 in connection with the perimeter model. Thus, a diagonal bond in cyclobutadiene or a 1,5 bond in octagonal cyclooctatetraene is expected to convert these perfect biradicals into heterosymmetric biradicaloids, to split the degeneracy of the nonbonding orbitals, and to lead toward an S_1–S_0 touching (Höweler et al., 1989). The polarizing perturbation δ introduced in this manner can be quite large; in going from cyclooctatetraene to pentalene, δ reaches δ_0 and then exceeds it, so the S_1 and S_0 states touch (Doehnert and Koutecký, 1980).

It is clear that the critical value δ_0 expected from the simple formula Equation 4.6 will not be quite right, since the 3 × 3 CI model ignores the effect of the second covalent VB structure. In a similar vein, one would not expect the analogous formula for the exact position of the state touching that is derived from the 2 × 2 VB model (Example 4.13) to be exactly right either, since the zwitterionic contributions to the D state are now left out. This inexactness should be particularly damaging when one considers the effects of electron-donating and electron-withdrawing substituents. Qualitatively, however, it is clear from the consideration of the states of H_4 that an energy decrease in S_1 and a S_1–S_0 touching are to be expected upon distortion of the square geometry of the initially diagonal interactions across the perimeter. Even in real organic molecules, the "bottom" of the pericyclic funnel is therefore to be sought at lower symmetry geometries containing such distortions. Since one or the other of a square's diagonals can be shortened, one can expect that two or possibly even a larger number of conical intersections will flank the "pericyclic minimum," producing rather intricate topological structure for the resulting "pericyclic funnel."

Example 4.13:
In the 2×2 VB model (Bernardi et al., 1988, 1990b) the ground- and excited-state energies are given as

$$E_0 = Q - T$$
$$E_1 = Q + T$$

where Q is the Coulomb energy and T the total exchange energy given by the London formula (cf. Eyring et al., 1944):

$$T = K_R^2 + K_P^2 + K_X^2 - K_R K_P - K_R K_X - K_P K_X)^{1/2}$$
$$= [\,(K_P - K_X)(K_R - K_X) + (K_P - K_R)^2]^{1/2}$$

K_R, K_P, and K_X are defined (Bernardi et al., 1988, 1990b) in terms of two center VB exchange integrals* as follows:

$$K_R = K_{12} + K_{34}$$
$$K_P = K_{13} + K_{24}$$
$$K_X = K_{14} + K_{23}$$

if orbitals 1,2 and 3,4 are coupled in the reactants while orbitals 1,3 and 2,4 are coupled in the product. (Cf. Scheme 1.)

At the conical intersection the two states are, by definition, degenerate ($E_1 = E_0$), and the total exchange energy T must vanish. Therefore the following two independent relations must hold:

$$K_R = K_P$$
$$K_R = K_X \text{ (or } K_P = K_X)$$

Since the behavior of the exchange integrals K_{ij} is easily predicted as a function of the geometrical coordinates, the structure corresponding to a conical intersection can be predicted by distorting the molecular structure of the system in such a way as to satisfy those two equations. (Cf. Section 6.2.1.)

In practical applications of the simple H_4 model there will be other bond-forming electrons in addition to the four electrons involved in the bond-switching or bond-crossing process. These additional bonds will prevent many of the total or partial fragmentation processes expected for H_4. For instance, the uninterrupted downhill slope of the D surface toward four separated atoms (cf. Figure 4.28) will not exist in the case of the $[2_s + 2_s]$ cycloaddition of two ethylenes, since the C—C σ bonds of the two ethylenes will prevent it. With due caution, however, it should be possible to transfer much of the information obtained from the 20×20 CI model to $[2_s + 2_s]$ and $x[2_s + 2_s]$ reactions in general, particularly information concerning the

---

*The two-center exchange integral K_{ij} has the same interpretation as in the Heitler-London treatment of H_2 and can be written as $K_{ij} = \langle ij|\hat{g}|ji\rangle + 2S_{ij}\langle i|\hat{h}|j\rangle$, where $\langle ij|\hat{g}|ji\rangle$ is the usual two-electron-exchange repulsion integral, $\langle i|\hat{h}|j\rangle$ is the one-electron integral, and S_{ij} the overlap integral between orbitals i and j.

physical nature of the various excited states, and qualitative understanding of the reasons for the ways in which their energies change with the molecular geometry. (Cf. Section 6.2.1; Chapter 7.)

4.4.2 Spectroscopic Nature of the States Involved in Pericyclic Reactions

In discussing certain photochemical reactions, such as photocycloadditions, it is of interest to relate the electronic states at the pericyclic minimum to the states of the two reacting fragments that come together side by side, in the present model case, to the states of two H_2 molecules. It then turns out that in the limit of infinite intermolecular separation, the G state corresponds to a combination of two ground-state fragments $H_2 + H_2$, the S state to a combination of a ground-state fragment H_2 with a singly excited fragment H_2^*, and the D state to an overall singlet coupling of two triplet-excited fragments $H_2^* + H_2^*$. Thus, at infinite separation, the S state can also be labeled *exciton state* and the D state *triplet–triplet annihilation state*. There actually are two exciton states, represented simply as $H_2H_2^* \pm H_2^*H_2$; S is the lower of these, and the higher one is labeled S′ in Figure 4.27. In the limit of infinite internuclear separation, two additional charge-transfer or ion-pair states can be identified, which can be represented simply as $H_2^{\oplus}H_2^{\ominus} \pm H_2^{\ominus}H_2^{\oplus}$. In Figure 4.27 they are labeled I and I′.

At finite separations, these four zero-order terms interact pairwise and produce states S, S′, I, and I′ of strongly mixed exciton–charge-transfer character, where the VB functions are ionic and contain terms such as $A\text{—}BC^{\oplus}D^{\ominus} \leftrightarrow A\text{—}BC^{\ominus}D^{\oplus}$ (exciton) or $A^{\oplus}B\text{—}CD^{\ominus}$ (CT). Of the four states, S is greatly stabilized compared to the others, through exciton as well as CT interactions—that is, exactly by the factors responsible for the stability of excimers and exciplexes. (Cf. Sections 5.4.2 and 5.4.3.) The S state has a minimum at the biradicaloid square geometry that may be referred to as the excimer minimum. A strong absorption band in the near-IR region corresponding to a transition between the S and S′ states of the fluorene excimer has been observed by photodissociation spectroscopy (Sun et al., 1993).

Because of the especially high symmetry and because all the electrons are involved in binding interactions, the conditions in H_4 are evidently optimal for the formation of an excimer minimum. The situation will be different for photocycloadditions in which π bonds are converted into σ bonds and were additional electrons in mutually repelling closed shells make an intimate approach of the two reactants difficult. In these cases, the excimer minimum will be flatter and will occur only at larger nuclear separations, where the molecules are barely touching. (Cf. Section 5.4.2 and Figure 6.7.)

The D state with its pericyclic funnel is not related to the four low-energy ionic states. It originates from a mixing of the triplet–triplet annihilation wave function with totally symmetric higher excited configurations and charge-transfer terms and acquires partial ionic character as the intermolecular separation decreases. In contrast to the excimer minimum, the pericyclic minimum will generally occur at a geometry at which the old bonds in the perimeter are half-broken, the new bonds in the perimeter half-established, and a strong diagonal interaction introduced—that is, roughly halfway along the reaction path. The relative location of the excimer minimum and the pericyclic funnel can thus vary over a wide range. (Cf. Section 6.2.3 and Figure 6.17.) They can represent two separate topological features in the S_1 surface, but it is also conceivable that they lie above each other and are not both present in the S_1 surface. (Cf. Figure 7.27.)

Supplemental Reading

General

Michl, J. (1972), "Photochemical Reactions of Large Molecules," *Mol. Photochem.* **4**, 243, 257, 287.

Michl, J. (1974), "Physical Basis of Qualitative MO Arguments in Organic Photochemistry," *Fortschr. Chem. Forsch.* **46**, 1.

Michl, J. (1978), "The Role of the Excited State in Organic Photochemistry," in *Excited States in Quantum Chemistry*; Nicolaides, C.A., Beck, D.R., Eds.; Riedel Publ.: Dordrecht.

Michl, J., Bonačić-Koutecký, V. (1990), *Electronic Aspects of Organic Photochemistry*; Wiley: New York.

Simons, J. (1983), *Energetic Principles of Chemical Reactions*; Jones and Bartlett Publ.: Boston.

Potential Energy Hypersurfaces

Klessinger, M. (1982), *Elektronenstruktur organischer Moleküle*; Verlag Chemie: Weinheim.

Mezey, P.G. (1987), *Potential Energy Hypersurfaces*; Elsevier: Amsterdam.

Salem, L. (1982), *Electrons in Chemical Reactions: First Principles*; Wiley: New York.

Spin-Orbit Coupling and Hyperfine Interactions

Gould, I.R., Turro, N.J., Zimmt, M.B. (1984), "Magnetic Field and Magnetic Isotope Effects on the Products of Organic Reactions"; *Adv. Phys. Org.* **20**, 1.

Khudyakov, I.V., Screbrennikov, Y.A., Turro, N.J. (1993), "Spin-orbit Coupling in Free Radical Reactions on the Way to Heavy Elements," *Chem. Rev.* **93**, 537.

Michl, J. (1991), "The States of an Electron Pair and Photochemical Reactivity," in *Theoretical and Computational Models for Organic Chemistry*; Formosinho, S.J., et al., Eds.; Kluwer: Dordrecht.

Richards, W.G., Trivedi, H.P., Cooper, D.L. (1981), *Spin-Orbit Coupling in Molecules*; Clarendon Press: Oxford.

Salem, L., Rowland, C. (1972), "The Electronic Properties of Diradicals," *Angew. Chem. Int. Ed. Engl.* **11**, 92.

Salikhov, K.M., Molin, Y. u. N., Sagdeev, R.Z., Buchachenko, A.L. (1984), *Magnetic and Spin Effects in Chemical Reactions*; Elsevier: Amsterdam.

Steiner, U.E., Ulrich, R. (1989), "Magnetic Field Effects in Chemical Kinetics and Related Phenomena," *Chem. Rev.* **89**, 51.

Avoided Crossings

Bonačić-Koutecký, V. (1983), "On Avoided Crossings between Molecular Excited States: Photochemical Implications," *Pure Appl. Chem.* **55**, 213.

Devaquet, A. (1975), "Avoided Crossings in Photochemistry," *Pure Appl. Chem.* **41**, 455.

Devaquet, A., Sevin, A., Bigot, B. (1978), "Avoided Crossings in Excited States Potential Energy Surfaces," *J. Am. Chem. Soc.* **100**, 479.

Salem, L., Leforestier, C., Segal, G., Wetmore, R. (1975), "On Avoided Surface Crossings," *J. Am. Chem. Soc.* **97**, 479.

Conical Intersections

Atchity, G.J., Xantheas, S.S., Ruedenberg, K. (1991) "Potential Energy Surfaces near Intersections", *J. Chem. Phys.* **95**, 1862.

Bernardi, F., Olivucci, M., Robb, M.A. (1990), "Predicting Forbidden and Allowed Cycloaddition Reactions: Potential Energy Topology and Its Rationalization," *Acc. Chem. Res.* **23**, 405.

Herzberg, G., Longuet-Higgins, H.C. (1963), "Intersection of Potential Energy Surfaces in Polyatomic Molecules," *Disc. Faraday Soc.* **25**, 77.

Longuet-Higgins, H.C. (1975), "Intersection of Potential Energy Surfaces in Polyatomic Molecules," *Proc. Roy. Soc. London* **A344**, 147.

Teller, E. (1937), "The Crossing of Potential Surfaces", *J. Phys. Chem.* **41**, 109.

Teller, E. (1969), "Internal Conversion in Polyatomic Molecules," *Israel J. Chem.* **7**, 227.

Correlation Diagrams

Bigot, B., Devaquet, A., Turro, N.J. (1981), "Natural Correlation Diagrams. A Unifying Theoretical Basis for Analysis of n Orbital Initiated Ketone Photoreactions," *J. Am. Chem. Soc.* **103**, 6.

Michl, J. (1974), "Photochemical Reactions: Correlation Diagrams and Energy Barriers," in *Chemical Reactivity and Reaction Paths*; Klopman, G., Ed.; Wiley: New York.

Pearson, R.G. (1976), *Symmetry Rules for Chemical Reactions*; Wiley: New York.

Woodward, R.B., Hoffmann, R. (1970), *The Conservation of Orbital Symmetry*; Verlag Chemie: Weinheim.

Biradicals and Biradicaloids

Bonačić-Koutecký, V., Koutecký, J., Michl, J. (1987), "Neutral and Charged Biradicals, Zwitterions, Funnels in S_1 and Proton Translocation: Their Role in Photochemistry, Photophysics and Vision," *Angew. Chem. Int. Ed. Engl.* **26**, 170.

Borden, W.T. (1982), *Diradicals*; Wiley: New York.

Johnston, L.J. (1993), "Photochemistry of Radicals and Biradicals," *Chem. Rev.* **93**, 251.

Johnston, L.J., Scaiano, J.C. (1989), "Time-Resolved Studies of Biradical Reactions in Solution," *Chem. Rev.* **89**, 521.

Michl, J. (1977), "The Role of Biradicaloid Geometries in Organic Photochemistry," *J. Photochem. Photobiol.* **25**, 141.

Michl, J. (1992), "Singlet and Triplet States of an Electron Pair in a Molecule—a Simple Model," *J. Mol. Struct. (Theochem)* **260**, 299.

Platz, M.S., Ed. (1990), *Spectroscopy of Carbenes and Biradicals*; Plenum: New York.

Gerhartz, W., Poshusta, R.D., Michl, J. (1976), "Excited Potential Energy Hypersurfaces for H_4 at Trapezoidal Geometries. Relation to Photochemical 2s + 2s Processes," *J. Am. Chem. Soc.* **98**, 6427.

Gerhartz, W., Poshusta, R.D., Michl, J. (1977), "Excited Potential Energy Hypersurfaces for H_4.2. 'Triply Right' (C_{2v}) Tetrahedral Geometries. A Possible Relation to Photochemical 'Cross-Bonding' Processes," *J. Am. Chem. Soc.* **99**, 4263.

TICT States

Lippert, E., Rettig, W., Bonačić-Koutecký, V., Heisel, F., Miehé, J.A. (1987), "Photophysics of Internal Twisting," *Adv. Chem. Phys.* **68**, 1.

Rettig, W. (1986), "Charge Separation in Excited States of Decoupled Systems," *Angew. Chem. Int. Ed. Engl.* **25**, 971.

CHAPTER

5

Photophysical Processes

Electronically excited states have only a short lifetime. In general, several processes are responsible for the dissipation of the excess energy of an excited state. These will be discussed in the following sections. For this purpose it is useful to distinguish between photophysical and photochemical pathways of deactivation, although such a distinction is not always unequivocal. (Cf. the formation of excimers, Section 5.4.2.) The present chapter deals with photophysical processes, which lead to alternative states of the same species such that at the end the chemical identity of the molecule is preserved. Photochemical processes that convert the molecule into another chemical species will be dealt with in later chapters.

5.1 Unimolecular Deactivation Processes

The excess energy taken up by light absorption can be dissipated through unimolecular processes either as radiation (emission) or by radiationless transitions. It can also be transferred to other molecules through bimolecular processes. The relative importance of these various processes depends on the molecular structure as well as on the surroundings of the molecules.

5.1.1 The Jablonski Diagram

The various *unimolecular photophysical processes* may be envisaged in a rather illuminating way with the help of the Jablonski diagram shown in Fig-

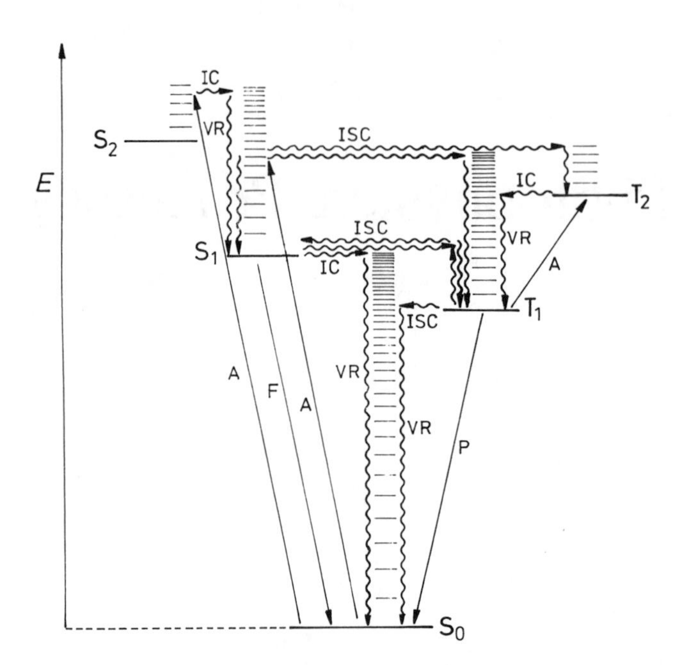

Figure 5.1. Jablonski diagram. Absorption (A) and emission processes are indicated by straight arrows (F = fluorescence, P = phosphorescence), radiationless processes by wavy arrows (IC = internal conversion, ISC = intersystem crossing, VR = vibrational relaxation).

ure 5.1. This diagram schematically displays the singlet ground state S_0, the excited singlet states S_1 and S_2, as well as the triplet states T_1 and T_2. For polyatomic molecules the spacing between the vibrational levels decreases rapidly with increasing energy, and the density of states increases very rapidly as the vibrational energy increases. Rotational levels of the various vibrational states have therefore been omitted for the sake of clarity. Standard convention shows absorption and emission processes as straight arrows and radiationless processes as wavy arrows. Bimolecular photophysical processes and photochemical processes are not shown in the Jablonski diagram. They provide different possible pathways for the deactivation of excited states and will be discussed in later sections.

Emission or *luminescence* is referred to as *fluorescence* or *phosphorescence,* depending on whether it corresponds to a spin-allowed or a spin-forbidden transition, respectively. Similarly, radiationless transitions between states of the same multiplicity and of a different multiplicity are known as *internal conversion* (IC) and *intersystem crossing* (ISC), respectively.

From Figure 5.1 it can be seen that a molecule can reach an excited vibrational level of the electronically excited state S_1 either by the absorption of a photon of appropriate energy or by internal conversion from one of the vibrational levels of a higher electronic state such as S_2. In liquid solutions *vibrational relaxation* (VR) to the vibrational ground state (or more accurately, to a Boltzmann distribution over the vibrational levels corresponding to thermal equilibrium) is very rapid, and the excess vibrational energy is converted into heat through collisions with solvent molecules. From the zero-vibrational level of the S_1 state the molecule can return to the ground state S_0 by fluorescence (F), or it can reach the triplet state T_1 by intersystem crossing (ISC), and after loss of excess vibrational energy it can return to the ground state S_0 by phosphorescence. Radiationless deactivation from S_1 to S_0 can occur via internal conversion (IC) and subsequent vibrational relaxation (VR). Radiationless deactivation from T_1 to S_0 can occur by intersystem crossing (ISC) followed by vibrational relaxation.

In many instances intersystem crossing is fast enough to compete with fluorescence or even with vibrational relaxation. As indicated in Figure 5.1, higher excited triplet states can be reached under such circumstances, and subsequent internal conversion and vibrational relaxation then proceeds in the triplet manifold. Higher excited triplet states are also accessible from T_1 via triplet–triplet absorption. Yet another pathway is shown in Figure 5.1. This corresponds to the thermal activation of T_1 and reverse intersystem crossing into S_1. This process gives rise to E-type *delayed fluorescence,* which derives its name from the circumstance that it was first detected for eosin, in contrast to P-type delayed fluorescence, initially observed for pyrene. The latter will be described in Section 5.4.5.5.

5.1.2 The Rate of Unimolecular Processes

If the spontaneous emission of radiation of the appropriate energy is the only pathway for a return to the initial state, the average statistical time that the molecule spends in the excited state is called the *natural radiative lifetime*. For an individual molecule the probability of emission is time-independent and the total intensity of emission depends on the number of molecules in the excited state. In a system with a large number of particles, the rate of decay follows a first-order rate law and can be expressed as

$$I = I_0 \, e^{-k_0 t} \tag{5.1}$$

where I_0 and I are the intensities of emitted radiation immediately after excitation and at a later time t, respectively (cf. Figure 5.2); k_0 is the rate constant and has the dimension of reciprocal time. The quantity

$$\tau_0 = 1/k_0 \tag{5.2}$$

is the mean natural lifetime of the excited state (in s). k_0 in Equation (5.2) is the rate constant for spontaneous emission and is given by the Einstein prob-

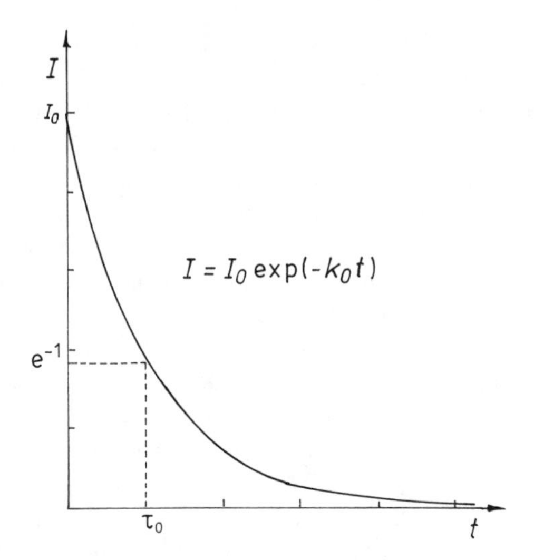

Figure 5.2. Exponential decay curve for emission intensity following the first-order rate law.

ability of spontaneous emission A_{mn}. From this relation the natural radiative lifetime in s can be estimated as

$$\tau_0 = \frac{3.47 \times 10^8}{\tilde{\nu}^2_{\max}} \frac{1}{\int \varepsilon(\tilde{\nu})\mathrm{d}\tilde{\nu}} \approx \frac{1.5}{\tilde{\nu}^2_{\max} f} \tag{5.3}$$

where $\tilde{\nu}_{\max}$ is the wave number of the absorption maximum and f the oscillator strength of the corresponding electronic transition (Strickler and Berg, 1962). Since emission shows the same dependence on the transition moment as absorption, it follows that the emission decay time is inversely proportional to the integrated intensity of the absorption. Using $\tilde{\nu} = 40{,}000\ \mathrm{cm}^{-1}$ and $f = 1$ or $f = 10^{-9}$ for a multiplicity-allowed transition and a singlet–triplet transition, respectively, the natural radiative lifetimes of an excited singlet and triplet state may be estimated from Equation (5.3) as $\tau_0 = 10^{-9}$ s and $\tau_0 = 1$ s, respectively.

The radiative lifetime is nearly temperature independent, but depends to some extent on the environment. Thus, the solvent can influence the radiative lifetime either by means of its effect on the transition moment (cf. Section 2.7.2) or through its refractive index. (See Förster, 1951.)

From the above estimate of the lifetime of a singlet state the rate constant of fluorescence $k_F = 1/\tau_0$ is obtained as $k_F = 10^6$–$10^9\ \mathrm{s}^{-1}$, depending on the value of the oscillator strength f. Phosphorescence, on the other hand, is a spin-forbidden process with a much smaller rate constant, $k_P = 10^{-2}$–10^4 s^{-1}.

Each process competing with spontaneous emission reduces the *observed lifetime* τ relative to the natural lifetime τ_0. In the case where only unimolecular processes i with rate constant k_i compete with emission, one has

$$\tau = 1/(k_F + \sum_i k_i) \tag{5.4}$$

According to Equation (5.3), the rate constant of IR emission is much smaller than k_F, since $\tilde{\nu}^2$ is correspondingly small. Vibrational energy is therefore ordinarily transferred to the surroundings through radiationless dissipation and converted to translational, librational, and rotational energy by collisions. At sufficiently high pressures the collisional frequency is 10^{13} s^{-1}, so the vibrational relaxation rate constant k_{VR} is expected to be $k_{VR} \leq 10^{13}$ s^{-1}. In fact, a value $k_{VR} = (4 \pm 1) \times 10^{12}$ s^{-1} has been measured by picosecond spectroscopy for the vibrational relaxation of the first excited singlet state of 9,10-dimethylanthracene (Rentzepis, 1970) and many similar values have been measured for other molecules. In solid solutions at low temperatures, vibrational relaxation may be slower by several orders of magnitude.

The rate of internal conversion (IC), a radiationless transition between isoenergetic levels of different states of the same multiplicity, may be of the same order of magnitude or even faster than vibrational relaxation. It depends, however, on the energy separation $\Delta E_{0,0}$ between the zero-vibrational levels of the electronic states involved (energy gap law, see Section 5.2.1). Similar relations hold for intersystem crossing transitions between states of different multiplicity, which are slower by 4–8 orders of magnitude.

In conclusion, the following rough estimates of rate constants may be given: For internal conversion for instance $k_{IC} \approx 10^{12}$–10^{14} s^{-1} for $S_n \rightsquigarrow S_1$ and $k_{IC} < 10^8$ s^{-1} for $S_1 \rightsquigarrow S_0$, whereas for intersystem crossing typical values are $k_{ST} \approx 10^6$–10^{11} s^{-1} for $S_1 \rightsquigarrow T_1$ and $k_{TS} \approx 10^4$–10^{-1} s^{-1} for $T_1 \rightsquigarrow S_0$. Intersystem crossing in carbonyl compounds is particularly fast. In biradicals, where T_1 and S_0 are very nearly degenerate, k_{TS} tends to be high, often in the range $10^6 - 10^8$ (Johnston and Scaiano, 1989).

5.1.3 Quantum Yield and Efficiency

A useful quantity in the description of photophysical and photochemical processes is the *quantum yield* Φ. The quantum yield Φ_j of a process j is defined as the number n_A of molecules A undergoing that process divided by the number n_Q of light quanta absorbed, that is,

$$\Phi_j = n_A/n_Q \tag{5.5}$$

On the other hand, the quantum yield can also be defined as

$$\Phi_j = (\mathrm{d}n_A/\mathrm{d}t)/I_{abs} \tag{5.6}$$

the ratio of the rate dn_A/dt of the process j to the intensity of the absorbed radiation $I_{abs} = dn_Q/dt$.* If the quantum yield is not constant over the entire reaction time, these two definitions of the *total* and the *differential quantum yield* agree only after extrapolation to the time $t = 0$.

For many purposes it will be useful to distinguish between quantum yield Φ_j related to the absorbed radiation and *efficiency* η_j related to the number of molecules in a given state. η_j may be defined by

$$\eta_j = n_j/n_{A^*} \tag{5.7}$$

which is the ratio of the number n_j of molecules that take a specific reaction path j to the number n_{A^*} of molecules in the precursor state A*. Analogously, the efficiency may be expressed as the ratio of a process involving a given excited state to the rate of production of that state. If $\sum_i k_i$ is the sum of rate constants for all processes under consideration, the probability that any one of the molecules will yield the product j is given by

$$\eta_j = k_j/\sum_i k_i \tag{5.8}$$

The quantum yield Φ_R of a reaction R is given as the product of the efficiencies of all steps necessary to reach the product R. Therefore, for a one-step reaction

$$\Phi_R = \eta_{abs}\eta_R \tag{5.9a}$$

whereas for a reaction with one metastable intermediate Z

$$\Phi_R = \eta_{abs}\eta_Z\eta_R \tag{5.9b}$$

For the selective excitation of the zero-vibrational level of the S_1 state $\eta_{abs} = 1$; in general this is also valid for higher energy radiation. For a two-photon process, where two photons are absorbed simultaneously, correspondingly $\eta_{abs} = 0.5$.

If only processes that obey a purely exponential rate law such as fluorescence, internal conversion, and intersystem crossing with rate constants k_F, k_{IC}, and k_{ISC} are involved in deactivating the singlet state S_1, the quantum yield of fluorescence may be written according to Equations (5.8) and (5.9) as

$$\Phi_F = \eta_{abs}\eta_F = 1 \times \frac{k_F}{k_F + k_{IC} + k_{ST}} = k_F\tau_S = \frac{\tau_S}{\tau_0^S} \tag{5.10}$$

where the natural lifetime τ_0^S and the observed lifetime τ_S of the singlet state are given by Equations (5.2) and (5.4). Thus, under these circumstances the

* If the absorbed intensity is measured per volume V, Equation (5.6) takes the form $\Phi_j = (dn_A/dt)/I_{abs} = (dc_A/dt)/(I_{abs}V)$, where c_A is the concentration of A.

Table 5.1 Relations Between Quantum Yield, Lifetime, and Rate Constant of Unimolecular Photophysical Processes

$$\Phi_F = k_F/(k_F + k_{ST} + k_{IC}) = k_F\tau_S$$
$$\eta_{IC} = k_{IC}/(k_F + k_{ST} + k_{IC}) = k_{IC}\tau_S$$
$$\eta_{ST} = k_{ST}/(k_F + k_{ST} + k_{IC}) = k_{ST}\tau_S$$
$$\Phi_P = \eta_{ST}k_P/(k_P + k_{TS}) = \eta_{ST}k_P\tau_T$$
$$\eta_{TS} = \eta_{ST}k_{TS}/(k_P + k_{TS}) = \eta_{ST}k_{TS}\tau_T$$

quantum yield of fluorescence is given as the ratio of the observed to the natural lifetime. Expressions for the quantum yield of internal conversion Φ_{IC} and of intersystem crossing Φ_{ST}, which populates the T_1 state, may be derived in a similar way.

If the T_1 state is deactivated only by first-order processes such as phosphorescence and intersystem crossing with rate constants k_P and k_{TS}, the quantum yield of phosphorescence is given according to Equation (5.8) by

$$\Phi_P = 1 \times \eta_{ST}\eta_P = \frac{k_{ST}}{k_F + k_{IC} + k_{ST}} \times \frac{k_P}{k_P + k_{TS}} = \eta_{ST}k_P\tau_T = \eta_{ST}\frac{\tau_T}{\tau_0^T} \tag{5.11}$$

where the appropriate expression for the efficiency η_{ST} for triplet state formation has been used. In the same way, an expression for the quantum yield Φ_{TS} of the radiationless deactivation of the T_1 state may be obtained. All these relationships are collected in Table 5.1.

Table 5.2 Observable Photophysical Parameters and their Relationship to Rate Constants of Various Photophysical Processes and Sources of their Information*

Photophysical Parameter	Symbol	Relation to Rate Constant	Source
Fluorescence quantum yield	Φ_F	$\frac{k_F}{k_{IC} + k_{ST} + k_F}$	Fluorescence spectrum
Phosphorescence quantum yield	Φ_P	$\frac{k_P}{(k_{TS} + k_P)} \frac{k_{ST}}{(k_{IC} + k_{ST} + k_F)}$	Phosphorescence spectrum
Triplet formation quantum yield	Φ_T	$\frac{k_{ST}}{k_{IC} + k_{ST} + k_F}$	T_1–T_n absorption
Singlet lifetime	τ_F	$\frac{1}{k_{IC} + k_{ST} + k_F}$	Fluorescence decay
Triplet lifetime	τ_P	$\frac{1}{k_{TS} + k_P}$	Phosphorescence decay

*These relations hold assuming that second-order processes such as quenching and photochemical reactions may be neglected.

The quantum yields of fluorescence and phosphorescence, Φ_F and Φ_P, may be determined experimentally by means of a fluorescent standard such as a rhodamine B solution whose Φ_F is independent of the exciting wavelength within a wide range. Lifetimes τ_F and τ_P are also experimentally accessible through time-resolved fluorescence measurements (phase method or single-photon counting) or by measuring the time dependence of phosphorescence. (Cf. Rabek, 1982.) In Table 5.2 the observable quantities and their relationship to rate constants are collected.

5.1.4 Kinetics of Unimolecular Photophysical Processes

From the five relationships collected in Table 5.2, the five rate constants k_F, k_P, k_{IC}, k_{ST}, and k_{TS} of unimolecular photophysical processes may in principle be obtained if all observable quantities are known. Simplifying assumptions are often possible. For instance, if internal conversion $S_1 \rightsquigarrow S_0$ is negligible one has $\Phi_F + \Phi_P + \Phi_{TS} \approx 1$ or $\Phi_{TS} = 1 - (\Phi_F + \Phi_P)$, so from $\Phi_{TS}/\Phi_P = k_{TS}/k_P$,

$$k_{TS} \approx k_P \frac{1 - (\Phi_F + \Phi_P)}{\Phi_P} \tag{5.12}$$

is obtained.

Example 5.1:
For benzene in EPA at 77 K $\Phi_F = 0.19$, $\Phi_P = 0.18$, and $\tau_P = 6.3$ s have been measured (Li and Lim, 1972); from the integrated area under the absorption curve $k_F = 1/\tau_0 = 2 \times 10^6\ s^{-1}$ is obtained. With the simplifying assumption $\Phi_F + \Phi_T \approx 1$ the relationships collected in Table 5.2 yield

$$\frac{\Phi_F}{\Phi_T} = \frac{k_F}{k_{ST}}$$

or

$$k_{ST} = k_F \frac{1 - \Phi_F}{\Phi_F}$$

from which the estimate

$$k_{ST} = \frac{1 - 0.19}{0.19} \times 2 \times 10^6\ s^{-1} = 8.5 \times 10^6\ s^{-1}$$

results. Using $\eta_{TS} = \Phi_T$

$$k_P = \frac{\Phi_P}{\Phi_T \tau_P} = \frac{0.18}{(1 - 0.19) \times 6.3} = 3.5 \times 10^{-2}\ s^{-1}$$

is obtained from Equation (5.11), and finally Equation (5.12) yields

$$k_{TS} = 3.5 \times 10^{-2} \frac{1 - (0.18 + 0.19)}{0.18} = 1.2 \times 10^{-1}\ s^{-1}$$

5.1.5 State Diagrams

Data obtained from spectroscopic measurements can be collected in a Jablonski diagram. In this way a state diagram is obtained for a molecule, and this can be extremely useful in discussing its photochemistry.

As an example the state diagram of biacetyl is displayed in Figure 5.3. From the 0–0 transition in the absorption or fluorescence spectrum, as well as in the phosphorescence spectrum, the energies $E(S_1)$ and $E(T_1)$ of the lowest singlet and triplet states are obtained. From the fluorescence quantum yield $\Phi_F = 0.01$ it is concluded that only 1% of the excited molecules emit; the remaining 99% are transferred into the T_1 state, assuming that radiationless deactivation $S_1 \rightsquigarrow S_0$ and photochemical reactions are negligible. From the fluorescence decay one has $k_F = 10^5\ s^{-1} = 1/\tau_0^S$, and the rate constant for intersystem crossing k_{ST} is obtained from the relationships collected in Table 5.2 as

$$k_{ST} = k_F \frac{1 - \Phi_F}{\Phi_F} = 10^5 \frac{0.99}{0.01} = 10^7\ s^{-1}$$

$\Phi_P = 0.25$ means that 25% of the molecules in the T_1 state phosphoresce with $k_P = 1.2 \times 10^2\ s^{-1}$. The remaining 75% return nonradiatively to the ground state, with

$$k_{TS} = k_P \frac{1 - (\Phi_P + \Phi_F)}{\Phi_P} = 3.5 \times 10^2\ s^{-1}$$

according to Equation (5.12).

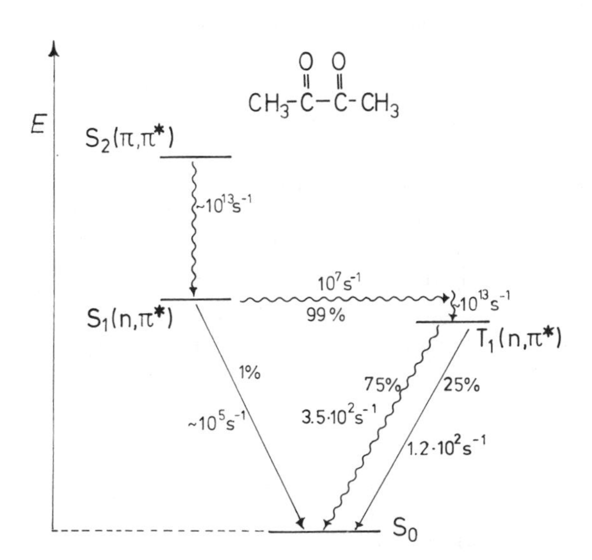

Figure 5.3. Jablonski energy diagram of biacetyl (adapted from Dubois and Wilkinson, 1963).

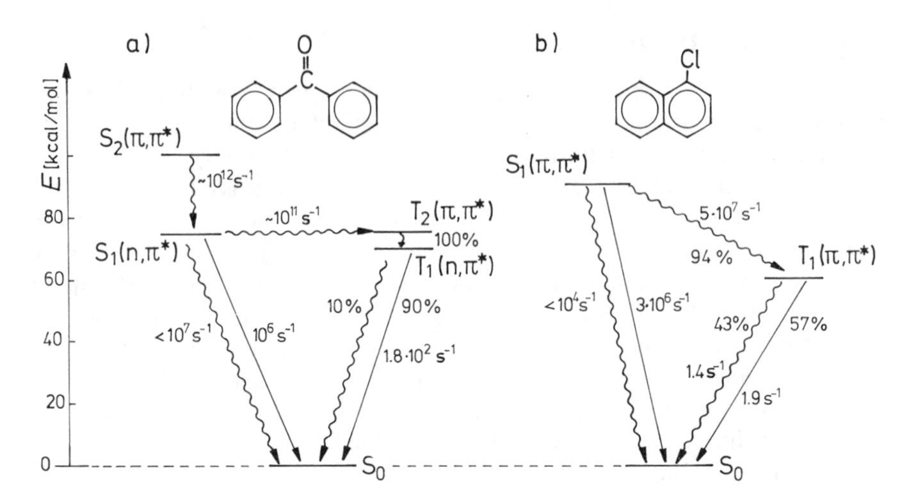

Figure 5.4. Jablonski energy diagram a) of benzophenone and b) of 1-chloronaphthalene (by permission from Turro, 1978).

Figure 5.4 displays state diagrams of other molecules of interest. (Cf. Example 6.6.)

5.2 Radiationless Deactivation

The electronic excitation of a molecule in general produces a state whose equilibrium geometry differs from the ground-state equilibrium geometry. If the excitation energy is sufficiently high, molecular vibrations of the higher state will also be excited. Reactions of "hot" molecules generated in this way will be discussed in Chapter 6.

Before a polyatomic molecule can move over potential energy barriers along a reaction path, the initially introduced vibrational energy is quickly distributed among the various normal vibrations and can be partially or completely dissipated by collisions. This is particularly true for condensed phases where the exchange of vibrational energy with the environment is so fast that thermal equilibrium can be reached in a time as short as 10^{-11} s. (Cf. Maier et al., 1977.)

5.2.1 Internal Conversion

According to Figure 5.1 the term internal conversion (IC) is used for a transition $S_n^v \rightsquigarrow S_m^{v'}$ or $T_n^v \rightsquigarrow T_m^{v'}$ between two isoenergetic vibrational levels denoted by v and v' of different electronic states n and m of the same multiplicity, which may have quite different energies at their respective equilibrium geometries. Often, however, the designation internal conversion is used in a wider sense encompassing vibrational relaxation as well. It then denotes radiationless transitions $S_n \rightsquigarrow S_1$ or $T_n \rightsquigarrow T_1$ from a higher excited singlet or triplet state into the lowest excited singlet or triplet state, respectively, as well as the radiationless transition $S_1 \rightsquigarrow S_0$ from the first excited singlet state into the vibrationally equilibrated ground state.

The raditionless processes $S_n \rightsquigarrow S_1$ and $T_n \rightsquigarrow T_1$ are usually so fast that lifetimes of higher excited states are very short and quantum yields of emission from higher excited states are very small. In the vast majority of cases luminescence is observed exclusively from the lowest excited state. This so-called *Kasha's rule* is of course relative in that what is observed depends on the sensitivity of the detector. For benzenoid aromatics, fluorescence from higher excited states in addition to fluorescence from the lowest excited state was observed for the first time in 1969 (Geldorf et al., 1969), whereas the fluorescence from the S_2 state of azulene had been known for quite some time. (See below.)

The radiationless transition $S_1 \rightsquigarrow S_0$, however, is in general much slower than $S_n \rightsquigarrow S_1$. For instance, most aromatic compounds fluoresce, and the internal conversion $S_1 \rightsquigarrow S_0$ contributes at most partially to the deactivation of the first excited singlet state. Thus, the cascade of nonradiative conversions of higher excited states normally ends at the S_1 state, and does not lead to the ground state S_0.

Example 5.2:
From the oscillator strength $f = 1.89$ of the absorption band of anthracene (**1**) at 39,700 cm^{-1} the natural lifetime of the excited state can be estimated from Equation (5.3) as $\tau_0 \approx 0.5 \times 10^{-9}$ s. Since no fluorescence from this higher excited state S_n is observed, the actual lifetime must be smaller at least by a factor of 10^{-4}, so one has

$$\tau < (0.5 \times 10^{-9}) \times 10^{-4} = 0.5 \times 10^{-13} \text{ s}$$

which gives

$$k_{IC}(S_n \rightsquigarrow S_1) = 1/\tau > 2 \times 10^{13} \text{ s}^{-1}$$

for the rate constant of internal conversion.

In the case of pyrene (**2**) the environment-dependent rate constant for the deactivation of the S_1 state is approximately 10^6 s^{-1}, and since deactivation occurs practically exclusively by fluorescence and intersystem crossing, one has

$$k_{IC}(S_1 \rightsquigarrow S_0) < 10^6 \text{ s}^{-1}$$

Thus, typical rate constants for the two nonradiative processes $S_n \rightsquigarrow S_1$ and $S_1 \rightsquigarrow S_0$ differ by a factor of 10^6–10^7 (Birks, 1970).

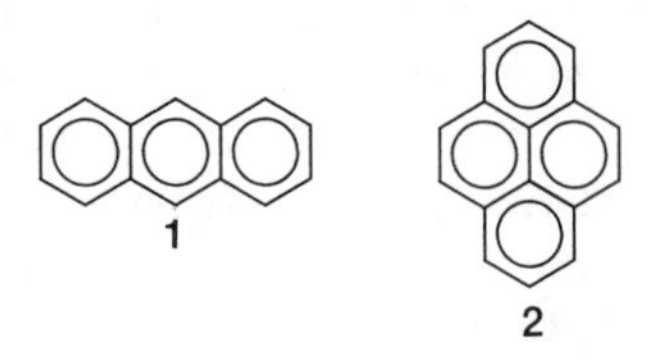

From experimental results it has been concluded that for aromatic hydrocarbons the radiationless transition $S_1 \rightsquigarrow S_0$ is negligible if the energy difference $\Delta E(S_1 - S_0)$ between S_1 and S_0 states is larger than 60 kcal/mol. For molecules with low-lying singlet states such as tetracene (**3**) ($\Delta E = 57$ kcal/mol) and its homologues, however, it becomes increasingly more important, and it accounts for more than 90% of the S_1 state deactivation in the case of hexacene (**4**) ($\Delta E = 40$ kcal/mol) (Angliker et al., 1982). These observations can be summarized by the relationship

$$k_{IC} = 10^{13}\,e^{-\alpha\Delta E} \tag{5.13}$$

which shows the dependence of the rate constant k_{IC} on the energy gap $\Delta E(S_1 - S_0)$ and which is referred to as the *energy-gap law* (Siebrand, 1967). α is a proportionality constant, approximately equal to 4.85 eV^{-1} for benzenoid aromatics.

Exceptions to the rules described are observed in the case of azulene (**5**) and its derivatives (Beer and Longuet-Higgins, 1952), where for the internal conversion $S_2 \rightsquigarrow S_1$ the rate constant $k_{IC}(S_2 \rightsquigarrow S_1) \approx 7 \times 10^8\ s^{-1}$ is exceptionally small. This may be at least in part due to the large energy gap, $\Delta E(S_2 - S_1) \approx 40$ kcal/mol. The radiationless transition from the first excited singlet state into the ground state, on the other hand, is extremely fast with $k_{IC}(S_1 \rightsquigarrow S_0) \approx 10^{12}\ s^{-1}$, although the energy gap $\Delta E(S_1 - S_0)$ is of the same order of magnitude as $\Delta E(S_2 - S_1)$. This is consistent with computational results which demonstrate the existence of an S_1–S_0 conical intersection that can be reached from the Franck–Condon region with almost no barrier (Bearpark et al., 1994). Another class of compounds that exhibit $S_2 \rightsquigarrow S_0$ fluorescence are the thiocarbonyl compounds (Maciejewski and Steer, 1993; cf. Section 6.1.5.3).

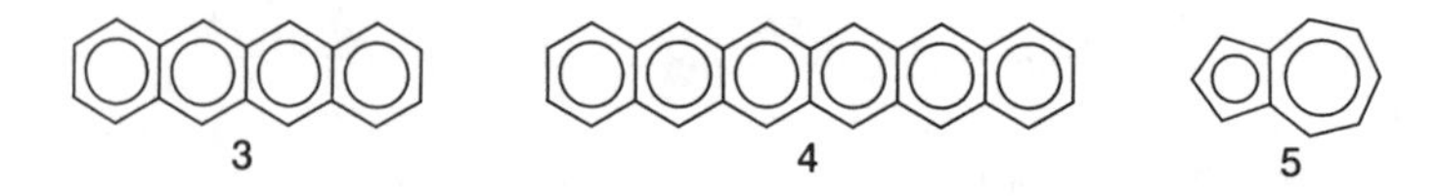

5.2.2 Intersystem Crossing

Transitions from singlet to triplet states and vice versa become possible through spin inversion. Of particular importance is the radiationless deactivation of the lowest excited singlet and triplet states, that is, the intersystem crossing $S_1 \rightsquigarrow T_1$ and $T_1 \rightsquigarrow S_0$, with rate constants k_{ST} and k_{TS}.

The transition $S_1 \rightsquigarrow T_1$ can take place either by direct spin-orbit coupling of S_1 to the higher vibrational levels of T_1 or by spin-orbit coupling to one of the higher states T_n followed by rapid internal conversion $T_n \rightsquigarrow T_1$. The rate-determining step is the spin inversion, and rate constant values k_{ST} are in the range 10^7 to 10^{11} s^{-1} and depend on the extent of spin-orbit coupling as well as on the energy gap between the states involved.

According to the selection rules for intersystem crossing known as *El Sayed's rules* (El Sayed, 1963) transitions

$$^1(n,\pi^*) \leftrightarrow {}^3(\pi,\pi^*) \qquad {}^3(n,\pi^*) \leftrightarrow {}^1(\pi,\pi^*) \tag{5.14a}$$

are allowed, while transitions

$$^1(n,\pi^*) \leftrightarrow {}^3(n,\pi^*) \qquad {}^1(\pi,\pi^*) \leftrightarrow {}^3(\pi,\pi^*) \tag{5.14b}$$

are forbidden.

These selection rules can be related to spin-orbit coupling with the help of the *Fermi golden rule*. (Cf. Section 5.2.3.) The values $k_{ST} \approx 10^6$ s^{-1} for naphthalene and $k_{ST} \approx 10^9$ s^{-1} for 1-bromonaphthalene (Birks, 1970), the difference of which can be explained through the heavy atom effect, also indicate clearly the influence of spin-orbit coupling. The fact that the rate

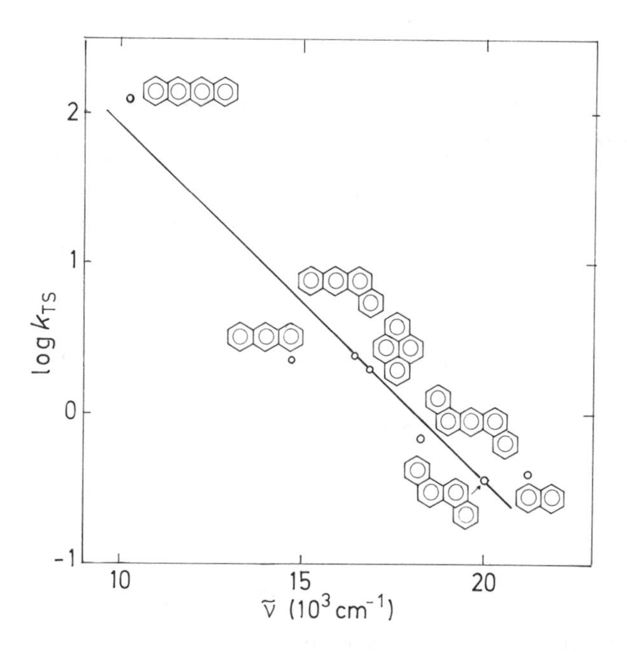

Figure 5.5. Relationship between the energy gap $\Delta E(T_1 - S_0)$ and the logarithm of the rate constant k_{TS} of intersystem crossing in aromatic hydrocarbons (data from Birks, 1970).

constants k_{ST} and k_{TS} can differ by a factor of up to 10^9, so that, for instance, $k_{TS} = 0.41\ s^{-1}$ for naphthalene, can also be related to the energy gap. From Figure 5.5 it can be seen that the k_{TS} values of aromatic hydrocarbons show a similar dependence on the energy gap $\Delta E(T_1 - S_0)$ between the T_1 and S_0 states to the one given in Equation (5.13) for k_{IC}.

Example 5.3:
Intersystem crossing is temperature dependent in some anthracene derivatives. Thus, the rate constant for 9,10-dibromoanthracene (**6**) may be written as

$$k_{ST} \approx 10^{12}\ e^{-E_a/RT}$$

with an activation energy $E_a \approx 4$ kcal/mol. Triplet–triplet absorption spectra show that T_2 is energetically higher than S_1 by 4–5 kcal/mol. It can be concluded that due to the large energy gap between S_1 and T_1, intersystem crossing does not occur into one of the higher vibrational levels of T_1, but with a small activation energy instead predominantly into T_2 (Kearvell and Wilkinson, 1971).

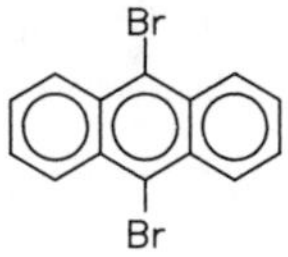

Br

Br

6

The observation that the rate constants k_{ST} for the $S_1 \rightsquigarrow T$ intersystem crossing in anthracene (**1**) and pyrene (**2**), k_{ST}(anthracene) $\approx 10^8\ s^{-1}$ and k_{ST}(pyrene) $\approx 10^6\ s^{-1}$, differ by a factor of 100 may be explained along similar lines. In pyrene a direct transition into a vibrationally excited level of T_1 occurs, the energy gap between the zero-point vibrational levels being $\Delta E(S_1 - T_1) \approx 30$ kcal/mol (Dreeskamp et al., 1975), whereas in anthracene the transition leads to the nearly isoenergetic T_2 state (Almgren, 1972).

The rate of intersystem crossing can be increased by the presence of paramagnetic molecules such as oxygen as well as by molecules with heavy atoms such as halogen or organometallic compounds. These are concentration-dependent effects. In the case of oxygen the rate constant of a radiationless singlet–triplet transition can be written as

$$k_{ST}^{obs} = k_{ST} + k_{ST}^{O_2}[O_2]$$

where $k_{ST}^{O_2} \approx 10^9$–$10^{10}$ mol^{-1} s^{-1}. For a 10^{-2} molar solution of O_2 one has $k_{ST}^{O_2}[O_2] \approx 10^7$–$10^8\ s^{-1}$, so the oxygen effect becomes noticeable when $k_{ST} \approx 10^8$ or smaller (Stevens and Algar, 1967).

5.2.3 Theory of Radiationless Transitions

The probability of radiationless transitions between different states is particularly great when the potential energy surfaces of the corresponding states touch or come at least very close to each other.

In the framework of the Born-Oppenheimer approximation, radiationless transitions from one surface to another are impossible. (See, e.g., Michl and Bonačić-Koutecký, 1990.) It is therefore necessary to go beyond the Born-Oppenheimer approximation and to include the interaction between different electronic molecular states through the nuclear motion in order to be able to describe such transitions. Using the time-dependent perturbation theory for the rate constant $k_{i \to f}$ of a transition between a pair of states one arrives at

$$k_{i \to f} = \frac{2\pi}{\hbar} <\Psi_f|\hat{H}'|\Psi_i>^2 \varrho_E \tag{5.15}$$

where $\hat{H}'$ is the perturbation operator, Ψ_i and Ψ_f are the wave functions of the initial and the final state, and ϱ_E is the density of states, given by the number of energy levels per energy unit in the final state at the energy of the initial state (Bixon and Jortner, 1968).* Equation (5.15) is referred to as the *Fermi golden rule* for the dynamics of transitions between states.

In the case of internal conversion between states of equal multiplicity $\hat{H}' = \hat{H}'_N$ is the kinetic energy operator of the nuclei. In the case of weak coupling the matrix element of the perturbation operator can be split into an electronic part β^{IC} and a contribution due to the vibrational terms which, with the help of further simplifying assumptions, can be written as the Franck-Condon overlap integral:

$$<\Psi_f|\hat{H}'_N|\Psi_i> \sim \beta^{IC} <\chi_f|\chi_i> \tag{5.16}$$

In the case of intersystem crossing transitions between states of different multiplicity, an additional spin-orbit coupling term $\hat{H}_{SO}$ has to be considered. From the perturbational expansion it follows that the contribution

$$<\Psi_f|\hat{H}_{SO}|\Psi_i> \sim \beta^{ISC} <\chi_f|\chi_i> \tag{5.17}$$

due to the spin-orbit interaction is dominant. Here Ψ_i and Ψ_f are wave functions of different multiplicity (singlet and triplet wave functions) and β^{ISC} is the electronic part of the interaction integral. The El-Sayed selection rules [Equation (5.14)] follow from Equation (5.17). (Cf. Example 1.8 and Section 4.3.4.)

---

* The density of states is approximately given by the number of ways of distributing ΔE over the normal modes of vibration. Because of many low-frequency modes, the number of these overtones and combinations is enormous: there are as many as ~3×10^5 states per cm^{-1} in the S_1–T_1 case ($\Delta E = 8{,}500\ cm^{-1}$) of benzene with 30 normal modes of vibration.

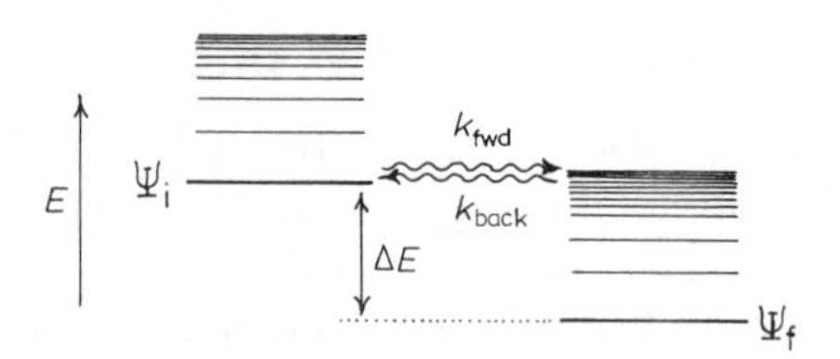

Figure 5.6. Nonradiative conversion in polyatomic molecules. Due to the difference in state density of the initial state Ψ_i and the final state Ψ_f, $k_{fwd} \gg k_{back}$, and the transition is practically irreversible.

The influence of the energy gap between the states involved on the rate constant of radiationless transitions may be clarified with the aid of Equations (5.15) and (5.16) as follows: According to Figure 5.6, the density of states ϱ_E in the final state increases with an increasing energy gap. The electronic excitation energy, which has to be converted into vibrational motion of the nuclei, increases at the same time. The larger the difference in vibrational quantum numbers, the smaller the overlap of the vibrational wave functions. Thus, the expected increase of the rate constant k of a radiationless transition with increasing energy gap ΔE due to the density of states is overcompensated by a decrease due to increasingly unfavorable Franck-Condon factors. Theoretical arguments lead to an exponential dependence of the Franck-Condon factors on the energy gap (Siebrand, 1966; Englman and Jortner, 1970).

According to Equation (5.15), the rate constant for a transition from the higher to the lower state is larger than that of the reverse transition. This is because for a given energy the density of states ϱ_E of the lower state is larger than the density of states ϱ_E' of the higher state, as can be seen from Figure 5.6. Also, the rate-determining transition between approximately isoenergetic levels of close-lying states is followed by a very fast dissipation of the vibrational energy. This prevents the back reaction and makes radiationless transitions into lower states virtually irreversible. (Cf. Figure 5.6.)

Example 5.4:

Figure 5.7 gives a schematic representation of the potential energy curves for two states of a diatomic molecule. Depending on the relative positioning of these curves different probabilities for radiationless transitions result because the Franck-Condon factors can differ appreciably. If the energy difference between the two states is large, as is generally the case for S_0 and S_1, the zero-vibrational level ($v' = 0$) of S_1 overlaps with a higher vibrational level (e.g., $v = 12$) of S_0 in a region close to the equilibrium geometry, where the kinetic energy of the nuclear motion is large and the probability χ_v^2 is small. The overlap integral $<\chi_v|\chi_{v'}>$ is therefore close to zero (Figure 5.7a) and the Franck-Condon factor makes the transition very unlikely. If the states are energetically closer to each other, such as S_2 and S_1, or if the potential energy curves cross,

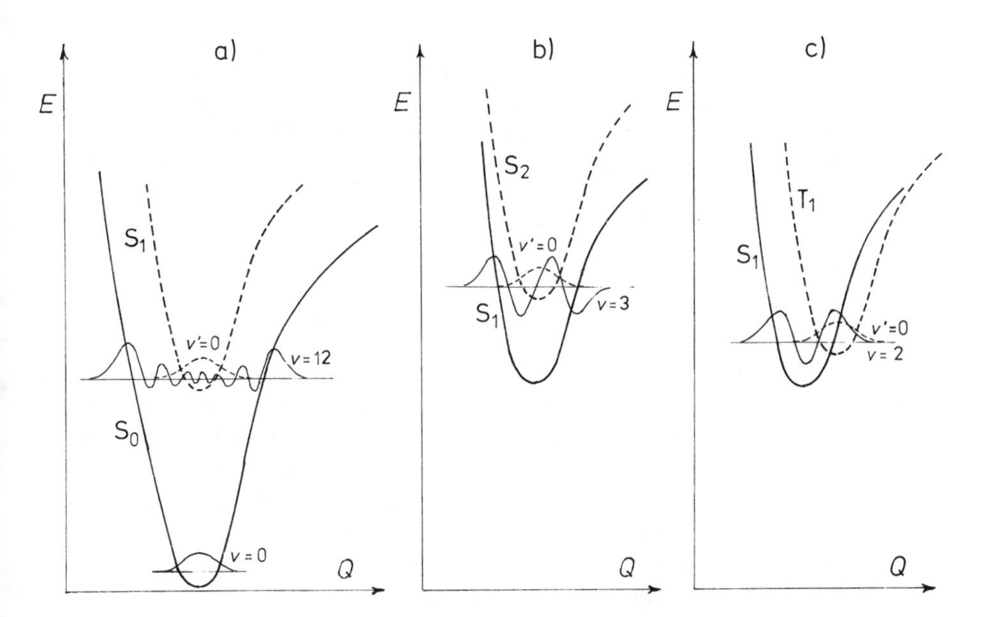

Figure 5.7. Franck-Condon factors for radiationless transitions between different potential energy curves of a diatomic molecule; a) for a large and b) for a small energy gap, such as those observed, for instance, between S_1 and S_0 or between S_2 and S_1, respectively, and c) for the case that the potential energy curves (e.g., S_1 and T_1) cross.

as may be the case for S_1 and T_1, the overlap $<\chi_v|\chi_{v'}>$ between the zero-vibrational level of the higher state and the isoenergetic vibrational level of the lower state will be much larger (Figure 5.7b and c). Radiationless transitions between these states are therefore much more likely.

According to Figure 5.7, the rate of radiationless transitions depends not only on the energy gap, but also on the equilibrium geometries of the states involved. Thus, rigid π systems such as condensed aromatic hydrocarbons possess only a relatively small energy gap between S_1 and S_0, but the bonding characteristics in the ground state and in the first excited (π,π^*) state differ so little that the potential energy surfaces of these states are nearly parallel. Since the amplitudes of the vibrational wave functions χ_v of the higher excited vibrational levels of S_0 oscillate very quickly around zero near the equilibrium geometry, integration results in very small Franck-Condon factors. Therefore, in these systems fluorescence can compete with radiationless deactivation.

Example 5.5:

In radiationless transitions from the triplet state to the ground state of aromatic hydrocarbons, the excess electronic energy goes predominantly into the CH stretching vibrations. Being of high frequency ($\tilde{\nu} \approx 3{,}000\ \mathrm{cm}^{-1}$), they are widely spaced and much smaller vibrational quantum numbers are required than for other normal modes of vibration whose wave numbers are at most half this size. Indeed, the triplet lifetime of naphthalene increases from $\tau \approx 2$ s to $\tau \approx 20$ s on perdeuteration. The energy gap $\Delta E(T_1 - S_0)$ is virtually the same for the deuterated and the undeuterated compound, but higher numbers of vibrational quanta are necessary to overcome this gap because the frequencies of the CD vibrations are smaller by roughly 30%. Hence, the Franck-Condon factors are much less favorable for the deuterated than for the protiated compounds (Laposa et al., 1965).

5.3 Emission

Emission of a photon from an electronically excited state is referred to as luminescence. Fluorescence and phosphorescence can be differentiated depending on whether the transition is between states of equal or different multiplicity and hence spin-allowed or spin-forbidden. (Cf. Section 5.1.1.) Thus, for molecules with singlet ground states fluorescence constitutes a pathway for deactivating excited singlet states whereas phosphorescence is observed in the deactivation of triplet states.

5.3.1 Fluorescence of Organic Molecules

According to the results presented in the last section, a fraction of the energy that a molecule acquires through absorption of a light quantum is dissipated in condensed phases very rapidly via radiationless deactivation and thermal equilibration. In general, the rate of energy loss by emission is comparable with the rate of radiationless deactivation only for the lowest excited singlet state S_1 or the lowest triplet state T_1. As a rule, therefore, the energy of the emitted radiation is lower than that of the absorbed radiation by the amount of energy that has been dissipated nonradiatively. The resulting emission is of longer wavelengths than the absorbed light. As a further consequence of thermal equilibration the intensity distribution in fluorescence and phosphorescence spectra is independent of the exciting wavelength.

The shapes of absorption and emission bands are determined in the same way by Franck-Condon factors. The shift of the emission maximum with respect to the absorption maximum, which is referred to as *Stokes' shift*, increases with the increasing difference between the equilibrium geometries of the ground and the excited states. This is schematically shown for a diatomic molecule in Figure 5.8; the maximum intensity of absorption will be observed for the vertical transition $v = 0 \rightarrow v' = n$, whereas emission will occur after vibrational relaxation, with the highest probability for the transition form $v' = 0$ to a different ground-state vibrational level, $v = m$.

Often the 0–0 transitions of absorption and emission do not coincide, and a "0–0 gap" results. This is referred to as *anomalous Stokes shift* and is due

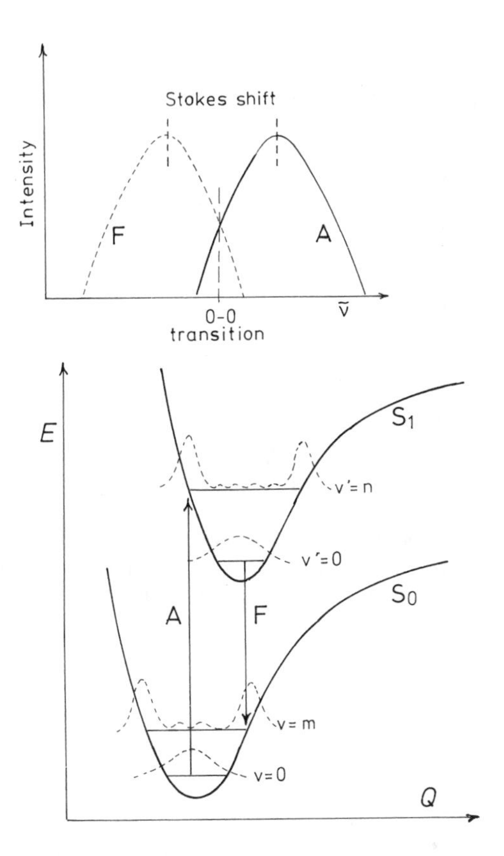

Figure 5.8. Stokes shift; a) definition and b) dependence on the difference in equilibrium geometries of ground and excited states. Shown is the probability distribution in various vibrational levels, which is proportional to the square of the vibrational wave function (adapted from Philips and Salisbury, 1976).

to different intermolecular interactions in the ground and excited states. The energy of the excited-state molecule decreases during its lifetime through reorientation of the surrounding medium, so fluorescence is shifted to longer wavelengths. As an example the difference of the 0–0 transitions in the absorption and in the fluorescence of *p*-amino-*p'*-nitrobiphenyl is shown in Figure 5.9 as a function of solvent polarity. In solid solutions at low temperatures the motion of solvent molecules can be slowed down to such an extent that no reorganization occurs and the anomalous Stokes shift disappears.

Frequently, the fluorescence spectrum is the mirror image of the absorption spectrum, as exemplified for perylene in Figure 5.10. This spectral symmetry is due to the fact that the excited-state vibrational frequencies, responsible for the fine structure of the absorption band, and the ground-state vibrational frequencies, which show up in the fluorescence band, are often

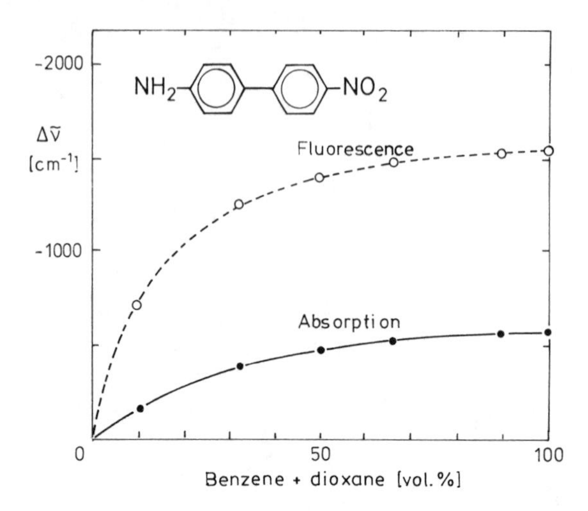

Figure 5.9. Anomalous Stokes shift illustrated by displacements of 0–0 transitions in absorption and fluorescence of *p*-amino-*p*′-nitrobiphenyl in benzene/dioxane as a function of the dioxane content (by permission from Lippert, 1966).

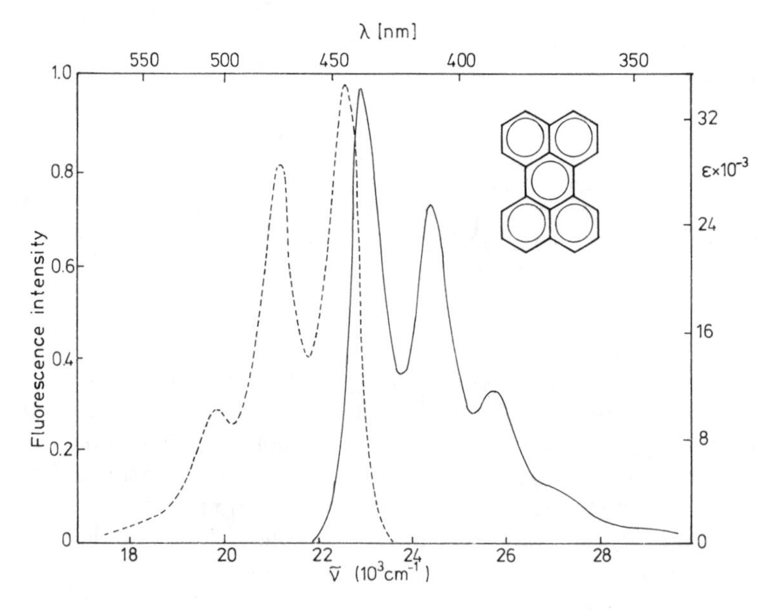

Figure 5.10. Absorption and fluorescence spectrum of perylene in benzene (by permission from Lakowicz, 1983).

quite similar. According to Figure 5.8 the intensity distribution given by the Franck-Condon factors for absorption and emission are also comparable. This is particularly true in those cases where electronic structures and equilibrium geometries of the ground and excited states differ very little, as for instance in $\pi \rightarrow \pi^*$ transitions of delocalized π systems. In biphenyl, however, which is less twisted around the central single bond in the excited than in the ground state (cf. Section 1.4.1), the absorption and fluorescence spectra differ appreciably, and vibrational structure is observed only in emission.

Another example of the mirror-image relation between absorption and fluorescence spectra is provided by anthracene (Figure 5.11). In this particular case the situation is complicated by vibronic coupling; the ^1L_a and ^1L_b bands overlap in the absorption spectrum, and the location of the ^1L_b origin has been inferred from the substituent effects evident from the MCD spectrum (Steiner and Michl, 1978).

The relationship between the quantum yield of fluorescence and molecular structure is determined to a large extent by the structural dependence of the competing photophysical and photochemical processes. Thus, for most rigid aromatic compounds fluorescence is easy to observe, with quantum yields in the range $1 > \Phi_F > 0.01$. This may be explained by the fact that the Franck-Condon factors for radiationless processes are very small because changes in equilibrium geometry on excitation are small; thus internal conversion becomes sufficiently slow. (Cf. Calzaferri et al., 1976.)

A fundamental factor that determines the fluorescence quantum yield is the nature of the lowest excited singlet state—that is to say, the magnitude of the transition moment between S_0 and S_1. If the $S_0 \rightarrow S_1$ transition is symmetry forbidden, as in benzene, k_F is small compared to $\sum_i k_i$ for the com-

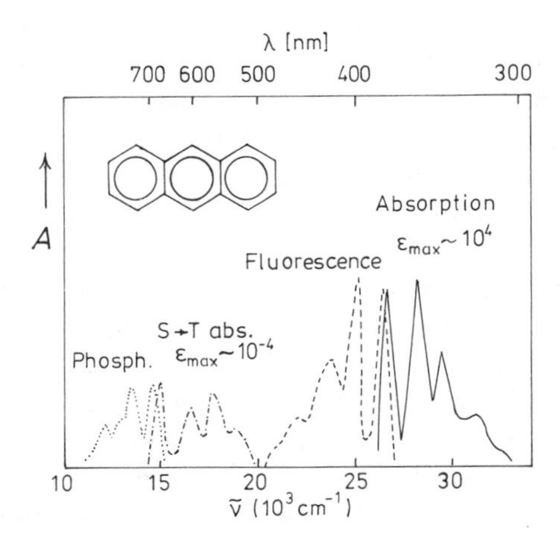

Figure 5.11. Absorption and emission spectra of anthracene (by permission from Turro, 1978).

peting processes and only minor quantum yields of fluorescence are observed. If due to substitution the intensity of the $S_0 \rightarrow S_1$ transition becomes larger, as in aniline, the fluorescence yield increases.

Most compounds whose lowest excited singlet state is an (n,π*) state exhibit only very weak fluorescence. The reason is that due to spin-orbit coupling, intersystem crossing into an energetically lower triplet state is particularly efficient. (Cf. Section 5.2.2.) Heavy atoms in the molecule (e.g., bromonaphthalene) or in the solvent (e.g., methyl iodide) may favor intersystem crossing among (π,π*) states to such an extent that fluorescence can be more or less completely suppressed. At low temperatures, photochemical deactivation and energy-transfer processes involving diffusion and collisions become less important, and low-frequency torsional vibrations that are particularly efficient for radiationless deactivation are suppressed, so fluorescence quantum yields increase. For instance, for *trans*-stilbene (**7**), $\Phi_F = 0.05$ at room temperature, but $\Phi_F = 0.75$ at 77 K. If the stilbene chromophore is fixed in a rigid structural frame as in **8**, Φ_F equals 1.0 independent of temperature (Sharafy and Muszkat, 1971; Saltiel et al., 1968).

7

8

The quantum yield of fluorescence from unsaturated compounds is independent of the exciting wavelength, unless photochemical reactions originating from higher excited singlet states or intersystem crossing compete with internal conversion. A good example is benzene in the gas phase at low pressure (less than 1 torr): on excitation of the S_1 state at $\lambda = 254$ nm fluorescence occurs with a quantum yield of $\Phi_F = 0.4$. This decreases if higher vibrational levels of this state are excited, and at $\lambda < 240$ nm no emission can be detected at all. It is assumed that a radiationless transition is possible from the higher vibrational levels of the S_1 state of benzene into very highly excited vibrational levels of the ground state of the isomeric benzvalene (**9**), as shown in the schematic representation in Figure 5.12. Benzvalene can either be stabilized by vibrational relaxation or can undergo a hot ground-state reaction and return to the benzene ground state (Kaplan and Wilzbach, 1968). Calculations show that a funnel in S_1 (conical intersection of S_1 and S_0), separated from the vertical geometry by a small barrier to isomerization, provides a mechanism for an ultrafast return to S_0, as soon as the molecule has sufficient vibrational energy to overcome the barrier (Palmer et al., 1993; Sobolewski et al., 1993). The so-called "channel 3" effect (see Riedle et al.,

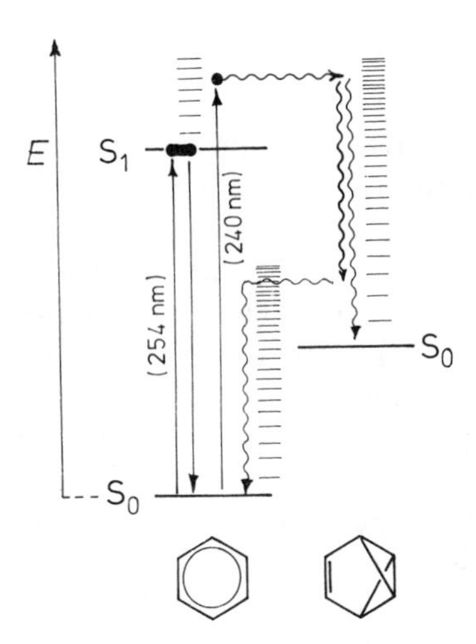

Figure 5.12. Qualitative state diagram for the fluorescence quenching of benzene by radiationless transition into one of the higher vibrational levels of the isomeric benzvalene. The back reaction is a hot ground-state reaction.

1990 for leading references) thus involves no special mechanism of nonradiative decay.

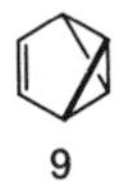

9

By measuring the fluorescence intensity of sufficiently dilute solutions as a function of the exciting wavelengths a *fluorescence excitation spectrum* is obtained. Such a measurement represents a remarkably sensitive method for investigating the absorption spectrum. With I_A and I_F being the intensities of the absorbed light and of the light emitted as fluorescence, respectively,

$$I_F = \Phi_F I_A = \Phi_F I_0(1 - e^{-2.303\varepsilon cd})$$

where I_0 is the intensity of the exciting light, ε the extinction coefficient, c the concentration of the sample, and d the optical path length. (Cf. Section 1.1.2.) If the absorbance $A = \varepsilon cd$ is smaller than say 0.2–0.3, a series expansion of the exponential function $e^{-x} = 1 - x + \ldots$, with neglect of the higher powers of x, yields

$$I_F = \Phi_F I_0 \, 2.303\varepsilon cd \tag{5.18}$$

and as a result, for a given Φ_F and I_0, the fluorescence intensity reflects the wavelength or wave-number dependence of the extinction coefficient ε.

5.3.2 Phosphorescence

Because internal conversion and vibrational relaxation are very fast, phosphorescence corresponds to a transition from the thermally equilibrated lowest triplet state T_1 into the ground state S_0 and the phosphorescence spectrum is approximately a mirror image of the $S_0 \rightarrow T_1$ absorption spectrum, which is spin forbidden and therefore difficult to observe because of the low intensity. This mirror-image symmetry is evident from the singlet–triplet absorption and phosphorescence spectra of anthracene shown in Figure 5.11. In general the T_1 state is energetically below the S_1 state, and phosphorescence occurs at longer wavelengths than fluorescence, as shown in Figure 5.11.

Since the transition moment of the spin-forbidden $T_1 \rightarrow S_0$ transition is very small, the natural lifetime τ_0^P of the triplet state is long. Consequently, radiationless processes can compete with phosphorescence in deactivating the T_1 state. Of particular importance are collision-induced bimolecular processes (cf. Section 5.4), and phosphorescence of gases and liquid solutions is relatively difficult to observe (Sandros and Bäckstrom, 1962). An exception is biacetyl with a very short T_1 lifetime τ_0^P. (Cf. Figure 5.3.) Phosphorescence spectra are commonly measured using samples in solvents or mixed solvents that form rigid glasses at 77 K (such as EPA = ether-pentane-alcohol mixture). (Cf., however, Example 5.6).

The natural lifetime of the triplet state $\tau_0^P = 1/k_P$ may be estimated from the observed lifetime and the quantum yields of fluorescence and phosphorescence. According to Equation (5.11)

$$\Phi_P = \eta_{ST} \frac{\tau_P}{\tau_0^P}$$

or

$$\tau_0^P = \frac{\eta_{ST}}{\Phi_P} \tau_P$$

If internal conversion and all energy-transfer processes and photochemical reactions are negligible, $\eta_{ST} = 1 - \Phi_F$, and one obtains

$$\tau_0^P = \tau_P \frac{1 - \Phi_F}{\Phi_P} \tag{5.19}$$

The natural lifetime τ_0^P varies between 10^{-6} s and many seconds.

The rate constant k_{ST} of intersystem crossing depends on the energy gap ΔE_{ST} between the singlet and triplet states and in particular on spin-orbit coupling. The difference in the magnitude of spin-orbit coupling contributes greatly to the fact that the quantum yield Φ_{ST} of triplet formation is small for aromatic hydrocarbons, but nearly unity for carbonyl compounds. (Cf. the El-Sayed rules, Section 5.2.2.) In discussing the energy gap dependence of the intersystem crossing rate one has to note that the triplet state closest to

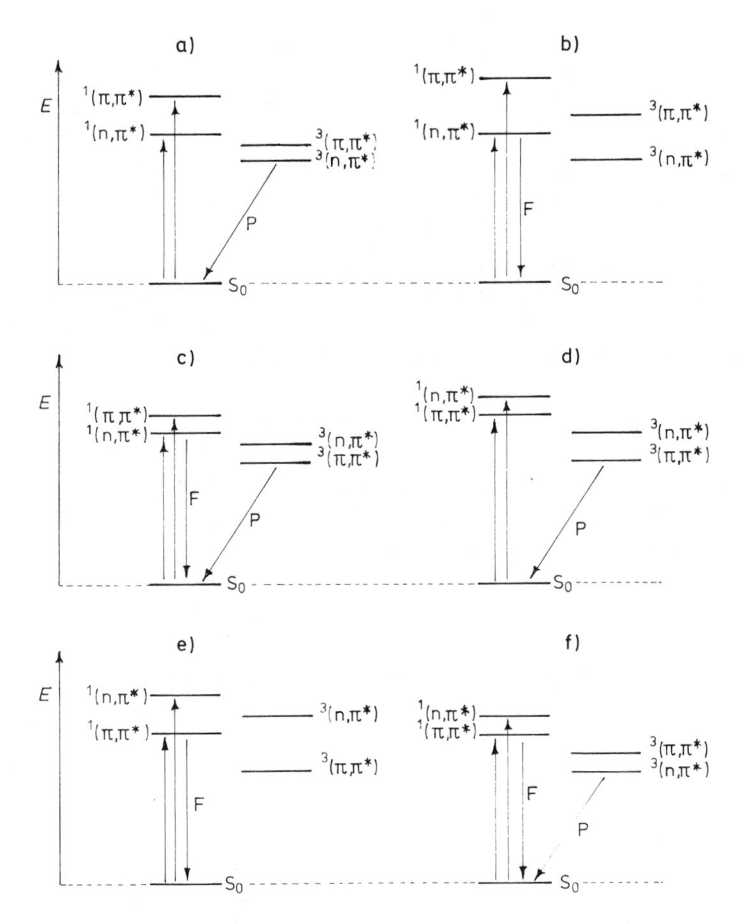

Figure 5.13. Various possiblities for the disposition of the lowest singlet and triplet (n,π*) and (π,π*) states of organic molecules (adapted from Wilkinson, 1968).

S_1 can be one of the higher triplet states T_n, such as is the case for anthracene (cf. Example 5.3), and that in molecules with lone pairs of electrons either T_1 or T_n may be of the same type as S_1. This is illustrated in Figure 5.13, where various possibilities for the energy order of the (π,π*) and (n,π*) states are displayed schematically. Not all the situations shown are equally probable; it is, for instance, very unlikely that the singlet–triplet splitting of the (n,π*) states could be appreciably larger than that of the (π,π*) states, as shown in Figure 5.13f.

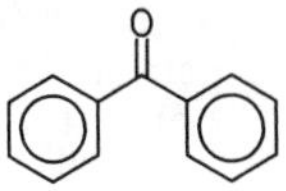

O
N

10 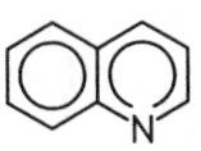 **11**

The relative position of the excited states of benzophenone (**10**) is shown in Figure 5.13a. Intersystem crossing leads from a $^1(n,\pi^*)$ state to a $^3(\pi,\pi^*)$ state; this type of transition is favored by spin-orbit coupling to such an extent that $\Phi_{ST} \approx 1$ and no fluorescence is observed. If, however, S_1 and T_1 are (n,π^*) states and are disposed as in Figure 5.13b, Φ_{ST} is so small that practically no phosphorescence can be observed although the molecule has $n \rightarrow \pi^*$ transitions. The relative disposition of (n,π^*) and (π,π^*) states may be changed by solvent effects and the rate constants of the various photophysical and photochemical processes can be drastically altered. Thus lone pairs of electrons in molecules such as quinoline (**11**) may be stabilized by hydroxylic solvents to such an extent that the (n,π^*) states become higher in energy than the (π,π^*) states, as shown in Figure 5.13d and e. Phosphorescence would predominate in the former case, fluorescence in the latter.

As a result of the heavy-atom effect or the effect of paramagnetic molecules such as O_2, which both enhance $S_0 \rightarrow T_n$ absorption (cf. Section 1.3.2), phosphorescence $T_1 \rightarrow S_0$ as well as the rate constants k_{ST} and k_{TS} of intersystem crossing will be favored. The consequence according to Equation (5.11) is an increase in η_{ISC}, whereas η_P will increase or decrease depending on which of the two processes, radiationless deactivation of the triplet state or phosphorescence, is more strongly favored. Frequently, an increase in the

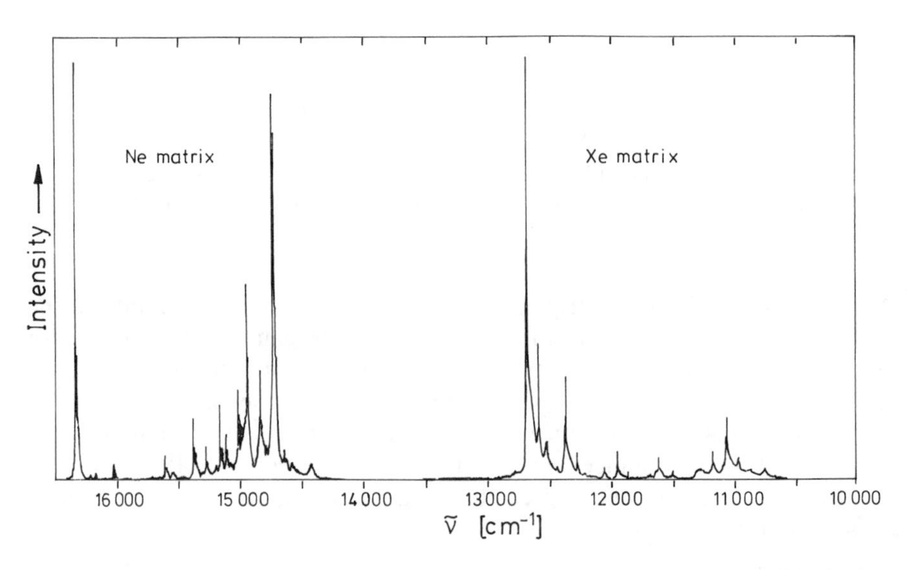

Figure 5.14. Luminescence of free-base porphine (**12**) in Ne (4 K) and Xe (12 K). The former is totally dominated by fluorescence (left) and the latter by phosphorescence (right) (by permission from Radziszewski et al., 1991).

quantum yield Φ_P of phosphorescence by the heavy-atom effect is observed. For example, free-base porphine (**12**) exhibits practically only fluorescence in a neon matrix and practically only phosphorescence in a xenon matrix (Figure 5.14). It has been concluded that the phosphorescence rate constant is enhanced by at least two orders of magnitude by the heavy-atom effect (Radziszewski et al., 1991).

12

Example 5.6:
The heavy-atom effect can be utilized for measuring phosphorescence in solution. In Figure 5.15 the luminescence spectrum of 1,4-dibromonaphthalene

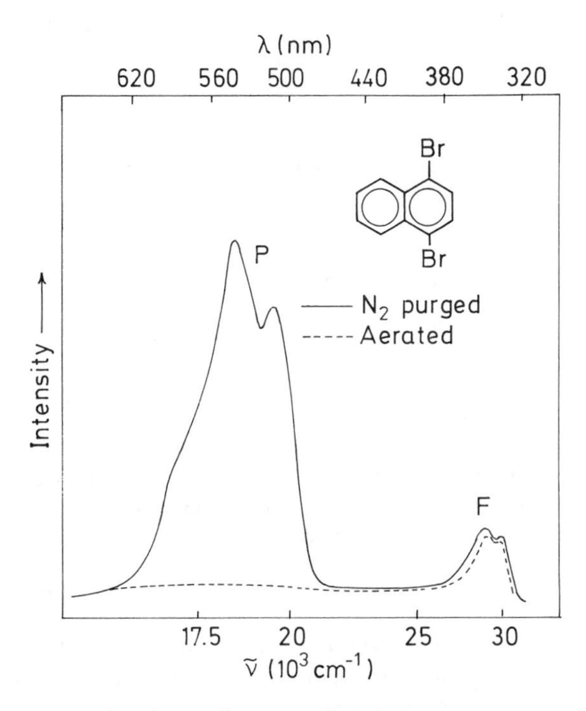

Figure 5.15. Luminescence spectrum of 1,4-dibromonaphthalene in acetonitrile with and without purging with N_2 (by permission from Turro et al., 1978).

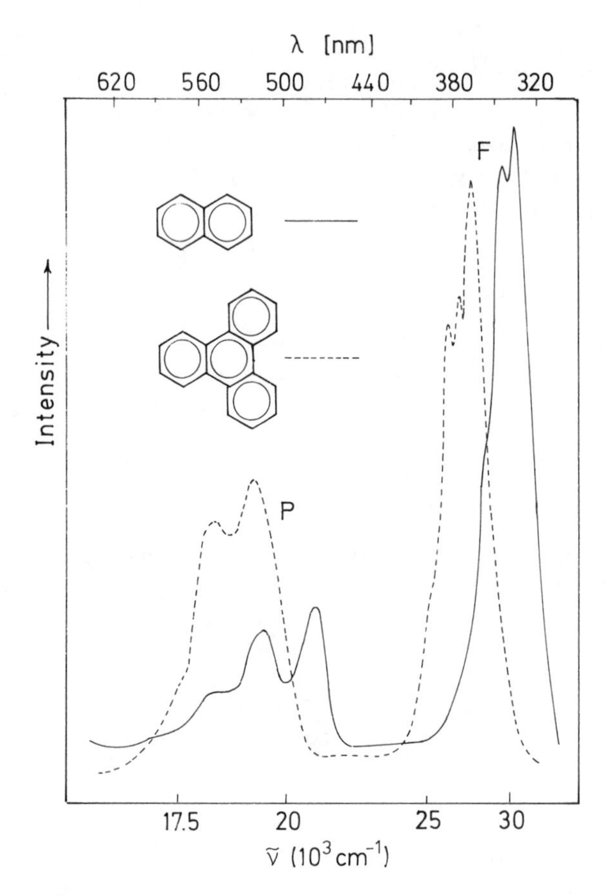

Figure 5.16. Luminescence spectra of naphthalene and triphenylene in 1,2-dibromoethane (by permission from Turro et al., 1978).

is shown as an example of the "internal heavy-atom effect." The spectra of naphthalene and triphenylene in Figure 5.16 demonstrate the "external heavy-atom effect" due to the solvent 1,2-dibromoethane. For such measurements it is important to use highly purified and deoxygenated solvents since otherwise phosphorescence is too weak to be detected, as is evident from Figure 5.15 (Turro et al., 1978).

When Φ_P is independent of the excitation wavelength the extreme sensitivity associated with emission spectroscopy can be utilized to obtain S_0–T absorption spectra by measuring *phosphorescence excitation spectra* (Marchetti and Kearns, 1967). The principle of the method is the same as for fluorescence excitation spectra and has been discussed in detail in Section 5.3.1. The use of a heavy-atom solvent such as ethyl iodide and of an intense light source permits the direct measurement of $S_0 \rightarrow T_1$ transitions by this method. This technique may be used even when the compound itself is non-phosphorescent by having present a phosphorescent molecule of lower triplet energy that is excited by energy transfer (see Section 5.4.5) from the nonphosphorescent triplets.

Example 5.7:
The phosphorescence excitation spectrum may be used to decide whether the lowest triplet state is a $^3(n,\pi^*)$ or a $^3(\pi,\pi^*)$ state, if measurements are carried out in two different solvents with or without the heavy-atom effect, respectively. The intensity of the phosphorescence excitation spectrum will in general increase due to the heavy-atom effect if T_1 is a $^3(\pi,\pi^*)$ state as in *p*-hydroxyacetophenone. However, no intensity increase is observed if T_1 is a $^3(n,\pi^*)$ state as in benzophenone, since spin-orbit coupling associated with a $n \rightarrow \pi^*$ transition is already so strong that the additional solvent effect is negligible. The spectrum in Figure 5.17 illustrates the intensity increase expected for a $^3(\pi,\pi^*)$ state.

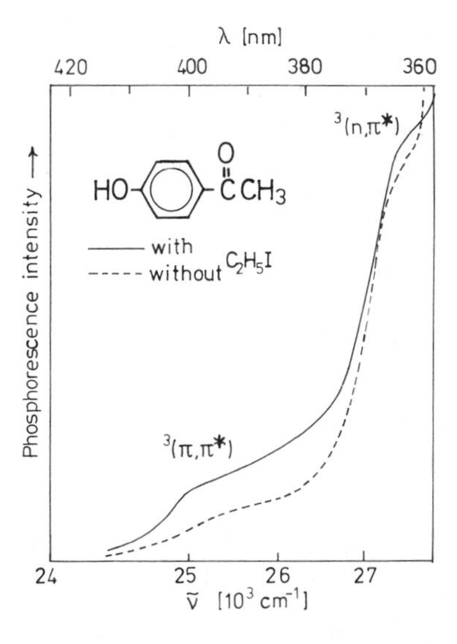

Figure 5.17. Phosphorescence excitation spectrum of *p*-hydroxybenzophenone in an ether-toluol-ethanol mixture with (—) and without (---) the addition of ethyl iodide. The curves are normalized in such a way that the $S \rightarrow T$ excitation of the $^3(n,\pi^*)$ state shows about the same intensity in both solvents (by permission from Kearns and Case, 1966).

5.3.3 Luminescence Polarization

A light quantum of appropriate energy can be absorbed by a molecule fixed in space only if the light electric field vector has a component parallel to the molecular transition moment. If the directions of the transition moment and of the electric field vector form an angle φ, the absorption probability is proportional to $\cos^2 \varphi$. (Cf. Section 1.3.5.) The light quanta of luminescence are also polarized, with the intensity again proportional to $\cos^2 \varphi$.

The polarization direction of an electronic transition may be determined by measurement of the absorption of polarized light by aligned molecules. (Cf. Michl and Thulstrup, 1986.) Orientation may be achieved in a number of ways. When single crystals are used or when the molecules of interest are incorporated into appropriate single crystals, a very high degree of orientation can be obtained if the crystal structure is favorable. Other methods accomplish orientation by embedding the molecules in stretched polymer films (polyethylene, PVA; Thulstrup et al., 1970) or in liquid crystals; further possibilities are orientation by an electric field (Liptay, 1963), or, in the case of polymers such as DNA, by a flow field (Erikson et al., 1985).

The relative polarization directions of the different electronic transitions of a molecule may be determined by exciting with polarized light and analyzing the degree of polarization of the luminescence, referred to briefly as *luminescence polarization*. When a solution of unoriented molecules is exposed to plane-polarized light of a wavelength appropriate for a specific electronic transition, only those molecules that have their transition moment oriented parallel to the electric field vector absorb with maximum probability. Using this selection process, known as *photoselection,* an effective alignment of the excited molecules is achieved. Only excited molecules can emit, and the directions of the transition moment of emission [in general $\boldsymbol{M}(S_1 \to S_0)$ or $\boldsymbol{M}(T_1 \to S_0)$] and the transition moment of absorption $\boldsymbol{M}(S_0 \to S_n)$ will form an angle α. If rotation of the excited molecules during their lifetimes is prevented by high viscosity of the solution, the emission will also be polarized.

The *degree of polarization* is defined as

$$P = \frac{I_{\parallel} - I_{\perp}}{I_{\parallel} + I_{\perp}} \tag{5.20a}$$

while the *degree of anisotropy,* which sometimes leads to simpler formulas (cf. Michl and Thulstrup, 1986), is defined as

$$R = \frac{I_{\parallel} - I_{\perp}}{I_{\parallel} + 2I_{\perp}} \tag{5.20b}$$

Here $I_{\parallel}$ and $I_{\perp}$ are the intensities of the components of the emitted light parallel and perpendicular to the electric vector of the exciting light, respectively. The curves $P(\lambda)$ or $P(\tilde{\nu})$ are called the *polarization spectrum*. Depending on whether the measurement is carried out with constant excitation wavelength λ_1 or at constant luminescence wavelength λ_2, different polarization spectra result. Using excitation light of fixed wavelength λ, a fluorescence polarization spectrum (FP) or a phosphorescence polarization spectrum (PP), respectively, is obtained. Measurement at a constant emission wavelength yields an absorption-wavelength-dependent polarization spectrum of either fluorescence [AP(F)] or phosphorescence [AP(P)].

The relationship between the degree of polarization P and the angle α between the transition moments of absorption and emission is given by

$$P = \frac{3\cos^2\alpha - 1}{\cos^2\alpha + 3} \tag{5.21}$$

Ideally, P can assume values ranging between $P = 0.5$ (corresponding to $\alpha = 0°$) and $P = -0.33$ (corresponding to $\alpha = 90°$). If a molecule possesses a symmetry axis of order $n > 2$ and absorption and emission are isotropically polarized in a plane perpendicular to this axis, $P = \text{const} = 0.14$ (Dörr and Held, 1960). The value of P is affected by the overlap of different electronic transitions and also reflects the orientational dependence of the transition moment on vibronic mixing with nonsymmetrical vibrations. Under favorable conditions it is therefore possible not only to determine the relative directions of electronic transition moments from the polarization spectra but also to locate bands hidden in the absorption spectrum and to ascertain the symmetry of the vibrations giving rise to the fine structure. Energy migration (repeated intermolecular energy transfer) causes depolarization; such measurements have therefore to be carried out in dilute solutions.

Example 5.8:

In Figure 5.18 the absorption and emission spectra of azulene are shown. The anomalous fluorescence of azulene from the S_2 state is easy to recognize. The AP(F) spectrum exhibits a deep minimum at 33,900 cm^{-1}. The small peak in the absorption spectrum at the same wave number is therefore not due to vibrational structure but rather to another electronic transition, the polarization of which had been predicted by PPP calculations. Figure 5.19 shows all four types of polarization spectra of phenanthrene. FP becomes negative at the vibrational maxima of the fluorescence; the most intense vibration is not totally symmetric, in contrast to the one which shows up weakly. For all absorption bands, AP(P) ≈ -0.3. The polarization direction of phosphorescence is perpendicular to the transition moments of all $\pi\rightarrow\pi^*$ transitions lying in the molecular plane and is therefore perpendicular to the molecular plane. One observes PP ≈ -0.3 as well, with a modulation due to vibrations.

The observed phosphorescence polarization direction may be accounted for by the fact that singlet–triplet transitions acquire their intensity by spin-orbit coupling of the S_0 state with triplet states and particularly, of the T_1 state with singlet states. (Cf. Section 1.3.2.) Under usual conditions the phosphorescence is an unresolved superposition of emissions from the three components of the triplet state, which in the absence of an external magnetic field are described

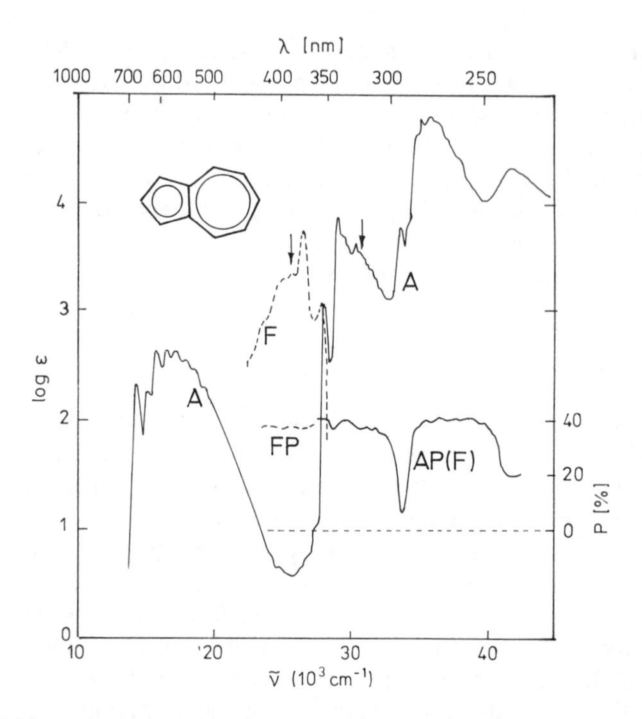

Figure 5.18. Absorption (A) and fluorescence spectrum (F) of azulene in ethanol at 93 K. FP and AP(F) denote the polarization spectrum of fluorescence and the excitation–polarization spectrum of fluorescence, respectively (by permission from Dörr, 1966).

by the spin functions Θ_x, Θ_y, and Θ_z [cf. Equation (4.7)] and have the same symmetry properties as the rotations $\hat{R}_x$, $\hat{R}_y$, and $\hat{R}_z$ about the molecular symmetry axes. In general they belong to different irreducible representations of the molecular point group and therefore mix with different singlet states under the influence of the totally symmetric spin-orbit coupling operator. The transition moments of the different admixed singlet states then determine the intensity and polarization direction of emission from the various components of the T_1 state. In the case of phenanthrene, the triplet component with the transition moment perpendicular to the molecular plane contributes most of the intensity, since spin-orbit coupling of this component to $^1(\sigma,\pi^*)$ states with polarization direction perpendicular to the plane predominates over the spin-orbit coupling of the other components. The $T_1 \rightarrow S_0$ transition steals its intensity from such $S_n \rightarrow S_0$ transitions.

In Figure 5.20, the emission spectrum of triphenylene is shown: $P = 0.14$ in the FP spectrum due to the threefold symmetry axis perpendicular to the molecular plane, which is also the polarization plane for all $\pi \rightarrow \pi^*$ transitions.

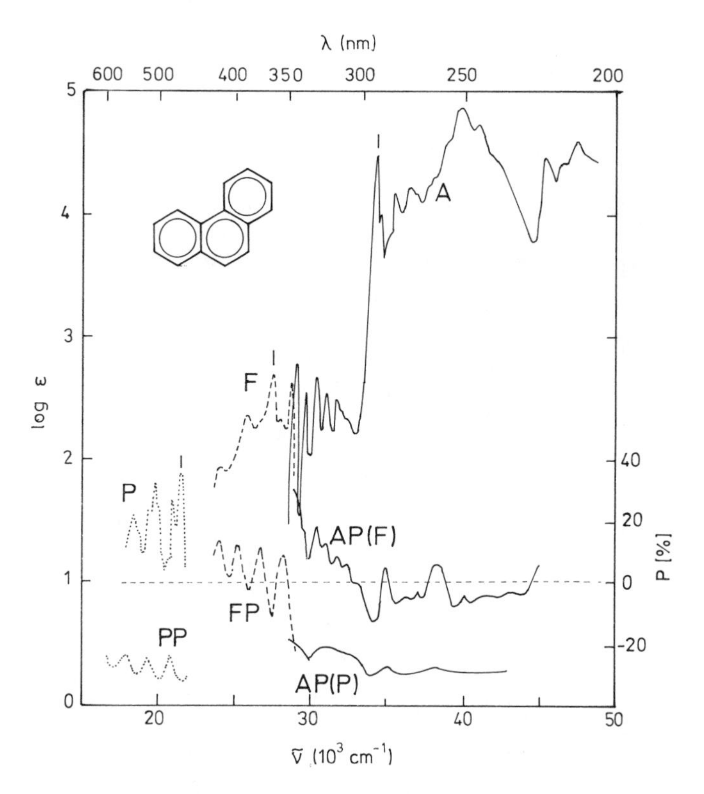

Figure 5.19. Absorption (A) and emission spectra (F and P) of phenanthrene in ethanol at 93 K. FP and PP denote the polarization spectra of fluorescence and phosphorescence, AP(F) and AP(P) the excitation polarization spectra of fluorescence and phosphorescence, respectively (by permission from Dörr, 1966).

The phosphorescence polarization direction is again perpendicular to the molecular plane and is modified by out-of-plane vibrations.

If the excited molecules merely return to their ground state by radiative or nonradiative processes, no permanent orientation remains after the photoselective irradiation is terminated. However, when the excited molecules undergo a permanent chemical change, photoselection leads to a lasting alignment of that portion of the reactant molecules that remain when the irradiation is interrupted. Sometimes, the product molecules are aligned as well, depending on the degree of correlation between the average orientation of the parent reactant and the daughter photoproduct molecules in space (Michl and Thulstrup, 1986).

Samples oriented by photoselection have been used for studies of molecular anisotropy by polarized absorption spectroscopy. For instance, the

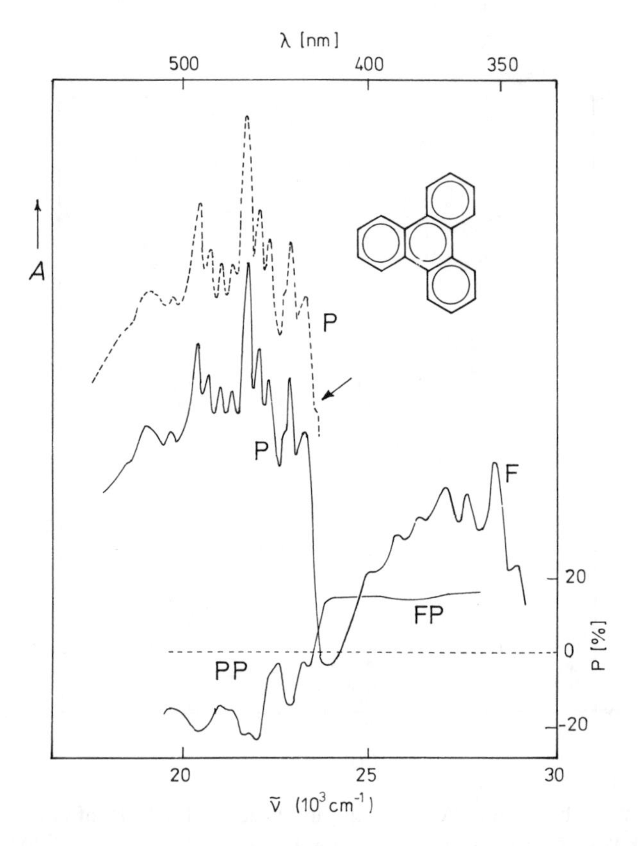

Figure 5.20. Emission spectra (F and P) of triphenylene in ethanol at 93 K. FP and PP denote the polarization spectra of fluorescence and phosphorescence, respectively (by permission from Dörr, 1966).

symmetries of all IR-active vibrations of free-base porphine (**12**) have been measured on a sample photooriented with visible light in a rare-gas matrix (Radziszewski et al., 1987, 1989). Photoorientation of the major and the minor conformer of 1,3-butadiene in rare-gas matrices was used to determine the directions of the IR-transition moments in both, which revealed that the latter is planar in these media (i.e., s-cis and not gauche), and yielded the average relation between the orientation of the s-cis reactant and that of the s-trans photoproduct (Arnold et al., 1990, 1991).

5.4 Bimolecular Deactivation Processes

In addition to monomolecular processes such as emission and radiationless deactivation there are very important bimolecular deactivation mechanisms that involve the transfer of excitation energy from one molecule to another. These processes are generally referred to as *quenching processes*. The suppression of emission by energy-transfer processes is in particular referred to as luminescence quenching (quenching in the strict sense). If it is not the deactivation that is of principal interest during a bimolecular process but rather the excitation of the energy acceptor molecule, the process is referred to as *sensitization*. States that otherwise would be accessible only with difficulty or even not at all may be populated through sensitized excitation.

5.4.1 Quenching of Excited States

Fluorescence quenching is a very general phenomenon that occurs through a variety of different mechanisms. All chemical reactions involving molecules in excited states can be viewed as luminescence quenching. Such photochemical reactions will be dealt with in later chapters.

Photophysical quenching processes that do not lead to new chemical species can in general be represented as

$$M^* \xrightarrow{Q} M'$$

where M′ is the ground state or another excited state of M. According to whether the quencher Q is a molecule M of the same kind or a different molecule *self-quenching* or *concentration quenching* can be distinguished from *impurity quenching* by some other chemical species.

Most intermolecular deactivation processes are based on collisions between an excited molecule M* and a quencher Q. They are subject to the Wigner–Witmer spin-conservation rule according to which the total spin must not change during a reaction (Wigner and Witmer, 1928).

Example 5.9:

In order that the products *C* and *D* lie on the same potential energy surface as the reactants *A* and *B*, the total spin has to be conserved during the reaction. The spins S_A and S_B of the reactants may be coupled according to vector addition rules in such a way that the total spin of the transition state can have the following values:

$$(S_A + S_B), (S_A + S_B - 1), \ldots, |S_A - S_B|$$

Similarly, the spins S_C and S_D of the products may be coupled to give one of the total spin values

$$(S_C + S_D), (S_C + S_D - 1), \ldots, |S_C - S_D|$$

A reaction is allowed according to the Wigner–Witmer spin-conservation rule if the reactants can form a transition state with a total spin that can also be obtained by coupling the product spins, that is, if the two sequences above have a number in common.

Thus, for $S_A = S_B = 0$ the reaction is allowed if $S_C = S_D = 0$, but not if $S_C = 1$, $S_D = 0$ since in the latter case S_C and S_D can be coupled only to the total spin $(S_C + S_D) = |S_C - S_D| = 1$. Therefore, the singlet–singlet energy transfer

$$^1D^* + {}^1A \rightarrow {}^1D + {}^1A^*$$

is allowed. Similarly, the triplet–triplet energy transfer

$$^3D^* + {}^1A \rightarrow {}^1D + {}^3A^*$$

is seen to be allowed. For the reaction of two molecules in their triplet states one has $S_A = S_B = 1$ and the total spin can take the values 2, 1, and 0. Triplet–triplet annihilation

$$^3A^* + {}^3A^* \rightarrow {}^1A + {}^{5,3,1}A^*$$

thus gives one molecule in a singlet state while the other one may be in a singlet, a triplet, or a quintet state.

Except for some long-distance electron-transfer and energy-transfer mechanisms bimolecular deactivation involves either an *encounter complex* (M* . . . Q), or an *exciplex* (MQ*) or *excimer* (MM)*. An exciplex or an excimer has a binding energy larger than the average kinetic energy $(3/2)kT$ and represents a new chemical species with a more or less well-defined geometrical structure corresponding to a minimum in the excited-state potential energy surface. This is not true for an encounter complex in which the components are separated by widely varying distances and have more random relative orientations.

Encounter complexes, exciplexes, and excimers can lose their excitation energy through either fluorescence or phosphorescence, by decay into M + Q* which corresponds to an energy transfer, by electron transfer to give $M^{\oplus} + Q^{\ominus}$ or $M^{\ominus} + Q^{\oplus}$, by internal conversion, and by intersystem crossing (Schulten et al., 1976b). All these processes lead to quenching of the excited state M* and are therefore referred to as quenching processes.

5.4.2 Excimers

Frequently, it is observed that an increase in the concentration of a fluorescent species such as pyrene is accompanied by a decrease in the quantum yield of its fluorescence. This phenomenon is called self-quenching or concentration quenching and is due to the formation of a special type of complex formed by the combination of a ground-state molecule with an excited-state molecule. Such a complex is called an excimer (*exc*ited di*mer*) (Förster and Kaspar, 1955). Whereas for a system M + M of two separate ground-state molecules all interactions are purely repulsive except for those yielding the van-der-Waals minima observed in the gas phase, stabilizing interactions are possible according to Figure 5.21 if one of the molecules is in an excited state and the HOMO and LUMO of the combined system are only singly

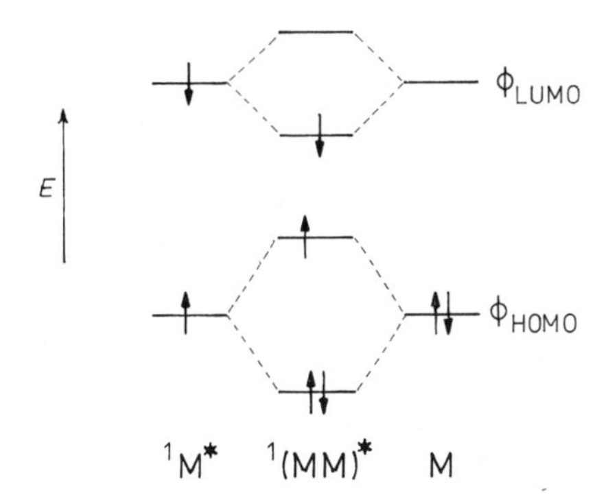

Figure 5.21. MO scheme of excimer and exciplex formation.

occupied. This gives rise to the relative minimum of the excimer $^1(MM)^*$ on the excited-state potential energy surface.

$$
\begin{array}{ccccc}
 & ^1M^* + M & \rightleftharpoons & ^1(MM)^* & \\
\textbf{(Monomer Fluorescence)} & -h\nu \downarrow \uparrow h\nu & & \downarrow -h\nu & \textbf{(Excimer Fluorescence)} \\
 & M + M & \longleftarrow & (MM) &
\end{array}
$$

In Section 4.4 it was shown for the $(H_2 + H_2)$ system that interaction between locally excited states described by Ψ_{MM^*} and Ψ_{M^*M} results in exciton

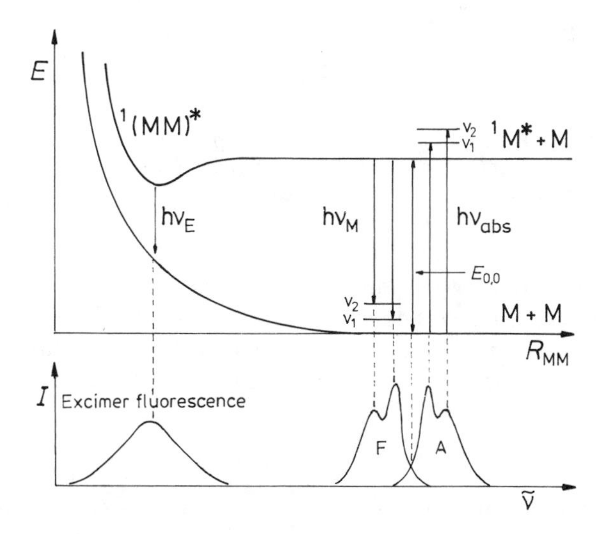

Figure 5.22. Schematic representation of the potential energy surfaces for excimer formation and of the difference between monomer fluorescence and excimer fluorescence (adapted from Rehm and Weller, 1970a).

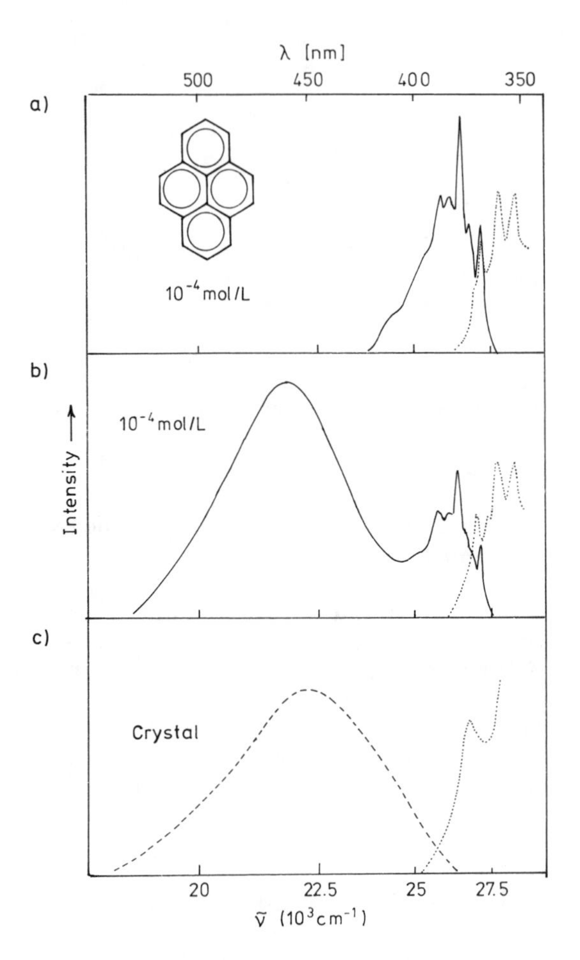

Figure 5.23. Absorption (···) and fluorescence spectrum (—) of pyrene a) 10^{-4} mol/L in ethanol, b) 10^{-2} mol/L in ethanol, and c) absorption and emission (---) of crystalline pyrene (adapted from Förster and Kaspar, 1955).

states. The magnitude of the interaction is a measure of the energy transfer from one molecule to the other. (Cf. the Förster mechanisms of energy transfer, Section 5.4.5.) Interaction with ion-pair or CT states further stabilizes the lower of the exciton states, so the excimer can be described by a wave function of the form

$$\Psi_{\text{Excim}} = c_1(\Psi_{MM^*} \pm \Psi_{M^*M}) + c_2(\Psi_{M^+M^-} \pm \Psi_{M^-M^+}) \tag{5.22}$$

Whether the stabilization is given by a plus or minus combination depends on the orientation of M* relative to M. (For a more detailed treatment see Michl and Bonačić-Koutecký, 1990.)

The potential energy surfaces of the ground state M + M and the excited state M* + M resulting from these arguments are represented schematically in Figure 5.22. From this diagram it is evident that excimer fluorescence is to be expected at longer wavelengths than monomer fluorescence and that the associated emission band should be broad and generally without vibrational structure, being due to a transition into the unbound ground state. The spectrum of pyrene in Figure 5.23 is the perfect confirmation of these expectations.

Excimer formation is observed quite frequently with aromatic hydrocarbons. Excimer stability is particularly great for pyrene, where the enthalpy of dissociation is ΔH = 10 kcal/mol (Förster and Seidl, 1965). The excimers of aromatic molecules adopt a sandwich structure, and at room temperature, the constituents can rotate relative to each other. The interplanar separation is 300–350 pm and is thus in the same range as the separation of 375 pm between the two benzene planes in 4,4′-paracyclophane (**13**), which exhibits the typical structureless excimer emission. For the higher homologues, such as 5,5′-paracylophane, an ordinary fluorescence characteristic of *p*-dialkylbenzenes is observed (Vala et al., 1965).

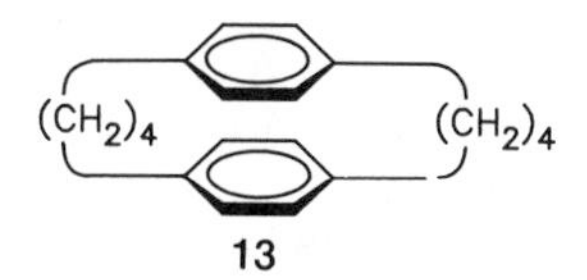

Calculations based on a wave function corresponding to Equation (5.22) also indicate a sandwich structure and an interplanar distance of 300–360 pm (Murrell and Tanaka, 1964).

5.4.3 Exciplexes

Two different molecules M and Q can also form complexes with a definite stoichiometry (usually 1:1). If complexation is present already in the ground state and leads to CT absorption (Section 2.6), which is absent in the individual components, the complex is referred to as a charge-transfer or donor–acceptor complex. If, however, the complex shows appreciable stability only in the excited state, it is called an exciplex (*exci*ted com*plex*):

$$\begin{array}{lcccl} & {}^1M^* + Q & \rightleftharpoons & {}^1(MQ)^* & \\ \textbf{(Monomer Fluorescence)} & -h\nu \downarrow\uparrow h\nu & & \downarrow -h\nu & \textbf{(Exciplex Fluorescence)} \\ & M + Q & \longleftarrow & (MQ) & \end{array}$$

The spectra of anthracene-dimethylaniline shown in Figure 5.24 exemplify the new structureless emission at longer wavelengths due to the formation of an exciplex. This fluorescence is very similar to excimer fluorescence.

Contrary to the situation for excimers, one component of the exciplex acts predominantly as the donor (D); the other one acts as the acceptor (A).

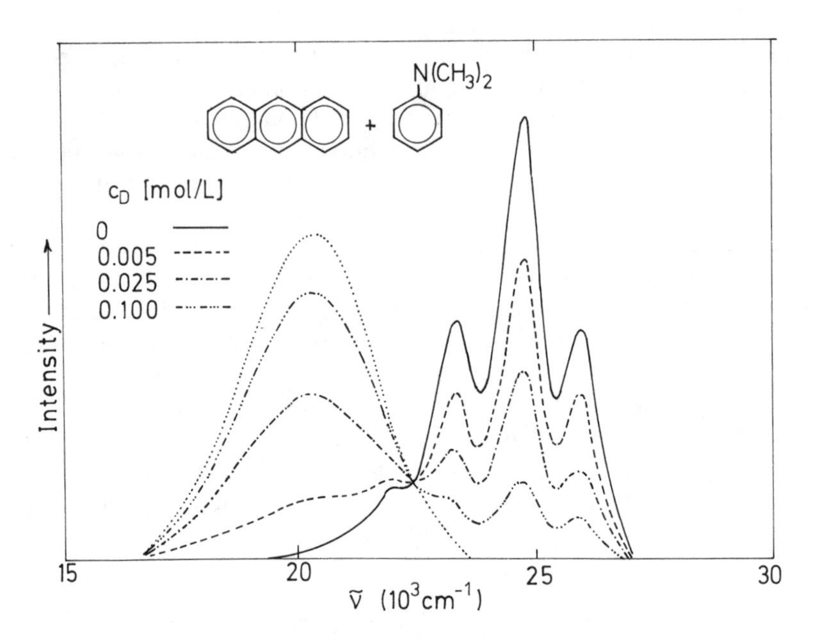

Figure 5.24. Fluorescence and exciplex emission from anthracene in toluene ($3.4 \cdot 10^{-5}$ mol/L) for various concentrations c_D of dimethylaniline (by permission from Weller, 1968).

If use is made of this fact in the notation, one obtains instead of Equation (5.22) the wave function

$$\Psi_{\text{Excipl}} = c_1\Psi_{DA^*} + c_2\Psi_{D^*A} + c_3\Psi_{D^+A^-} + c_4\Psi_{D^-A^+} \tag{5.23}$$

where $c_1 \neq c_2$ and $c_3 \neq c_4$; the third term corresponds to a charge-transfer (CT) excited state (cf. Section 4.4) and is by far the most important one, and the exciplex corresponds to a contact ion pair. Experimentally the charge-transfer character is revealed by the high polarity, with exciplexes from aromatic hydrocarbons and aromatic tertiary amines having dipole moments $\mu_{(AD)^*} > 10$ D (Beens et al., 1967; cf. Section 1.4.2).

From a simple MO treatment it follows that the electron transfer leading to exciplex formation can occur either from an excited donor to an acceptor or from a donor to an excited acceptor. (See Figure 5.25.) In both cases, the singly occupied orbitals of the resulting exciplex correspond to the HOMO of the donor and the LUMO of the acceptor. Neglecting solvent effects, the energy of exciplex emission is therefore given by

$$h\nu_{\text{Excipl}} = \text{IP}_D - \text{EA}_A + C \tag{5.24}$$

where IP_D and EA_A are the ionization potential of the donor and the electron affinity of the acceptor, and C describes the Coulomb attraction between the components of the ion-pair $D^{\oplus}A^{\ominus}$.

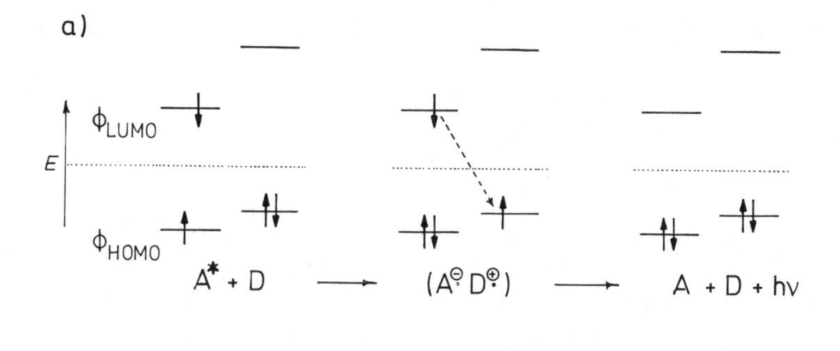

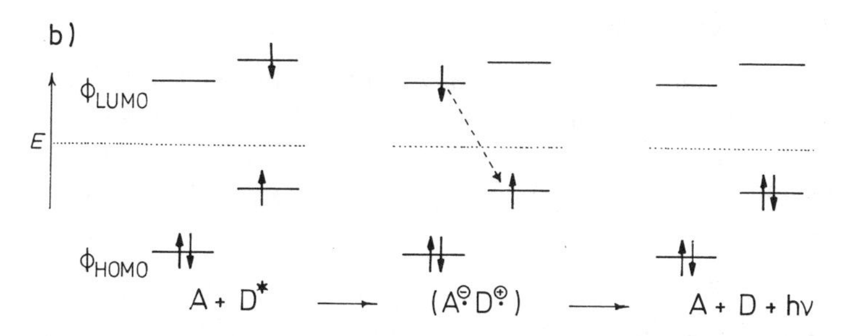

Figure 5.25. Exciplex formation by charge transfer a) from the donor to the excited acceptor and b) from the excited donor to the acceptor. Exciplex emission is indicated by a broken arrow (by permission from Weller, 1968).

Exciplexes and excimers appear to be involved in many photochemical processes; in particular, they are probably involved in many quenching and charge-transfer processes and in many photocycloadditions.

5.4.4 Electron-Transfer and Heavy-Atom Quenching

According to the following scheme, electron transfer, intersystem crossing and energy transfer can all compete with fluorescence in deactivating an exciplex $^1(DA)^*$ formed from the molecules A and D.

$$
\begin{array}{ccccc}
 & & ^3(DA)^* & & D + A + \Delta \\
^1D^* + {}^1A & \rightleftharpoons & \uparrow k_{ISC} & \nearrow k_D & \\
 & & ^1(DA)^* & & \\
^1D + {}^1A^* & \rightleftharpoons & \downarrow k_e & \searrow k_F & \\
 & & D^{\ominus\cdot} + A^{\ominus\cdot} & & D + A + h\nu
\end{array}
$$

In polar solvents, polar exciplexes (contact ion pairs) dissociate into nonfluorescent radical ions (loose ion pairs or free ions) due to the stabilization of the separated ions by solvation. It has been observed that exciplex emission decreases with increasing polarity of the solvent and that at the same time the free-radical ions $D^{\oplus}$ and $A^{\ominus}$ can be identified by flash spectroscopy (Mataga, 1984).

Assuming the existence of a quasi-stationary state the rate constant of an exothermic electron-transfer reaction can be written as

$$k_{obs} = k_{diff}/(1 + k_{-diff}/k_e)$$

k_{diff}, k_e, k_{-diff}, and k_{-e} are the rate constants for diffusion, forward electron transfer from D to A, for the dissociation of the encounter complex to A and D, and for the back electron transfer from $A^{\ominus}$ to $D^{\oplus}$, respectively. The terms with k_{-e} have been neglected since $k_{-e} \ll k_e$ can be assumed for exothermic reactions. According to the Marcus theory (1964) a relationship of the form

$$\Delta G^{\ddagger} = \lambda/4\,(1 + \Delta G/\lambda)^2 \tag{5.25}$$

exists for adiabatic *outer-sphere* electron-transfer reactions* between the free enthalpy of activation $\Delta G^{\ddagger}$ and the free enthalpy of reaction ΔG. λ in Equation (5.25) is the reorganization energy essentially due to changes in bond distances and solvation. As a consequence of this relation, the rate constant

$$k_e \sim \exp(-\Delta G^{\ddagger}/RT) \tag{5.26}$$

should first increase with increasing exothermicity until the value $\Delta G = -\lambda$ is reached and then decrease again. The dependence of log k_e on ΔG obtained in this way is shown in Figure 5.26 by the dashed curve; the region of decreasing rate for strongly exergonic electron-transfer reactions ($\Delta G < \lambda$) is referred to as the *Marcus inverted region*.

The temperature dependence of electron-transfer rate constants is interesting. In the normal region, it shows an activation energy as predicted from simple Marcus theory. In the inverted region, the activation energy is very small or zero. This agrees with the quantum mechanical version of the theory (Kestner et al., 1974; Fischer and Van Duyne, 1977), which makes it clear that the transition from the upper to the lower surface behaves just like ordinary internal conversion.

Studies of the fluorescence quenching in acetonitrile have shown that the electron-transfer reaction

$$^1A^* + {}^1D \rightarrow {}^1({}^2A^{\ominus}\,{}^2D^{\oplus})$$

---

* Redox processes between metal complexes are divided into outer-sphere processes and inner-sphere processes that involve a ligand common to both coordination spheres. The distinction is fundamentally between reactions in which electron transfer takes place from one primary bond system to another (outer-sphere mechanism) and those in which electron transfer takes place within a primary bond system (inner-sphere mechanism) (Taube, 1970).

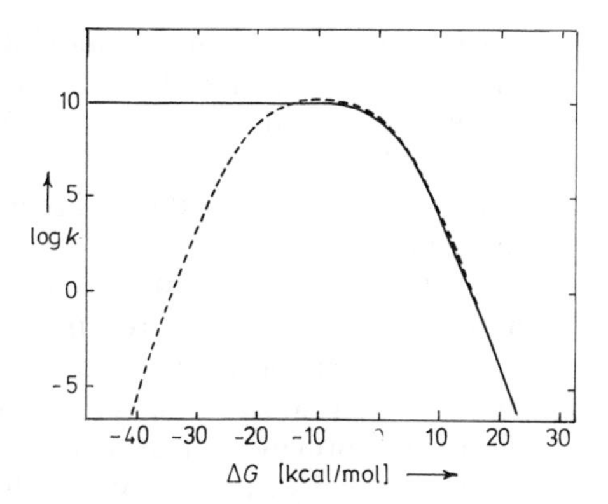

Figure 5.26. Dependence of the rate constant log k for electron-transfer processes on the free reaction enthalpy ΔG, Rehm-Weller plot (—) and Marcus plot (---) for $\lambda = 10$ kcal/mol (adapted from Eberson, 1982).

leading to a radical ion pair is diffusion controlled if the free enthalpy of this reaction is $\Delta G \leq -10$ kcal/mol, and that in contrast to the Marcus theory it remains diffusion-controlled even for very negative values of ΔG. Therefore, Rehm and Weller (1970b) proposed the empirical relationship

$$\Delta G^{\ddagger} = \Delta G/2 + [(\Delta G/2)^2 + (\lambda/4)^2]^{1/2} \tag{5.27}$$

which adequately describes the experimentally observed data, as can be seen from the solid curve in Figure 5.26. The free enthalpy of electron transfer ΔG can be estimated according to Weller (1982a) from the equation

$$\Delta G = E^{\text{ox}}_{1/2}(\text{D}) - E^{\text{red}}_{1/2}(\text{A}) - \Delta E_{\text{exc}}(\text{A}) + \Delta E_{\text{Coul}} \tag{5.28}$$

where $E^{\text{ox}}_{1/2}(\text{D})$ and $E^{\text{red}}_{1/2}(\text{A})$ are the half-wave potentials of the donor and acceptor, $\Delta E_{\text{exc}}(\text{A})$ is the (singlet or triplet) excitation energy of the acceptor, and ΔE_{Coul} is the Coulombic energy of the separated charges in the solvent in question. The authors suggested that the disagreement with the Marcus theory is due to fast electron transfer occurring via exciplex formation (Weller, 1982b), as shown in the following reaction scheme:

Encounter complex ($d \approx 700$ pm) **Exciplex** ($d \approx 300$ pm)

$$^1\text{A}^* + {}^1\text{D} \underset{k_{-\text{diff}}}{\overset{k_{\text{diff}}}{\rightleftharpoons}} {}^1({}^1\text{A}^* + {}^1\text{D}) \xrightarrow{k_{\text{S}}} {}^1(\text{A}^*\text{D})$$

$$^1({}^1\text{A}^* + {}^1\text{D}) \xrightarrow{k_{\text{rad}}} {}^1({}^2\text{A}^{\ominus} + {}^2\text{D}^{\oplus}) \qquad {}^1(\text{A}^*\text{D}) \xrightarrow{k^{\text{Exc}}_{\text{IC}}} {}^1(\text{A}^{\ominus}\text{D}^{\oplus})$$

$$^1({}^2\text{A}^{\ominus} + {}^2\text{D}^{\oplus}) \xleftarrow{k^{\text{Exc}}_{\text{diss}}} {}^1(\text{A}^{\ominus}\text{D}^{\oplus})$$

Experimental values of k_{diss}^{Exc} indicate that dissociation of weakly solvated dipolar exciplexes into a strongly solvated radical ion pair requires charge separation against the Coulomb attraction as well as diffusion of solvent molecules during resolvation (Weller, 1982b).

More recent investigations of rigidly fixed donor–acceptor pairs, for example, of type **14** with different acceptors A (Closs et al., 1986), and also of free donor–acceptor systems (Gould et al., 1988) have shown, however, that rates of strongly exothermic electron-transfer reactions in fact decrease again, as is to be expected for the Marcus inverted region. The deviation of the original Weller–Rehm data from Marcus theory at very large exothermicities may well be due to the formation of excited states of one of the products, for which the exothermicity is correspondingly smaller, and to compensating changes in the distance of intermolecular approach at which the electron transfer rate is optimized.

Recently a number of covalently linked porphyrin–quinone systems such as **15** (Mataga et al., 1984) or **16** (Joran et al., 1984) have been synthesized in order to investigate the dependence of electron-transfer reactions on the separation and mutual orientation of donor and acceptor. These systems are also models of the electron transfer between chlorophyll a and a quinone molecule, which is the essential charge separation step in photosynthesis in green plants. (Cf. Section 7.6.1.) Photoinduced electron transfer in supramolecular systems for artificial photosynthesis has recently been summarized (Wasielewski, 1992).

A

14

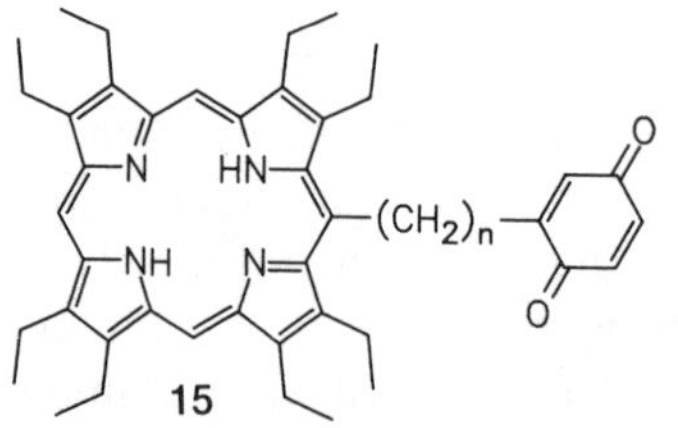

N HN NH N $(CH_2)_n$ O O

15

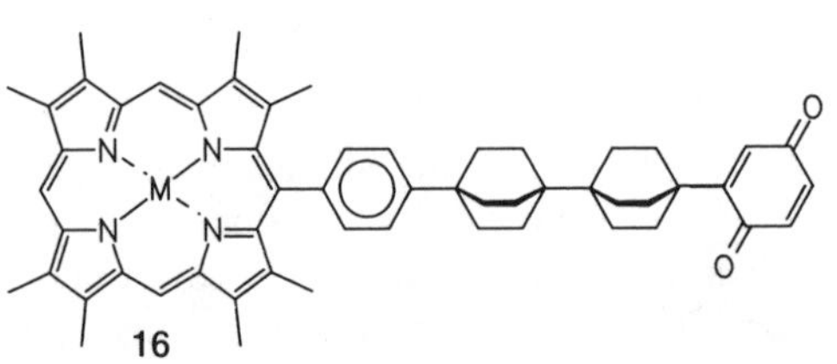

N N M N N O O

16

Heavy-atom quenching occurs if the presence of a heavy-atom–containing species enhances the intersystem crossing $^1(MQ)^* \xrightarrow{ISC} {}^3(MQ)^*$ to such an extent that it becomes the most important deactivation process for the exciplex. Since the triplet exciplex is normally very weakly bound and dissociates into its components, what one actually observes in such systems is

$$^1M^* + Q \rightarrow {}^3M^* + Q$$

Luminescence quenching by oxygen appears to be a similar process with $Q = {}^3O_2$. The process is diffusion-controlled, and it may be thought of as occurring via a highly excited triplet exciplex $^3(MO_2)^{**}$ that undergoes internal conversion to the lowest triplet exciplex $^3(MO_2)^*$ and decays into the components:

$$^1M^* + {}^3O_2 \xrightarrow{k_{diff}} {}^3(MO_2)^{**} \xrightarrow{IC} {}^3(MO_2)^* \rightarrow {}^3M^* + {}^3O_2$$

The net reaction consists of a catalyzed intersystem crossing which is spin allowed as opposed to the simple intersystem crossing (Birks, 1970). Triplet states may also be quenched by oxygen, but triplet–triplet annihilation (cf. Section 5.4.5.5) seems to be the predominant mechanism.

5.4.5 Electronic Energy Transfer

If an excited donor molecule D* reverts to its ground state with the simultaneous transfer of its electronic energy to an acceptor molecule A, the process is referred to as *electronic energy transfer*:

$$D^* + A \rightarrow D + A^*$$

The acceptor can itself be an excited state, as in triplet–triplet annihilation. (Cf. Section 5.4.5.5.) The outcome of an energy-transfer process is the quenching of the emission or photochemical reaction associated with the donor D* and its replacement by the emission or photochemical reaction characteristic of A*. The processes resulting from A* generated in this manner are said to be *sensitized*.

Energy transfer can occur either radiatively through absorption of the emitted radiation or by a nonradiative pathway. The nonradiative energy transfer can also occur via two different mechanisms—the Coulomb or the exchange mechanism.

5.4.5.1 Radiative Energy Transfer

Radiative energy transfer is a two-step process and does not involve the direct interaction of donor and acceptor:

$$D^* \rightarrow D + h\nu$$
$$h\nu + A \rightarrow A^*$$

The efficiency of radiative energy transfer, frequently described as "trivial" because of its conceptual simplicity (Förster, 1959), depends on a high quantum efficiency of emission by the donor in a region of the spectrum where the light-absorbing power of the acceptor is also high. It may be the dominant energy transfer mechanism in dilute solutions, because its probability decreases with the donor–acceptor separation only relatively slowly as compared with other energy-transfer mechanisms. When donor and acceptor are identical and emission and absorption spectra overlap sufficiently, *radiative trapping* may occur through repeated absorption and emission that increases the observed luminescence lifetime.

5.4.5.2 Nonradiative Energy Transfer

The nonradiative energy transfer

$$D^* + A \rightarrow D + A^*$$

is a single-step process that requires that the transitions $D^* \rightarrow D$ and $A \rightarrow A^*$ be isoenergetic as well as coupled by a suitable donor–acceptor interaction. If excited-state vibrational relaxation is faster than energy transfer, and if energy transfer is a vertical process as implied by the Franck-Condon principle, the spectral overlap defined by

$$J = \int_0^\infty \bar{I}_D(\tilde{\nu})\bar{\varepsilon}_A(\tilde{\nu})d\tilde{\nu} \tag{5.29}$$

is proportional to the number of resonant transitions in the emission spectrum of the donor and the absorption spectrum of the acceptor. (Cf. Figure 5.27). The spectral distributions $\bar{I}_D(\tilde{\nu})$ and $\bar{\varepsilon}_A(\tilde{\nu})$ of donor emission and accep-

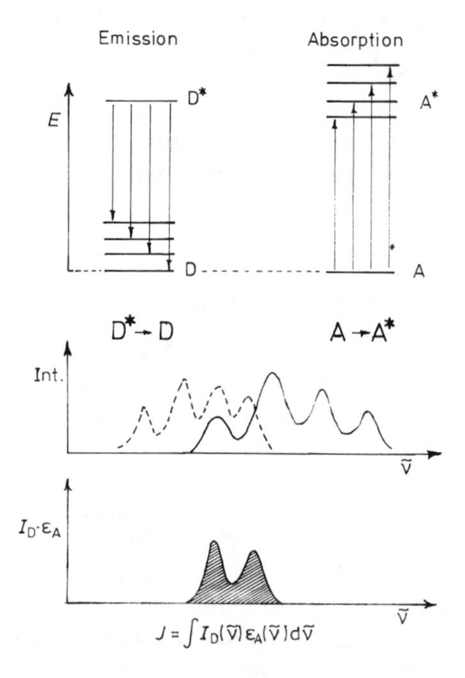

Figure 5.27. Schematic representation of the spectral overlap J and its relation to the emission and absorption spectrum.

tor absorption, respectively, are normalized to a unit area on the wave-number scale, that is

$$\int_0^\infty \bar{I}_D(\tilde{\nu})d\tilde{\nu} = \int_0^\infty \bar{\varepsilon}_A(\tilde{\nu})d\tilde{\nu} = 1$$

This clearly reflects the fact that J is not connected to the oscillator strengths of the transitions involved.

The coupling of the transitions is given by the interaction integral

$$\beta = \langle\Psi_f|\hat{H}'|\Psi_i\rangle = \int \Psi_f\hat{H}'\Psi_i d\tau \tag{5.30}$$

where $\hat{H}'$ involves the electrostatic interactions of all electrons and nuclei of the donor with those of the acceptor, and $\Psi_i = \hat{A}\Psi_{D*}\Psi_A$ and $\Psi_f = \hat{A}\Psi_D\Psi_{A*}$ are antisymmetrized product wave functions of the initial and the final state. The total interaction β may be written as the sum of a Coulomb and an exchange term; thus in the two-electron case $\Psi_i = [\psi_{D*}(1)\psi_A(2) - \psi_{D*}(2)\psi_A(1)]/\sqrt{2}$ and $\Psi_f = [\psi_D(1)\psi_{A*}(2) - \psi_D(2)\psi_{A*}(1)]/\sqrt{2}$, and

$$\beta = (\beta^C - \beta^E) = [\int\psi_{D*}(1)\psi_A(2)\hat{H}'\psi_D(1)\psi_{A*}(2)d\tau_1 d\tau_2 - \int\psi_{D*}(1)\psi_A(2)\hat{H}'\psi_D(2)\psi_{A*}(1)d\tau_1 d\tau_2] \tag{5.31}$$

where the second integral, the exchange term, differs from the Coulomb term in that the variable 1 and 2 are interchanged on the right.*

The functions ψ_i are spin orbitals and contain a space factor ϕ_i and a spin factor α or β. (Cf. Section 1.2.2.)

The Coulomb term represents the classical interaction of the charge distributions $Q_1(1) = |e|\ \phi_{D*}(1)\phi_D(1)$ and $Q_2(2) = |e|\ \phi_A(2)\phi_{A*}(2)$ and may be expanded into multipole terms: dipole-dipole, dipole-quadrupole, etc. With $\boldsymbol{M}_D$ and $\boldsymbol{M}_A$ denoting the transition moments of the two molecules, and at not too small distances R_{AD} between the donor and the acceptor, the dipole term, which dominates for allowed transitions, may be written as

$$\beta^C(\text{dipole-dipole}) \sim \boldsymbol{M}_D\boldsymbol{M}_A/R_{AD}^3 \tag{5.32}$$

and is thus related to experimentally measurable quantities (Förster, 1951).

The exchange interaction, which is responsible for example for the singlet–triplet splitting, is a purely quantum mechanical phenomenon and does not depend on the oscillator strengths of the transitions involved. The exchange integral

$$\beta^E = \int\psi_{D*}(1)\psi_A(2)\frac{e^2}{r_{12}}\psi_D(2)\psi_{A*}(1)d\tau_1 d\tau_2 \tag{5.33}$$

* It should be noted that here ψ_{A*} and ψ_{D*} denote wave functions of the excited states of A and D, respectively, and not the complex conjugate of ψ_A and ψ_D.

representing the interaction of the charge densities $Q_1^E(1) = |e| \, \phi_{D^*}(1)\phi_{A^*}(1)$ and $Q_2^E(2) = |e| \, \phi_D(2)\phi_A(2)$ vanishes if the spin orbitals ψ_A and ψ_D or ψ_{A^*} and ψ_{D^*}, respectively, contain different spin functions. Since the charge densities Q_1^E and Q_2^E depend on the spatial overlap of the orbitals of D and A, the exchange interaction decreases exponentially with increasing internuclear distances, similarly as the overlap.

For forbidden donor and acceptor transitions the Coulomb term vanishes and the exchange term will predominate. If both transitions are allowed and the distance is not too small, the dipole–dipole interactions will prevail. The higher multipole terms are important only at very short distances, where, however, the essential contributions arise again from the exchange interaction, unless it vanishes due to the spin symmetry.

Time-dependent perturbation theory yields for the rate constant of nonradiative energy transfer

$$k_{D^* \to A} = \frac{2\pi}{\hbar} \beta^2 \varrho_E$$

(Cf. the Fermi golden rule, Section 5.2.3.) The density of states ϱ_E (number of states per unit energy interval) is related to the spectral overlap J, and using the relations for β given above the expressions derived by Förster (1951) and Dexter (1953) for the rate constant of energy transfer by the Coulomb and the exchange mechanism, respectively, may be written as

$$k_{ET}(\text{Coulomb}) \sim \frac{f_D f_A}{R_{DA}^6 \tilde{\nu}^2} J \tag{5.34}$$

and

$$k_{ET}(\text{exchange}) \sim e^{-2R_{DA}/L} J \tag{5.35}$$

f_D and f_A are the oscillator strengths of the donor and acceptor transition, respectively, L is a constant related to an effective average orbital radius of the electronic donor and acceptor states involved, and J is the spectral overlap.

5.4.5.3 The Coulomb Mechanism of Nonradiative Energy Transfer

Energy transfer according to the Coulomb mechanism, which is also referred to as the Förster mechanism, is based on classical dipole–dipole interactions. From Equation (5.32) the interaction energy is seen to be proportional to R_{AD}^{-3}, with R_{AD} being the donor–acceptor separation, and is significant at distances up to the order of 10 nm, which is large but less than the range of radiative energy transfer. Introduction of reasonable numerical values into the Förster equation for the rate constant [Equation (5.34)] leads to the expectation that k_{ET} can be much larger than the diffusion rate constant k_{diff}.

From Equation (5.31) it is seen that spin integration yields a nonvanishing Coulomb interaction only if there is no change in spin in either component. Thus

$$^1D^* + {}^1A \rightarrow {}^1D + {}^1A^*$$

and

$$^1D^* + {}^3A \rightarrow {}^1D + {}^3A^*$$

are fully allowed.

However, triplet–triplet energy transfer

$$^3D^* + {}^1A \rightarrow {}^1D + {}^3A^*$$

is forbidden. Nevertheless, it is sometimes observed, because the spin-selection rule, which strongly reduces the magnitude of k_{ET}, also prolongs the lifetime of $^3D^*$ to such an extent that the probability of energy transfer can still be high compared with the probability of deactivation of $^3D^*$ (Wilkinson, 1964).

5.4.5.4 The Exchange Mechanism of Nonradiative Energy Transfer

According to Equation (5.35), energy transfer by the exchange mechanism is a short-range phenomenon since the exchange term decreases exponentially with the donor–acceptor separation R_{AD}. Since it requires the intervention of an encounter complex ($D^* \cdots A$) it is also called the overlap or collision mechanism. The Wigner–Witmer spin-selection rules (Section 5.4.1) apply, and the spin-allowed processes

$$^1D^* + {}^1A \rightarrow {}^1D + {}^1A^*$$

and

$$^3D^* + {}^1A \rightarrow {}^1D + {}^3A^*$$

are referred to as singlet–singlet and triplet–triplet energy transfer, respectively. The singlet–singlet energy transfer is also allowed under the Coulomb mechanism, which, as a long-range process, in general predominates. Collisional singlet–singlet energy transfer is therefore likely to be rare and observable only under special conditions—for example, with biacetyl as a quencher, since this shows only a very weak absorption in the UV/VIS region (Dubois and Van Hemert, 1964), and when it is intramolecular (Hassoon et al., 1984).

Triplet–triplet energy transfer, on the other hand, is a very important type of energy transfer observed with solutions of sufficient concentration. It has been established as occurring over distances of 1–1.5 nm, comparable with collisional diameters. Steric effects were shown to be significant; for instance, the introduction of *gem*-dimethyl groups into the diene used as a quencher reduces the rate constant of fluorescence quenching of diazabicyclooctene (**17**) by a factor of 3–4. Apparently, it is important which regions of the molecules touch in the encounter complex (Day and Wright, 1969).

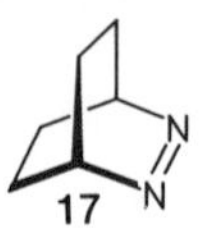

Intramolecular triplet–triplet energy transfer in compounds of the type D—Sp—A, where Sp is an appropriate spacer, such as *trans*-decalin or cyclohexane, has also been studied and a relation between charge transfer and energy transfer has been shown to exist. As depicted in Figure 5.28, electron transfer can be symbolized as electron exchange between the LUMOs of donor and acceptor, hole transfer as electron exchange between the HOMOs, and triplet–triplet energy transfer as a double electron exchange involving both the HOMOs and the LUMOs. This simplified view suggests that the probability of triplet–triplet transfer should be proportional to the product of the probabilities of hole transfer and electron transfer. Indeed, a remarkably good proportionality between rate constants k_{ET} of triplet–triplet energy transfer and the product $k_e k_h$ of rate constants of electron transfer and hole transfer has been observed experimentally (Closs et al., 1989). This proportionality has also been found in ab initio calculations indicating that the observed distance dependence of electron transfer, hole transfer, and triplet–triplet energy transfer is determined not only by the number of intervening CC bonds but also by an angular dependence of the through-bond coupling (Koga et al., 1993).

Triplet–triplet energy transfer is most important in photochemical reactions. It is utilized to specifically excite the triplet state of the reactant. This process is referred to as *photosensitization* and the donor $^3D^*$ is called a *triplet sensitizer* $^3Sens^*$. For efficient triplet sensitization the sensitizer must absorb substantially in the region of interest, its intersystem crossing effi-

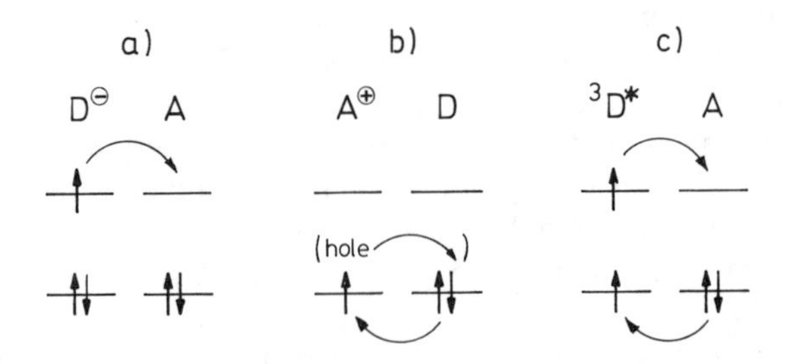

Figure 5.28. Frontier orbital representation of electron exchange in a) electron transfer, b) hole transfer, and c) triplet–triplet energy transfer (adapted from Closs et al., 1989).

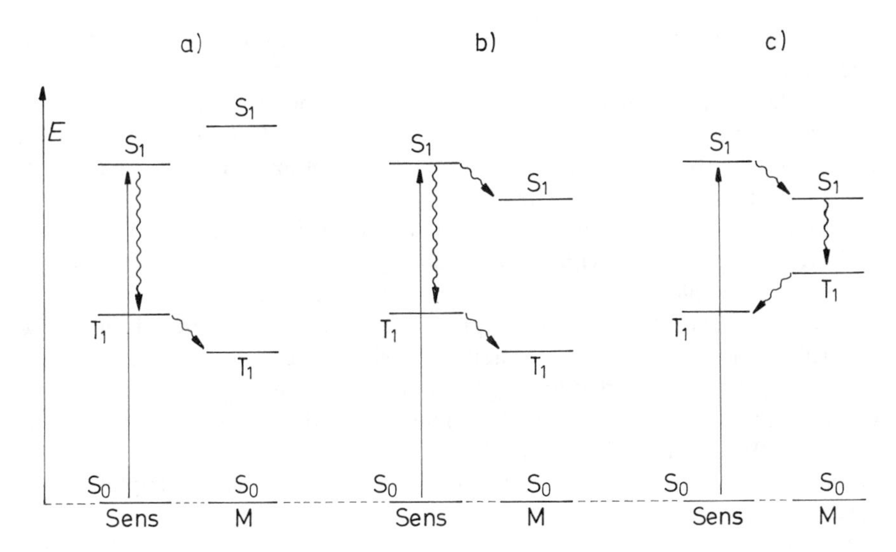

Figure 5.29. Three possible situations for luminescence quenching: a) only triplet excitation can be transferred from the sensitizer Sens to the molecule M (triplet quenching), b) singlet as well as triplet transfer is possible, and c) M can capture singlet excitation from Sens and transfer triplet excitation back to Sens.

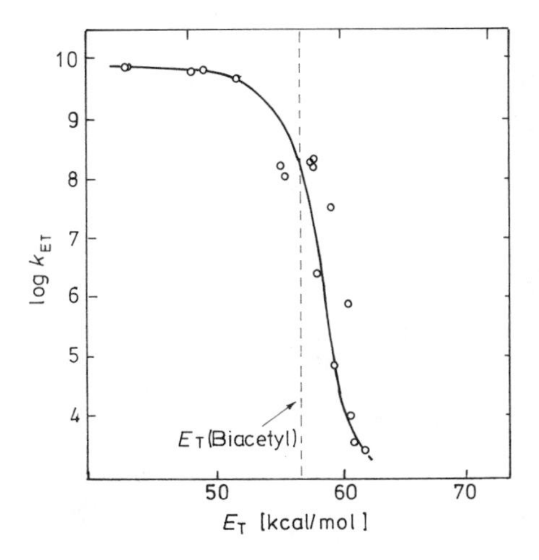

Figure 5.30. Triplet–triplet energy transfer from biacetyl in benzene to various acceptors. Rate constant k_{ET} as a function of the triplet energy E_T of the acceptor (by permission from Lamola, 1968).

ciency must be high, and its triplet energy must be higher than that of the acceptor.

The desired relationship of the energy levels of the sensitizer Sens and the acceptor molecule M is depicted in Figure 5.29a. If the energy levels are disposed as shown in Figure 5.29b, singlet–singlet and triplet–triplet energy transfer are both possible. The situation shown in Figure 5.29c, however, would allow singlet–singlet energy transfer from the sensitizer to the molecule M and subsequent triplet–triplet energy transfer back to the sensitizer.

When the triplet energy E_T of the acceptor is about 3.5 kcal/mol or more below that of the donor, triplet–triplet energy transfer is generally diffusion controlled, that is, $k_{ET} = k_{diff}$. When both triplet energies are the same, then only the 0–0 bands overlap and k_{ET} is smaller by a factor of 10^2; it decreases even further with increasing E_T. This is evident from the results of Lamola (1968) shown in Figure 5.30, where the rate constant k_{ET} of the triplet–triplet energy transfer of biacetyl is plotted against the triplet energy E_T of various acceptors.

Example 5.10:

Terenin and Ermolaev (1956) were the first to observe triplet–triplet energy transfer by measuring the benzophenone-sensitized phosphorescence of naphthalene in a rigid glassy solution at 77 K. The relative energies of the lowest singlet and triplet states of these molecules are shown in Figure 5.31. Excitation of the $^1(\pi,\pi^*)$ state and subsequent intersystem crossing produce the T_1 state of benzophenone, which may transfer the excitation energy to the T_1 state

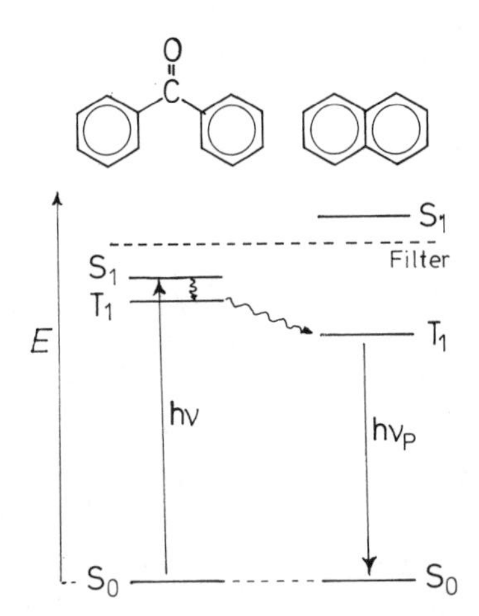

Figure 5.31. Relative energies of the lowest singlet and triplet states of benzophenone and naphthalene, and triplet sensitization of naphthalene; direct excitation of the higher-lying naphthalene S_1 state is prevented by a filter.

of naphthalene. The use of a filter prevented direct population of the T_1 state of naphthalene via excitation of its S_1 state followed by intersystem crossing.

5.4.5.5 *Triplet–Triplet Annihilation*

Energy transfer can also occur between two molecules in their excited states; this phenomenon is most common for two triplet states due to their relatively long lifetimes. According to the Wigner–Witmer spin-selection rules the following triplet–triplet annihilation processes are allowed by the exchange mechanism:

$$^3D^* + {}^3A^* \rightarrow {}^1D + {}^1A^{**}$$
$$^3D^* + {}^3A^* \rightarrow {}^1D + {}^3A^{**}$$
$$^3D^* + {}^3A^* \rightarrow {}^1D + {}^5A^{**}$$

where the double asterisk is used to denote higher excited states. The first process is most important and for most organic molecules the combined triplet energy is sufficient to excite one of them into an excited singlet state. The donor and the acceptor molecules are identical in many cases, and, as shown for the $(H_2 + H_2)$ system in Section 4.4, the acceptor molecule will undergo internal conversion and finally reach the lowest excited singlet state, so triplet–triplet annihilation may be summarized by the equation

$$^3M^* + {}^3M^* \rightarrow {}^1M^* + M$$

If $^1M^*$ is fluorescent, triplet–triplet annihilation produces *delayed fluorescence,* reflecting the long lifetime of $^3M^*$ (Parker, 1964). This phenomenon was first studied for pyrene, and was therefore dubbed "P-type delayed fluorescence" in contrast to the "E-type delayed fluorescence" discussed in Section 5.1.1.

Example 5.11:
Triplet–triplet annihilation (TTA) between different molecules A and X (hetero-TTA) may produce the excited singlet state of either A or X. This reaction is utilized in a general and synthetically useful method of generating singlet oxygen according to

$$\text{Sens} + h\nu \rightarrow {}^1\text{Sens}^* \rightarrow {}^3\text{Sens}^*$$
$$^3\text{Sens}^* + {}^3O_2 \rightarrow \text{Sens} + {}^1O_2$$

Strongly absorbing dyes such as Rose Bengal or methylene blue are usually used as photosensitizers. (Cf. Section 7.6.3.)

Singlet oxygen generation is used in photodynamic tumor therapy—for instance, with porphyrins as photosensitizers. (Cf. Dougherty, 1992.) The disadvantage of natural porphyrins such as hematoporphyrin is that they are of low chemical stability and have an absorption spectrum similar to that of hemoglobin. Therefore, expanded porphyrins such as [22]coproporphyrin II tetramethylester (**18**) (Beckman et al., 1990) and octaethyl[22]porphyrin(2.2.2.2) (**19**) (Vogel et al., 1990) have been synthesized.

MeCO2 CO2Me NH HN N MeCO2 CO2Me

18

19

Delayed fluorescence from a very-short-lived upper excited singlet state S_k^A populated by hetero-TTA has been observed for the first time using the system A = anthracene and X = xanthone (Nickel and Roden, 1982). An energy-level diagram for this system is shown in Figure 5.32, and the corrected spectrum of the delayed fluorescence of anthracene and xanthone in trichlorotrifluoroethane is depicted in Figure 5.33. The band at 36,000–40,000 cm^{-1} has been assigned to the $S_6^A \rightarrow S_0^A$ delayed fluorescence of anthracene produced by T_1^A +

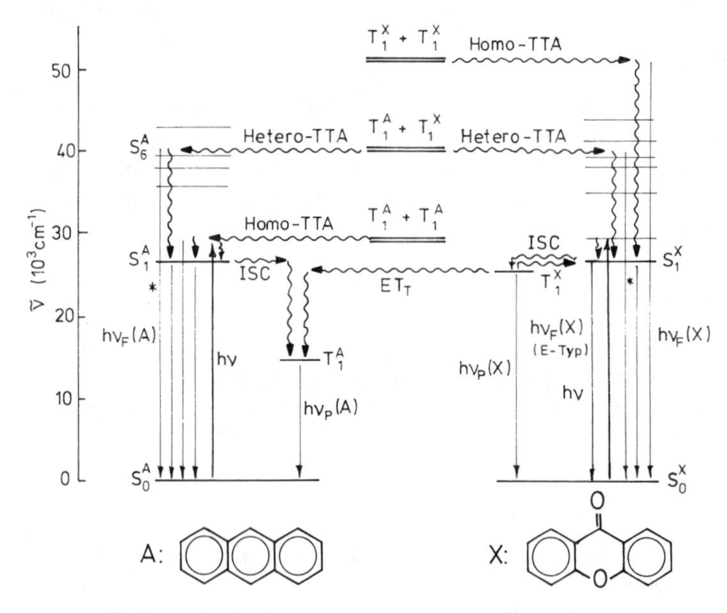

Figure 5.32. Energy-level diagram of anthracene (A) and xanthone (X). Double lines denote the three different triplet pairs for which TTA processes are indicated; asterisks mark the delayed fluorescence (DF) resulting from hetero-TTA (by permission from Nickel and Roden, 1982).

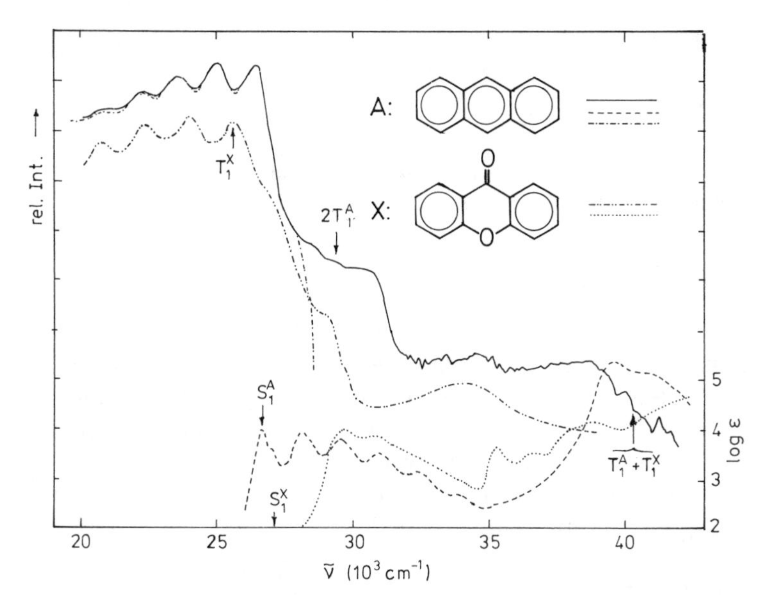

Figure 5.33. Corrected delayed fluorescence spectrum (——) of a solution of anthracene (A, [A] ≈ 7×10^{-6} mol/L) and xanthone (X, [X] ≈ 2×10^{-5} mol/L) in 1,1,2-trichlorotrifluoroethane at 243 K. For comparison spectra of delayed fluorescence of A alone (-·-·-) and of X alone (-··-··-) as well as absorption spectra of A (------) and X(······) are also shown (by permission from Nickel and Roden, 1982).

T_1^X. It is almost exactly a mirror image of the $S_0^A \rightarrow S_6^A$ absorption band in the anthracene spectrum.

5.4.6 Kinetics of Bimolecular Photophysical Processes

In the absence of photochemical reactions an excited-state molecule M* can be deactivated by emission, by radiationless decay, or by quenching:

$$\mathrm{M} \xrightarrow[I_{abs}]{h\nu} \mathrm{M}^* \begin{cases} \xrightarrow{k_F} \mathrm{M} + h\nu \\ \xrightarrow{k_D} \mathrm{M} \\ \xrightarrow[\mathrm{Q}]{k_q[\mathrm{Q}]} \mathrm{M} + \mathrm{Q} \end{cases}$$

The quantum yield of fluorescence is then

$$\Phi_F^Q = \frac{k_F}{k_F + k_D + k_q[Q]} \tag{5.36}$$

while in the absence of quencher, one has from Equation (5.10)

$$\Phi_F = \frac{k_F}{k_F + k_D}$$

Hence, the ratio Φ_F/Φ_F^Q of the quantum yield of fluorescence without quencher to that with quencher is given by

$$\frac{\Phi_F}{\Phi_F^Q} = \frac{k_F + k_D + k_q[Q]}{k_F + k_D} = 1 + \frac{k_q[Q]}{k_F + k_D} = 1 + \tau_F k_q[Q] \quad (5.37)$$

where τ_F is the excited-state lifetime in the absence of quencher. Equation (5.37) is known as the *Stern-Volmer equation* (Stern and Volmer, 1919). If Φ_F/Φ_F^Q is plotted against the quencher concentration [Q] a straight line of slope $k_q\tau$ will result. Hence if τ is known, the quenching rate constant k_q is immediately obtained, and if k_q is known τ may be determined.

Because the direct measurement of lifetimes is easily performed nowadays, it is also useful to transform the Stern–Volmer relation into the form

$$\frac{\tau_F}{\tau_F^Q} = 1 + \tau_F k_q[Q] \quad (5.38)$$

which is obtained from the expression for τ_F given in Table 5.2 and from the corresponding expression that takes into account bimolecular quenching.

For many systems k_q is of the order 10^{10} L mol^{-1} s^{-1}, that is, close to the diffusion-controlled rate constant k_{diff}. This suggests that in these cases quenching is so rapid that the rate-determining step is the actual diffusion of the molecules to form an encounter complex. Thus,

$$M^* + Q \underset{k_{-1}}{\overset{k_{diff}}{\rightleftharpoons}} (M^* \cdots Q) \xrightarrow{k_q} M + Q$$

where k_q is the actual quenching rate constant within the complex. Under steady-state conditions

$$k_{diff}\,[M^*][Q] = [(M^* \cdots Q)](k_q + k_{-1})$$

and from

$$k_q\,[(M^* \cdots Q)] = \frac{k_q k_{diff}}{k_q + k_{-1}}\,[M^*][Q] = k_q(\text{obs})[M^*][Q]$$

the observed quenching rate constant

$$k_q(\text{obs}) = \frac{k_q k_{diff}}{k_q + k_{-1}}$$

is obtained.

If $k_q \gg k_{-1}$, then $k_q(\text{obs}) \approx k_{diff}$ and the observed quenching rate that is equal to the diffusion rate will be dependent on solvent viscosity.

If $k_q \ll k_{-1}$, then $k_q(\text{obs}) \approx k_q k_{\text{diff}}/k_{-1} = k_q K$, where K is the equilibrium constant for the formation of the complex; k_q(obs) will be independent of solvent viscosity.

Finally, if k_q and k_{-1} are of the same order, then k_q(obs) will be less than k_{diff}.

Quenching of two excited states, one of which is inaccessible to the quencher, will yield a Stern–Volmer plot that is not a straight line but rather a curve bent toward the x axis. An upward curvature, concave toward the y axis as shown in Figure 5.36, is observed if *combined dynamic and static quenching* occurs. Static quenching results from the formation of a nonfluorescent ground-state complex between fluorophore and quencher or from the statistical presence of the quencher next to the excited molecule (Sun et al., 1992a). In the case of purely static quenching the Stern–Volmer plot yields a straight line with a slope equal to the equilibrium constant $K = [\text{MQ}]/[\text{M}][\text{Q}]$, whereas $\tau_F/\tau_F^Q = 1$, since the lifetime of the uncomplexed molecules is unperturbed by the presence of the quencher.

Example 5.12:
Figure 5.34 shows the Stern–Volmer plot for triplet quenching of the photochemical addition of benzaldehyde to 2,3-dimethyl-2-butene (cf. Section 7.4.3) by piperylene. The ratio of the quantum yield Φ of oxetane formation without quencher to that with quencher is plotted against the quencher concentration [Q]. From the resulting straight line it may be concluded that the reaction proceeds via a single reactive state which is assigned as $^3(n,\pi^*)$.

In the case of diffusion-controlled quenching, k_q should depend on solvent viscosity in the same way as k_{diff}, that is, $k_q = \alpha T/\eta$. A plot of k_q(obs) for triplet

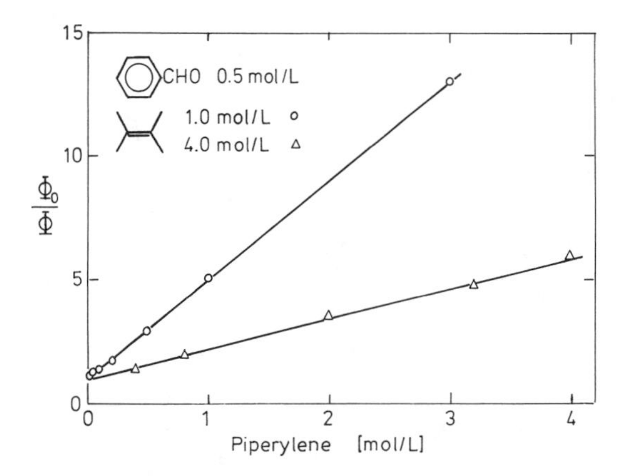

Figure 5.34. Stern-Volmer plot for triplet quenching of the oxetane formation from benzaldehyde and 2,3-dimethyl-2-butene by piperylene for different olefin concentrations (by permission from Yang et al., 1967).

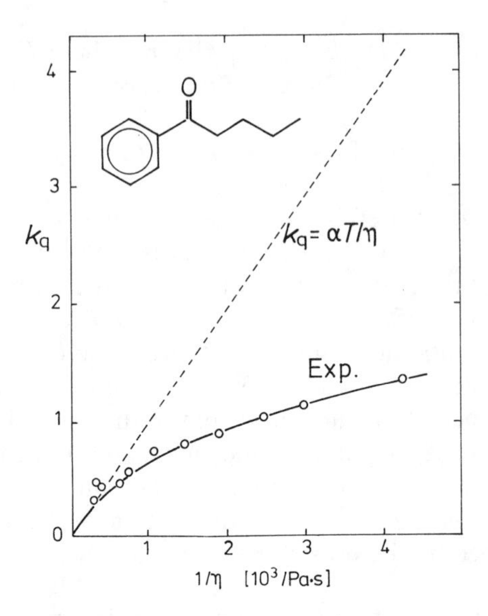

Figure 5.35. Comparison of the theoretical and the observed relation between k_q and $1/\eta$ for triplet quenching of valerophenone by 2,5-dimethyl-2,4-hexadiene (adapted from Wagner and Kochevar, 1968).

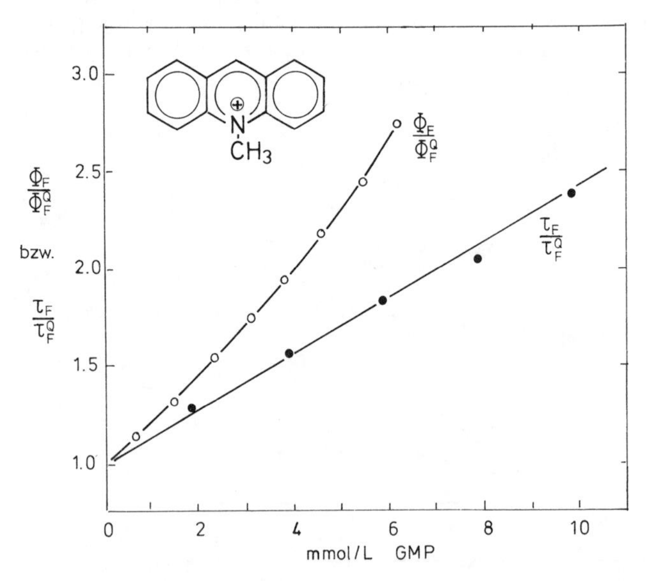

Figure 5.36. Fluorescence quenching of 10-methylacridinium chloride by guanosine-5′-monophosphate and separation of the static and dynamic quenching contributions (by permission from Kubota et al., 1979).

quenching of valerophenone by 2,5-dimethyl-2,4-hexadiene against $1/\eta$ does not give a straight line (Figure 5.35), so in this case energy transfer cannot be solely diffusion-controlled; that is, not every encounter complex between an excited donor and an acceptor molecule results in energy transfer.

The Stern–Volmer plot of fluorescence quantum yields for the quenching of 10-methylacridinium chloride by the nucleotide guanosine-5′-monophosphate shown in Figure 5.36 provides an excellent example in which both static and dynamic quenching occur. Since lifetime measurements were available, the contribution of static quenching could be separated off; the Stern–Volmer plot of the purely dynamic quenching was obtained using τ_F/τ_F^Q.

5.5 Environmental Effects

For a deeper understanding of photophysical processes it is sometimes very useful to study the influence of the environment. The results of some such studies will be briefly mentioned in the following sections.

5.5.1 Photophysical Processes in Gases and in Condensed Phases

In solution vibrational relaxation is extraordinarily fast, and usually, thermal equilibrium is reached before emission or other photophysical or photochemical processes can come into play. In low-pressure vapors, however, time intervals between collisions will be large enough for other processes to be able to compete with vibrational relaxation. "Hot" reactions in excited states and particularly in the ground state, with the molecule excited to a

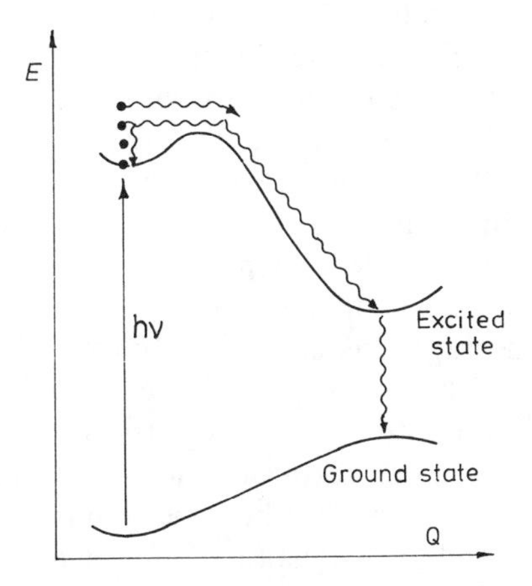

Figure 5.37. Schematic representation of the potential energy surfaces of "hot" excited-state reactions.

higher vibrational state than would correspond to thermal equilibrium, will become important. The excess excitation energy may be sufficient to overcome barriers to chemical reactions and to enable the molecule to undergo processes not accessible in thermal equilibrium. This situation is illustrated in Figure 5.37 by means of schematic potential energy surfaces. It is easy to see from this representation that hot excited-state reactions are dependent on the excitation wavelength; a reaction is possible only if excitation is into a vibrational level higher in energy than the barrier. The conversion of benzene into benzvalene discussed in Section 5.3.1 is an example of such a wavelength-dependent photoreaction.

5.5.2 Temperature Dependence of Photophysical Processes

Photochemical reactions competing with photophysical processes often require thermal activation in order to cross small barriers in the S_1 or T_1 state. (See Section 6.1.) In general, such reactions may be suppressed at low temperatures and photophysical processes would then be favored. This is by far the most important temperature effect on photophysical processes although the following effects should also be considered.

In discussing photophysical processes conformers can be considered as different species since in the excited state the conversion of one conformer into another is frequently slower than many of the competing processes. (Cf. Example 7.13.) Also, conformational changes are often much slower in the excited state than in the ground state, because formal single bonds of conjugated systems can show higher π-bond orders in the excited state than in the ground state. This is true, for instance, for polyenes. The conformational distribution in the ground state is temperature dependent, and rates of photophysical or photochemical processes characteristic for a particular conformer will become temperature dependent as well. There is, however, also evidence for just the opposite case: in the (σ,σ^*) excited state of certain oligosilanes, conformational changes are faster than in the ground state and are driven by large differences in the energies of conformers that are of nearly equal energy in the ground state (Sun et al., 1992b).

Another type of temperature dependence that may be thought of as competition between temperature-dependent and temperature-independent processes has been observed for substituted anthracenes. It is due to the fact that the T_2 state of anthracene is energetically very close to the lowest excited singlet state S_1. When T_2 is just above S_1 the intersystem crossing $S_1 \rightsquigarrow T_2$ will be associated with a barrier E_a. The rate constant of this transition can be expressed in the form of a normal Arrhenius equation $k = Ae^{-E_a/kT}$ and the intersystem crossing becomes temperature dependent. This situation is found in 9- and 9,10-substituted anthracenes. (Cf. Example 5.3.) The fluorescence quantum yield Φ_F increases strongly with decreasing temperature. In other substituted anthracenes as well as in anthracene itself T_2

is just below S_1; intersystem crossing is facile and without a barrier, resulting in the quantum yield of fluorescence being lower and temperature independent.

A case in which the fluorescence rate constant increases with temperature has been described recently (Van Der Auweraer et al., 1991). The fluorescence from TICT (twisted internal charge transfer) states of donor–acceptor pairs that may be considered as strongly heterosymmetric biradicaloids (see Section 4.3.3) is strongly forbidden since the donor and acceptor orbitals avoid each other in space nearly perfectly at the orthogonally twisted geometry and the overlap density is nearly exactly zero. Vibrational activation populates levels in which the twist angle deviates from 90° and that carry larger transition moments to the ground state. A similar observation was made for highly polar intramolecular exciplexes.

5.5.3 Solvent Effects

Polar solvents and in particular hydrogen bonds are able to stabilize polar molecules. This is also true for the excited states of a molecule that may have quite different polarities, and solvent variation may possibly change the energy order of the excited states as well as transition moments. (Cf. Section 2.7.2.) As an example we recall quinoline, whose solvent-dependent fluorescence quantum yield has been discussed in Section 5.3.2.

Example 5.13:
p-Dimethylaminobenzonitrile (**20**), which has been studied by Lippert (1969), is a classical example of the solvent effect on the ordering of excited states.

$(CH_3)_2N-C_6H_4-CN$

20

Whereas the ground state and locally excited benzene states are of lower polarity, there is another excited state of **20** consisting of an NR_2 group as donor and the C_6H_4CN group as acceptor that has a dipole moment $\mu = 12$ D and that can be described as a TICT state with intramolecular charge transfer (Grabowski and Dobkowski, 1983) stabilized by twisting around the single bond. (Cf. Section 4.3.3.) In nonpolar solvents fluorescence from a locally excited state is observed. In polar solvents, however, the TICT state is stabilized to such an extent that it becomes the lowest excited state, and fluorescence from this state is observed. By means of picosecond spectroscopy it has been shown that in nonpolar solvents, relaxed fluorescence sets in 7–20 ps after the exciting flash due to vibrational relaxation, whereas relaxed fluorescence from the TICT state is characterized by an appreciably longer inhibition time of $\approx$ 40 ps associated with a geometry change and reorientation of the polar solvent molecules in the field of the large excited-state dipole moment (Struve et al., 1973).

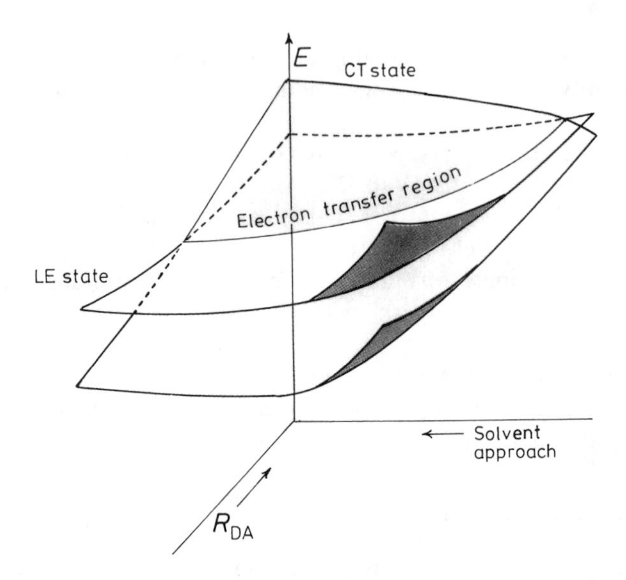

Figure 5.38. Dependence of the light-induced electron transfer on donor–acceptor separation R and upon solvation characterized by the distance between solvent molecules and donor–acceptor system (by permission from Ramunni and Salem, 1976).

Solvent effects may be discussed in a particularly illuminating way in the case of electron-transfer reactions. Figure 5.38 gives a schematic three-dimensional representation of the potential energy surfaces responsible for photoinduced electron transfer and their dependence upon donor–acceptor separation and upon the approach of solvent molecules (Ramunni and Salem, 1976). The diagram is based on model calculations for a system consisting of NH_3 as the donor, cyanoethylene as the acceptor, and two H_2O molecules. These molecules were arranged in such a way that the locally

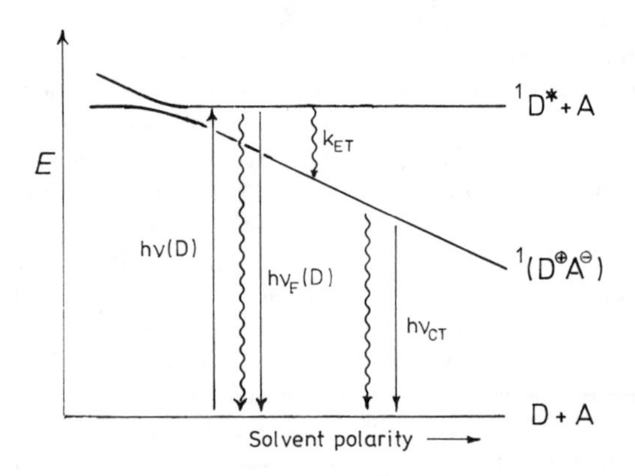

Figure 5.39. Schematic representation of the fluorescence of donor–acceptor systems as a function of solvent polarity (by permission from Pasman et al., 1985).

excited state and the CT state have different symmetries so that the crossing of the potential energy surfaces is not avoided.

If variations in the distance between H_2O molecules and the donor–acceptor system are taken to be characteristic for solvation changes and hence for variations in solvent polarity, the schematic diagram of Figure 5.39 is obtained which shows the solvent dependence of the fluorescence of donor–acceptor pairs. According to this diagram the energy of the exciplex fluorescence decreases with increasing solvent polarity.

Example 5.14:
Using rigid bichromophoric systems with donor and acceptor components sterically fixed by spacers, the dependence of electron transfer on the geometry as well as on solvation has been studied. According to Figure 5.40 exciplex fluorescence to be expected from the diagram in Figure 5.39 can be observed for system **22a** in polar solvents, whereas electron transfer does not occur in solvents of low polarity, and only fluorescence of the donor component is seen,

X Y   X Y

CH_3O **21** CH_3O **22** CH_3O **23**

a: X = H, Y = $COOC_2H_5$ **a**: X = COOMe, Y = H
b: X = Y = CN   **b**: X = Y = CN

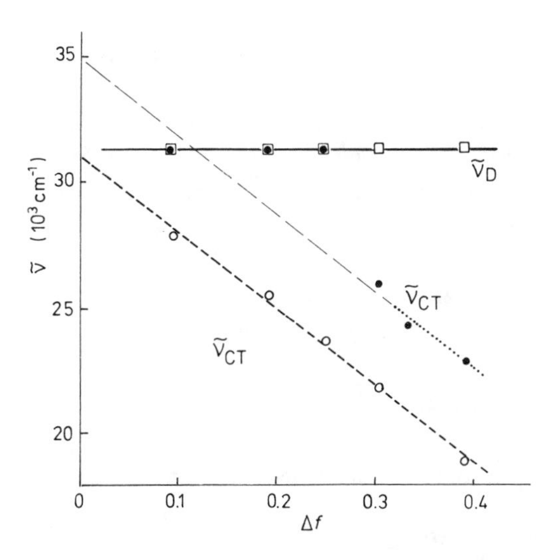

Figure 5.40. Fluorescence maxima of compounds 22a (●), 22b (○), and 23 (□) as a function of solvent polarity (by permission from Pasman et al., 1985).

and this agrees with the fluorescence from compound **23**. For larger donor–acceptor separations, as in **21a**, only donor fluorescence independent of solvent polarity is observed.

Irrespective of solvent polarity and donor–acceptor separation, systems **21b** and **22b**, with inherently better acceptor moieties, undergo extremely fast electron transfer ($k_e > 10^{11}$ s^{-1}) and show the expected exciplex fluorescence (Figure 5.40). Solvent reorganization is believed to provide a significant barrier for long-range electron transfer from the donor to the weaker but not to the stronger acceptor (Pasman et al., 1985). A relation has been derived that provides useful design criteria for systems to display optimally fast charge separation across a given distance that is also virtually independent of both medium effects and temperature (Kroon et al., 1991). From this relation it is predicted that a D—bridge—A system which is barrierless in a nonpolar solvent like *n*-hexane will stay barrierless in any other solvent. If it is not barrierless in *n*-hexane it will never become so whatever more polar solvent is used.

Other possible solvent effects depend on the heavy-atom effect, which may favor intersystem crossing, and on the viscosity, which may change the rate of diffusion and hence the collisional triplet–triplet energy transfer. These effects have already been mentioned in previous sections.

Supplemental Reading

Barltrop, D.A., Coyle, J.D. (1975), *Excited States in Organic Chemistry*; Wiley: London.

Birks, J.B. (1970), *Photophysics of Aromatic Molecules*; Wiley: New York.

Birks, J.B., Ed. (1973/75), *Organic Molecular Photophysics,* Vol. **1–2**, Wiley: New York.

Gilbert, A., Baggot, J. (1991), *Essentials of Molecular Photochemistry*; CRC Press: Boca Raton.

Lamola, A.A. (1971–76), *Creation and Detection of Excited States,* Vol. **1–4**; Marcel Dekker: New York.

Rabek, J.F. (1982–91), *Experimental Methods in Photochemistry and Photophysics,* Part 1 and 2, *Photochemistry and Photophysics;* CRC Press: Boca Raton, FL.

Simons, J.P. (1971), *Photochemistry and Spectroscopy*; Wiley: London.

Turro, N.J. (1978), *Modern Molecular Photochemistry*; Benjamin: Menlo Park.

Time-resolved Laser Spectroscopy

Fleming, G.R. (1986), *Chemical Applications of Ultrafast Spectroscopy*; Oxford University Press: Oxford.

Eisenthal, K.B., Ed. (1983), *Applications of Picosecond Spectroscopy to Chemistry,* NATO ASI Series; Reidel: Dordrecht.

Rentzepis, P.M. (1982), "Advances in Picosecond Spectroscopy," *Science* **218**, 1183.

Radiationless Deactivation

Avouris, P., Gelbart, W.M., El-Sayed, M.A. (1977), "Nonradiative Electronic Relaxation under Collision-Free Conditions," *Chem. Rev.* **77**, 793.

Caldwell, R.A. (1984), "Intersystem Crossing in Organic Photochemical Intermediates," *Pure Appl. Chem.* **56**, 1167.

Freed, K.F. (1978), "Radiationless Transitions in Molecules," *Acc. Chem. Res.* **11**, 74.

Robinson, G.W., Frosch, R.P. (1962), "Theory of Electronic Energy Relaxation in the Solid Phase," *J. Chem. Phys.* **37**, 1962.

Robinson, G.W., Frosch, R.P. (1963), "Electronic Excitation Transfer and Relaxation," *J. Chem. Phys.* **38**, 1187.

Siebrand, W. (1967), "Radiationless Transitions in Polyatomic Molecules," *J. Chem. Phys.* **46**; 440; **47**, 2411.

Teller, E. (1960), "Internal Conversion in Polyatomic Molecules," *Israel J. Chem.* **7**, 227.

Luminescence

Lakowicz, J.R. (1983), *Principles of Fluorescence Spectroscopy*; Plenum Press: New York, London.

Parker, C.A. (1969), *Photoluminescence in Solution*; Elsevier: London.

Turro, N.J., Liu, K.-C., Chow, M.-F., Lee, P. (1978), "Convenient and Simple Methods for the Observation of Phosphorescence in Fluid Solutions. Internal and External Heavy Atom and Micellar Effects," *J. Photochem. Photobiol.* **27**, 523.

Polarization

Dörr, F., Held, M. (1960), "Ultraviolet Spectroscopy using Polarized Light," *Angew. Chem.* **72**, 287.

Michl, J., Thulstrup, E.W. (1986), *Spectroscopy with Polarized Light*; VCH Inc.: New York.

Thulstrup, E.W., Michl, J. (1989), *Elementary Polarization Spectroscopy*; VCH Inc.: New York.

Excimers and Exciplexes

Beens; H., Weller, A. (1975), in *Organic Molecular Photophysics,* Vol. **2**, Birks, J.B., Ed.; Wiley: London.

Caldwell, R.A., Creed, D. (1980), "Exciplex Intermediates in [2 + 2] Photocycloadditions," *Acc. Chem. Res.* **13**, 45.

Davidson, R.S. (1983), "The Chemistry of Excited Complexes: a Survey of Reactions," *Adv. Phys. Org. Chem.* **19**, 1.

Förster, Th. (1969), "Excimers," *Angew. Chem. Int. Ed. Engl.* **8**, 333.

Gordon, M., Ware, W.R., Eds. (1975), *The Exciplex*; Academic Press: New York.

Lim, E.C. (1987), "Molecular Triplet Excimers," *Acc. Chem. Res.* **20**, 8.

Weller, A. (1982), "Exciplex and Radical Pairs in Photochemical Electron Transfer," *Pure Appl. Chem.* **54**, 1885.

Electron Transfer

Closs, G.L., Calcaterra, L.T., Green, N.J., Penfield, K.W., Miller, J.R. (1986), "Distance, Stereoelectronic Effects, and the Marcus Inverted Region in Intramolecular Electron Transfer in Organic Radical Anions," *J. Phys. Chem.* **90**, 3673.

Eberson, L. (1987), *Electron Transfer Reactions in Organic Chemistry*; Springer: Berlin, Heidelberg, New York.

Fox, M.A., Ed. (1992), "Electron Transfer: A Critical Link between Subdisciplines in Chemistry," thematic issue of *Chem. Rev.* **92**, 365–490.

Fox, M.A., Chanon, M., Eds. (1988), *Photoinduced Electron Transfer,* Vol. **1–4**, Elsevier: Amsterdam.

Gould, I.R., Farid, S. (1988), "Specific Deuterium Isotope Effects on the Rates of Electron Transfer within Geminate Radical-Ion Pairs," *J. Am. Chem. Soc.* **110**, 7883.

Kavarnos, G.J. (1993), *Fundamentals of Photoinduced Electron Transfer;* VCH Inc.: New York.

Kavarnos, G.J., Turro, N.J. (1986), "Photosensitization by Reversible Electron Transfer: Theories, Experimental Evidence, and Examples," *Chem. Rev.* **86**, 401.

Marcus, R.A., Sutin, N. (1985), "Electron Transfers in Chemistry and Biology," *Biochim. Biophys. Acta* **811**, 265.

Mattay, J., Ed. (1990–92), "Photoinduced Electron Transfer, Vol. I–IV," *Topics of Current Chemistry,* **156**, **158**, **159**, **163**; Springer: Heidelberg.

Salem, L. (1982), *Electrons in Chemical Reactions: First Principles*; Wiley: New York.

Weller, A. (1968), "Electron Transfer and Complex Formation in the Excited State," *Pure Appl. Chem.* **16**, 115.

Energy Transfer

Closs, G.L., Johnson, M.D., Miller, J.R., Piotrowiak, P. (1989), "A Connection between Intramolecular Long-Range Electron, Hole, and Triplet Energy Transfers," *J. Am. Chem. Soc.* **111**, 3751.

Lamola, A.A. (1969), "Electronic Energy Transfer in Solution: Theory and Applications" in *Techniques of Organic Chemistry* **14**; Weisberger, A., Ed.; Wiley: New York.

Turro, N.J. (1977), "Energy Transfer Processes," *Pure Appl. Chem.* **49**, 405.

Yardley, J.T. (1980), *Introduction to Molecular Energy Transfer*; Academic Press: New York.

CHAPTER

6

Photochemical Reaction Models

Ground-state reactions are easily modeled using the absolute reaction-rate theory and the concept of the activated complex. The reacting system, which may consist of one or several molecules, is represented by a point on a potential energy surface. The passage of this point from one minimum to another minimum on the ground-state surface then describes a ground-state reaction, and the saddle points between the minima correspond to the activated complexes or transition states.

For a theoretical discussion of photochemical reactions at least two potential energy surfaces are required: the ground-state surface with reactant and product minima as well as the excited-state surface on which the photoreaction is launched. Furthermore, the theoretical model must be capable of describing what happens between the time of light absorption by a molecule in its electronic ground state and the appearance of the product molecule, also in its electronic ground state, and in thermal equilibrium with its surroundings.

6.1 A Qualitative Physical Model for Photochemical Reactions in Solution

A starting point for the discussion of experimental results in mechanistic photochemistry is the knowledge of the shapes of the ground-state (S_0) and first excited-state (S_1) singlet surfaces and the lowest triplet-state surface (T_1). Three steps can then be distinguished:

- First, minima and funnels in S_1 and T_1 have to be located.
- Then, it must be estimated which minima (funnels) are accessible, given the reaction conditions, and which ones will actually be populated with significant probabilities.
- Finally, from the shape of the ground-state surface S_0 it must be determined what the products of return from these important minima (funnels) in S_1 and T_1 will be.

This simple model ignores molecular dynamics problems as well as information on additional excited states, density of vibrational levels, vibronic coupling matrix elements, etc., which are required in more sophisticated advanced applications.

6.1.1 Electronic Excitation and Photophysical Processes

According to the Franck-Condon principle, light absorption is a "vertical" process. Consequently, immediately after brief excitation with broad-band light the geometry of the system is virtually identical with that just before excitation. Subsequently, however, the motions of the nuclei are suddenly governed by a new potential energy surface, so the geometry of the system will change in time. In solution, the surrounding medium will act as a heat bath and efficiently remove excess vibrational energy. In a very short time, on the order of a few (5–50) picoseconds, thermal equilibrium will be established and the molecule will be sitting in one or another of the numerous minima in the excited-state surface. (Cf. Section 4.1.3.) If, due to the initial kinetic energy of the nuclei, the molecule can cross barriers before reaching this minimum, the process is referred to as a reaction of "hot molecules in an excited state" or simply as a *hot excited-state reaction*. (See Section 6.1.3.)

If the initial excitation was not into the lowest excited state of given multiplicity, a fast crossing to that state will occur via internal conversion, that is, typically to the S_1 or T_1 state (Kasha's rule, cf. Section 5.2.1). In some cases a funnel in S_1 is accessible and internal conversion from S_1 to S_0 can be so fast that the first thermal equilibration of the vibrational motion in these molecules is achieved in a minimum in the S_0 state. Such a process is referred to as a *direct reaction*. Here as well, the excess kinetic energy of the nuclei may take the molecule over barriers in the S_0 state into valleys other than the one originally reached; analogous to the above-mentioned reactions, such processes are referred to as *hot ground-state reactions*.

Finally, in the presence of heavy atoms or in other special situations (cf. Section 4.3.4) intersystem crossing may proceed so fast that it is able to compete with vibrational relaxation. The first thermally equilibrated species formed may then be found in a minimum on the T_1 surface, even if the initial excitation was into S_n. T_1 may also be reached via sensitization, that is, by excitation energy transfer from another molecule in the triplet state. (Cf. Section 5.4.5.)

In one way or another, picoseconds after the initial excitation, the molecule will typically find itself thermally equilibrated with the surrounding medium in a local minimum on the S_1, T_1, or S_0 surfaces: in S_1 if the initial excitation was by light absorption, in T_1 if it was by sensitization or if special structural features such as heavy atoms were present, and in S_0 if the reaction was direct. Frequently, the initially reached minimum in S_1 (or T_1) is a spectroscopic minimum located at a geometry that is close to the equilibrium geometry of the original ground-state species, so no net chemical reaction can be said to have taken place so far.

Slower processes are the next to come into play. The most important among these are: First, thermally activated motion from the originally reached minimum over relatively small barriers to other minima or funnels, representing the adiabatic photochemical reaction proper. Second, intersystem crossing that takes the molecule from the singlet to the triplet manifold and thus eventually to a new minimum in T_1. Third, fluorescence or phosphorescence that return the molecule to the ground-state surface S_0. Fourth, radiationless conversion from S_1 or T_1 to S_0, which is usually not competitive unless the energy gap $\Delta E(S_1 - S_0)$ or $\Delta E(T_1 - S_0)$ between the S_1 or T_1 and S_0 states in the region of the minimum is small. (Cf. Section 5.2.3.) In this case, a "hot" molecule is produced, and a hot ground-state reaction can take place before removal of the excess energy. Other processes are also possible. These are, for example, further photon absorption (cf. Example 6.1), or either excitation or deexcitation via energy transfer, such as triplet–triplet annihilation or quenching. (Cf. Section 5.4.) A complex reaction mechanism that consists either of a two- or three-quantum process, depending on the temperature, will be discussed in Example 7.18.

Example 6.1:

The photochemical electrocyclic transformation of the cyclobutene derivative **1** to pleiadene (**2**) does not proceed from the thermalized S_1 and T_1 states. At room temperature, the reaction cannot compete with fluorescence or intersystem crossing (ISC) due to high intervening barriers. However, these can be overcome by excitation to higher states, populated starting from S_0 either by excitation with a single photon of sufficiently high energy ($\lambda < 214$ nm) or in a rigid glass at 77 K by subsequent two-photon excitation. In the latter case UV absorption followed by intersystem crossing yields the long-lived T_1 state. This intermediate can absorb visible light corresponding to a $T_1 \rightarrow T_n$ transition (Castellan et al., 1978). The situation is illustrated in Figure 6.1. Experimentally, it was not possible to distinguish between a reaction from one of the higher ex-

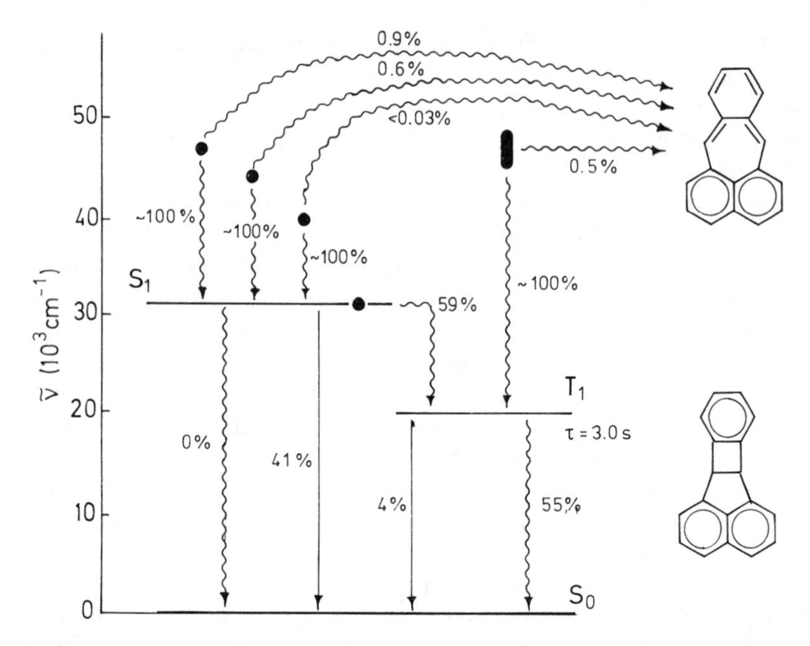

Figure 6.1. Photophysical processes and photochemical conversion of 6b,10b-dihydrobenzo[3,4]cyclobut[1,2-*a*]acenaphthylene. The numbers given are quantum yields of processes starting at levels indicated by black dots (by permission from Castellan et al., 1978).

cited S_n or T_n states and a reaction of vibrationally "hot" molecules in one of the lower excited states, possibly even S_1 or T_1.

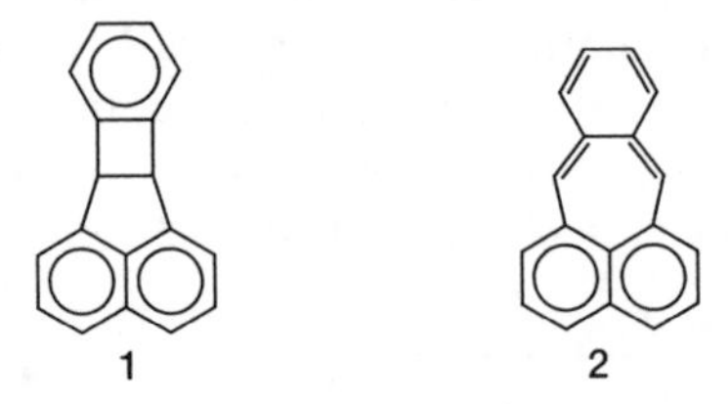

No matter whether the eventual return to S_0 will be radiative or radiationless, its most important characteristic is the geometry of the species at the time of the return, that is, the location of the minimum in S_1 or T_1 from which the return occurs. If the returning molecule reaches a region of the S_0 surface that is sloping down to the starting minimum as shown in Figure 6.2a, the whole process is considered photophysical, since there is no net chemical change. If the return is to a region of S_0 that corresponds to a "continental divide" (cf. Figure 6.2b) or that clearly slopes downhill to some other minimum in S_0, a net chemical reaction will have occurred as a result of the initial

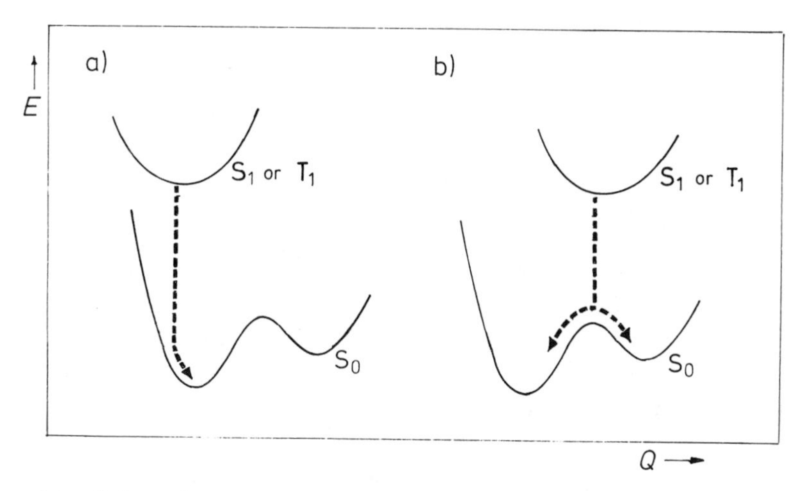

Figure 6.2. Schematic representation of the jump from an excited-state surface (S_1 or T_1) to the ground-state surface a) without chemical conversion and b) in the region of a "continental divide" with partial conversion.

excitation at least for some of the molecules, and the process is considered photochemical.

The description given so far is best suited for unimolecular photochemical reactions. As mentioned above, if the process is bimolecular, both components together must be considered as a "supermolecule." The description given here remains valid except that some motions on the surface of the supermolecule may be unusually slow since they are diffusion-limited.

6.1.2 Reactions with and without Intermediates

Figure 6.3 shows a schematic representation of two surfaces of the ground state (S_0) and of an excited state (S_1 or T_1) and of various processes following initial excitation. The thermal equilibration with the surrounding dense medium requires a sojourn in some local minimum on the surface for approximately 1 ps or longer. This may happen for the first time while the reacting species is still in an electronically excited state (Figure 6.3, path c), or it may occur only after return to the ground state (Figure 6.3, path a). In the former case, the reaction mechanism can be said to be "complex," in the latter, "direct." Direct reactions such as direct photodissociations cannot be quenched. The first ground-state minimum in which equilibrium is reached need not correspond to the species that is actually isolated. Instead, it may correspond to a pair of radicals or to some extremely reactive biradical, etc. (Figure 6.3, path k). Nevertheless, this seems to be a reasonable point at which the photochemical reaction proper can be considered to have been

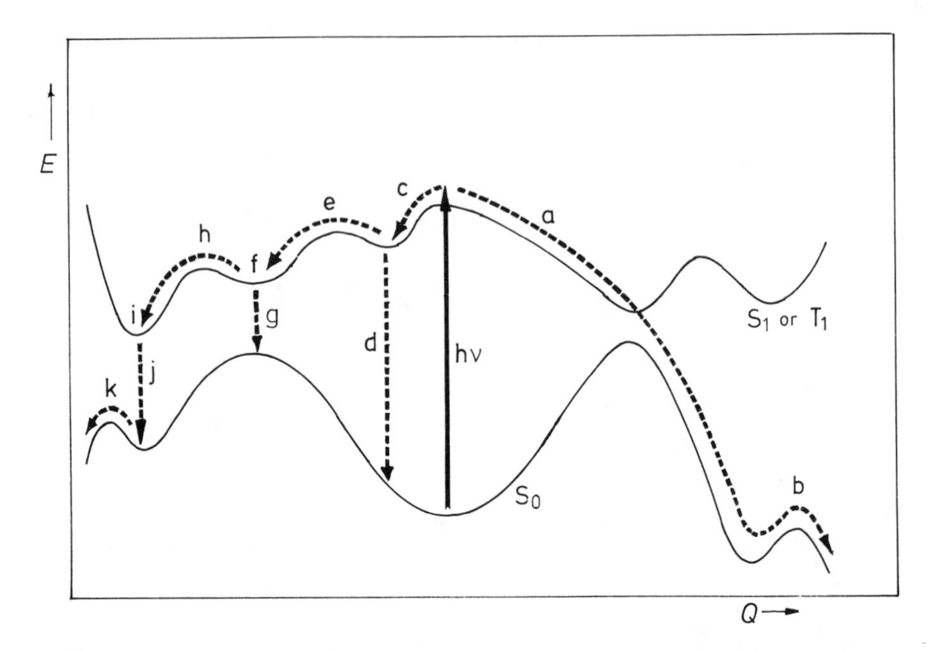

Figure 6.3. Schematic representation of potential energy surfaces of the ground state (S_0) and an excited state (S_1 or T_1) and of various processes following initial excitation (by permission from Michl, 1974a).

completed, even though the subsequent thermal reactions may be of decisive practical importance.

If return to S_0 from the minimum in S_1 or T_1 originally reached by the molecule is slow enough for vibrational equilibration in the minimum to occur first, the reaction can be said to have an *excited-state intermediate*. The sum of the quantum yields of all processes that proceed from such a minimum, that is, from an intermediate, cannot exceed one.

Those minima in the lowest excited-state surface that permit return to the ground-state surface so rapidly that there is not enough time for thermal equilibration are termed "funnels." (Usually, they correspond to conical intersections, cf. Section 4.1.2.) By definition, direct photochemical reactions without an intermediate proceed through a funnel. Sometimes, a funnel may be located on a sloping surface and not actually correspond to a minimum. Since different valleys in S_0 may be reached through the same funnel depending on which direction the molecule first came from, the sum of quantum yields of all processes proceeding from the same funnel can differ from unity. The reason for the difference between an intermediate and a funnel is that a molecule in a funnel is not sufficiently characterized by giving only the positions of the nuclei—the directions and velocities of their motion are needed as well (Michl, 1972).

Example 6.2:
Passage of a reacting system to the S_0 surface through a funnel in the S_1 surface is common in organic photochemical reactions, but the quantitative description of such an event is not easy. Until recently, it was believed that the funnels mostly correspond to surface touchings that are weakly avoided, but recent work of Bernardi, Olivucci, Robb, and co-workers (1990–1994) has shown that the touchings are actually mostly unavoided and correspond to true conical intersections. This discovery does not have much effect on the description of the dynamics of the nuclear motion, since the unavoided touching is merely a limiting case of a weakly avoided touching. Only when the degree of avoidance becomes large, comparable to vibrational level spacing, does the efficiency of the return to S_0 suffer much. The expression "funnel" is ordinarily reserved for regions of the potential energy surface in which the likelihood of a jump to the lower surface is so high that vibrational relaxation does not compete well.

The simplest cases to describe are those with only one degree of freedom in the nuclear configuration space. In such a one-dimensional case, the probability P for the nuclear motion to follow the nonadiabatic potential energy surface

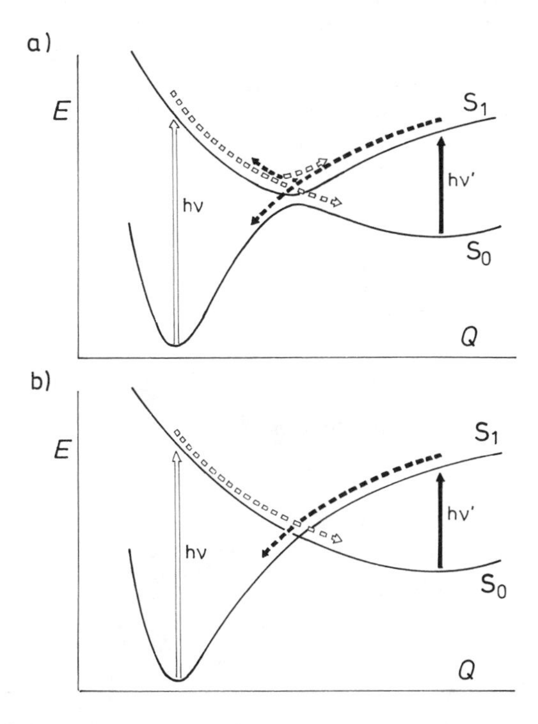

Figure 6.4. Schematic representation of potential energy surfaces S_0 and S_1 as well as $S_0 \rightarrow S_1$ excitation (solid arrows) and nuclear motion under the influence of the potential energy surfaces (broken arrows) a) in the case of an avoided crossing, and b) in the case of an allowed crossing (adapted from Michl, 1974a).

(i.e., to perform a jump from the upper to the lower adiabatic surface) is given approximately by the relation of Landau (1932) and Zener (1932):

$$P = \exp\left[(-\pi^2/h)\,(\Delta E^2/v\Delta S)\right]$$

Here, ΔE is the energy gap between the two potential energy surfaces at the geometry of closest approach, ΔS is the difference of surface slopes in the region of the avoided crossing, and v is the velocity of the nuclear motion along the reaction coordinate. Thus, the probability of a "jump" from one adiabatic Born-Oppenheimer surface to another increases, on the one hand, with an increasing difference in the surface slopes and increasing velocity of the nuclear motion, and on the other hand, with decreasing energy gap ΔE. The less avoided the crossing, the larger the jump probability; when the crossing is not avoided at all, $\Delta E = 0$ and therefore $P = 1$. This is indicated schematically in Figure 6.4.

In the many-dimensional case, the situation is more complicated (Figure 6.5). The arrival of a nuclear wave packet into a region of an unavoided or weakly avoided conical intersection (Section 4.1.2) still means that the jump to the lower surface will occur with high probability upon first passage. However, the probability will not be quite 100%. This can be easily understood qualitatively, since the entire wave packet cannot squeeze into the tip of the cone when viewed in the two-dimensional branching space, and some of it is forced to experience a path along a weakly avoided rather than an unavoided crossing, even in the case of a true conical intersection.

An actual calculation of the $S_1 \rightarrow S_0$ jump probability requires quantum mechanical calculations of the time evolution of the wave packet representing the initial vibrational wave function as it passes through the funnel (Manthe and

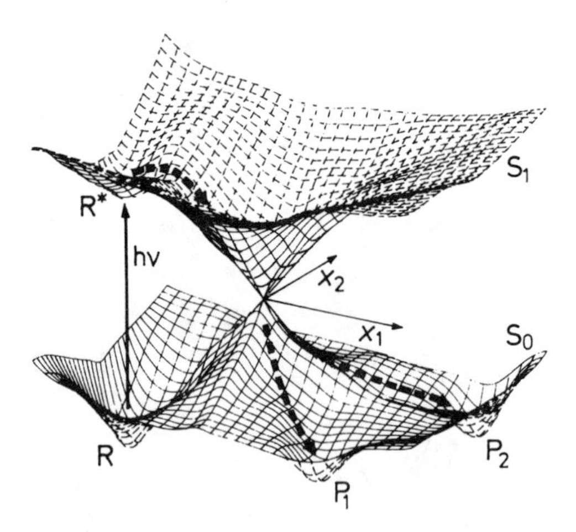

Figure 6.5. Conical intersection of two potential energy surfaces S_1 and S_0; the coordinates x_1 and x_2 define the branching space, while the touching point corresponds to an $(F - 2)$-dimensional "hyperline." Excitation of reactant R yields R*, and passage through the funnel yields products P_1 and P_2 (by permission from Klessinger, 1995).

Köppel, 1990), or a more approximate semiclassical trajectory calculation (Herman, 1984). In systems of interest to the organic photochemist, simultaneous loss of vibrational energy to the solvent would also have to be included, and reliable calculations of quantum yields are not yet possible. It is perhaps useful to provide a simplified description in terms of classical trajectories for the simplest case in which the molecule goes through the bottom of the funnel, that is, the lowest energy point in the conical intersection space.

The trajectories passing exactly through the "tip of the cone" (Figure 6.5) proceed undisturbed. They follow the typically quite steep slope of the cone wall, thus converting electronic energy into the energy of nuclear motion. This acceleration will often be in a direction close to the $\boldsymbol{x}_1$ vector, that is, the direction of maximum gradient difference of the two adiabatic surfaces. (Cf. Example 4.2.) Trajectories that miss the cone tip and go near it only in the two-dimensional branching space have some probability of staying on the upper surface and continuing to be guided by its curvature, and some probability of performing a jump onto the lower surface and being afterward guided by it. However, in this latter case, an amount of energy equal to the "height" of the jump is converted into a component of motion in a direction given by the vector $\boldsymbol{x}_2$ (direction of the maximum mixing of the adiabatic wave functions, cf. Example 4.2), which is generally not collinear with $\boldsymbol{x}_1$ and is often approximately perpendicular to it (Dehareng et al., 1983; Blais et al., 1988). After the passage through the funnel, motion in the branching space, defined by the $\boldsymbol{x}_1$, $\boldsymbol{x}_2$ plane, is most probable. Of course, momentum in other directions that the nuclei may have had before entry into the funnel will be superimposed on that generated by the passage through the funnel.

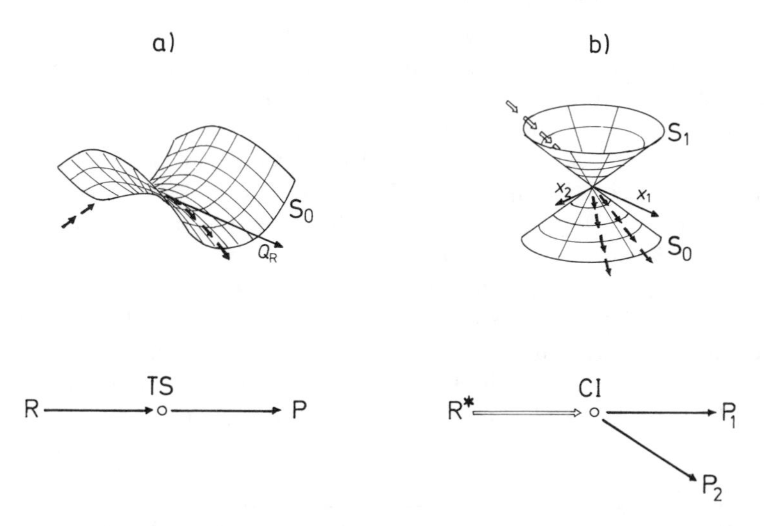

Figure 6.6. Schematic representation a) of the transition state of a thermal reaction and b) of the conical intersection as a transition point between the excited state and the ground state in a photochemical reaction. Ground- and excited-state reaction paths are indicated by dark and light arrows, respectively (adapted from Olivucci et al., 1994b).

A bifurcated reaction path will probably result, and several products are often possible as a result of return through a single funnel. (Cf. Figure 6.5.) In summary, the knowledge of the arrival direction, the molecular structure associated with the conical intersection point, and the resulting type of molecular motion in the x_1, x_2 plane centered on it provide the information for rationalizing the nature of the decay process, the nature of the initial motion on the ground state, and ultimately, after loss of excess vibrational energy, the distribution of product formation probabilities. Depending on the detailed reaction dynamics, the sum of the quantum yields of photochemical processes that proceed via the same funnel could be as low as zero or as high as the number of starting points.

To appreciate the special role of a conical intersection as a transition point between the excited and the ground state in a photochemical reaction, it is useful to draw an analogy with a transition state associated with the barrier in a potential energy surface in a thermally activated reaction (Figure 6.6). In the latter, one characterizes the transition state with a single vector that corresponds to the reaction path through the saddle point. The transition structure is a minimum in all coordinates except the one that corresponds to the reaction path. In contrast, a conical intersection provides two possible linearly independent reaction path directions.

Often, the minimum in S_1 or T_1 that is originally reached occurs near the ground-state equilibrium geometry of the starting molecule (Figure 6.3, path c); the intermediate corresponds to a vibrationally relaxed approximately vertically excited state of the starting species, which can in general be identified by its emission (fluorescence or phosphorescence, Figure 6.3, path d). Quenching experiments can help decide whether the emitting species lies on the reaction path or represents a trap that is never reached by those molecules that yield products. However, often the minimum that is first reached is shallow and thermal energy will allow the excited species to escape into other areas on the S_1 or T_1 surface before it returns to S_0 (Figure 6.3, path e). This is particularly true for the T_1 state due to its longer lifetime. In the case of intermolecular reactions the rate also depends on the frequency with which diffusion brings in the reaction partner. The presence of a reaction partner may provide ways leading to minima that were previously not accessible, for example, by exciplex formation. Other possibilities available to a molecule for escaping from an originally reached minimum are classical energy transfer to another molecule, absorption of another photon (cf. Example 6.1), triplet–triplet annihilation (cf. Example 6.4), and similar processes.

Example 6.3:

For reactions proceeding from the S_1 state, intersystem crossing is frequently a dead end. Thus, irradiation of *trans*-2-methylhexadiene (**4**) in acetone (**3**) yields the oxetanes **5** and **6** through stereospecific addition of the ketone in its

$^1(n,\pi^*)$ state (S_1) to the CC double bond, whereas intersystem crossing into the $^3(n,\pi^*)$ state (T_1) prevents oxetane formation. The $^3(n,\pi^*)$ state is quenched by triplet energy transfer to the diene, which then undergoes a sensitized trans-cis isomerization to **7** (Hautala et al., 1972).

Example 6.4:

A well-studied example of a photoreaction involving excimers is anthracene dimerization (Charlton et al., 1983). Figure 6.7 shows part of the potential energy surfaces of the supermolecule consisting of two anthracene molecules. Singlet excited anthracene ($^1A^* + {}^1A$) can either fluoresce (monomer fluores-

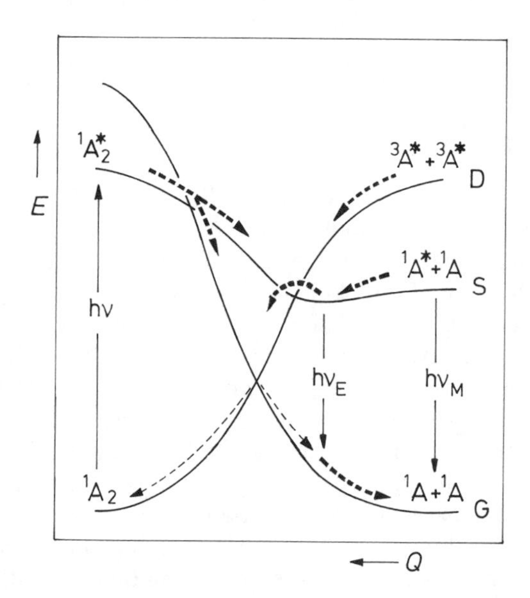

Figure 6.7. Schematic potential energy curves for the photodimerization of anthracene (adapted from Michl, 1977).

cence $h\nu_M$), intersystem cross to the triplet state ($^3A^* + {}^1A$), or undergo a bimolecular reaction to form an excimer $^1(AA)^*$, which can be identified by its fluorescence ($h\nu_E$). Starting from the excimer minimum and crossing over the barrier, the molecule can reach the pericyclic funnel and proceed to the ground-state surface leading to the dimer (24%) or to two monomers (76%). In this case intersystem crossing to the triplet state does not necessarily represent a dead end, because two triplet-excited anthracene molecules may undergo a bimolecular reaction to form an encounter pair whose nine spin states are reached with equal probability. (Cf. Section 5.4.5.5.) One of these is a singlet state $^1(AA)^{**}$ that can reach the pericyclic funnel without forming the excimer intermediate first. The experimentally observed probability for the dimer formation via triplet–triplet annihilation agrees very well with the spin-statistical factor (Saltiel et al., 1981).

From Example 6.4 it can be seen that the molecule may also end up in a minimum or funnel in S_1 or T_1 that is further away from the geometry of the starting species. This then corresponds to a "nonspectroscopic" minimum or funnel (Figure 6.3, minimum f) such as the pericyclic funnel of the anthracene dimerization in Figure 6.7, or even to a spectroscopic minimum of another molecule or another conformer of the same molecule (Figure 6.3, minimum i). Reactions of the latter kind can sometimes be detected by product emission (Figure 6.3, path j). (Cf. Example 6.5.)

In many photochemical reactions return to the ground-state surface S_0 occurs from a nonspectroscopic minimum (Figure 6.3, path g) that may be reached either directly or via one or more other minima (Figure 6.3, sequence c,e).

6.1.3 "Hot" Reactions

Even in efficient heat baths, there are usually several short periods of time during a photochemical reaction when the reactant vibrational energy is much higher than would be appropriate for thermal equilibrium. One of these occurs at the time of initial excitation, unless the excitation leads into the lowest vibrational level of the corresponding electronic state. The amount of extra energy available for nuclear motion is then a function of the energy transferred to the molecule in the excitation process, that is, a function of the exciting wavelength. Further, such periods occur during internal conversion (IC) or intersystem crossing (ISC), when electronic energy is converted into kinetic energy of nuclear motion, or after emission that can lead to higher vibrational levels of the ground state.

The length of time during which the molecule remains "hot" for each of these periods clearly depends on the surrounding thermal bath. In ordinary liquids at room temperature it appears to be of the order of picoseconds. (Cf. Section 5.2.1.) Although short, this time period permits nuclear motion during 10^2–10^3 vibrational periods. During this time, a large fraction of the molecules can move over potential energy barriers that would be prohibitive at thermal equilibrium. Since the vibrational energy may be concentrated in specific modes of motion, a molecule can move over barriers in these favored directions and not move over lower ones in less-favored directions. Chemical reactions (i.e., motions over barriers) that occur during these periods are called "hot."

Example 6.5:
Reactions of hot molecules are known for the ground state as well as for excited states. Thus, after electronic excitation of 1,4-dewarnaphthalene (**8**) in a glass at 77 K emission of **8** as well as of naphthalene (**9**) produced in a photochemical electrocyclic reaction is observed. The intensity of fluorescence from **9** relative to that from **8** increases if the excitation wavelength is chosen such that higher vibrational levels of the S_1 state of **8** are reached. This suggests that there is a barrier in S_1. (Cf. Example 6.1.) If some of the initial vibrational energy is utilized by the molecule to overcome the reaction barrier before it is lost to the environment, **9** is formed and can be detected by its fluorescence. If the energy is not sufficient or if thermal equilibrium is reached so fast that the barrier can no longer be overcome, fluorescence of **8** is observed. The situation is even more complicated because at very low temperatures a quantum mechanical tunneling through the small barrier in S_1 of **8** occurs (Wallace and Michl, 1983).

Thermal or *quasi*-equilibrium reactions, as opposed to hot ones, can be usefully discussed in terms of ordinary equilibrium theory as a motion from one minimum to another, using concepts such as activation energy, activation entropy, and transition state, as long as the motion remains confined to a single surface. "Leakage" between surfaces can be assigned a temperature-independent rate constant. In this way jumps from one surface to another through a funnel may be included in the kinetic scheme in a straightforward manner.

Example 6.6:
For the photoisomerization of 1,4-dewarnaphthalene (**8**) to naphthalene already discussed in Example 6.5 the following efficiencies could be measured in an N_2 matrix at 10 K (Wallace and Michl, 1983): $\eta_F(\mathbf{8}) = 0.13$, $\eta_{ST}(\mathbf{8}) = 0.29$, $\eta_{Ra} = 0.14$, $\eta_{Rd} + \eta_{ret} = 0.44$, $\eta_F(\mathbf{9}) = 0.047$, and $\eta_{TS}(\mathbf{9}) = 0.95$. Here, the indices Ra and Rd refer to the adiabatic and diabatic reaction from the singlet state, ret refers to the nonvertical return to the ground state of **8**, and TS refers to the intersystem crossing to the ground state of **9**. Making the plausible assumption that triplet **8** is converted to triplet **9** with 100% efficiency, these data

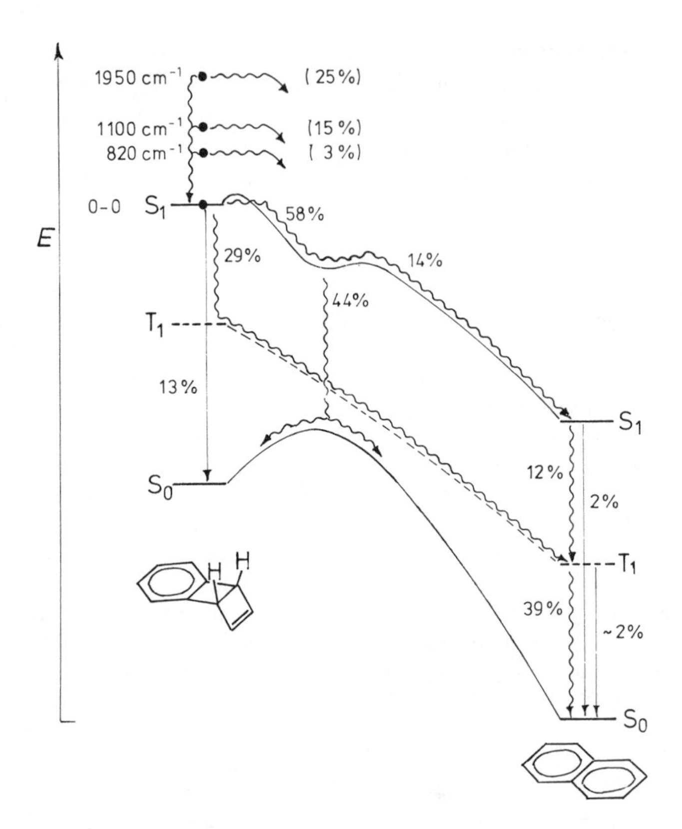

Figure 6.8. Schematic representation of the potential energy surfaces relevant for the photochemical conversion of 1,4-dewarnaphthalene to naphthalene. Radiative and nonradiative processes postulated are shown and probabilities with which each path is followed are given (by permission from Wallace and Michl, 1983).

may be combined to produce the graphical representation shown in Figure 6.8. Here the probabilities with which each path is followed after initial excitation of a molecule of **8** into the lowest vibrational level of its ^1L_b (S_1) state are shown. Using the values $\tau_F(\mathbf{8}) = 14.3 \pm 1$ ns and $\tau_F(\mathbf{9}) = 177 \pm 10$ ns for the fluorescence lifetimes, rate constants can be derived from Equation (5.8) for the various processes. Thus, one obtains $k_F = \eta_F/\tau_F = 9.1 \times 10^6\ s^{-1}$ for the fluorescence of **8** and $k_R = (\eta_{Ra} + \eta_{Rd} + \eta_{ret})/\tau_F = 4.1 \times 10^7\ s^{-1}$ for the rate of passing the barrier.

6.1.4 Diabatic and Adiabatic Reactions

The description of the course of photochemical processes outlined so far implies that there is a continuous spectrum of reactions from unequivocally *diabatic* ones that involve a nonradiative jump between surfaces to unequivocally *adiabatic* ones that proceed on a single surface (Förster, 1970). The *diabatic limit* is represented by reactions in which return to the ground state occurs at the geometry of the starting material. The rest of the reaction then occurs in the ground state and the reaction must be a hot ground-state one; otherwise the ground-state relaxation would regenerate the starting material, resulting in no net chemical change. The other extreme, a *purely adiabatic reaction,* is represented by reactions that proceed in an electronically excited state all the way to the ground-state equilibrium geometry of the final product. Here the excited state converts to the ground state via emission of light or by a radiationless transition. In both these limiting cases, the return to the ground state occurs from a spectroscopic minimum at an ordinary geometry.

In addition to such minima the lowest excited states tend to contain numerous minima and funnels at biradicaloid geometries, through which return to the ground state occurs most frequently. Most photochemical reactions then proceed part way in the excited state and the rest of the way in the ground state, and the fraction of each can vary continuously from case to case. (Cf. Figure 6.3, path a.) It is common to label "adiabatic" only those reactions that produce a "spectroscopic" excited state of the product (cf. Figure 6.3, path h), so distinction between diabatic and adiabatic reactions would appear to be sharp rather than blurred. But this is only an apparent simplification, since it is hard to unambiguously define a spectroscopic excited state.

A second obvious problem with the ordinary definition of adiabatic reactions is the vagueness of the term "product." If the product is what is actually isolated from a reaction flask at the end, few reactions are adiabatic. (Cf. Example 6.7.) If the product is the first thermally equilibrated species that could in principle be isolated at sufficiently low temperature, many more can be considered adiabatic. A triplet Norrish II reaction is diabatic if an enol and an olefin are considered as products. It would have to be considered adiabatic, however, if the triplet 1,4-biradical, which might easily be observed, were considered the primary photochemical product. (See Section 7.3.2.)

Example 6.7:

For a photochemical conversion to be adiabatic, the excited-state surface has to exhibit an overall downhill slope from reactant to product geometries and must not contain unsurmountable barriers or local minima by which the reacting molecules get trapped and funneled off to the ground-state surface before reaching the product geometry. For instance, the photochemical conversion of 1,4-dewarnaphthalene to naphthalene is so strongly exothermic that the pericyclic minimum is without doubt shallow enough to facilitate efficient escape on the excited surface. (Cf. Figure 6.8; for a semiempirical calculation of the potential energy curves, see Jug and Bredow, 1991.) The result is that a portion of the reactants proceeds along the adiabatic path. An example of this behavior is the photochemical decomposition of the benzene dimers **10** and **11**, which yields excited benzene efficiently. Since the benzene excited state lies at 110 kcal/mol and the wavelength of the exciting light was 335 nm, corresponding to 85.3 kcal/mol, the result clearly demonstrates that a part of the chemical energy stored in the reactant is utilized in the adiabatic process to generate electronic excitation in the product (Yang et al., 1988).

10 **11**

The situation is different for the valence isomerization of the metacyclophanediene **12** to the methano-*cis*-dihydropyrene **13**, which also proceeds adiabatically (Wirz et al., 1984). In this case reactant and product HOMOs correlate with each other in the same way as the corresponding LUMOs. The reaction is therefore allowed in the ground state. Nevertheless, the excited state has no correlation-imposed barrier, although the photochemical reaction should be forbidden by the simple Woodward–Hoffmann rules. (See also Example 7.18.) Other examples for adiabatic reactions may be found among triplet reactions, as funnels do not exist in the triplet surface, and return to the singlet ground state requires spin inversion. The ring opening of 1,4-dewarnaphthalene to naphthalene is an example of such a case. (See Figure 6.8.)

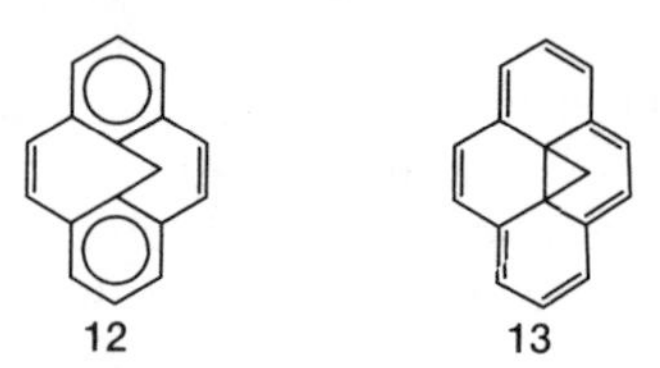

6.1.5 Photochemical Variables

In addition to concentration there are essentially four reaction variables that can be relatively easily controlled and that may have a considerable effect on the course of a photochemical reaction; these are the reaction medium and temperature, and the wavelength and intensity of the exciting light. In addition, magnetic field and isotope effects may come into play.

6.1.5.1 The Effect of the Reaction Medium

Medium effects can be divided into two classes: those that directly modify the potential energy surfaces of the molecule, such as polarity or hydrogen bonding capacity, affecting through strong solvation in particular the (n,π^*) as opposed to the (π,π^*) state energies, and those that operate in a more subtle manner. Examples of the latter are microscopic heat conductivity, which determines the rate of removal of excess vibrational energy, the presence of heavy atoms, which enhance rates of spin-forbidden processes, or viscosity, which affects diffusion rates and thus influences the frequencies of bimolecular encounters. Through these, it may also control triplet lifetimes or the competition between monomolecular and bimolecular processes.

Very high solvent viscosity, encountered in crystalline and glassy solids, also effectively alters the shape of potential energy surfaces by making large changes in molecular geometry either difficult or impossible. This increases the probability that the excited molecule will not greatly change the initial geometry and will eventually emit light rather than react.

The structural dependence of biradicaloid minima discussed in Section 4.3.3 on an example of twisting of a double bond *A*=*B* can be extended to take solvent effects into account. Not only the nature of the atoms *A* and *B* but also polar solvents and counterions affect the stability of zwitterionic states and states of charged species. Then, depending on the solvent, a biradicaloid minimum can represent either an intermediate or a funnel for a direct reaction.

Example 6.8:
If light-induced electron transfer is the crucial step in a photochemical reaction, the solvent dependence expected for this process (cf. Section 5.5.3) may carry over to the whole reaction. An example is the reaction of 1-cyanonaphthalene (**14**) with donor-substituted acetic acids such as *p*-methoxyphenylacetic acid (**15**).

$[14]^* + RCH_2COOH$ (15) → 14 + $RCH_2COOH + h\nu$

$[14]^* + RCH_2COOH$ (15) → $[14]^{\ominus\bullet} + [RCH_2COOH]^{\oplus\bullet}$

↓

(1-cyano-1,4-dihydronaphthyl radical) + $R\dot{C}H_2 + CO_2$

↓

16 + $(RCH_2)_2$ (17) + 18

R = CH_3O–C_6H_4–

In polar solvents such as acetonitrile, electron transfer occurs followed by proton transfer from the radical cation to the radical anion with concurrent loss of CO_2. The radicals collapse to addition products such as **16** or **17**. Alternatively, the radical pair may escape the solvent cage to give, after hydrogen abstraction from a suitable hydrogen source, the reduction product **18**. In nonpolar solvents such as benzene, however, electron transfer is not possible; only exciplex emission and no chemical reaction are observed (Libman, 1975).

6.1.5.2 Temperature Effects

Temperature can have an essential effect on the course of a photochemical reaction because it can affect the rates at which molecules escape from minima in S_1 or T_1. At very low temperatures, even barriers of a few kcal/mol are sufficient to suppress many photochemical processes more or less completely. Processes such as fluorescence, which were too slow at room temperature, may then be able to compete. (Cf. Example 6.5.)

At times, the excited molecule has the choice of reacting in two or more competing ways, and the competition between them can also be temperature dependent. An example is the temperature dependence of the diastereoselectivity and regioselectivity of the cycloaddition of menthyl phenylglyoxylate **19** to tetramethylethylene (**20**) and to ketene acetal **21** (Buschmann et al., 1991):

19 + **20** → **22** (54 % de); **19** + **21** → **23** (41 % de) + **24** (28 % de), 5 : 1

Figure 6.9 shows a plot of ln k/k' versus T^{-1}, where k and k' are the overall formation rate constants for the major and the minor oxetane isomer, respectively. The diastereomeric excess (*% de*) can either increase or decrease with decreasing temperature.

Changes in temperature generally affect ratios of conformer concentration of the starting ground-state molecules and also their distribution among individual vibrational levels, thus codetermining the nuclear configuration of the average molecule just after excitation. The equilibrium between valence tautomers is also affected by temperature changes, and a temperature effect on the photochemical reactivity may result. Thus, irradiation of benzene oxide (**25**) at room temperature produces the furan **27**, whereas irradiation at $-80°C$ gives 11% **27**, 74% phenol and 15% benzene. At lower

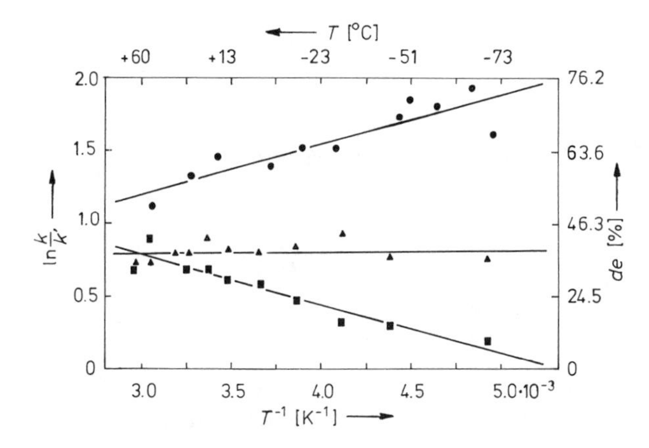

Figure 6.9. Temperature dependence of the diastereoselectivity of oxetane formation (**22**: ●, **23**: ▲, **24**: ■) by cycloaddition of menthyl phenylglyoxalate (**19**) to tetramethylethylene (**20**) and ketene acetal (**21**) (by permission from Buschmann et al., 1991).

temperatures the benzene oxide predominates in the benzene oxide–oxepin equilibrium, and only the photochemistry of this tautomer is observed. At higher temperatures, however, the photochemical reactions of the oxepin (**26**), present in the equilibrium, prevail due to its considerably higher extinction coefficient ε at the wavelength of irradiation (Holovka and Gardner, 1967).

OH; $h\nu$, -80 °C; **25**; **26**; $h\nu$, RT; **27**

6.1.5.3 Effects of Wavelength and Intensity of the Exciting Light

Changes in light wavelength determine the total amount of energy initially available to the excited species. As pointed out in Section 5.2.1, in dense media part of this energy is rapidly lost and after 10^{-11}–10^{-12} s the molecule reaches either one of the minima on the S_1 or T_1 surfaces, or in the case of a direct reaction through a funnel, one of the minima on the S_0 surface. Since internal conversions need not be "vertical" the probability that one or another minimum is reached may change drastically as the energy of the starting point changes. In general, one can imagine that a higher initial energy will allow the molecule to move above barriers that were previously forbidden, and that additional minima will become available. Whether the additional energy is actually used for motion toward and above such barriers or whether it is used for motion in "unproductive" directions and eventually lost as heat could be a sensitive function of the electronic state and vibra-

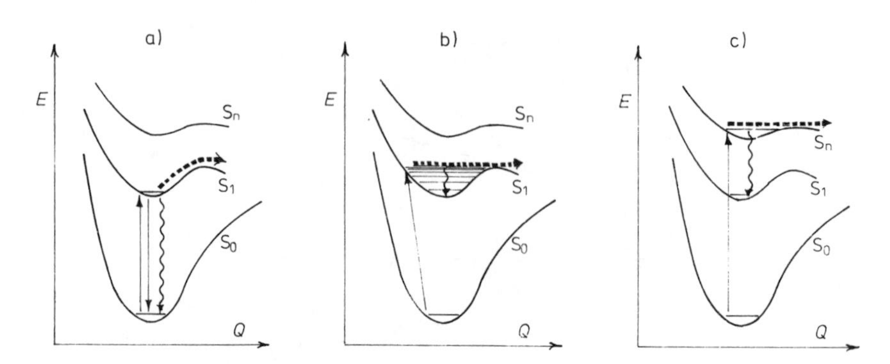

Figure 6.10. General state diagrams for reactions a) from a thermalized S_1 state, b) from a "hot" S_1 state, and c) from a higher excited state S_n.

tional level reached through initial excitation. According to Figure 6.10 three cases can be distinguished in principle. The simple model does not, however, warrant a differentiation between, for example, a hot S_1 and an isoenergetic but less hot S_2 state in a large molecule. In these large molecules the density of vibronic states is high, the Born–Oppenheimer approximation poor, and mixing of states likely to be extremely efficient.

The effect of barriers on the S_1 surface has already been discussed in Example 6.5 for the conversion of 1,4-dewarnaphthalene to naphthalene. A different kind of wavelength dependence is observed for the photochemical reaction of thiobenzophenone (**28**) with acrylonitrile (**29**). Depending on the wavelength of the exciting light, products are obtained that are produced either from the T_1 state reached by ISC from the S_1 state ($\lambda > 500$ nm) or from the S_2 state ($\lambda = 366$ nm). These results were demonstrated by appropriate quenching experiments (De Mayo and Shizuka, 1973). In this case, bimolecular reactions in the S_2 state are possible because the energy gap between the first two singlet states, the (n,π^*) and the (π,π^*) state, amounts to $\approx$ 50 kcal/mol for thiones, which is unusually large; internal conversion $S_2 \rightsquigarrow S_1$ is therefore slow. (Cf. Section 5.2.1.)

28 29

$\lambda = 366$ nm

$\lambda > 550$ nm

Δ

If rotational barriers are sufficiently high in the ground state as well as in the excited state, different conformers can exhibit different photochemistry. Thus, if the spectra of conformers are sufficiently different, as in the case of substituted hexatrienes, the photoreactions of one or another of the conformers can be observed depending on the excitation wavelengths.

Example 6.9:
The photocyclization of (*Z*)-1,3,5-hexatriene represents a well-studied example of the influence of ground-state conformational equilibrium on the product composition (Havinga, 1973). The unsubstituted triene exists mainly in the s-trans-s-trans conformation and therefore cannot undergo photocyclization; only *Z*-*E* isomerization has been observed. With the 2,5-dimethyl derivative the s-cis-s-cis conformation predominates and photocyclization yields dimethylcyclohexadiene and some dimethylvinylcyclobutene. Finally, with the 2-methyl derivative the s-cis-s-trans conformation is preferred, and a relatively good yield of methylbicyclo[3.1.0]hexene as well as methylvinylcyclobutene and methylhexatriene-1,2,4 is found. These results are collected in Scheme 1.

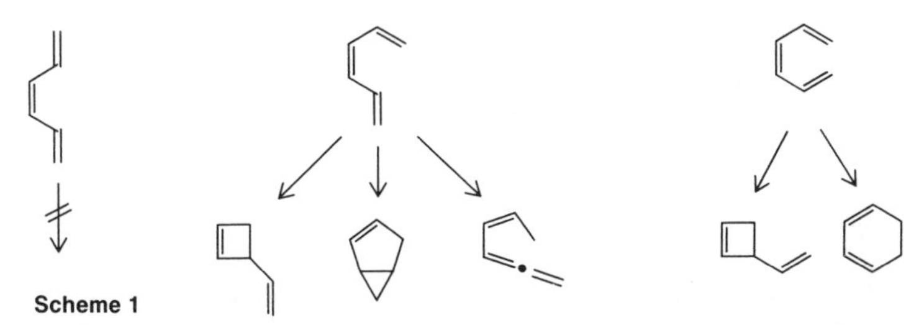

Scheme 1

The product composition may also be determined by the wavelength dependence of a photostationary state. If bicyclononadiene (**30**) is irradiated with 254-nm light, cyclononatriene (**31**) is formed. On further irradiation, it converts slowly to the tricyclic product **32** at the expense of **30** and **31**. If 300-nm light is used, no cyclononatriene (**31**) is found; only slow formation of the tricycle **32** is observed. This wavelength dependence has been explained by a spatial orientation of the π system of triene **31** as a Möbius array of orbitals (**33**) that favors the conrotatory recyclization to form **30** to such an extent that the quantum yield for closure of **31** is about 20 times that for the opening of **30**. This explains why at 254 nm the photostationary state consists of about 40% **30** and 60% **31**, although the extinction coefficient of **30** is about 30 times that of **31**. At 300 nm the ratio of extinction coefficients drops to about 0.02, and the photostationary state is greatly displaced in favor of the bicyclononadiene, thus allowing the less efficient formation of tricyclononene **32** to proceed from the diene (Dauben and Kellog, 1971).

A third example, the photochemical valence isomerization of the tricycle **34** to form 1-cyanoheptalene (**35**), is worthy of mention. It proceeds at 254 nm with a quantum yield nearly 100 times that at 365 nm. In this case, the rate of reaction depends on the initial concentration, but not on the instantaneous concentration. This is an indication that the isomerization proceeds via an exciplex with an electronically excited contaminant, whose concentration remains constant. It is quite possible that similar effects occur also with other wavelength-dependent photoreactions (Sugihara et al., 1985).

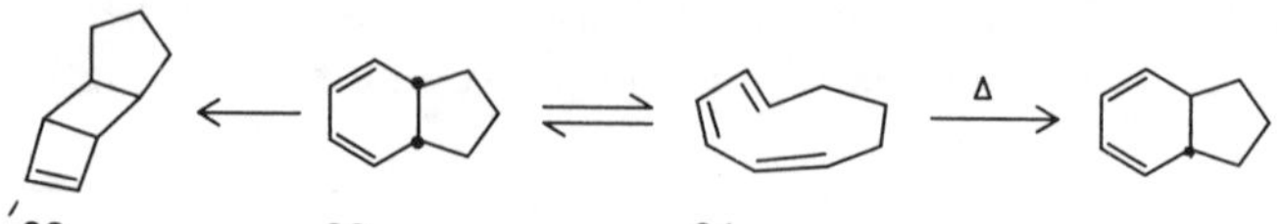

32

30

$\varepsilon_{254} = 3800$
$\varepsilon_{300} = 50$

31

$\varepsilon_{254} = 130$
$\varepsilon_{300} = 2000$

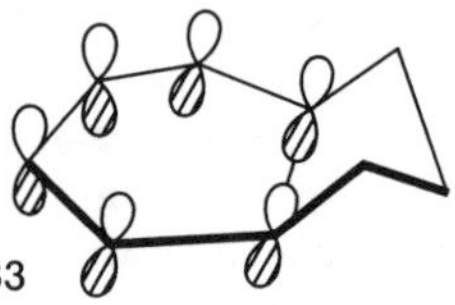

33

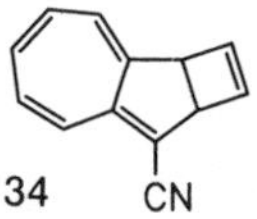

34 CN

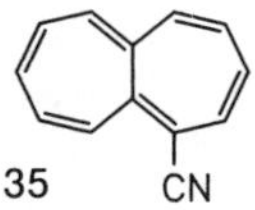

35 CN

Light intensity at the usual levels seldom has an effect on the primary photochemical step if all other variables are kept constant, although overall results may be considerably affected since it can control the concentration of the reactive intermediates. However, it will affect the outcome of a competition between primary one-photon and two-photon processes. These are particularly relevant in rigid glasses where triplets have a long lifetime and quite a few of them are likely to absorb a second photon. The additional energy can permit motion to new minima on the excited-state surfaces, thus leading to new products.

Example 6.10:
Irradiation of diphenyldiazomethane (**36**) with a laser at 249 nm yields fluorene (**37**), 9,10-diphenylanthracene (**38**), and 9,10-diphenylphenanthrene (**39**) in varying amounts depending on the laser intensity or the initial concentration. With conventional lamp excitation these products are not observed. These findings have been explained by the following scheme:

With increasing laser intensity the relative yields of products formed by second-order reactions (**38–40**) increase with respect to the production of fluorene (**37**) from a first-order reaction. It is assumed that fluorene and diphenylanthracene originate from the singlet carbene, whereas diphenylphenanthrene (**39**) arises from either a triplet–singlet carbene dimerization or a triplet–triplet

$Ph_2C=N_2$ (36) ⟶ $^1[Ph_2C:]$ ⇌ $^3[Ph_2C:]$ ⟶ $Ph_2C=CPh_2$ (40)

37 ↑ $^1[Ph_2C:]$; ↓ 38; $^3[Ph_2C:]$ ↓ 39

carbene dimerization. Tetraphenylethylene (**40**), which also is the major product under flash-photolysis conditions, appears to result predominantly from dimerization of two triplet carbenes (Turro et al., 1980a).

6.1.5.4 Magnetic-Field Effects

Reactions involving radical pairs or biradicals can be affected significantly by a magnetic field imposed from the outside. In these intermediates, the singlet and triplet levels can lie so close together that even the very weak perturbation represented by the magnetic field can be sufficient to affect their mixing, and thus the outcome of the photochemical reaction. The fundamentals needed for the understanding of these phenomena have been discussed in Section 4.3.4. (For recent reviews, see Steiner and Ulrich, 1989, and Khudyakov et al., 1993; see also Salikhov et al., 1984.)

6.1.5.5 Isotope Effects

Different isotopes of an element differ in their weight and possibly in other nuclear properties, such as magnetic moment. The mechanism of action of the ponderal effects is generally familiar from ground-state chemistry, and can be understood on the basis of differences in zero-point vibrational energies. (Cf. also Example 5.5.) In rare cases involving tunneling, the mass of the tunneling particle is affected by isotopic substitution. (Cf. Melander and Saunders, 1980.) In photochemical processes, contributions from both of these mechanisms can be quite large, since the reactions often involve very small barriers.

In addition, those photochemical reactions that proceed by radical pair or biradical intermediates offer an opportunity for very large isotope effects based on differences in the magnetic moments of the individual isotopes. As outlined in Section 4.3.4, in these intermediates the singlet and triplet levels can be very close in energy, and the hyperfine coupling mechanism of singlet–triplet coupling plays a role in determining the nature of the products.

This mechanism is exquisitely sensitive to nuclear magnetic moments. (Cf. Salikhov et al., 1984; Doubleday et al., 1989.)

6.2 Pericyclic Reactions

Pericyclic minima and funnels that can be easily predicted by means of correlation diagrams are of great importance in concerted pericyclic photoreactions. However, there may be additional minima and barriers on the excited-state surfaces that affect or even determine the course of the photoreaction.

6.2.1 Two Examples of Pericyclic Funnels

The recognition of a leak in the lowest excited singlet potential energy surface halfway along the path of ground-state–forbidden pericyclic reactions is due to Zimmerman (1966). He argued at the Hückel level of theory, at which configurational wave functions do not interact. The correlation diagram drawn for the high-symmetry reaction path then exhibits a touching of S_0, S_1, and S_2 surfaces (Figure 6.11a), since Φ_0, $\Phi_{1\to1}$, and $\Phi_{1\to1'}^{1\to1'}$ are all degenerate at the pericyclic geometry.

Van der Lugt and Oosterhoff (1969) took into account electron repulsion at the PPP level for the specific example of the conversion of butadiene to cyclobutene and noted that along the high-symmetry path the surface touching becomes avoided (Figure 6.11b), with a minimum in the S_1 surface rather than the S_2 surface as drawn in the older Longuet-Higgins–Abrahamson (1965) and Woodward–Hoffmann (1969) correlation diagrams. This "pericyclic minimum" results from an avoided crossing of the covalent G and D

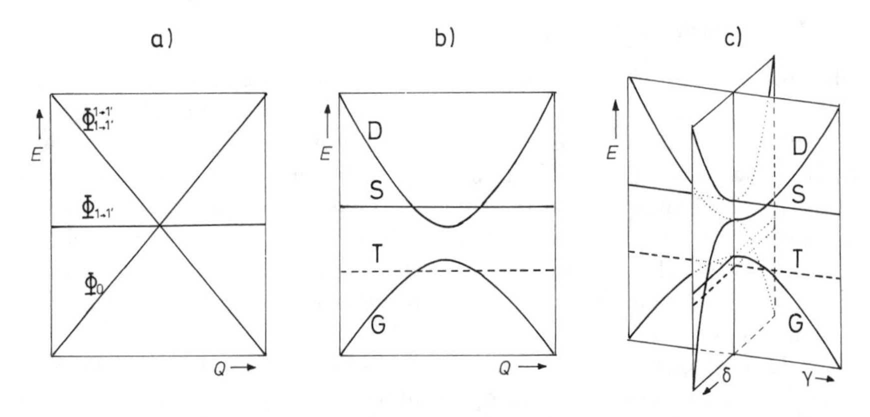

Figure 6.11. Schematic correlation diagrams for ground-state–forbidden pericyclic reactions; a) HMO model of Zimmerman (1966), b) PPP model of van der Lugt and Oosterhoff (1969), and c) real conical intersection resulting from diagonal interactions. The two planes shown correspond to the homosymmetric (γ) and heterosymmetric (δ) case. Cf. Figure 4.20.

configurations (Φ_0 and $\Phi_{1\to 1'}^{1\to 1'}$). Van der Lugt and Oosterhoff pointed out that it will provide an efficient point of return to S_0 and a driving force for the reaction.

Subsequently, from an ab initio study of a simple four-electron–four-orbital model problem (H_4), Gerhartz et al. (1977) concluded that the pericyclic "minimum" that results if symmetry is retained in the correlation diagram is not a minimum along other directions, specifically the symmetry-lowering direction that increases the diagonal interactions in the perimeter, stabilizing the S_1 state and destabilizing the S_0 state (Section 4.4.1). They identified an actual S_1–S_0 surface touching (conical intersection) in H_4, in a sense returning to the original concept of Zimmerman (1966), but at a less symmetrical geometry. They proposed that analogous diagonal bonding in the pericyclic perimeter is a general feature in organic analogues of the simple H_2 + H_2 reaction they studied, in which the simple minimum actually was not a minimum but a saddle point (transition structure) between two conical intersections (Figure 6.11c). Assuming, however, that in the low-symmetry organic analogues the surface touching probably would be weakly avoided, they continued to refer to the global region of the pericyclic funnel, including the diagonally distorted structure, as "pericyclic minimum." They proposed that because of the diagonal distortion at the point of return to the S_0 surface, the mechanism now accounted not only for the ordinary $[2_s + 2_s]$ pericyclic reactions, such as the butadiene-to-cyclobutene conversion, but also for $x[2_s + 2_s]$ reactions, such as the butadiene-to-bicyclobutane conversion. Although the exact structure at the bottom point of the pericyclic funnel remained unknown, it was clear from the experiments that it is a funnel in the sense of being extremely efficient in returning the excited molecules to the ground state.

Finally, in a remarkable series of recent papers, Bernardi, Olivucci, Robb and their collaborators (1990–1994) demonstrated that the S_1–S_0 touching actually is not avoided even in the low-symmetry case of real organic molecules, and they confirmed the earlier conjectures by computing the actual geometries of the funnels (conical intersections) in the S_1 surface at a reasonable level of ab initio theory. They also pointed out that still additional reactions can proceed through the same pericyclic funnel, such as the cis-trans isomerization of butadiene.

Sections 6.2.1.1 and 6.2.1.2 describe the pericyclic funnel in more detail for two particularly important and illuminating examples: the cycloaddition of two ethylene molecules and the isomerization of butadiene. We rely heavily on the results of recent ad initio calculations (Bernardi et al., 1990a; Olivucci et al., 1993).

6.2.1.1 The Cycloaddition of Two Ethylene Molecules

First, we consider the face-to-face addition of two ethylene molecules. The orbital correlation diagram for the high-symmetry path analogous to the rectangular $H_2 + H_2$ path discussed in Section 4.4.1 was given in Figure 4.15. The resulting configuration and state correlation diagrams are essentially the same as those for the electrocyclic ring closure of butadiene discussed in Section 4.2.3. At the halfway point (rectangular geometry, four equal resonance integrals along the perimeter), the situation corresponds to a perfect biradical, analogous to the π system of cyclobutadiene. (Cf. Section 4.3.2.) The order of the states is T below G (S_0), followed by D (S_1) and S (S_2) at higher energies. As discussed in Section 4.4.1 for the model case of H_4, one can use either of the simplified approaches, 3 × 3 CI or 2 × 2 VB, to understand the nature of the distortion that leads from the rectangular geometry to the conical intersection of S_1 with S_0. The use of the former permits a discussion of substituent effects, and we shall consider it first.

From Figure 4.20b it is apparent that a perturbation δ produces a heterosymmetric biradicaloid and promises to lead to the critically heterosymmetric situation ($\delta = \delta_0$), where S_0 and S_1 are degenerate and a real conical intersection results. From Figure 6.12 it is seen that a diagonal interaction differentiates the otherwise degenerate energies of the localized nonbonding orbitals of the cyclobutadiene-like perfect biradical and thus produces a δ perturbation. This suggests that the conical intersection may be reached by a rhomboidal distortion of the pericyclic geometry, which decreases the length of one diagonal and increases that of the other and therefore corresponds to either an enhanced 1,3 or an enhanced 2,4 interaction. For unsubstituted ethylenes, these two cases are equivalent by symmetry, and the pericyclic funnel consists of two conical intersections separated by a transition state at the rectangular geometry as shown in Figure 6.13 (Bernardi et al., 1990a,b; Bentzien and Klessinger, 1994).

In the case of the [2 + 2] photocycloaddition of substituted ethylenes, the two situations remain equivalent for head-to-head addition (1,2 arrangement of the substituents) but are no longer equivalent for head-to-tail addition (1,3 arrangement of the substituents), since the latter affects the energy difference between the two localized forms of the nonbonding orbitals. Ac-

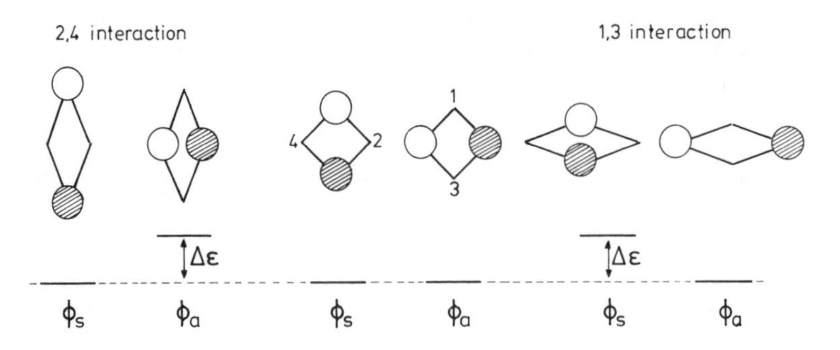

Figure 6.12. The effect of diagonal interactions on the energy of the degenerate localized nonbonding orbitals of a cyclobutadiene-like perfect biradical.

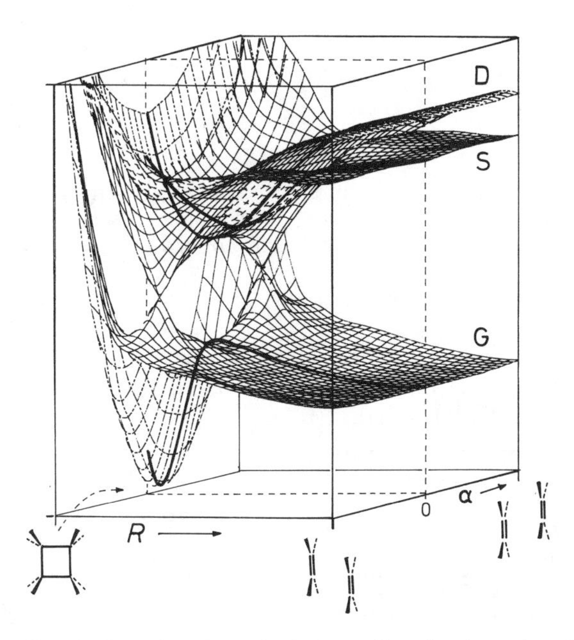

Figure 6.13. Pericyclic funnel region of ethylene dimerization, showing two equivalent conical intersections corresponding to 1,3 and 2,4 diagonal interactions and the transition state region at rectangular geometry ($\alpha = 0$). The curves shown for $\alpha = 0$ correspond to the van der Lugt–Oosterhoff model (by permission from Klessinger, 1995).

cess to one of the otherwise equivalent conical intersections will thus be favored. Electron-donating substituents will reinforce the effect of the diagonal interaction in the case of 1,3 substitution and will counteract it in the case of 2,4 substitution. The opposite will be true of electron-withdrawing substituents (Bonačić-Koutecký et al., 1987). The overall effect of substituents on the kinetics and product distribution in [2 + 2] photoaddition of olefins is however complicated by the possible intermediacy of an excimer or an exciplex, which will typically be more stable for the head-to-head arrangement of substituents. This is discussed in more detail in Section 7.4.2. Depending on the extent of diagonal interaction, the rhomboidal distortion may lead to preferential diagonal bonding in contrast to pericyclic bonding after return to the ground state, resulting in an $x[2_s + 2_s]$ addition product rather than a $[2_s + 2_s]$ product. Examples of $x[2 + 2]$ cycloaddition reactions also are discussed in Section 7.4.2.

In terms of the 2 × 2 VB model, the argument for rhomboidal geometries as favored candidates for a conical intersection (Bernardi, 1990a,b) goes as follows. (Cf. Figure 6.14.) The exchange integral K_R (cf. Example 4.13), which corresponds to the spin coupling in the reactants, will decrease rapidly with decreasing 1,2 and 3,4 overlap as the intrafragment CC distance

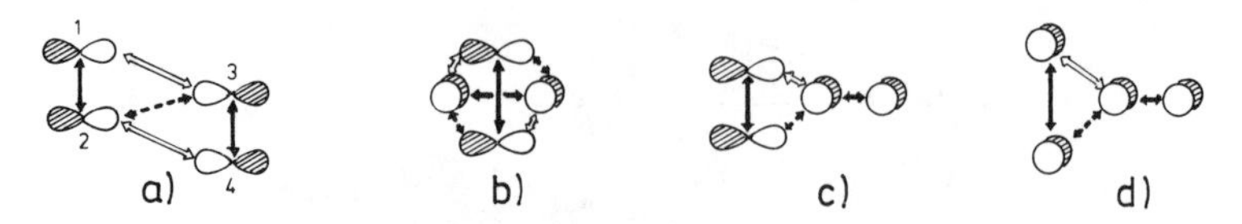

Figure 6.14. Schematic representation of possible geometries at which the conical intersection conditions may be satisfied. Dark double-headed arrows represent K_R, light arrows K_P, and broken arrows K_X contributions (adapted from Bernardi et al., 1990a).

increases. The exchange integral K_P, which corresponds to the product spin coupling, will increase rapidly with increasing 1,3 and 2,4 overlap, as the interfragment distance is decreased. K_X will be quite small for rectangular geometries. While one might satisfy $K_R = K_P$ at these geometries, $K_R = K_X$ can never be satisfied. Figure 6.14 gives a schematic representation of possible geometries on the $(F - 2)$-dimensional hyperline, where the conditions imposed by the equations of Example 4.13 may be satisfied and $E_1 = E_0 = Q$. The Coulomb energy Q will be repulsive in the region of structures b through d in Figure 6.14 because of interference of the methylene hydrogen atoms. Thus, in agreement with the argument based on the two-electron–two-orbital model, the favored region of conical intersection for the [2 + 2] cycloaddition is found at the rhomboidal geometries.

6.2.1.2 The Isomerization of Butadiene

Monomolecular photochemistry of butadiene is rather complex. Direct irradiation in dilute solution causes double-bond cis-trans isomerization and rearrangements to cyclobutene, bicyclo[1.1.0]butane, and 1-methylcyclopropene (Srinivasan, 1968; Squillacote and Semple, 1990). Matrix-isolation studies have established another efficient pathway, s-cis-s-trans isomerization (Squillacote et al., 1979; Arnold et al., 1990, 1991).

Scheme 2 illustrates the phototransformations of *s*-cis-butadiene. Until recently, it was not clear whether the reaction paths leading from the two conformers to the various observed photoproducts are different already on the S_1 surface or whether they do not diverge until after return to S_0. The latter alternative is now believed to be correct. As suggested by the results obtained for the H_4 model system (20 × 20 CI model, see Section 4.4), a simultaneous twist about all three CC bonds provides the combination of perimeter and diagonal interactions (Figure 6.15) needed to reach a funnel (S_1–S_0 degeneracy). Return to S_0 through this geometry accounts for the formation of both cyclobutene and of bicyclo[1.1.0]butane (Gerhartz et al., 1977). The twist around the central CC bond was first invoked by Bigwood and Boué (1974). Both the 3 × 3 CI and the 2 × 2 VB models account for the presence of this pericyclic funnel in S_1. Recent extensive calculations

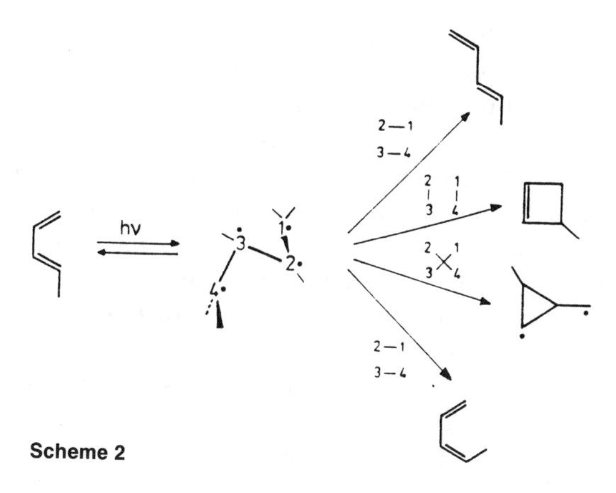

(Olivucci et al., 1993, 1994b; Celani et al., 1995) put these conjectures on firm ground and suggest very strongly that all the different photochemical transformations of butadienes listed in Scheme 2, including cis-trans isomerization, cyclization, bicyclization, and s-cis-s-trans-isomerization, involve passage through a common funnel located in a region of geometries characterized by a combination of peripheral and diagonal bonding (Figure 6.15c–f). Moreover, contrary to previous expectations, these calculations demonstrate that in the funnel region the S_1–S_0 gap not only is very small but vanishes altogether in a large region of geometries. The funnel therefore

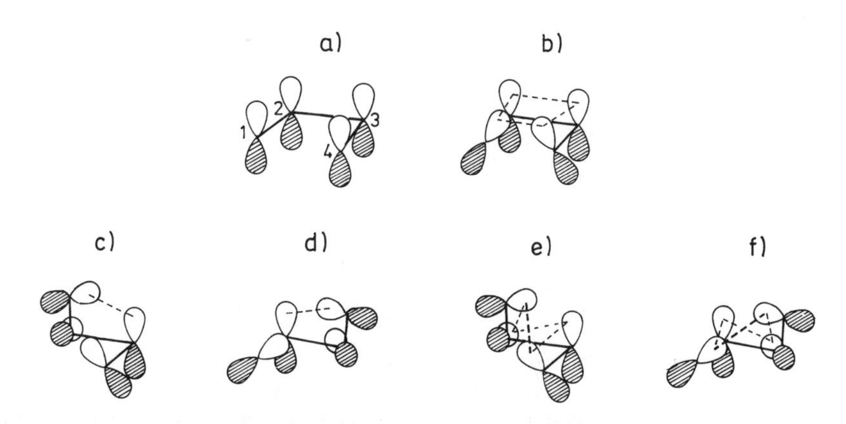

Figure 6.15. π-Orbital interactions in butadiene; a) planar geometry, p AOs, b) high-symmetry disrotatory pericyclic geometry, peripheral interactions along the perimeter, and c)–f) geometries of two equivalent funnels; diagonal interactions are shown in c) and d); peripheral interactions in e) and f).

corresponds to a true S_1–S_0 touching, that is, a conical intersection of the kind expected for a critically heterosymmetric biradicaloid geometry.

For unsubstituted butadienes, the 1,3- and 2,4-diagonal interactions are equivalent (Figure 6.15c and d), as are the two possible movements of the terminal methylene groups above or below the molecular plane. Thus, the pericyclic funnel consists of a total of four equivalent conical intersection regions, communicating pairwise through the two high-symmetry disrotatory "pericyclic minima" investigated by van der Lugt and Oosterhoff (1969). The funnel can be reached from either of the vertically excited geometries, *s-cis-* and *s-trans*-butadiene, after efficient crossing from the initially excited optically allowed B symmetry (S) state to the much more covalent A symmetry (D) state. Three points of minimal energy within the funnel, dubbed s-trans, s-cis, and central conical intersection, have been identified within the conical intersection region, as shown in Figure 6.16. However, the lowest-energy geometries within the funnel region are probably never reached in the S_1 state, since rapid return to S_0 most likely ensues

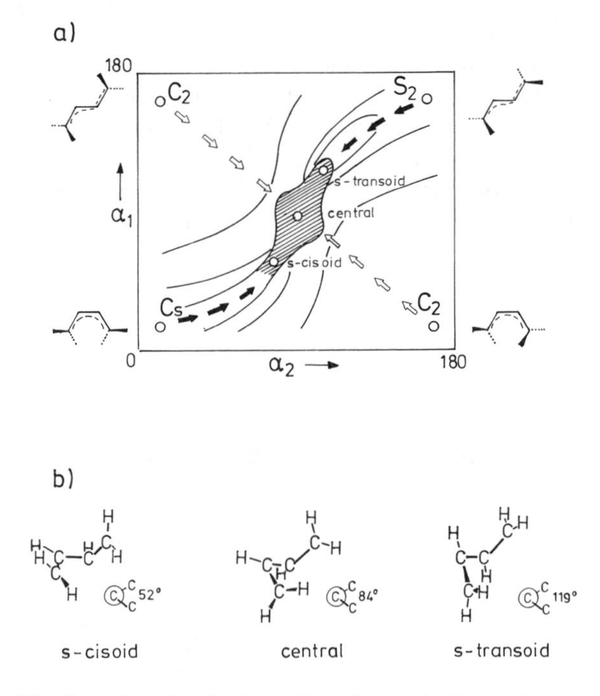

Figure 6.16. The funnel region for butadiene isomerization; a) cross section of the excited-state potential energy surface for $\beta = \alpha_2$, where α_1 and α_2 correspond to rotations about the two C═C bonds of butadiene and β to rotation about the single bond and b) optimized geometries of the three points of minimal energy within the funnel (by permission from Olivucci et al., 1993).

once the edge of the funnel region has been reached. There are two stereochemically different pathways connecting the spectroscopic minima with the central funnel region. As expected from the classical Woodward–Hoffmann correlation diagrams (Section 4.2.1), the disrotatory pathway is energetically more favorable than the conrotatory one. (Cf. Sections 7.1.3 and 7.5.1.)

In the absence of molecular dynamics calculations, it is impossible to make quantitative statements about the probability with which the various product geometries will be reached on the S_0 surface once the molecule has passed through the funnel. Bonding is possible both by increasing further the strength of the diagonal interaction to yield bicyclobutane (δ perturbation, $\boldsymbol{x}_1$ in the branching space) and by increasing the strength of either pair of opposite-side pericyclic interactions to yield either cyclobutene or butadiene (γ perturbation, $\boldsymbol{x}_2$ in the branching space). In the latter case, an s-cis or an s-trans product may result, depending on the direction of twist along the central CC bond, and either a cis or a trans geometry may result at the double bonds in suitably labeled butadienes. This ambiguity accounts for the lack of stereospecificity in the ring opening of labeled cyclobutenes, which is believed to occur through the same funnel (Bernardi et al., 1992c).

6.2.2 Minima at Tight and Loose Geometries

In Section 4.1.3 the notion of *loose* and *tight* biradicaloid geometries has been introduced using the H_2 molecule as an example. For a pericyclic reaction both of the cyclic conjugated orbitals, which are nonbonding for the biradicaloid geometry, are localized in the same region of space; the biradicaloid geometry of a pericyclic minimum is thus a tight one. In the case of ethylene dimerization it can be represented by formula **41**, which is isoconjugate with cyclobutadiene and has two nonbonding orbitals. However, in addition there are a number of open-chain biradicaloid geometries (e.g., **42** and **43**) for which the nonbonding orbitals (a and b) are localized in different regions of space, corresponding to loose biradicaloid geometries. They can be obtained from the pericyclic geometry by disrupting the cyclic conjugation through suitable bond stretching and twisting, and are favored in the S_0 and T_1 states but not in S_1. Here, the pericyclic minimum is evidently at tight geometries unless steric effects prevent it. In addition, substituents can stabilize other geometries, for example, by conjugative interactions. From the pericyclic minimum in S_1 all conformational changes will lead toward unfavorable loose geometries, thus requiring considerable energy. The molecule has little freedom for motion such as bond rotation and the reaction will be stereospecific. Besides, the duration of its sojourn in this minimum is undoubtedly very short, particularly if a diagonal distortion of the type discussed in Section 4.4.1 permits an approach to an area of S_1–S_0 touching, and in the limit, the molecule could pass through such a funnel during a single vibrational period. It will then find itself high on the S_0 surface at an unfavorable biradicaloid geometry, and the steepest slope is likely to lead to a minimum that corresponds either to the starting material or to the pericyclic product. It is also conceivable that the relaxation on the S_0 surface will actually first complete the diagonal bond, producing a biradical that can then close a ring ("crossed cycloaddition," cf. Section 7.4.2).

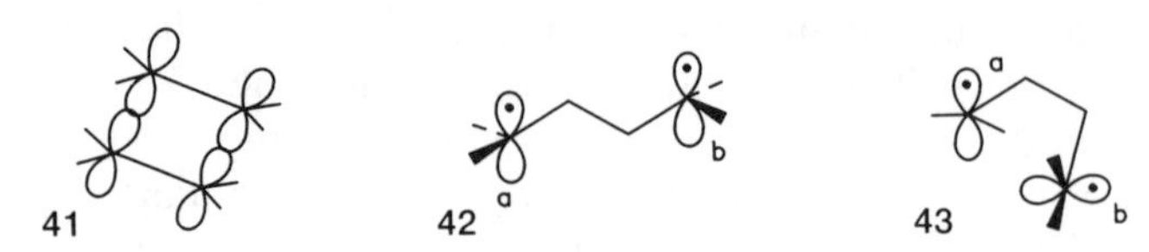

Similar but reversed arguments are also valid for the T_1 surface, if the pericyclic process involves an uncharged perimeter (i.e., if the number of electrons and AOs in the perimeter are equal). In this case loose geometries are more favorable, since there is now no need for a cyclic array of orbitals or any other rigid geometrical requirement. Biradicaloid minima at these geometries will typically allow considerable freedom of motion such as bond rotation, in contrast to the situation in S_1. Also, return to S_0 is spin forbidden and may be relatively slow, possibly permitting detection by direct observation. The ground state is then reached at a biradicaloid geometry, preferentially at geometries at which ISC is fast. (Cf. Section 4.3.4.) Subsequently, little if any additional stereochemical information is lost before a new bond is formed, or before the biradical can fall apart, as indicated in the following scheme:

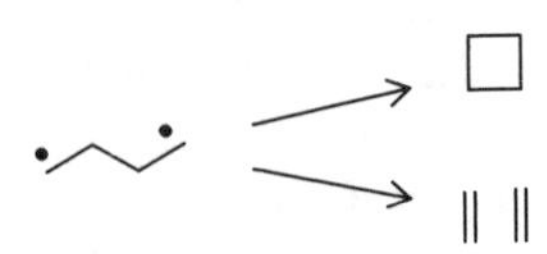

The electron-repulsion arguments for the preference of the triplet state for a loose biradicaloid geometry over a tight geometry, and for the resulting nonconcerted reaction course, only apply if the number of atomic orbitals involved in the pericyclic perimeter is the same as the number of electrons delocalized in the perimeter. In charged perimeters (cf. Figure 2.20c), the two differ, both the singlet and the triplet prefer the tight cyclic geometry, and both processes are expected to be concerted and stereospecific, as in the ring opening of oxirane or the ring closure in diphenylamine (Section 7.5.1, cf. Michl and Bonačić-Koutecký, 1990: Section 5.2.3.)

As long as loose biradicaloid geometries are more favorable, pericyclic processes will not occur. For example, the T_1 surface in Figure 4.8 goes downhill from the cyclobutene to the butadiene geometry along the disrotatory reaction path. What is not shown in this two-dimensional diagram is its path going even further downhill in a direction toward the loose biradicaloid geometry of twisted butadiene. Thus, after ISC, triplet-excited cyclobutene should reach the S_0 surface via the geometry of twisted butadiene and finally relax to ordinary ground-state butadiene in a nonstereospecific manner.

Example 6.11:
The photochemistry of butadiene (Hammond et al., 1964; Srinivasan, 1968) serves as a good example for the demonstration that minima at tight and loose geometries lead to different products. On direct irradiation in cyclohexane or ether, a singlet reaction takes place, believed to pass through the diagonally distorted pericyclic minimum at tight geometries, and gives cyclobutene and bicyclobutane, depending on the mode of relaxation on the S_0 surface. (Cf. Section 4.4.1.) The sensitized reaction, however, proceeds via the triplet minimum at loose geometries and produces the addition products **44–46**. (Cf. Section 7.4.1.)

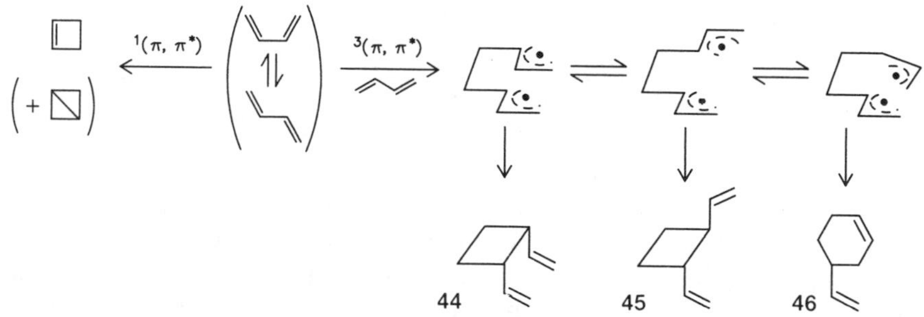

$^1(\pi, \pi^*)$ $^3(\pi, \pi^*)$

44 45 46

6.2.3 Exciplex Minima and Barriers

Photodimerizations and photocycloadditions are important examples of bimolecular reactions. For such reactions an encounter complex has to be first formed, which in the following will be treated as a supermolecule. Correlation diagrams can be constructed for this supermolecule in the usual manner and can be utilized to discuss the course of the reaction. This was demonstrated in Chapter 4 for the exploration of pericyclic minima using H_4 as an example.

The formation of the encounter complex will in general be diffusion controlled. When two (or more) molecules D and A form a sufficiently stable aggregate in the ground state, this aggregate will be a new chemical species commonly referred to as a CT complex and can be identified by its characteristic long-wavelength absorption. (Cf. Section 2.6.)

Stable complexes can form in excited states even when this is not possible in the ground state. These exciplexes or excimers can frequently be identified by their characteristic fluorescence. (Cf. Sections 5.4.2 and 5.4.3.) In view of the formation of an exciplex, ground-state–forbidden photocycloadditions typically involve an intermediate, E*, and return through a funnel,

P*. A representative correlation diagram for this case is that of the photodimerization of anthracene shown in Figure 6.4. The flat excimer minimum (E*) will occur at larger intermolecular separations, with the molecules barely touching. It corresponds to the minimum in the surface of the S state in the case of H_4 shown in Figure 4.27. Its displacement to rectangular geometries is due to the existence of additional bonds not present in H_4. The pericyclic funnel (P*), however, will still occur halfway through the reaction, that is, at a diagonally distorted geometry in which the old bonds are half broken and the new bonds half established. If the initial components A can approach in two different orientations, leading to two possible ground-state products B_1^0 and B_2^0 (e.g., syn and anti, or head-to-tail and head-to-head, see Section 7.4.2), the following kinetic scheme will result:

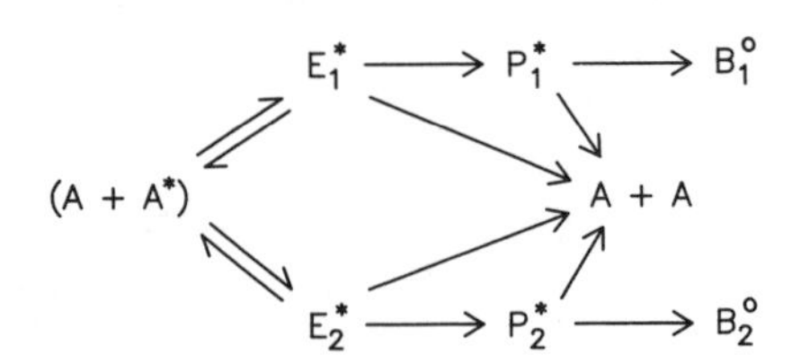

Scheme 3

Which of several possible products will be formed and whether photocycloaddition will occur at all will depend on the depth of the various minima on the excited surface S_1, on the height of the barriers between them, and on the partitioning between starting materials and products upon return to S_0. Additional complications result if the diagonally distorted funnels give rise to products of crossed cycloadditions.

Formation of a bound excimer ordinarily proceeds without an activation barrier other than that imposed by the need for diffusion to bring the reaction partners together in a medium of nonzero viscosity. It is then reasonable to assume preferential formation of the most stable excimer. This is also reasonable if an equilibrium is established among the various possible excimers. The relative stabilities of the excimers or exciplexes can be estimated from orbital interactions or from experimental data (exciton splitting and CT interactions, cf. Section 5.4.2). In the absence of steric complications, the syn head-to-head excimer is usually preferred. (Cf. Section 7.4.2.)

According to Figure 6.17, the height of the barrier between the excimer minimum and the pericyclic funnel depends both on their depths and on the relative placement of the two excited-state surfaces S and D. The depth of the diagonally distorted pericyclic funnel is determined by the nature of the biradical; its dependence on molecular structure, on the head-to-head and head-to-tail orientation of the components, and on reaction medium can be discussed using the principles outlined in Section 4.4.1.

A qualitative model for predicting the height of the barrier as a function of the location of the avoided crossing on the reaction coordinate was proposed by Caldwell (1980). It is based on an estimate of the crossing between the correlation line of the doubly excited configuration D and the ground

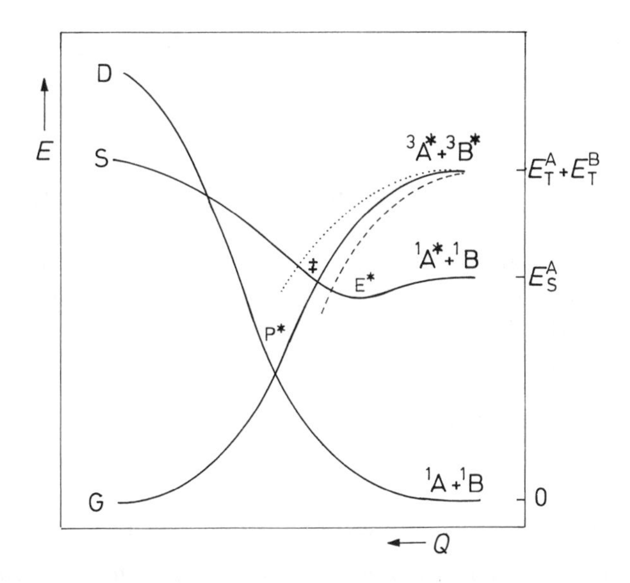

Figure 6.17. Schematic correlation diagram for photodimerization, showing the crossing (‡) that determines the barrier between excimer minimum (E*) and pericyclic minimum (P*), an earlier crossing (---), and a later crossing (···) (by permission from Caldwell, 1980). Effects of a likely diagonal distortion are not shown.

state G with the correlation line of the singly excited state S. (See Figure 6.17.) Due to the interpretation of the D state as $^3A^*$ + $^3B^*$ (cf. Section 4.4.2), the energy gap ΔE_∞ between the potential energy surfaces D and S at infinite separation of the reactants A and B may be expressed as

$$\Delta E_\infty = E_T^A + E_T^B - E_S^A \tag{6.1}$$

where $E_S^A \leq E_S^B$ has been assumed for the singlet excitation energies. The initial slope with which the D state descends from the separated partners toward the pericyclic minimum can be estimated using PMO arguments for the interaction of two triplet states. Assuming that only HOMO–HOMO and LUMO–LUMO interactions are important, the energy change during an addition reaction is given by

$$\Delta E_p(r) = 2\sum_{\varrho\sigma} (a_{\varrho \mathrm{HO}} b_{\sigma \mathrm{HO}} + a_{\varrho \mathrm{LU}} b_{\sigma \mathrm{LU}})\beta(r) \tag{6.2}$$

where a and b are the HMO coefficients of the HOMO (subscript HO) and the LUMO (subscript LU) of molecule A and B, respectively, and the new bonds formed are between positions ϱ in A and σ in B. According to Figure 6.17 an early crossing with the S–S correlation line will bring about a low barrier, and therefore high reactivity, due to the large distance r_C between the adducts at the crossing and the correspondingly small bonding interaction. The lateness of the crossing may be conveniently defined with respect to the E_S^A level instead of the S–S line, since in that case the crossing condition is simply $\Delta E_p(r_C) = -\Delta E_\infty$. This condition allows combination of Equations (6.1) and (6.2) into the relationship

$$\beta(r_C) = -\tfrac{1}{2}(E_T^A + E_T^B - E_S^A)/\sum_{\varrho\sigma}(a_{\varrho HO}b_{\sigma HO} + a_{\varrho LU}b_{\sigma LU}) \tag{6.3}$$

for the resonance integral $\beta(r_C)$ of the newly formed bonds. The conditions for $|\beta(r_C)|$ to be small, that is, for the reaction to start early and for the barrier to be low, are: low triplet energies, high singlet excitation energy E_S^A, and large HMO coefficients of HOMO and LUMO at reacting positions. In agreement with this argument, photodimerizations of olefins are quite facile when a value $|\beta(r_C)| < 20$ kcal/mol is obtained for the resonance integral from Equation (6.3), whereas for $|\beta(r_C)| > 24$ kcal/mol no dimerization has been observed. (Cf. Section 7.4.2.)

6.2.4 Normal and Abnormal Orbital Crossings

The initial problem in constructing a correlation diagram for a thermally forbidden pericyclic reaction is in the identification of those orbitals that are most responsible for the ground-state energy barrier. In particular, it is necessary to actually identify the originally bonding orbital of the reactant, say ϕ_i, which becomes antibonding (or at least nonbonding) in the product, and the originally antibonding reactant orbital, say ϕ_k, which becomes bonding in the product, in order to completely specify the "characteristic configuration." The second step in the argument consists of an assignment of states of both the reactant and the product to specific configurations, and may require a consultation of spectroscopic data and some CI calculations.

In many cases ϕ_i is the HOMO and ϕ_k the LUMO of the reactant, and as they cross along the reaction coordinate, ϕ_i becomes the LUMO and ϕ_k the HOMO of the product. This kind of crossing will be referred to as "normal orbital crossing." Although the crossing may be avoided in systems of low symmetry, the arguments do not change. An "abnormal orbital crossing" occurs if at least one of the crossing orbitals ϕ_i and ϕ_k is neither the HOMO nor LUMO of the reactant. In this case, already at the stage of the configuration correlation diagram, more or less significant barriers have to be expected on the S_1 and T_1 surfaces.

This becomes apparent from the following argument: The characteristic configuration $\Phi_{i \to k}$ of the reactant will correlate with the configuration $\Phi_{k \to i}$ of the product and this will be the lowest-energy configuration likely to remain at approximately constant energy along the path. If the crossing is of the normal type, this configuration most often predominates in the lowest excited singlet and triplet states S_1 and T_1 of both the reactant and the product, with the result that no barriers due to correlation are imposed starting on either side. If instead, either due to abnormal orbital crossing or due to configuration interaction effects, it predominates in one of the higher excited states, other configurations necessarily represent the lowest excited state, S_1 or T_1. This then rises along the reaction path until it meets the state primarily represented by the characteristic configuration, and barriers are imposed on the excited-state surfaces.

In the case of many aromatics, the lowest singlet excited state is not represented by the HOMO–LUMO excited configuration (^1L_a) but rather by a mixture of $\Phi_{i \to k+1}$ and $\Phi_{i-1 \to k}$ (^1L_b, cf. Section 2.2.2), and a barrier is expected in the excited state S_1. Since the difference in the S_1 and S_2 energies is usually quite small the resulting barrier will be small. (See Example 6.12.) In the triplet state, no barriers of this origin are expected since even in these cases the triplet represented by the HOMO→LUMO excitation is lowest.

Example 6.12:
Figure 6.18a gives a schematic representation of the orbital correlation diagram for the thermally forbidden conversion of one alternant hydrocarbon into another one. The following configuration correlations are easily verified from this diagram:

$$\Phi_0 \to \Phi'^{1 \to 1'}_{1 \to 1'}$$
$$\Phi_{1 \to 1'} \to \Phi'_{1 \to 1'}$$
$$\Phi_{1 \to 2'} \to \Phi'^{1 \to 1'}_{1 \to 2'}$$
$$\Phi_{2 \to 1'} \to \Phi'^{1 \to 1'}_{2 \to 1'}$$

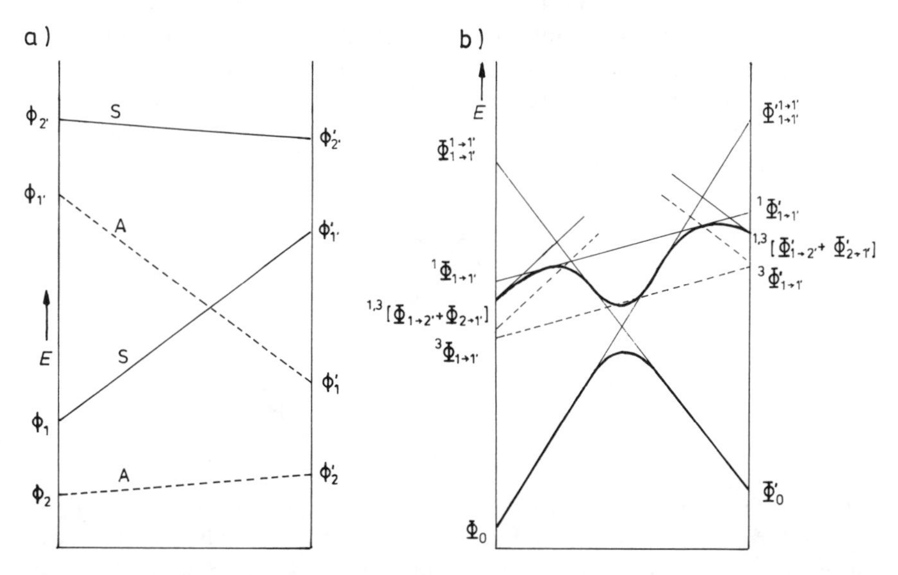

Figure 6.18. Excited-state barriers: a) orbital correlation diagram for a thermally forbidden conversion of an alternant hydrocarbon; b) the corresponding configuration and state correlation diagram for the case that the HOMO→LUMO excitation does not represent the longest-wavelength absorption.

where Φ and Φ' may refer either to reactant or to product. The ground state as well as the mixture of $\Phi_{1\to2'}$ and $\Phi_{2\to1'}$, which in aromatic hydrocarbons represents the ^1L_b state, correlates with high-lying doubly excited states. Assuming that both in the reactant and the product the ^1L_b state is below the ^1L_a state, which corresponds to the HOMO→LUMO excitation, the state correlation diagram shown in Figure 6.18b is obtained, which already takes into consideration the avoided crossings between singlet states of equal symmetry. From the derivation of this diagram it is evident that a barrier in the S_1 state is to be expected in all those cases in which the lowest excited singlet state is not represented by the HOMO→LUMO excitation, but is symmetrical with respect to the symmetry element that is being conserved during the correlation, like the ground state. An example of the situation discussed here is given by the conversion of dewarnaphthalene into naphthalene. (Cf. Example 6.5.) The corresponding correlation diagram was already shown in Figure 6.8.

From the general form of the correlation diagram in Figure 6.18 it is evident that since barriers can be overcome with the aid of thermal energy the initial excitation need not be into that state that is represented by the characteristic configuration. In this case low temperature can prevent the reaction from taking place. Excitation of higher vibrational levels of one-and-the-same absorption band can enable the nuclei to move over the barrier in a hot reaction.

In the case of an abnormal orbital crossing, a straightforward consideration of frontier orbitals would be misleading. Examples of high barriers due to abnormal orbital crossing have been observed in the electrocyclic ring opening of cyclobutenoacenaphthylene (**47**) and similar compounds (cf. Example 6.1 and Figure 6.1) (Michl and Kolc, 1970; Meinwald et al., 1970). An example of a low barrier due to abnormal orbital crossing that can be overcome using thermal energy has been reported for the cycloreversion reaction of heptacyclene **48** to two acenaphthylene molecules (**49**) (Chu and Kearns, 1970).

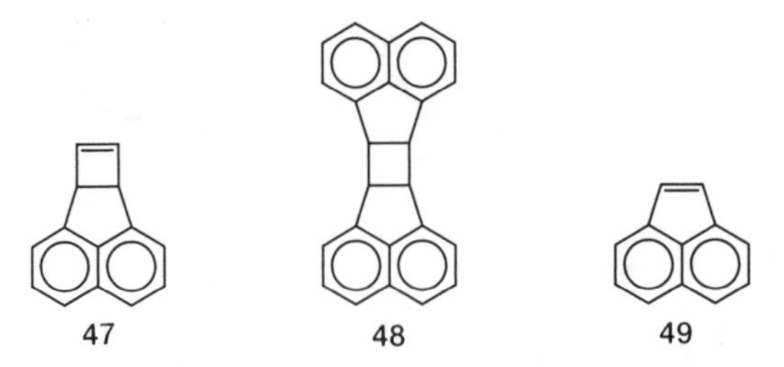

In deciding whether a normal or an abnormal orbital crossing is to be expected for an electrocyclic process, the two-step procedure for constructing qualitative orbital correlation diagrams described in Section 4.2.2 has proven very useful. The least-bonding MO of the reactant that interacts strongly with the bonding combination (σ) of the AOs on the two carbons originally joined by a single bond will be ϕ_i, and the least-antibonding MO that interacts strongly with the antibonding combination (σ^*) will be ϕ_k. If ϕ_i is the HOMO of the original molecule and ϕ_k the LUMO, the orbital crossing is normal; otherwise it is abnormal.

Example 6.13:
For cyclobutenophenanthrene (**50**) two different photochemical reaction pathways are conceivable (Michl, 1974b); the electrocyclic opening of the cyclobutene ring to form **51** and the cycloreversion reaction to give phenanthrene and acetylene. The HMO coefficients of the 2 and 2′ positions in biphenyl have opposite signs in the HOMO and equal signs in the LUMO. Thus, the HOMO

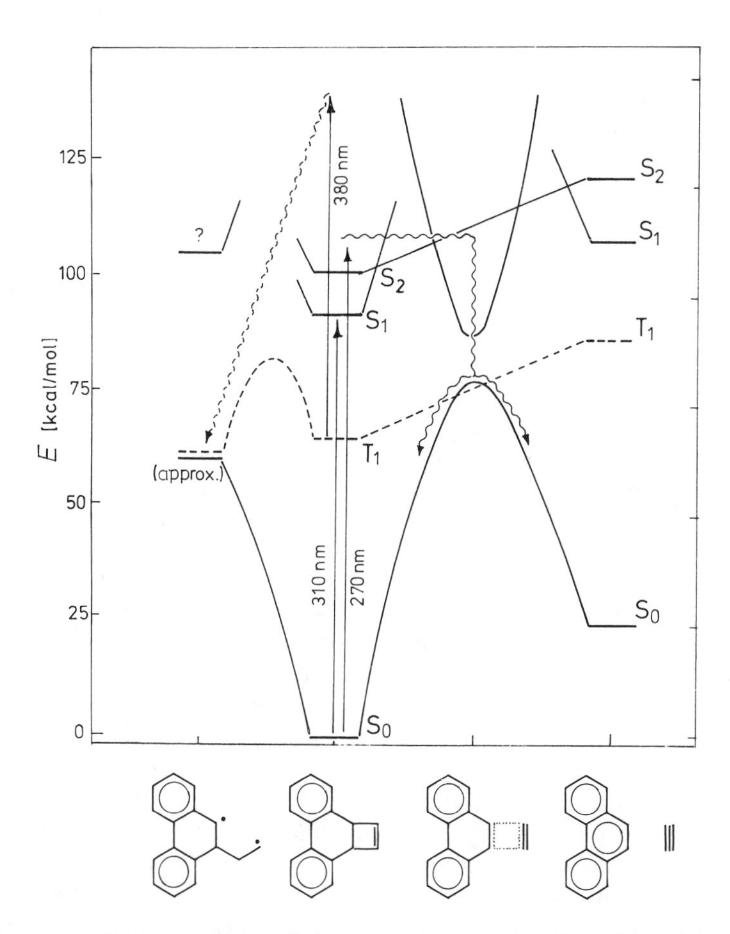

Figure 6.19. State correlation diagram for the fragmentation of cyclobutenophenanthrene. The straight arrows indicate absorption of light of a given wavelength; the wavy lines indicate how the barriers in S_1 (full lines) and in T_1 (broken lines) can be overcome when sufficient energy is available (by permission from Michl, 1974b).

cannot interact with the bonding combination and the LUMO cannot interact with the antibonding combination of the AOs of the original σ bond during the disrotatory opening of the cyclobutene ring in **50**. This results in an abnormal orbital crossing. This reaction has in fact so far not been detected.

Experimentally known, however, is the cycloreversion to give phenanthrene and acetylene. This is assumed to be concerted in the singlet state and stepwise in the triplet state as indicated in Figure 6.19. The orbital correlation diagram for the concerted reaction has been derived in Example 4.5 and exhibits a normal orbital crossing. (See Figure 4.16.) Since the ^1L_a band that is represented by the HOMO→LUMO transition corresponds to an excitation from S_0 to S_2 in biphenyl as well as in phenanthrene, a barrier in S_1 results, as shown for the general case in Figure 6.18. However, the triplet reaction is expected to be endothermic and the molecules will sooner or later again collect in the T_1 minimum for the starting geometry. Also indicated in Figure 6.19 is the nonconcerted pathway for which a barrier in T_1 is to be expected, since the T_1 is of a $\pi \rightarrow \pi^*$ nature while the characteristic configuration for reaching the open-chain minimum is of a $\sigma \rightarrow \sigma^*$ nature. (Cf. Figure 6.20.)

50

51

6.3 Nonconcerted Photoreactions

6.3.1 Potential Energy Surfaces for Nonconcerted Reactions

Most known photochemical processes are not pericyclic reactions. Even in many of these cases correlation diagrams can be helpful in estimating the location of minima and barriers on excited-state surfaces. (Cf. Section 4.2.2.) The derivation of these correlation diagrams, however, is often more difficult, not only because of lack of symmetry, but also because it may be difficult to identify any one excited state as the characteristic state, particularly in large molecules. For example, many of the $\sigma \rightarrow \sigma^*$ excited states of toluene will have some contribution from the $\sigma \rightarrow \sigma^*$ bond orbital excitation in one of the three C—H bonds in the methyl group.

Often, however, it is sufficient to distinguish between σ^* and π^* orbitals in order to decide whether a barrier is likely to occur. For instance, the T_1 surface of toluene along the path of nuclear geometries that leads to dissociation to $C_6H_5CH_2\cdot$ + H· can be viewed as originating from interaction of a locally excited $^3(\pi,\pi^*)$ configuration of the benzene chromophore and a locally excited $^3(\sigma,\sigma^*)$ configuration of the CH bond. The former is of lower energy at the initial geometries, as is the latter at the final geometries; somewhere along the way they intend to cross, but the crossing is avoided and results in a barrier separating two minima on the T_1 surface (Figure 6.20).

The minimum, which is essentially represented by the $^3(\pi,\pi^*)$ configuration, is a spectroscopic minimum, into which initial excitation occurs, either by energy transfer (triplet sensitization) or indirectly by intersystem crossing (ISC) from the singlet manifold. The minimum represented by the $^3(\sigma,\sigma^*)$ configuration, on the other hand, is a reactive minimum, from which the actual dissociation takes place. Similarly, the characteristic state for an α cleavage in a ketone triplet is of (σ,σ^*) nature, whereas the excitation is into the (n,π^*) state.

The spectroscopic minimum preserves the excited molecule until it can escape toward the reactive minimum. Its role is particularly crucial in bimolecular processes, where this escape has to wait for diffusion to introduce a reaction partner. A high barrier separating the spectroscopic minimum and the reactive minimum is to be expected if the orbitals ϕ_i and ϕ_k, which represent the characteristic configuration, do not interact with the orbitals involved in the electronic transition into the lowest excited state; the crossing between the potential energy curve that goes up in energy along the reaction coordinate and the one that comes down in energy would then not be avoided. Examples of this situation are the dissociation of an aromatic CH bond in toluene or the cleavage of a CC bond in a ketone that is rather far away from the carbonyl group. No doubt such reactions will show minima in the S_1 or T_1 surface. But they are separated by unsurmountable barriers, so these reactions cannot be observed.

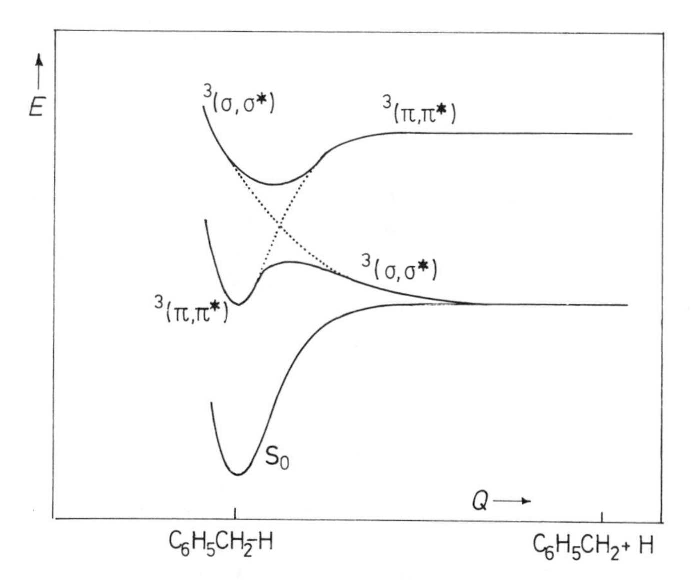

Figure 6.20. Energies of selected states during dissociation of a benzylic CH bond in toluene as a function of the reaction coordinate (by permission from Michl, 1974a).

However, when the orbitals of the characteristic configuration can interact with the orbitals involved in excitation, as is the case for the cleavage of the benzylic C—H bond in toluene or the α CC bond in a ketone, the crossing will be avoided and the barrier will be lowered depending on the interaction. The magnitude of the interaction is rather difficult to predict without a detailed calculation.

In the particular case of the photolytic dissociation of the benzylic CH bond in toluene, the barrier separating the two minima is so high that the reaction proceeds on absorption of a single photon only if light of sufficiently short wavelengths is used. With light of longer wavelengths another photon is needed to overcome the barrier. (Cf. Johnson and Albrecht, 1968 as well as Michl, 1974a.)

Example 6.14:
Except for the relief of ring strain, similar barrier heights are to be expected for the triplet-sensitized photochemical ring opening of benzocyclobutene (**52**) to form *o*-xylylene (**53**) and for the dissociation of the benzylic C—C bond in ethylbenzene leading to a benzyl and methyl radical. Since the energy of the characteristic configuration may be lowered by additional substituents, this may favor the interaction with the orbitals involved in the initial excitation and thus reduce the barrier height. In fact, for $\alpha,\alpha,\alpha'\alpha'$-tetraphenylbenzocyclobutene (**54**) a photochemical ring opening has been observed, whereas under similar conditions of low-temperature irradiation benzocyclobutene is inert (Quinkert et al., 1969; Flynn and Michl, 1974).

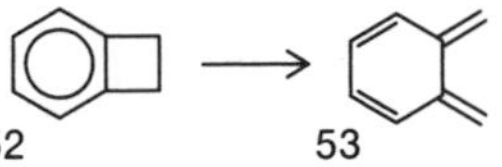

52 → 53 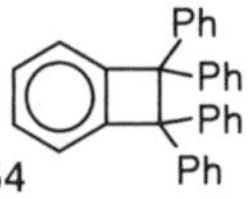   54 (Ph, Ph, Ph, Ph)

In the case of dissociation of the various methyl-substituted anthracenes to form the corresponding anthrylmethyl radicals the relative magnitude of the interaction between the σ orbitals of the characteristic configuration and the π and π^* orbitals involved in the excitation may be estimated quite easily. Assuming that the interaction is proportional to the square $c_{\mu i}^2$ of the LUMO coefficient of the carbon atom adjacent to the methyl group, the lowest barrier is expected for the 9-methyl derivative and the highest one for the 2-methyl derivative, as may be seen from the squared coefficients given in formula **55**.

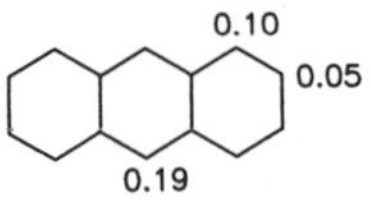

0.10, 0.05, 0.19
55

In Section 4.2.2 it has been shown that due to intended or natural orbital correlations, crossings may occur in the configuration correlation diagram that are avoided in the state correlation diagram. If no avoided crossing is to be expected on the basis of orbital symmetry, the interaction between these configurations is likely to be strong, so the barrier resulting from the avoided crossing may not be high.

The results of ab initio calculations for hydrogen abstraction by ketones according to

$$R^1C(=O)R^2 + H{-}R \xrightarrow{h\nu} R^1\dot{C}(OH)R^2 + \bullet R$$

represented in Figure 6.21, in fact show small barriers in the $^1(n,\pi^*)$ and $^3(n,\pi^*)$ excited states, which indicate that the MOs apparently "remember" the natural or intended correlation. (Cf. Figure 4.18.)

Another barrier on the T_1 surface is observed for hydrogen abstraction by ketones whose $^3(\pi,\pi^*)$ state is of lower energy than the $^3(n,\pi^*)$ state. From Figure 4.17 it is seen that the $^3(\pi,\pi^*)$ state goes up in energy while the $^3(n,\pi^*)$ state comes down in energy along the reaction coordinate or at least remains more or less constant (Figure 6.21), and a crossing of the corresponding potential energy surfaces will occur at a geometry intermediate between that of the reactant and a biradical. For nonplanar arrangements of the nuclei this crossing will be avoided and will produce a barrier. This amounts to 5 kcal/mol in the case of the hydrogen abstraction in the naphthyl

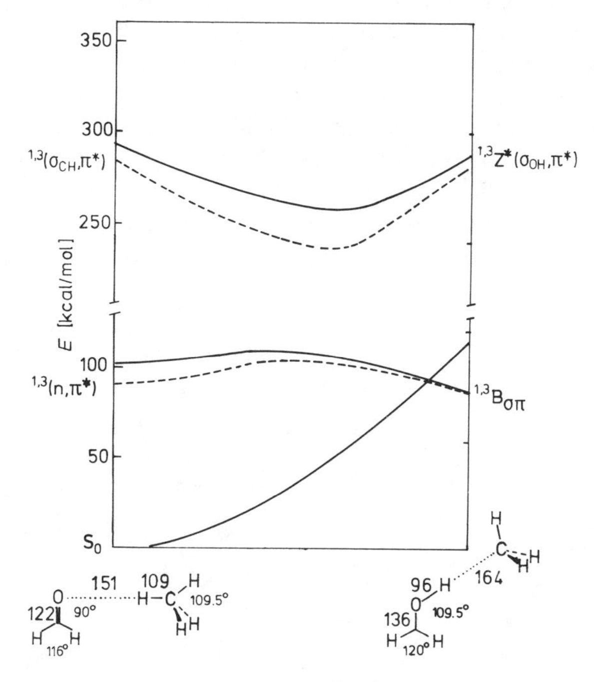

Figure 6.21. State correlation diagram for the photochemical hydrogen abstraction, as calculated for the system formaldehyde + methane (by permission from Devaquet et al., 1978).

ketone **56** to form **57**, an example in which the $^3(n,\pi^*)$ is higher in energy by 9 kcal/mol than the $^3(\pi,\pi^*)$ state. Besides, this is an abstraction of the δ-H atom that makes a nonplanar reaction path very likely (De Boer et al., 1973).

56 57

The picture becomes somewhat more complex if the α cleavage of ketones is considered.

$$R^1C(=O)R^2 \xrightarrow{h\nu} R^1\dot{C}=O + \cdot R^2$$

If all atoms involved in the reaction lie in the same plane, the unpaired electron of the acyl radical may be either in an orbital that is symmetric with respect to this plane or in an orbital that is antisymmetric—that is, either in a σ or in a π orbital, whereas only a σ orbital is available for the unpaired electron of radical R. Instead of just one singlet and one triplet covalent biradicaloid structure (Figure 4.5), there are now two of each, which may be denoted as $^{1,3}B_{\sigma,\sigma}$ and $^{1,3}B_{\sigma,\pi}$, respectively. Similarly, there are also different zwitterionic structures to be expected. The increase in complexity and the number of states that results from the presence of more than two active orbitals on the atoms of a dissociating bond has been formalized and used for the development of a classification scheme for photochemical reactions ("topicity"), as is outlined in more detail in Section 6.3.3.

As shown in Figure 6.22 in the example of the formyl radical, the σ acyl radical prefers a bent geometry with the unpaired electron in an approxi-

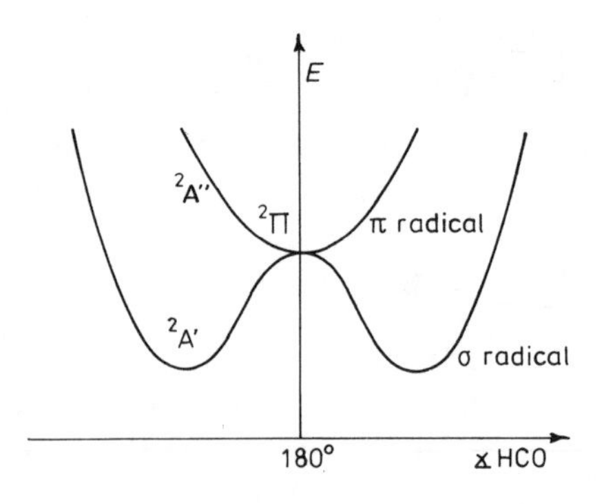

Figure 6.22. Qualitative representation of the energy of the σ and π formyl radical as a function of the HCO angle.

mately sp^2 hybrid AO that has some s character and thus is energetically more favorable than a pure p AO. In contrast, the π acyl radical prefers a linear geometry, since it has its unpaired electron in the π^* orbital; the p_C AO starts empty and overlaps with the doubly occupied n_O AO to form a doubly occupied MO. Since at linear geometries three electrons in one π system are equivalent to three electrons in two p AOs orthogonal to this π system, the σ and the π acyl radical states have the same energy, that is, are degenerate.

In discussing the reaction it is helpful to use bond dissociation and bond angle variation in the reaction product as independent coordinates. In this way the potential energy surfaces of the type shown in Figure 6.23 for the α-cleavage reaction of formaldehyde are obtained. Such a three-dimensional diagram is difficult to construct without calculations, and an initial analysis can be based on a consideration of the reactions to linear and bent products separately as shown in Figure 6.24. Sometimes the two reaction paths are plotted superimposed, with only the σ states of the acyl radical shown for the bent species and only the π states shown for the linear species (Salem, 1982), but such plots can be easily misunderstood and we avoid them. We shall return to these issues in connection with applications of these diagrams

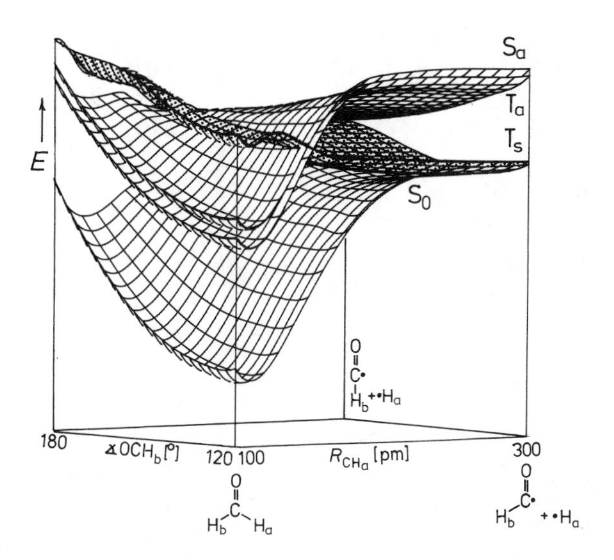

Figure 6.23. Potential energy surfaces for the α-cleavage reaction of formaldehyde as a function of the CH_a distance and the OCH_b angle; formaldehyde states are in the front left corner, correlation to the front right corner corresponds to cleavage of the α bond with bond angles kept constant, and correlation from here to the right rear corner corresponds to linearization of the formyl radical. Correlation between states of formaldehyde and of the linear biradical results from a cross section through these surfaces approximately along the diagonal from the front left to the rear right corner (by permission from Reinsch et al., 1987).

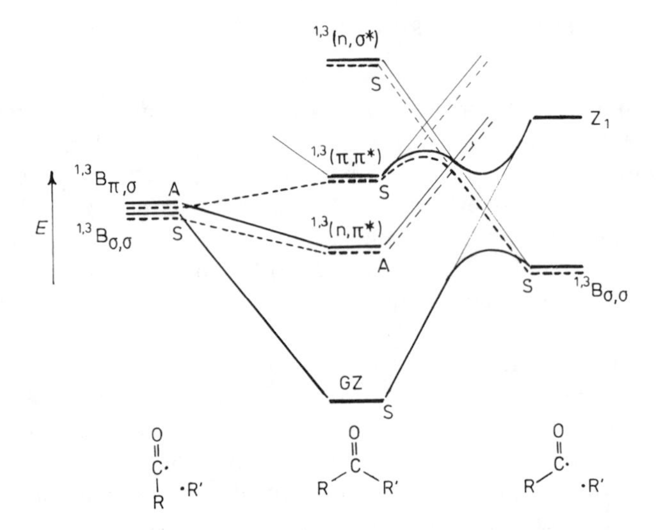

Figure 6.24. State correlation diagram for the α cleavage of saturated ketones. The path to a bent acyl radical is shown on the right, that to a linear acyl radical to the left. The right-hand part of the diagram corresponds to the front face of the three-dimensional representation in Figure 6.23; the left-hand part corresponds to the cross section along the diagonal in Figure 6.23.

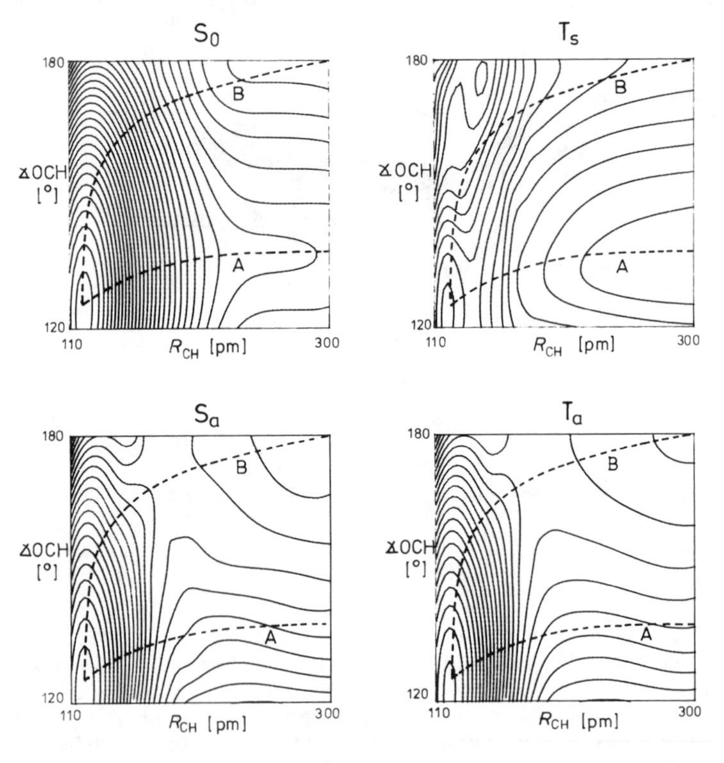

Figure 6.25. Contour diagrams of the potential energy surfaces for the α-cleavage reaction of formaldehyde shown in Figure 6.23. The broken lines indicate the reaction paths to the bent (A) and the linear (B) acyl radical (by permission from Reinsch et al., 1987).

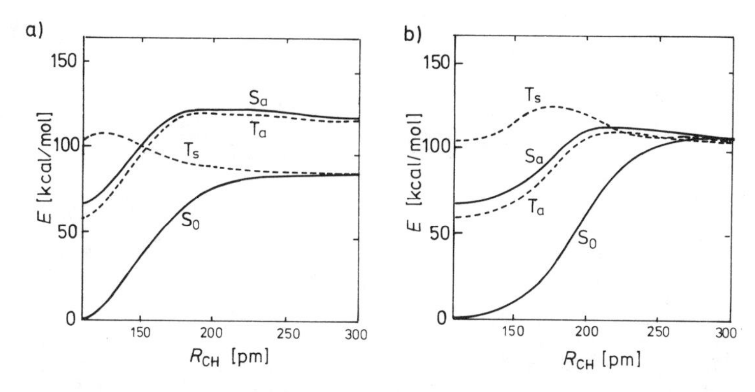

Figure 6.26. Potential energy curves for the α-cleavage reaction of formaldehyde. Part a) represents a cross section through the surfaces of Figure 6.23 approximately along the ∢OCH = 130° line, forming a bent acyl radical, and part b) a cross section along the minimum energy path on the (n,π*) excited surfaces S_a and T_a, forming a linear acyl radical.

in Section 7.2.1. It is important to remember that the two parts of Figure 6.24 do not represent different reactions but rather correspond to different reaction paths on the same potential energy surfaces. This becomes particularly evident from the contour diagrams shown in Figure 6.25. For the potential energy surfaces of Figure 6.23 these diagrams show the reaction path to the bent acyl radical, which is the minimum energy path in the ground state and in the $^3(\pi,\pi^*)$ state (S_0 and T_s), and the alternative reaction path to the linear acyl radical, which is more favorable in (n,π*) states (S_a and T_a). Cross sections through the surfaces of Figure 6.23 along these reaction paths are depicted in Figure 6.26. Figure 6.26a corresponds to the right-hand side of the correlation diagram shown in Figure 6.24, and Figure 6.26b to the left-hand side.

6.3.2 Salem Diagrams

In favorable cases state correlation diagrams of the type shown in Figure 6.21 and Figure 6.24 may be obtained simply from the leading VB structures of the reactant and the biradicaloid product. The molecular plane is chosen as the symmetry element—that is, one considers coplanar reactions, and any deviation from coplanarity may possibly be taken into account as an additional perturbation. As σ orbitals are symmetric and π orbitals antisymmetric with respect to the molecular plane, symmetries of the various states may be obtained by simply counting the number of σ and π electrons in a VB structure. A single structure is in general not sufficient to describe an electronic state accurately, but all contributing VB structures have the same symmetry, so the inclusion of just some of them is sufficient to establish a correlation diagram. The use of VB structures in the consideration of photochemical reaction paths was pioneered by Zimmerman (1969). The use of

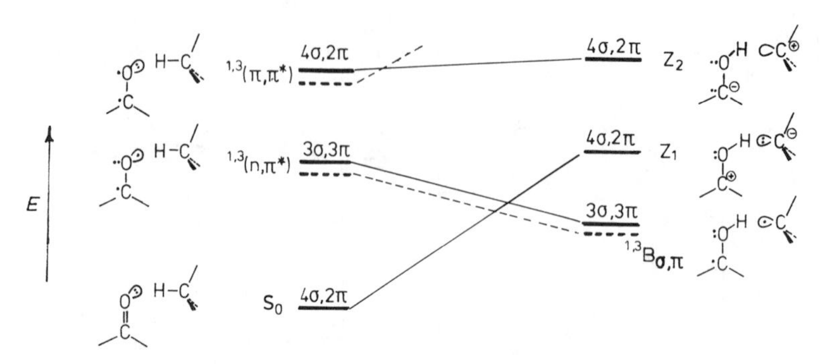

Figure 6.27. Salem diagram for hydrogen abstraction by a carbonyl compound. The biradicaloid product states are denoted by ^{1,3}B (dot–dot, covalent) and $Z_{1,2}$ (hole-pair, zwitterionic), respectively (by permission from Dauben et al., 1975).

VB correlation diagrams for this purpose was developed by Salem (1974) and collaborators (Dauben et al., 1975), and they are commonly referred to as *Salem diagrams*.

The photochemical hydrogen abstraction by carbonyl compounds, which has already been discussed in the last section, will be used to illustrate the procedure. The left-hand side of Figure 6.27 shows the dominant VB structures of the ground state and of the excited (n,π*) and (π,π*) states of the reactants. Only the electrons directly involved in the reaction are considered. In the ground and the (π,π*) states, the numbers of σ and π electrons are 4 and 2, respectively, and these states are symmetric relative to reflection in the mirror plane. In the (n,π*) states, both numbers are equal to 3, and these states are antisymmetric. The right-hand side shows the analogous VB structures for the various states of the biradicaloid system formed as a primary product. Dot–dot (covalent) states are denoted by ^{3}B and ^{1}B and hole-pair (zwitterionic) states by Z_1 and Z_2, using the nomenclature introduced in Section 4.3.1. The energy of these states depends on the nature of radical centers *A* and *B* as has also been discussed in Section 4.3.1.

In the present case both radical centers are C atoms, so the dot–dot states are clearly lower in energy than the charge-separated hole-pair ones. Since the electron count gives 3σ, 3π for the dot–dot and 4σ, 2π for the hole-pair states, correlation lines can be drawn immediately as shown in Figure 6.27. Thus, a diagram is obtained that is basically identical to the results represented in Figures 4.18 and 6.21 except for the barriers revealed by natural orbital correlations.

6.3.3 Topicity

The differences between the correlation diagrams for the dissociation of the H—H or Si—Si single bond (Figure 4.5) and for the dissociation of the C—C single bond in the Norrish type I process (Figure 6.24) are striking. In particular, in the former case, only the triplet state is dissociative, while in the latter case, a singlet state is as well. We have already seen in Section 4.3.3 that the correlation diagram for the dissociation of a single bond can change dramatically when the electronegativities of its termini begin to differ significantly. Now, however, both bonds to be cleaved are relatively nonpolar. The contrast is clearly not due to the differences in the properties of silicon and carbon, either. Rather, it is due to the fact that in one case a double bond is present at one of the termini of the bond that is being cleaved. This causes an increase in the number of low-energy states and changes the correlation diagram. This role of the additional low-energy states has been formalized in the concept of topicity (Salem, 1974; Dauben et al., 1975), which was generalized by its originators well beyond simple bond-dissociation processes. We believe that this concept is most powerful and unambiguous in the case of reactions in which only one bond dissociates, and shall treat this situation only.

Simple bond-dissociation reactions are classified as bitopic, tritopic, tetratopic, etc., according to the total number of active orbitals at the two terminal atoms of the bond. Active orbitals at an atom (AOs or their hybrids) are those whose occupancy is not always the same in all VB structures that are important for the description of the low-energy states that enter the correlation diagram for the bond-dissociation process. These are obviously the two orbitals needed to describe the bond to be cleaved, so the lowest possible topicity number is two, but possibly also other orbitals located at the two atoms: those containing lone pairs out of which excitation is facile (e.g., oxygen 2p), those that are empty such that excitation into them is facile (e.g., boron 2p), and those participating in multiple bonds adjacent to the bond being cleaved (e.g., carbonyl in the α-cleavage reaction).

It is now clear that the dissociation of the H—H bond and of the Si—Si bonds in saturated oligosilanes are bitopic processes. The Norrish type I α cleavage is an example of a tritopic process, sometimes subclassified further as a $\sigma(\sigma,\pi)$ tritopic process to specify that one terminal carries an active orbital of σ symmetry and the other two active orbitals, one of σ and one of π symmetry. Other examples of tritopic reactions are the dissociation of the C—N bond in a saturated amine and the dissociation of the C—O bond in a saturated alcohol (the oxygen 2s orbital is very low in energy, doubly occupied in all important states, and not counted as active). Figure 6.28 shows the correlation diagram for the dissociation of the C—O bond in methanol. Diagrams for such simple dissociations are not necessarily of use in themselves, not only since these reactions require high-energy excitation and are relatively rarely investigated by organic photochemists, but primarily because the lowest excited states of fully saturated organic molecules are usually of Rydberg character while only valence states are shown in the diagrams. Thus, at least one side of the diagram is quite unrealistic. Experimental (Keller et al., 1992; Jensen et al., 1993) and theoretical (Yar-

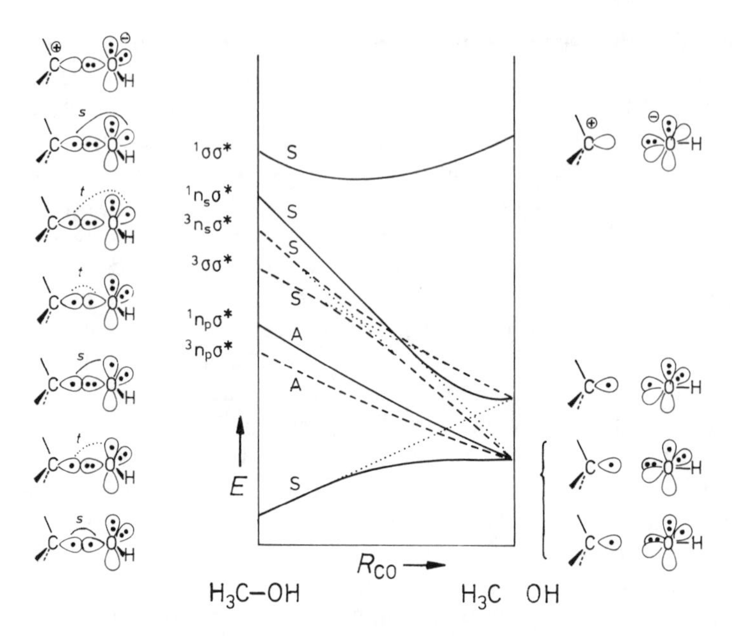

Figure 6.28. VB structure (dotted lines) and state (full lines) correlation diagram for the dissociation of the C—O bond in methanol (by permission from Michl and Bonačić-Koutecký, 1990).

kony, 1994) investigations of Rydberg-valence interactions in the processes $CH_3SH \rightarrow CH_3 + SH$ and $CH_3SH \rightarrow CH_3S + H$ exemplify this situation.

Although diagrams such as that of Figure 6.28 are useless for the understanding of molecules such as CH_3OH or CH_3SH themselves, they are still useful for the derivation of orbital correlation diagrams for bond dissociation reactions of molecules with unsaturated chromophores, using the stepwise procedure described in Section 4.2.2. Thus, Figure 6.28 forms the basis for the understanding of reactions such as the photo-Fries and photo-Claisen rearrangements discussed in Section 7.2.1; it applies to the ring opening reactions of oxiranes, etc.

Examples of tetratopic reactions are the C—N bond dissociation in azo compounds, discussed in Section 7.2.2, C—X bond dissociation in alkyl halides, and the O—O bond dissociation in peroxides. Examples of pentatopic reactions are the dissociation of the C—X bond in vinyl halides, of the C═C bond in ketenes, and of the C═N bond in diazoalkanes. An example of a hexatopic bond dissociation is the fragmentation of an alkyl azide to a nitrene and N_2. A verification of the topicity rules at a semiempirical level was reported (Evleth and Kassab, 1978), and a detailed description of the electronic structure aspects of bond dissociations characterized by various topicity numbers, with references to the original literature, has appeared recently (Michl and Bonačić-Koutecký, 1990).

Supplemental Reading

Dauben, W.G., Salem, L., Turro, N.J. (1975), "A Classification of Photochemical Reactions," *Acc. Chem. Res.* **8**, 41.

Dougherty, R.C. (1971), "A Perturbation Molecular Orbital Treatment of Photochemical Reactivity. The Nonconservation of Orbital Symmetry in Photochemical Pericyclic Reactions," *J. Am. Chem. Soc.* **93**, 7187.

Förster, Th. (1970), "Diabatic and Adiabatic Processes in Photochemistry," *Pure Appl. Chem.* **24**, 443.

Gerhartz, W., Poshusta, R.D., Michl, J. (1976), "Excited Potential Energy Hypersurfaces for H_4 at Trapezoidal Geometries. Relation to Photochemical 2s + 2s Processes," *J. Am. Chem. Soc.* **98**, 6427.

Michl, J. (1974), "Physical Basis of Qualitative MO Arguments in Organic Photochemistry," *Fortschr. Chem. Forsch.* **46**, 1.

Michl, J., Bonačić-Koutecký, V. (1990), *Electronic Aspects of Organic Photochemistry*; Wiley: New York.

Salem, L. (1976), "Theory of Photochemical Reactions," *Science* **191**, 822.

Salem, L. (1982), *Electrons in Chemical Reactions: First Principles*; Wiley: New York.

Turro, N.J., McVey, J., Ramamurthy, V., Cherry, W., Farneth, W. (1978), "The Effect of Wavelength on Organic Photoreactions in Solution. Reactions from Upper Excited States," *Chem. Rev.* **78**, 125.

Zimmerman, H.E. (1976), "Mechanistic and Explaratory Organic Photochemistry," *Science* **191**, 523.

Zimmerman, H.E. (1982), "Some Theoretical Aspects of Organic Photochemistry," *Acc. Chem. Res.* **15**, 312.

CHAPTER

7

Organic Photochemistry

Examples of photoreactions may be found among nearly all classes of organic compounds. From a synthetic point of view a classification by chromophore into the photochemistry of carbonyl compounds, enones, alkenes, aromatic compounds, etc., or by reaction type into photochemical oxidations and reductions, eliminations, additions, substitutions, etc., might be useful. However, photoreactions of quite different compounds can be based on a common reaction mechanism, and often the same theoretical model can be used to describe different reactions. Thus, theoretical arguments may imply a rather different classification, based, for instance, on the type of excited-state minimum responsible for the reaction, on the number and arrangement of centers in the reaction complex, or on the number of active orbitals per center. (Cf. Michl and Bonačić-Koutecký, 1990.)

Since it is not the objective of this chapter to give either a complete review of all organic photoreactions or an exhaustive account of the applicability of the various theoretical models, neither of these classifications is followed strictly. Instead, the interpretation of experimental data by means of theoretical models will be discussed for selected examples from different classes of compounds or reaction types in order to elucidate the influence of molecular structure and reaction variables on the course of a photochemical reaction.

7.1 Cis-trans Isomerization of Double Bonds

Cis-trans photoisomerizations have been studied in great detail and can serve as an instructive example for the use of state correlation diagrams in discussing photochemical reactions. They have been observed for olefins, azomethines, and azo compounds.

7.1.1 Mechanisms of cis-trans Isomerization

Possible cis-trans isomerization mechanisms of isolated double bonds can be discussed using ethylene as an example. A state correlation diagram for this conversion is shown in Figure 4.6. (Cf. also Figure 2.2.) This diagram clearly displays the strongly avoided crossing that gives rise to a biradicaloid minimum in the S_1 surface (Cf. Section 4.3.2). This is surely not a minimum with respect to distortions other than pure twisting (see below). The diagram also displays the biradicaloid minimum in the T_1 surface, where T_1 and S_0 are nearly degenerate.

According to the MO model, both the HOMO and the LUMO are singly occupied in the T_1 state by electrons of parallel spin. The minimum in this state then arises from the fact that the destabilizing effect of the π^* MO (LUMO) is stronger than the stabilizing effect of the π MO (HOMO) if overlap of the AOs is taken into account. For the biradicaloid geometry, however, the overlap vanishes and its destabilizing effect disappears. (Cf. Section 4.3.2.) From the state energies of a homosymmetric biradicaloid shown in Figure 4.20, the energy difference between the S_1 and T_1 states is expected to be constant; Figure 4.6 indicates that this is true to a good approximation. Since for a twisting of the double bond the internuclear distance is more or less fixed by the σ bond, there are in this case no loose geometries that could favor the T_1 state.

Similar arguments apply to the S_1 state. Since twisted ethylene is a homosymmetric biradicaloid, this state is described by configurations that correspond to charge-separated structures in the VB picture, such as

$$\overset{\oplus}{\text{C}}-\overset{\ominus}{\text{C}} \longleftrightarrow \overset{\ominus}{\text{C}}-\overset{\oplus}{\text{C}}$$

(cf. Section 4.3.1). It has been proposed that charge separations of this type, which may bring about very polar excited states if the symmetry is perturbed (*sudden polarization effect,* cf. Section 4.3.3), may also be important in photochemical reactions (Bruckmann and Salem, 1976), but no experimental evidence supports this so far.

At the level of the 3 × 3 CI model, heterosymmetric perturbations δ that introduce an energy difference between the two localized nonbonding orbitals of the twisted double bond may reduce the S_1–S_0 gap to zero and provide a very efficient relaxation path. (Cf. Section 4.3.3.) Calculations show that pyramidalization of one of the methylene groups alone does not represent a

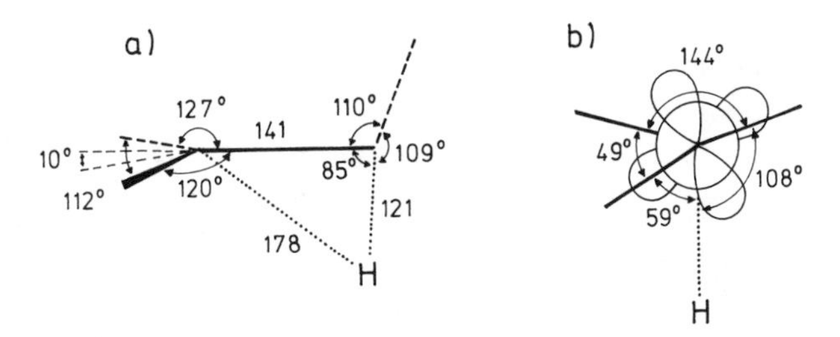

Figure 7.1. Geometry of the S_1–S_0 conical intersection of ethylene calculated at the CASSCF level, indicating the possibility of cis-trans isomerization and [1,2] hydrogen shift; a) side view, b) Newman projection with localized nonbonding orbitals (Freund and Klessinger, 1995).

perturbation δ sufficient for reaching a critically heterosymmetric biradicaloid geometry (real conical intersection). However, this is accomplished by an additional distortion of one of the CH bonds toward the other carbon atom (Ohmine, 1985; Michl and Bonačić-Koutecký, 1990). The geometry of the resulting conical intersection calculated at the CASSCF level is shown in Figure 7.1. Thus it is most likely that ethylene cis-trans isomerization and [1,2] hydrogen shift occur via the same funnel. (Cf. Section 6.2.1.)

Two essentially different cis-trans isomerization mechanisms may be derived from the state correlation diagram of ethylene:

1. Through absorption of a photon, the molecule reaches the biradicaloid minimum in S_1 via one of the excited singlet states; return from this minimum to the ground state S_0 at the pericyclic funnel region close to the orthogonal geometry can then lead to either of the planar geometries. Heterosymmetric perturbations that introduce an energy difference between the two localized nonbonding orbitals of the twisted form reduce the S_0–S_1 gap, possibly to zero, and provide very efficient relaxation paths. (Cf. Section 4.3.3.)
2. Energy transfer from a triplet sensitizer produces the T_1 state; return to the ground state S_0 occurs again for a twisted geometry. T_1 has a minimum near $\theta = 90°$, but spin-orbit coupling is inefficient at this geometry (cf. Section 4.3.4), and the return to S_0 most likely occurs at a smaller twist angle. The advantage of sensitization is that it is readily applicable to monoolefins, which require very high energy radiation for singlet excitation. Furthermore, competing reactions of the S_1 state such as valence isomerizations, hydrogen shifts, and fragmentations are avoided.

Given suitable reaction partners, the T_1 state of the olefin may also be reached via an exciplex and a radical ion pair (see Section 7.6.1), which may undergo ISC and subsequent reverse electron transfer (Roth and Schilling, 1980).

Additional mechanisms that have been established may be termed photocatalytic. One of these is the *Schenck mechanism,* which involves an addition of the sensitizer (Sens) to the double bond and formation of a biradicaloid intermediate that is free to rotate about the double bond and subsequently collapses to the sensitizer and olefin:

$$\text{Sens} \xrightarrow{h\nu} \text{Sens}^*$$

$$\text{Sens}^* + \text{RHC=CHR} \longrightarrow \bullet\text{Sens}-\text{CHR}-\text{CHR}\bullet \rightleftharpoons \bullet\text{Sens}-\text{CHR}-\text{CHR}\bullet \xrightarrow{-\text{Sens}} \text{RHC=CHR}$$

Obviously, the reaction can proceed according to this mechanism even if the triplet energy of the sensitizer is below that of the olefin (Schenck and Steinmetz, 1962).

Instead of the sensitizer, a photolytically generated halogen atom can also add to the olefin and produce a radical, which may then rotate about the double bond:

$$\frac{1}{2}\text{X}_2 \xrightarrow{h\nu} \text{X}\bullet \xrightarrow{\text{RCH=CHR}} \text{X}-\text{CHR}-\text{CHR}\bullet$$

7.1.2 Olefins

Cis-trans isomerization of simple nonconjugated olefins is difficult to achieve by direct irradiation because such high energy is required for singlet excitation ($\lambda < 200$ nm). Cis-trans isomerization via the S_1 state has been observed for 2-butene (**1**) (Yamazaki and Cvetanović, 1969); at higher concentrations a photochemical [2 + 2] cycloaddition comes into play. (See Section 7.4.1.) Cis-trans isomerizations of cycloalkenes clearly demonstrate the influence of ring strain: *trans*-cycloalkenes are impossible to make with unsaturated small rings; with six- and seven-membered rings the trans products can be detected at low temperatures (cf. Bonneau et al., 1976; Wallraff and Michl, 1986; Squillacote et al., 1989), whereas larger rings give *trans*-cycloalkenes that are stable at room temperature. *trans*-Cyclooctene (**2**) has been obtained on direct as well as on sensitized irradiation (Inoue et al., 1977). Enantioselective cis-trans isomerization with very high optical purities (64%) has been obtained for cyclooctene (**2**) by triplex-forming sensitizers (Inoue et al., 1993).

1 2

In protic solvents, strained *trans*-cycloalkenes such as methylcyclohexene (**3**) give an adduct according to the following scheme (Kropp et al., 1973):

CH_3 $h\nu$ CH_3 CH_3OD $\oplus$ CH_3 D + $CH_3O^{\ominus}$ H_3C OCH_3 D

3

From the state correlation diagram of cis-trans isomerization in Figure 4.6, it is seen that the crossing is strongly avoided with a large energy difference ΔE between S_0 and S_1 for the biradicaloid geometry ($\theta = 90°$). The energy gap ΔE will be reduced at less symmetrical geometries, where the electronegativities of the two termini of the double bond will differ. (Cf. Section 4.3.3.) Still, in simple olefins the result could well be a biradicaloid minimum rather than a funnel, and this suggests the occurrence of an intermediate. In the earlier literature this intermediate is referred to as *phantom state* $^1P^*$ (Saltiel et al., 1973). Therefore the fraction β of molecules in the biradicaloid minimum that reach the ground state of the *trans*-olefin is independent of the initial isomer. At a given wavelength λ the cis and the trans isomers c and t are formed in a constant ratio, and after a certain period of time a *photostationary state* (PSS) is reached. The composition of the photostationary state is given by

$$([c]/[t])_{PSS} = [\varepsilon_t(\lambda)/\varepsilon_c(\lambda)](\Phi_{t\rightarrow c}/\Phi_{c\rightarrow t}) \tag{7.1}$$

where $\varepsilon(\lambda)$ is the extinction coefficient at wavelength λ and Φ is the quantum yield of the photochemical conversion. (See Example 7.1.) If there are no competing reactions one has $\Phi_{c\rightarrow t} = \beta$ and $\Phi_{t\rightarrow c} = 1 - \beta$, and

$$([c]/[t])_{PSS} = [\varepsilon_t(\lambda)/\varepsilon_c(\lambda)][(1 - \beta)/\beta] \tag{7.2}$$

Complete conversion may be achieved only if it is possible to choose solvent and excitation wavelength such that only one of the isomers absorbs and $\varepsilon \approx 0$ for the other one.

Example 7.1:
The rate of a simple photoreaction of species X is given by

$$d[X]/dt = \Phi_X I_X^a$$

where the fraction I_X^a is obtained by multiplying the total absorbed radiation intensity I_{tot}^a by the ratio $A_X(\lambda)/A_{tot}(\lambda)$ of the absorbances at the corresponding wavelength λ. Photostationary equilibrium is reached if the rates of cis-trans and trans-cis conversion are equal, and for direct irradiation

$$\Phi_{c\rightarrow t}\frac{A_c(\lambda)}{A_{tot}(\lambda)}I_{tot}^a = \Phi_{t\rightarrow c}\frac{A_t(\lambda)}{A_{tot}(\lambda)}I_{tot}^a$$

From Lambert-Beer's law, $A_X(\lambda) = \varepsilon_X(\lambda)[X]d$, and one has

$$\Phi_{c\rightarrow t}\varepsilon_c(\lambda)[c] = \Phi_{t\rightarrow c}\varepsilon_t(\lambda)[t]$$

which then yields Equation (7.1). In the case of a sensitized reaction, the rate of formation of triplet-excited molecules $^3X^*$ is proportional to the product $k_{ET}^X[^3S^*][X]$ of the rate constant for energy transfer and the concentrations of sensitizer and molecule X in the ground state, and k_{ET}^c and k_{ET}^t replace ε_c and ε_t in the foregoing equations.

Sensitization of alkenes by carbonyl compounds, which is likely to proceed via an exciplex, can be accompanied by oxetane formation. (See Section 7.4.4.)

7.1.3 Dienes and Trienes

The singlet cis-trans isomerization mechanisms of conjugated olefins appear to be quite different from those of isolated olefins.

Direct irradiation of butadienes is known to yield a mixture of cis-trans isomerization and cyclization photoproducts. (Squillacote and Semple, 1990; Leigh, 1993) All these photochemical transformations are now believed to involve passage through a common funnel. (Cf. Scheme 2 in Section 6.2.1.2.) In agreement with expectations from correlation diagrams (Figure 4.6; see also Figure 7.3 below), the initially excited optically allowed state of B symmetry (S) is depopulated in ~10 fs owing to fast internal conversion to the nearby state of A symmetry (D) (Trulson and Mathies, 1990). Calculations (Olivucci et al., 1993; Celani et al., 1995) in fact yield a $1B_2$–$2A_1$ conical intersection near the bottom of the B state minimum. The energy of the $2A_1$ state drops rapidly as one proceeds further toward the pericyclic funnel at a disrotatory geometry in which all three CC bonds are considerably twisted, and pericyclic as well as diagonal interactions are present (Section 6.2.1 and Figures 6.15 and 6.16). The calculations suggest that after crossing from the B state it might be possible, but 10 kcal/mol more costly energetically, to reach the funnel region on the A surface by an initially conrotatory motion followed by a backward twist of one of the double bonds. This path, however, appears to have no chance to compete with the barrierless disrotatory approach toward the funnel region described earlier.

It is probably reasonable to assume that the excited-state motion is initially dominated by the slope of the B and A surfaces, which points in the disrotatory way and toward the diagonally-bonded pericyclic funnel, and to assume that the acquired momentum is kept after the jump to S_0. This would point in the direction of bicyclobutane. Unless the surface jump occurs right at the cone tip, it generates an additional momentum in the x_2 direction, that is, along the γ perturbation coordinate, toward cyclobutene and the original as well as cis-trans isomerized butadiene. Whether further facile motions on the S_0 surface follow, in particular s-cis-s-trans interconversion, before all excess vibrational energy is lost, is hard to tell.

The case of *s-trans*-butadiene has been investigated computationally in less detail. The disrotatory pathway is again favored and enters the pericyclic funnel region of conical intersections at a much larger twist angle along the central C—C bond. This pathway is again barrierless and is steeper than the s-cis pathway. After the jump to S_0, diagonal bonding to bicyclobutane appears more likely and peripheral bonding to cyclobutene quite unlikely. It seems probable that the molecule will enter the S_0 state with a higher velocity along the s-trans isomerization path, suggesting a higher efficiency for the s-trans→s-cis than for the s-cis→s-trans isomerization. In fact, it is known that in 2,3-dimethylbutadiene the s-trans→s-cis isomerization is 10 times more efficient than the s-cis→s-trans isomerization (Squillacote and Semple, 1990).

Triplet-sensitized cis-trans isomerization is frequently more efficient. Thus, the direct excitation of 1,3-pentadiene (piperylene), which has been studied in great detail, results in cis-trans isomerization with very low quantum yields ($\Phi_{c\to t} = 0.09$, $\Phi_{t\to c} = 0.01$) and very small quantities of dimethylcyclopropene as a side product, whereas when benzophenone is used as sensitizer, $\Phi_{c\to t} = 0.55$ and $\Phi_{t\to c} = 0.44$. This efficient quenching of higher-energy triplet states by conjugated dienes is utilized in mechanistic studies for identifying the excited state responsible for a photochemical reaction.

Computational results similar to those for butadiene have been obtained for the singlet excited-state cis-trans isomerization of *cis*-hexatriene around its single and double CC bonds (Olivucci et al., 1994a). For each of these torsional modes, a reaction pathway has been found on the D excited state (A symmetry). Both lead over a small barrier to funnels that are reached already at about 60° twist angles. In fact, the fluorescence excitation spectrum of very cold *cis*-hexatriene in supersonic jet is consistent with the existence of two very fast radiationless decay channels with a barrier below 1 kcal/mol (Petek et al., 1992). As in butadiene, the funnels are located at biradicaloid geometries with pericyclic and diagonal interactions, as indicated in Figure 7.2, which also schematizes some of the possible bond-forming processes that are expected to occur along different ground-state relaxation paths after the S_1–S_0 return.

The reaction pathways both for single- and double-bond isomerization enter the funnel region at less than one-third of the way toward the products, suggesting that the majority of excited-state molecules should decay back to the ground-state reactant. In addition, the excited-state barrier for single-bond isomerization is smaller than that for double-bond isomerization, reducing the quantum yield for cis-trans isomerization (since the product of s-cis-s-trans isomerization is just a different conformer, not a new cis-trans isomer of the reactant). Thus the low experimental value of $\Phi = 0.034$ reported for the quantum yield of *trans*-hexatriene from *cis*-hexatriene (Jacobs

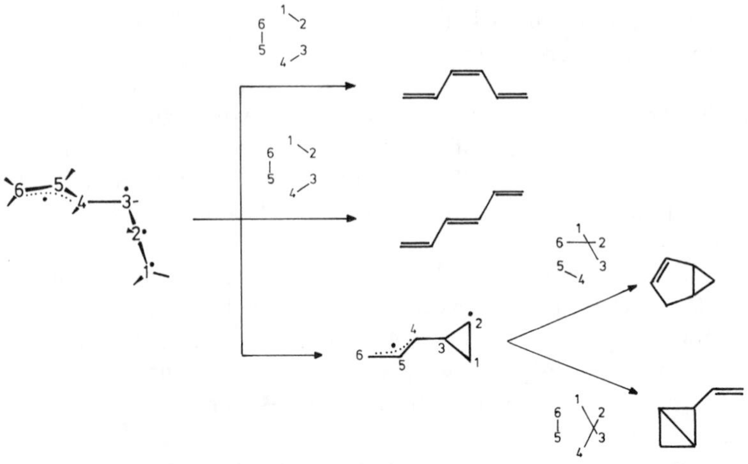

Figure 7.2. Schematic representation of the conical intersection structure for cis-trans isomerization of *cis*-hexatriene and of some of the possible bond-making processes that might occur along different ground-state relaxation paths (by permission from Olivucci et al., 1994a).

and Havinga, 1979) is easily rationalized. 2-Vinylbicyclo[1.1.0]butane (Datta et al., 1971) and bicyclo[3.1.0]hexene (Jacobs and Havinga, 1979) are also found in small yields, as might be suspected from the geometry of the funnel (Figure 7.2) and by analogy to butadiene.

Example 7.2:

From the quantum yields of the direct cis-trans photoisomerization of 2,4-hexadienes it is seen that a one-step conversion of the trans-trans isomer into the cis-cis isomer or vice versa does not occur. A common intermediate can therefore be excluded, and at least two different intermediates have to be assumed as indicated in Scheme 1 (Saltiel et al., 1970). In contrast to this singlet reaction with very fast return to S_0, the triplet reaction is characterized by intermediate lifetimes long enough to allow for interconversion. The result of the benzophenone-sensitized photoisomerization of 2,4-hexadienes is therefore in agreement with a common triplet intermediate for the isomerization of both double bonds (Saltiel et al., 1969).

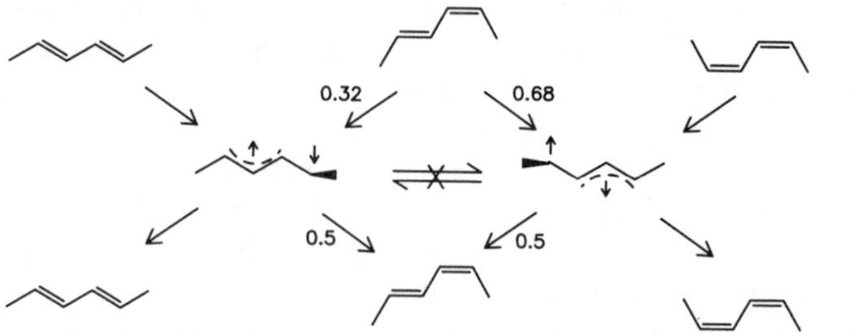

0.32 0.68

0.5 0.5

Scheme 1

7.1.4 Stilbene

The cis-trans isomerization of stilbene (**4**) has been thoroughly studied. The potential-energy diagram for a rotation about the CC double bond that preserves a twofold symmetry axis is shown in Figure 7.3.

4

As in ethylene, at a twist angle of about 90° the doubly excited state, responsible for the biradicaloid minimum in S_1, lies below the singly excited state. In Figure 4.6 it is fairly widely separated from S_0; in olefins with more extensive conjugation, however, the gap ΔE at the twisted geometry is smaller and the possibilities for reaching a critically heterosymmetric biradicaloid geometry richer (cf. Section 4.3.3). As a result, the biradicaloid minimum may actually turn into a funnel: In the case of stilbene, any possible intermediate in the cis→trans process is found to have a lifetime of less than 150 fs (Abrash et al., 1990). Nevertheless, a transient spectrum different from those of *cis*- and *trans*-stilbene has been observed upon excitation of

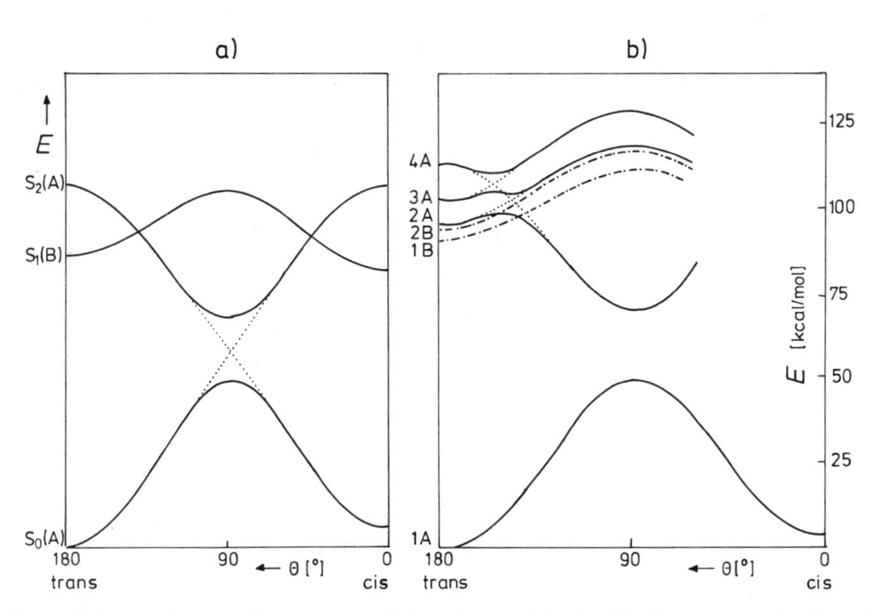

Figure 7.3. a) Schematic state correlation diagram for the cis-trans isomerization of stilbene along the symmetric path (by permission from Orlandi and Siebrand, 1975); b) modified Orlandi-Siebrand diagram (by permission from Hohlneicher and Dick, 1984).

cis-stilbene (Doany et al., 1985), presumably due to vibrationally hot excited *trans*-stilbene.

In contrast to the situation in ethylene, the model proposed by Orlandi and Siebrand (1975) suggests a maximum at $\theta = 90°$ along the symmetrical reaction path for the singly excited state because of conjugative stabilization of the planar configurations ($\theta = 0°$ and $\theta = 180°$), absent in ethylene. The result is a barrier in S_1 separating the spectroscopic minimum of the trans isomer ($\theta = 180°$), but not that of the less stable cis isomer (Greene and Farow, 1983), from the biradicaloid minimum at $\theta = 90°$. The situation is similar in the case of long-chain polyenes, which also show fluorescence (Hudson and Kohler, 1973). As a consequence of the barrier in the S_1 state, the quantum yield for the trans-cis isomerization of stilbene is temperature dependent; at lower temperatures fluorescence becomes increasingly important as a competing process; and Φ_F becomes nearly unity at temperatures below 100 K (Figure 7.4). If intersystem crossing can be neglected, one has

$$\Phi_{t \to c} = (1 - \Phi_F)(1 - \beta) \tag{7.3}$$

The rate constant for the isomerization of *trans*-stilbene in the S_1 state is also affected by solvent viscosity and has served as a favorite prototype for the investigation of solvent dynamics in fast monomolecular kinetic processes (Saltiel and Sun, 1990).

Quantum chemical calculations essentially confirm the above simple model for the trans-cis isomerization of stilbene. Two-photon excitation spectra and more recent calculations indicate, however, that contrary to what is shown in Figure 7.3a, it is not the lowest but rather a higher excited A state of *trans*-stilbene that correlates with the lowest excited singlet state

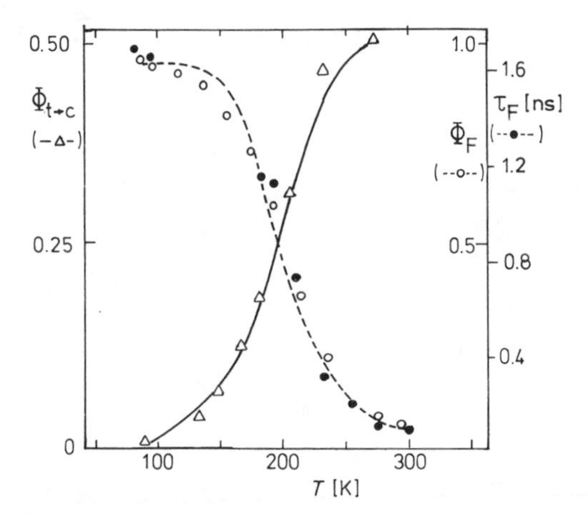

Figure 7.4. Temperature dependence of the quantum yield $\Phi_{t \to c}$ for trans-cis isomerization (△), of the quantum yield Φ_F of fluorescence (○), and of the fluorescence lifetime (●) of stilbene (by permission from Saltiel and Charlton, 1980).

at the twisted biradicaloid minimum. This results in several avoided crossings and a barrier in the lowest excited singlet state as shown in Figure 7.3b (Hohlneicher and Dick, 1984). Other authors have proposed that the barrier in S_1 does not result from an avoided crossing, and that the predominantly doubly excited state does not become S_1 until after the barrier has been passed and the perpendicular geometry is nearly reached (Troe and Weitzel, 1988). It is difficult to make any definitive a priori statements about the potential energy surfaces of a molecule of this size at this time, particularly since the optimal reaction path almost certainly preserves no symmetry elements.

Substituted stilbenes have also been extensively studied (Saltiel and Charlton, 1980; see also Saltiel and Sun, 1990).

On direct irradiation the formation of dihydrophenanthrene (**5**, DHP) accompanies the cis-trans isomerization of stilbene. Thus, one has

$$\Phi_{c \to t} = (1 - \Phi_{\mathrm{DHP}})\beta \tag{7.4}$$

where $\beta \approx 0.4$ is the fraction of molecules in the biradicaloid minimum that reach the ground state of *trans*-stilbene.

H
H′
5

Using azulene as a triplet quencher it has been shown that triplet states are not involved in the cis-trans isomerization of stilbene on direct excitation. Triplet-sensitized cis-trans isomerization, however, is observed and proceeds in both directions through a minimum in the triplet potential energy surface at a twisted geometry, often referred to as the triplet "phantom" state $^3P^*$.

From Example 7.1 the photostationary state may be written as

$$([c]/[t])_{\mathrm{PSS}} = (k^t_{\mathrm{ET}}/k^c_{\mathrm{ET}})(\Phi_{t \to c}/\Phi_{c \to t})$$

with quantum yields

$$\Phi_{t \to c} = 1 - \gamma$$

and

$$\Phi_{c \to t} = \gamma$$

where γ is the fraction of molecules in the triplet minimum that form the trans isomer. Since for stilbene β and γ are both close to 0.4, it is probable that the singlet and the triplet minima are located at similar geometries (Saltiel and Charlton, 1980). Insertion of the expressions for $\Phi_{t \to c}$ and $\Phi_{c \to t}$ yields

$$([c]/[t])_{\mathrm{PSS}} = (k^t_{\mathrm{ET}}/k^c_{\mathrm{ET}})\ [(1 - \gamma)/\gamma] \tag{7.5}$$

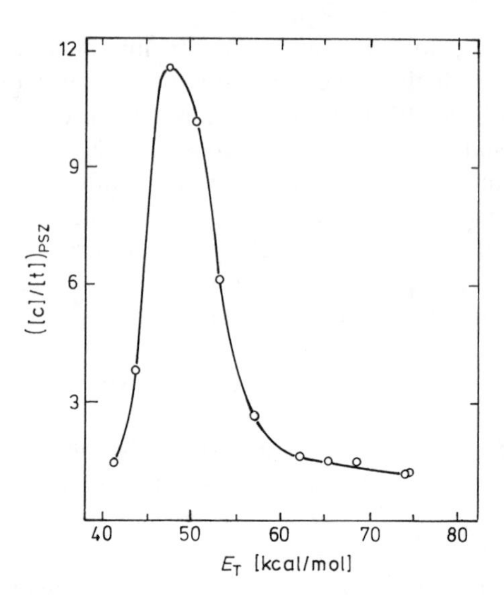

Figure 7.5. Variation of the photostationary state in cis-trans isomerization of stilbene with triplet energy E_T of the sensitizer (by permission from Saltiel and Charlton, 1980).

The ratio k^t_{ET}/k^c_{ET} may vary over a wide range. Vertical excitation of *trans*-stilbene and of *cis*-stilbene requires sensitizer triplet energies of $E_T > 50$ kcal/mol and $E_T > 57$ kcal/mol, respectively. When E_T of the sensitizer is sufficiently higher than the triplet energies of both the cis and trans olefin, triplet energy transfer is diffusion-controlled; that is, practically every encounter results in energy transfer. Sensitizers with a triplet energy above that of *trans*-stilbene but below that of *cis*-stilbene transfer triplet energy only to the trans isomer, and k^c_{ET} is very small. According to Equation (7.5), $([c]/[t])_{PSS}$ then becomes very large. The dependence of $([c]/[t])_{PSS}$ on E_T of the sensitizer shown in Figure 7.5 is thus easy to understand; the fact that sensitization occurs even for $E_T < 50$ kcal/mol has been explained by "nonvertical" energy transfer into the triplet minimum $^3P^*$ (Hammond and Saltiel, 1963), which presumably corresponds to transitions originating from a vibrationally excited ground state. Other mechanisms have been discussed as well (Saltiel and Sun, 1990).

7.1.5 Heteroatom, Substituent, and Solvent Effects

As indicated in Section 4.3.3, substituent as well as environmental effects on photochemical cis-trans isomerization can be discussed in terms of the dependence of the energy gap and consequently also of the $S_1 \rightarrow S_0$ transition rate on the electronegativity difference δ. This is determined by the nature of the atoms on the double bond, by substituents, and by the solvent. If δ is

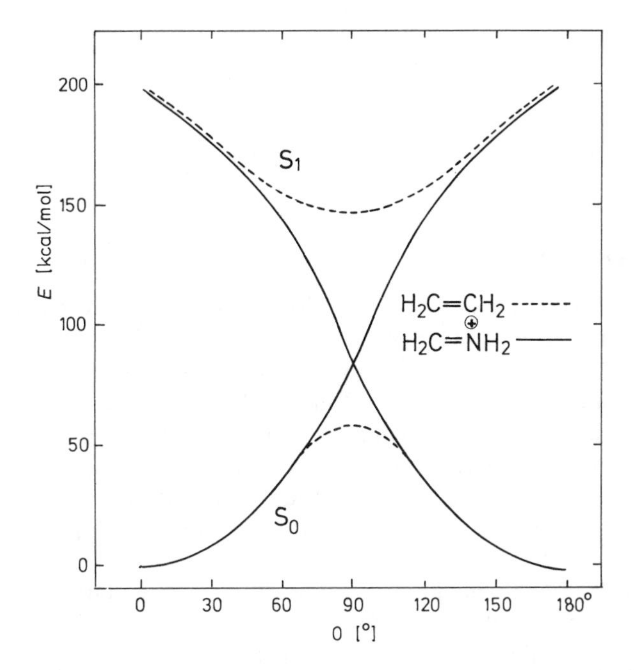

Figure 7.6. Calculated potential energy curves for a twist in the double bond in $CH_2{=}CH_2$ (---) and $CH_2{=}NH_2^{\oplus}$ (—) (by permission from Bonačić-Koutecký et al., 1984).

equal to the critical quantity δ_0 the energy gap disappears and the S_1 and S_0 surfaces touch. According to Figure 7.6 this situation is just attained for the twisted formaldiminium ion $CH_2{=}NH_2^{\oplus}$. In this case a thermal equilibrium in S_1 is not to be expected at the biradicaloid minimum, which corresponds to a funnel instead. Both the conversion of the trans excited state to the cis ground state and the conversion of the cis excited state to the trans ground state would then proceed with dynamical memory and the sum of the quantum yields $\Phi_{c\rightarrow t}$ and $\Phi_{t\rightarrow c}$ might therefore reach the limiting values of zero and 2 (Michl, 1972; see also the conversion of azomethines below). These processes are very important for the understanding of the rapid deactivation of excited states of triphenylmethane and rhodamine dyes (Rettig et al., 1992). In cyanine dyes (Momicchioli et al., 1988), for example, δ can go beyond the critical value δ_0, and it can become very large in TICT molecules. (Cf. Section 5.5.2 and 5.5.3.)

In the case of the retinal Schiff base (**6**) the efficiency of cis-trans isomerization of the double bond between C-11 and C-12 is considerably enhanced by polar solvents on the one hand and by protonation of the Schiff base on the other hand (Becker and Freedman, 1985). This is rather important because **6** is the chromophore of rhodopsin and this isomerization represents one of the primary steps in vision.

6

7.1.6 Azomethines

Syn-anti isomerization about a C=N double bond is intrinsically more complicated than cis-trans isomerization of a C=C double bond. This is due to the fact that n→π* excitations have to be discussed in addition to π→π* excitations, and because syn-anti isomerization can be effected by either of two linearly independent kinds of motion or their linear combination, namely twisting and in-plane inversion at the nitrogen atom. (See Figure 7.7.) It is believed that in simple azomethines thermal isomerizations occur through inversion, while photochemical isomerizations proceed along a twisting path (Paetzold et al., 1981).

This has been confirmed by quantum chemical calculations of the potential energy surfaces of the ground state and the lowest excited states of formaldimine in the two-dimensional subspace defined by the twisting and linear inversion motions (Bonačić-Koutecký and Michl, 1985a). Selected cuts through these surfaces for different dihedral angles are displayed in Figure 7.8. Whereas the ground state prefers planar geometries

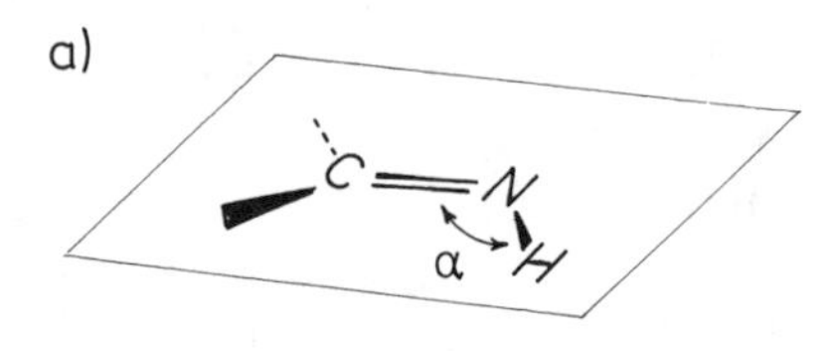

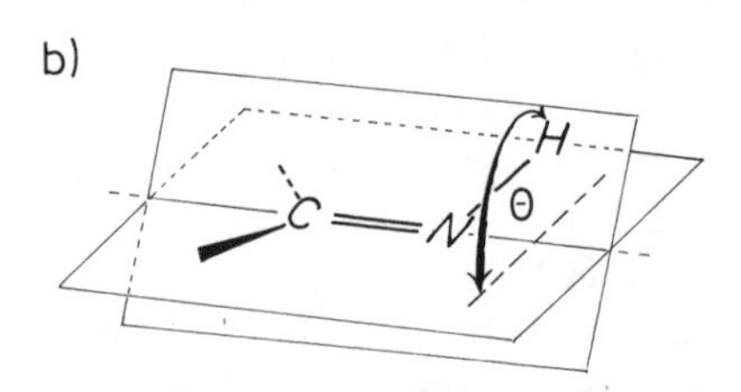

Figure 7.7. The syn-anti isomerization of formaldimine a) through in-plane inversion and b) by rotation. α is the CNH valence angle and θ the torsional angle.

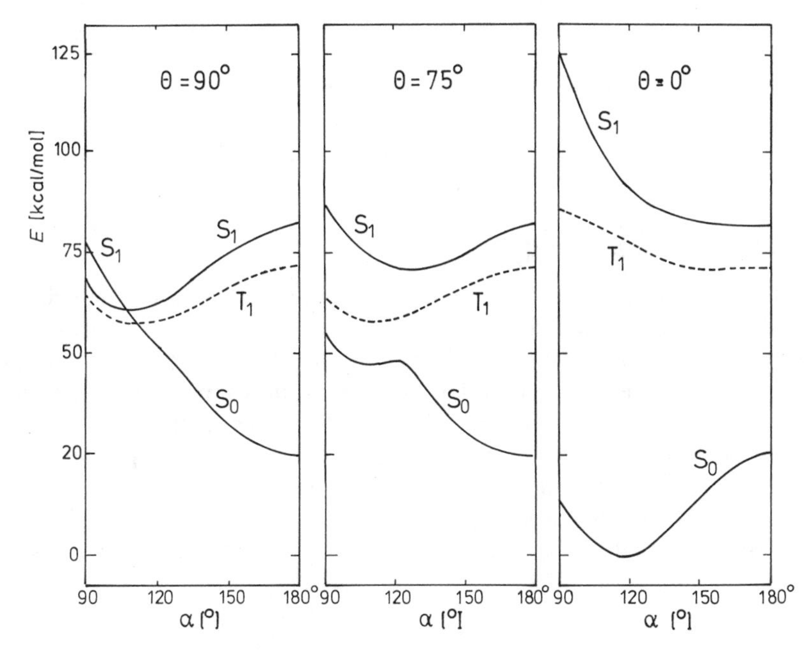

Figure 7.8. Dependence of the energy of the lowest states of formaldimine on the valence angle α, shown for selected values of the twist angle θ (by permission from Bonačić-Koutecký and Michl, 1985a).

($\theta = 0°$ or 180°), orthogonal geometries ($\theta = 90°$) are preferred by the T_1 and S_1 states that correspond to $n \rightarrow \pi^*$ excitation. The very small energy gap between S_1 and S_0 for orthogonal geometries in the region $100° < \alpha < 120°$ can be viewed as a consequence of a conical intersection at a valence angle $\alpha = 106.5°$. Thus, vertical excitation into the S_1 state should be followed by vibrational relaxation to an orthogonal twisted geometry and to the funnel in S_1, followed by a very rapid radiationless relaxation to S_0, leaving little opportunity for fluorescence or intersystem crossing. Back in the S_0 state, the molecule should vibrationally relax rapidly to one of the two symmetry-equivalent planar forms of the imine ("syn" and "anti") with equal probability, and one should have $\Phi_{syn \rightarrow anti} = \Phi_{anti \rightarrow syn} = 0.5$. However, if a significant fraction of the excited molecules reaches the region of the conical intersection without having lost dynamical memory of their original geometry, syn or anti, both quantum yields of isomerization may deviate from 0.5, even in the absence of other competing processes.

Relatively little is known about the *E-Z* isomerization of *N*-alkylimines (**7a**). The reversible photoisomerization of anils (**7b**), however, has been studied in some detail. Since the quantum yield of intersystem crossing Φ_{ISC}

is relatively large, it is assumed to be a triplet reaction. (Cf. Paetzold et al., 1981.)

$R^1R^2C{=}NR$ **7** a: R = Alkyl b: R = Aryl

7.1.7 Azo Compounds

Finally, azoalkanes (**8**) have lone pairs of electrons on both nitrogen atoms, and additional n→π* transitions and additional kinds of motion have to be considered in discussing cis-trans isomerization. The effect of the n orbitals is apparent from the orbital correlation diagram shown in Figure 7.9. In constructing this diagram use has been made of the fact that the orbitals n_1 and n_2 of the lone pairs of electrons on the two nitrogen atoms split due to an appreciable interaction, and that the orbital ordering is the natural one in the cis isomer, with the combination $n_+ = (n_1 + n_2)/\sqrt{2}$ below the combination $n_- = (n_1 - n_2)/\sqrt{2}$, while in the trans isomer orbital interaction produces the opposite ordering.

R–N=N–R **8** a b

From PE spectroscopy results for azomethane (Haselbach and Heilbronner, 1970) the energy of the π MO is known to lie between those of the n_+ and n_- orbitals. Thus, a HOMO-LUMO crossing results and the trans-cis

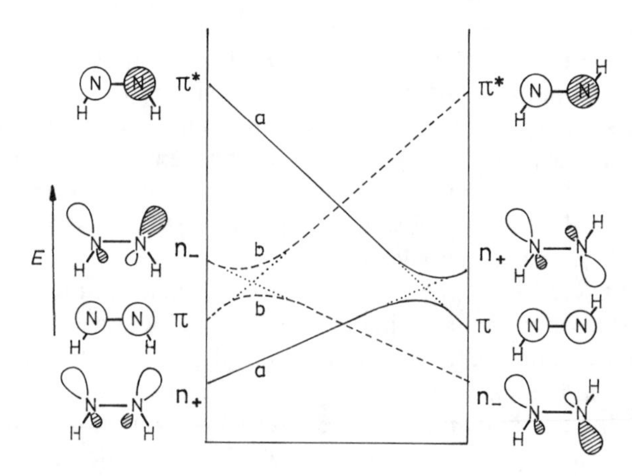

Figure 7.9. Orbital correlation diagram for the cis-trans isomerization of diimide HN=NH.

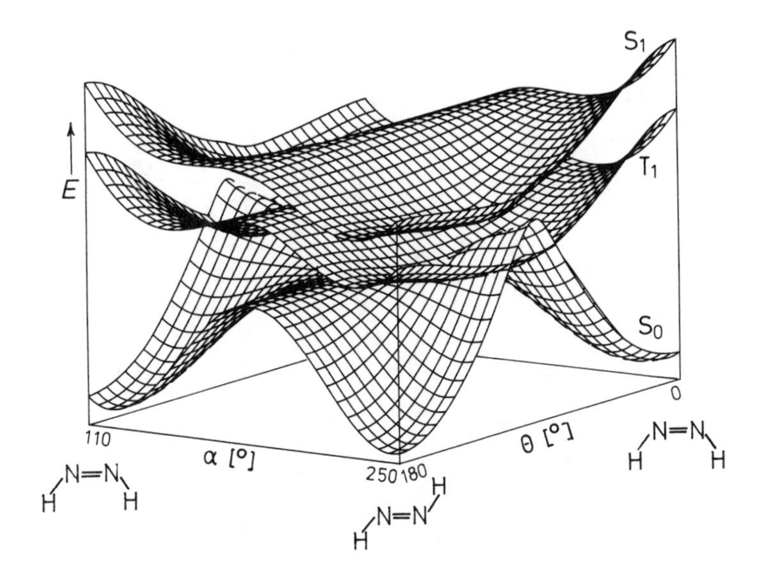

Figure 7.10. Computed potential energy surfaces of the ground state S_0 and the (n,π*) excited states T_1 and S_1 for the cis-trans isomerization of diimide as a function of the twist angle θ and the valence angle α at one of the nitrogen atoms.

isomerization is forbidden in the ground state and allowed in the first excited singlet and triplet states, as in the case of olefins. In contrast to the olefins, however, the lowest excited states of azoalkanes are the (n,π*) excited states. On the (π,π*) excited singlet and triplet surfaces the reaction encounters a correlation-imposed barrier. For instance, the configuration $(n_-)^2(\pi)^1(n_+)^2(\pi^*)^1$ of the trans isomer correlates with the doubly excited configuration $(n_+)^1(\pi)^2(n_-)^1(\pi^*)^2$ of the cis isomer. The computed potential energy surfaces of Figure 7.10 confirm these concepts. From these calculations it is also seen that isomerization by motion of the substituent in the molecular plane (variation of α) is energetically preferable to twisting the N=N bond (variation of θ) in the ground state, while the opposite is true in the ^{1,3}(n,π*) excited states.

In agreement with the theoretical results, the photoisomerization of simple azoalkanes is found to be rather effective. For azomethane in benzene at 25°C quantum yields of $\Phi_{t\rightarrow c} = 0.42$ and $\Phi_{c\rightarrow t} = 0.45$ have been observed (Thompson et al., 1979). *Cis*-azo compounds are moderately stable. Only tertiary *cis*-azoalkanes are thermally unstable and decompose to nitrogen and radicals. (See Section 7.2.2.)

Example 7.3:
In azobenzene the cis-trans isomerization in the ^{1}(π,π*) state apparently proceeds along a twisting path whereas in the ^{1}(n,π*) state it proceeds along the inversion path. This has been suggested by the fact that for azobenzenes such

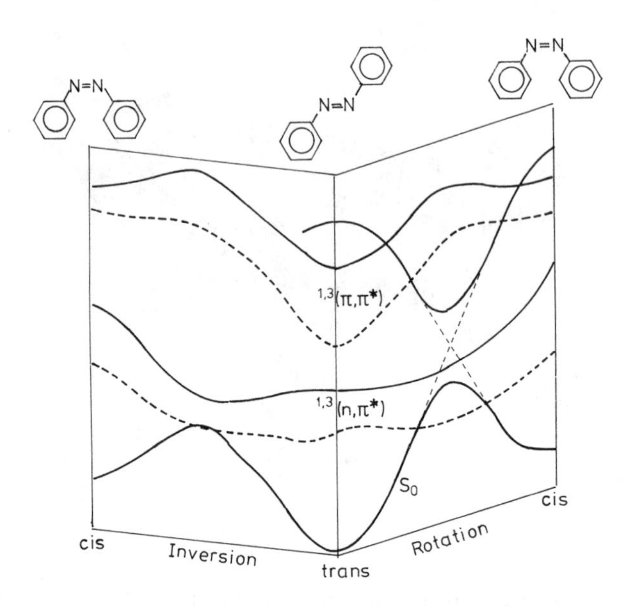

Figure 7.11. Schematic state correlation diagram for the cis-trans isomerization of azobenzene for two reaction paths that correspond to a twist mechanism and an inversion mechanism, respectively (adapted from Rau, 1984).

as the bridged crown ether **9**, for which twisting is inhibited for steric reasons, the quantum yield $\Phi_{t \to c}$ of trans-cis isomerization is independent of the exciting wavelength. In azobenzene itself, however, n→π^* excitation (λ = 436 nm) and $\pi \to \pi^*$ excitation (λ = 313 nm) give different quantum yields $\Phi_{t \to c}$. According to the schematic state correlation diagram of Figure 7.11, inversion, which is preferred in the $^1(n,\pi^*)$ state, is forbidden in the (π,π^*) state. However, from this state a funnel for a 90° twisted geometry may be reached, followed by a rapid transition to the ground-state surface (Rau, 1984).

9

7.2 Photodissociations

These reactions are initiated by the cleavage of a single or a double bond. A double-bond dissociation is normally possible only if the bond is unusually weak, that is, if one or both of the fragments are unusually stable. The C=N_2 and N=N_2 bonds in diazo compounds and azides are the best-known examples.

Single-bond dissociation is the simplest when it takes place between atoms that carry no multiple bonds, lone pairs, or empty orbitals. In that case it is bitopic (two active orbitals). If the single bond is covalent and uncharged, the $^1(\sigma,\sigma^*)$ state (which often represents S_1) is weakly bound and often fluorescent, and the $^3(\sigma,\sigma^*)$ state (which often represents T_1) is purely dissociative, as shown schematically in Figure 4.5 for the dissociation of H_2. Another example is the photodissociation of the Si—Si bond, mentioned in Section 7.2.3. Bitopic photodissociation normally takes place in the triplet state, which correlates with a radical pair, and not in the singlet state, which correlates with an ion pair.

If one of the termini of the single bond is charged (onium cation or ate anion), both the singlet and the triplet (σ,σ^*) configurations are dissociative (cf. Figure 4.24a), and the single bond can be cleaved in the singlet as well as the triplet state. Ammonium and sulfonium salts are the best-known substrates for such dissociations. If the single bond is of the dative type, which is uncommon in organic chemistry, neither the singlet nor the triplet (σ,σ^*) state is dissociative (both correlate with an ion pair).

Single-bond dissociation reactions of higher topicity are more common. In these, at least one of the atoms originally connected by the cleaved bond carries a lone pair, an empty orbital, or a multiple bond. The number of active orbitals on the two reaction centers then determines the topicity of the reaction (Section 6.3.3). For instance, the cleavage of the carbon–heteroatom bond in amines and alcohols is tritopic, the cleavage of the saturated carbon–halogen and of the oxygen–oxygen bond is tetratopic, the cleavage of a vinylic carbon–halogen bond is pentatopic, etc. Already for tritopic bond-dissociation reactions, and more so for those of higher topicity, there is at least one configuration of each multiplicity that is dissociative (Figure 6.28), and homolytic photochemical bond cleavage is not restricted to the triplet state. Correlation diagrams of the kind discussed in Section 6.3.1 are useful for the construction of qualitative potential energy diagrams for these reactions.

Often, the singlet or triplet (σ,σ^*) configuration does not enter the S_1 or T_1 state with significant weight at the initial geometry, and these are represented by other lower-energy configurations, such as (π,π^*). This situation normally leads to a barrier in the reaction surface that needs to be overcome before dissociation can take place (Figure 6.20, cf. "intended" correlations in Section 4.2.2). An example of tritopic reactions of this kind are the singlet-state dissociations of the benzylic C—X bonds in compounds containing the ArX—C moiety (photo-Fries, photo-Claisen, and others), in which the (π,π^*) excited state of the arene chromophore lies below the (σ,σ^*) configuration, and for which the principles embodied in Figures 6.20 and 6.28 are relevant (Grimme and Dreeskamp, 1992).

Examples of correlation diagrams for a series of reactions of higher topicity, up to hexatopic, have been discussed in some detail elsewhere (Michl and Bonačić-Koutecký, 1990). Here, we shall only outline the results for two important cases: Norrish type I C—C bond cleavage in carbonyl compounds (tritopic), and C—N bond cleavage in azo compounds (tetratopic).

7.2.1 *α* Cleavage of Carbonyl Compounds (Norrish Type I Reaction)

α Cleavage of a carbonyl compound is known as a Norrish type I reaction. It gives an acyl and an alkyl radical in the initial step. According to Scheme 2 the acyl radical can lose CO (path a); the resulting radicals can undergo recombination or disproportionation. Alternatively, the acyl radical can react via hydrogen abstraction (path b) or by hydrogen loss and formation of a ketene (path c).

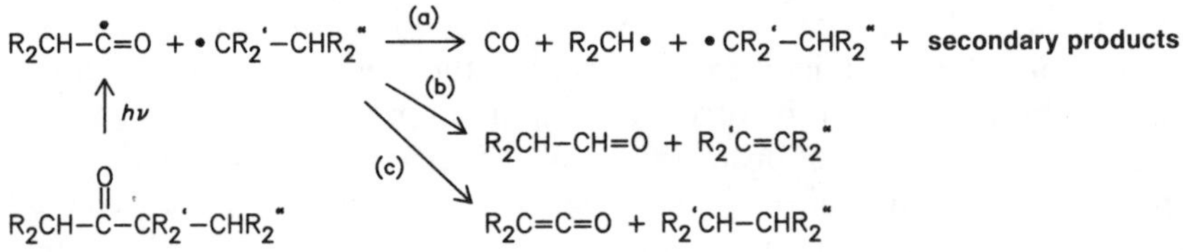

$R_2CH{-}\dot{C}{=}O + \bullet CR_2'{-}CHR_2'' \xrightarrow{(a)} CO + R_2CH\bullet + \bullet CR_2'{-}CHR_2''$ + secondary products

$\xrightarrow{(b)} R_2CH{-}CH{=}O + R_2'C{=}CR_2''$

$\xrightarrow{(c)} R_2C{=}C{=}O + R_2'CH{-}CHR_2''$

$\uparrow h\nu$

$R_2CH{-}C(=O){-}CR_2'{-}CHR_2''$

Scheme 2

The correlation diagrams for the *α*-cleavage reaction have been derived in Section 6.3.1 for the path to a bent acyl radical as well as for that to a linear acyl radical and have been discussed with the aid of potential energy surfaces drawn as a function of the distance R of the dissociating bond and of the OCR angle of the acyl radical. In Figure 7.12 the potential energy surfaces of the photochemical *α* cleavage of the hydrogen atom of acetaldehyde and benzaldehyde are displayed in a similar way. The symmetry with respect to the molecular plane is indicated by a subscript s (symmetric) or a (antisymmetric), respectively. Detailed quantum chemical ab initio calculations for acetaldehyde have shown that reaction profiles for CH_3 cleavage and H cleavage are very similar in the $^3(n,\pi^*)$ state (Yadav and Goddard, 1986), and the surfaces in Figures 7.12a and b may be taken as characteristic for the *α* cleavage of saturated and unsaturated carbonyl compounds, respectively.

In agreement with the results from Section 6.3.1, Figure 7.12a shows that in saturated ketones the lowest singlet and triplet states correlate directly with the biradicaloid states of the acyl radical at its linear geometry. Reaching this geometry, although allowed, is usually endothermic to such an extent that it is practically negligible. On the other hand, formation of the bent acyl radical is markedly less endothermic and occurs via a correlation-induced barrier. This barrier stems from a crossing of the T_a and T_s surfaces.

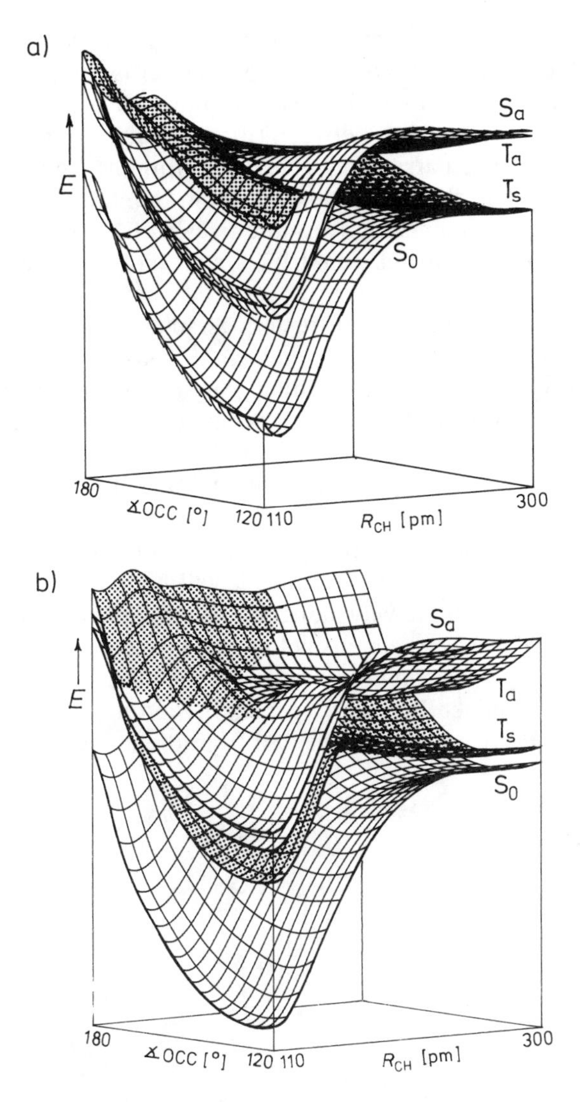

Figure 7.12. Potential energy surfaces for the α-cleavage reaction a) of acetaldehyde and b) of benzaldehyde. Formation of a bent acyl radical corresponds approximately to a reaction path along the front face ($\sphericalangle$OCC $\approx$ 130°) of the diagrams, whereas formation of the linear acyl radical proceeds on a path leading from the left front corner to the right rear corner (by permission from Reinsch and Klessinger, 1990).

For the initial geometry, these correspond to n→π^* and π→π^* excitation, respectively. This crossing can also be discerned in the correlation diagram for the path to a bent acyl radical in Figure 6.24 and will be avoided when the arrangement of the reacting centers is no longer coplanar; it therefore represents a conical intersection.

In the absence of intersystem crossing (ISC) the $^1(n,\pi^*)$ state can react only to give the linear acyl radical. Intersystem crossing followed by conversion into the bent acyl radical is favored, however, since it is associated with a transition of an electron between two mutually orthogonal p orbitals. (Cf. Example 1.8.) The barrier for α cleavage from the $^3(n,\pi^*)$ state is considerably smaller than that for α cleavage from the $^1(n,\pi^*)$ state. This is in agreement with experimental activation energies of 17 kcal/mol and 6 kcal/mol observed for the singlet and triplet states of acetone, respectively (Turro, 1978).

The singlet ground state of the resulting radical pair is of much higher energy than that of the initial ketone and can therefore easily be deactivated back to the reactant. The corresponding reaction of the triplet radical pair with conservation of spin, however, is symmetry forbidden and in general also endothermic. Dissociation into free radicals is thus preferred. This is another reason why α cleavage is much more efficient from the $^3(n,\pi^*)$ state than from the $^1(n,\pi^*)$ state.

From Figure 7.12b it is seen that in conjugated ketones the $^3(\pi,\pi^*)$ state can be stabilized to such an extent that it will be energetically below the $^{1,3}(n,\pi^*)$ states. Then, all the way along the reaction path leading to the bent acyl radical, the T_s surface remains below the T_a and S_a surfaces, which for the initial geometry correspond to an $n\rightarrow\pi^*$ excitation. In this case, the reaction rate is determined by the T_s barrier. According to the discussion in Section 6.3.1, a barrier is to be expected from the fact that excitation does

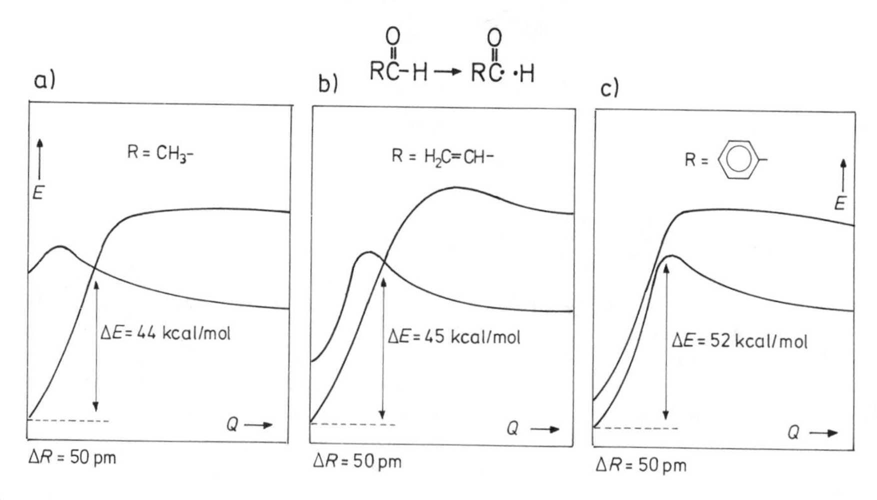

Figure 7.13. Cross sections through the T_a (n,π^*) and T_s (π,π^*) triplet surfaces for the α cleavage of a) acetaldehyde, b) acrolein, and c) benzaldehyde, leading to the bent acyl radicals. In a) and b), the surface crossing that is avoided for nonplanar geometries determines the barrier height and the location of the transition state (by permission from Reinsch et al., 1988).

not reside primarily in the σ bond to be cleaved but mostly in the carbonyl group. (Cf. the natural correlations shown in Figure 6.24.)

These results also follow from the cross sections through the potential energy surfaces calculated in the semiempirical all-valence electron approximation by the MNDOC-CI method for acetaldehyde, acrolein, and benzaldehyde, which are displayed in Figure 7.13. Stabilization of the $^3(\pi,\pi^*)$ state by conjugation is seen to increase slightly the barrier, and this should reduce the reactivity. The α cleavage of *t*-butyl phenyl ketone is in fact slower by a factor of 10^3 than the same reaction of *t*-butyl methyl ketone (Yang et al., 1970). If the $^3(\pi,\pi^*)$ state is below the $^3(n,\pi^*)$ state, a considerably higher barrier is to be expected from Figure 7.13. Biphenyl *t*-butyl ketone (**10**), whose lowest triplet state is (π,π^*), is in fact photostable (Lewis and Magyar, 1972).

10

In Figure 7.14a the calculated barrier heights of the lowest triplet state are displayed for a number of alkyl methyl ketones with different alkyl groups R. For cleavage of the methyl group the barrier is independent of R, whereas for cleavage of the R group it is determined by the degree of branching at the α carbon. For primary alkyl groups —CH_2R' the calculated activation energy $E_a \approx 30$ kcal/mol is independent of R′, for secondary alkyl

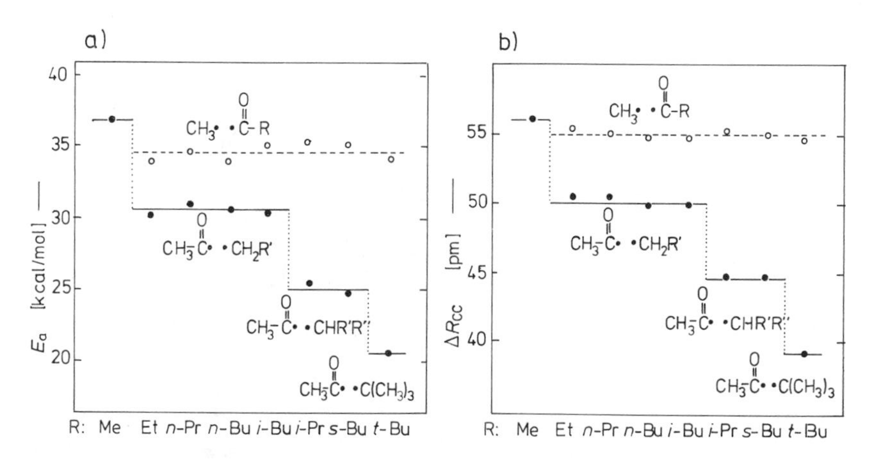

Figure 7.14. α cleavage of saturated methyl ketones CH_3COR; a) activation energy E_a and b) location of the transition state for different groups R. ΔR_{CC} is the increase in the length of the CC bond in the transition state (by permission from Reinsch and Klessinger, 1990).

groups —CHR′R″, $E_a \approx 25$ kcal/mol, and for —$CH(CH_3)_3$, $E_a \approx 20$ kcal/mol. This is in good agreement with experimental data that indicate that the rate and efficiency of the α-cleavage reaction increase with increasing stability of the resulting radicals. In unsymmetrical ketones the weakest CC bond is cleaved; methyl ethyl ketone thus preferentially yields ethyl and acetyl radicals, but the selectivity of the α cleavage decreases with increasing energy of the exciting light (Turro et al., 1972a).

Finally, from Figure 7.14b it is seen that similar regularities are found for the transition-state geometries: the smaller the activation energy E_a, the smaller the elongation ΔR_{CC} of the bond to be cleaved. This is exactly what is to be expected from the correlation diagram in Figure 6.24, if for a given acyl radical the energy of the ${}^3B_{\sigma,\sigma}$ state is determined essentially by the nature of the radical center at the cleaved group R.

Further information about the mechanism of the α-cleavage reaction is available from the study of photochemical reactions in micelles (cf. Figure 7.15), formed in aqueous solutions of detergents (Turro et al., 1985). According to the Wigner–Witmer spin-conservation rules (Section 5.4.1) a triplet radical pair is formed at first. In homogeneous solutions intersystem crossing (ISC) is much slower than diffusion ($k_{TS} < k_{diff}$) and free radicals will form. In micelles, however, k_{diff} is appreciably smaller and of the same order as the escape rate from the micelle (10^6–10^7 s^{-1}), and singlet radical pairs and their recombination products may well form.

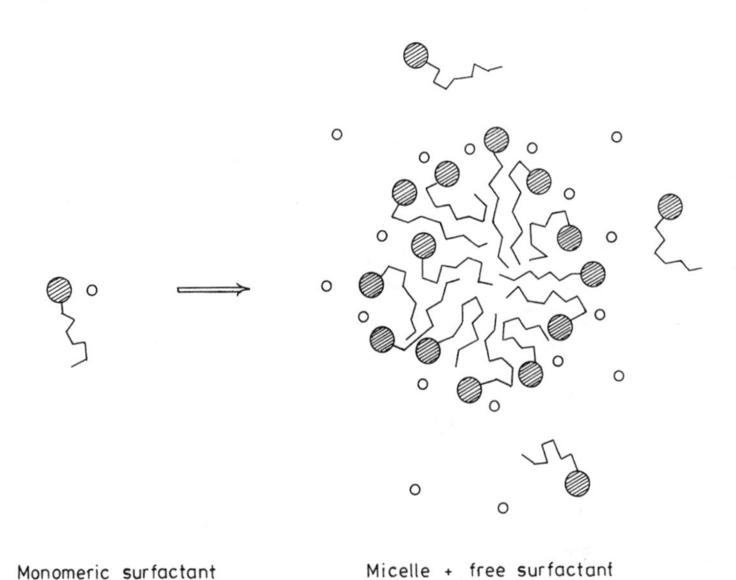

Figure 7.15. Schematic representation of a micelle (by permission from Bunton and Savelli, 1986).

Thus, upon photolysis of dibenzyl ketone (**11**) in micelles, 4-methylphenyl benzyl ketone (**12**) is formed as a side product according to Scheme 3.

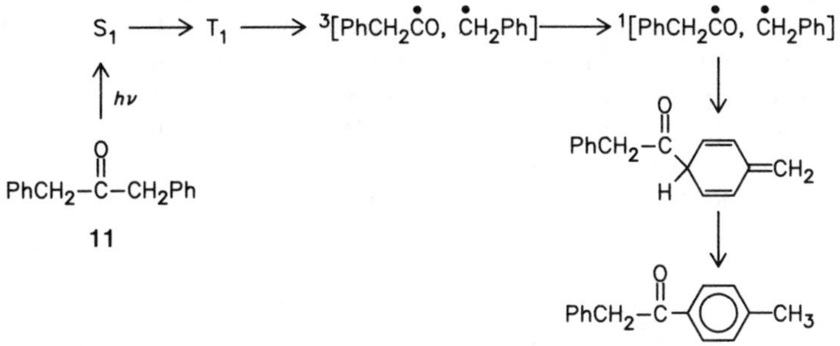

$$S_1 \longrightarrow T_1 \longrightarrow {}^3[PhCH_2\dot{C}O,\ \dot{C}H_2Ph] \longrightarrow {}^1[PhCH_2\dot{C}O,\ \dot{C}H_2Ph]$$

$PhCH_2{-}CO{-}CH_2Ph$ (**11**) $\xrightarrow{h\nu}$ S_1

Scheme 3 **12**

Due to the cage effect in micelles, unsymmetrically substituted dibenzyl ketones such as **13** yield predominantly the unsymmetrical diphenylethanes on photodecarbonylation, whereas in homogeneous solution all three possible products are formed in the statistical ratio 1:2:1 (Turro and Kraeutler, 1978).

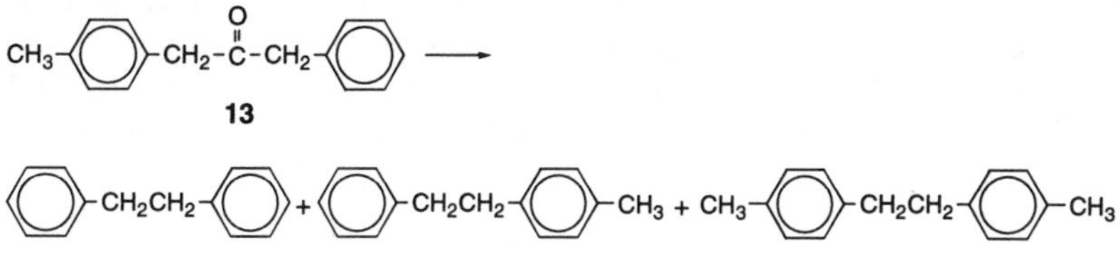

13

Homogeneous Solution	1	:	2	:	1
Micelle	1	:	6	:	1

Photolysis of ketones in micelles with simultaneous application of an external magnetic field permits a ^{13}C isotope enrichment. (Cf. Section 6.1.5.5.) This is the case because ^{13}C nuclei have a magnetic moment and thus accelerate the spin inversion by the hyperfine interaction mechanism. (Cf. Example 4.9.) Due to the more efficient recombination of radicals containing ^{13}C, the initial product formed after photolysis in a back reaction is ^{13}C enriched (Turro et al., 1980b).

In spite of the occasionally rather low efficiency, decarbonylation has been used to generate strained ring systems. An important example is the synthesis of tetra-*t*-butyltetrahedrane (**14**) (Maier et al., 1981).

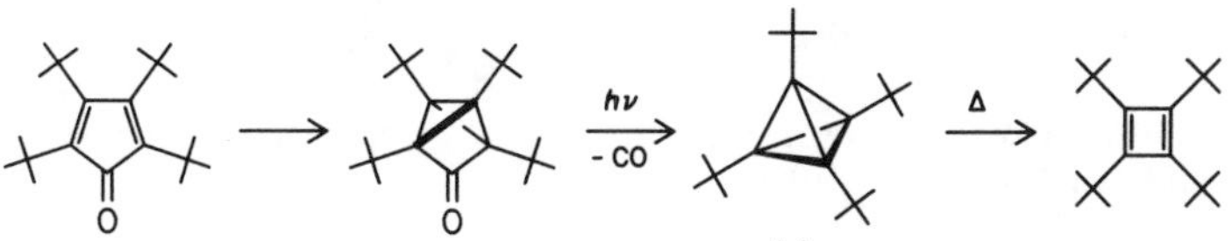

$\xrightarrow{h\nu,\ -CO}$ **14** $\xrightarrow{\Delta}$

Example 7.4:
The photochemical behavior of cyclobutanone (**15**) contrasts sharply with that of other ketones. Cyclobutanone undergoes α cleavage also from the $^1(n,\pi^*)$ state, with subsequent fragmentation to ketene and olefin, decarbonylation to cyclopropane or cyclization to oxacarbene (**16**), whose concerted formation has also been proposed on the basis of stereochemical observations (Stohrer et al., 1974). In contrast, cyclohexanone cleaves exclusively from the triplet state and undergoes disproportionation reactions. The photochemical activity of cyclobutanone persists even at low temperatures (77 K) where cyclohexanone is photostable.

15 $\xrightarrow{h\nu}$ → CO + ·CH₂CH₂CH₂·; → $CH_2{=}C{=}O + CH_2{=}CH_2$; → 16 $\xrightarrow{ROH}$

16

These results may be rationalized by means of the correlation diagram in Figure 7.16, which was constructed using thermochemical and spectroscopic data to estimate state energies. The (n,π^*) states of the initial ketone are higher in energy than the B states of the linear acyl radical that is an unstrained open-chain species, and the reaction corresponding to the direct correlation will be exothermic. Furthermore, it has been proposed that the resulting linear acyl

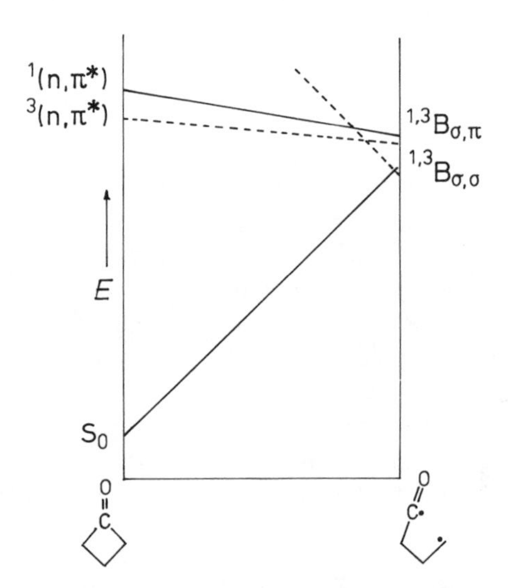

Figure 7.16. Correlation diagram for the α cleavage of cyclobutanone (adapted from Turro et al., 1976).

radical is formed in an excited state and can dissociate into CO and $\cdot CH_2—CH_2—CH_2\cdot$ (Turro et al., 1976).

α-Cleavage reactions are also observed for esters and amides. The photochemical rearrangement of aryl esters that proceeds via α cleavage and yields a mixture of *o*- and *p*-hydroxyaryl ketones according to Scheme 4 is known as the photo-Fries reaction:

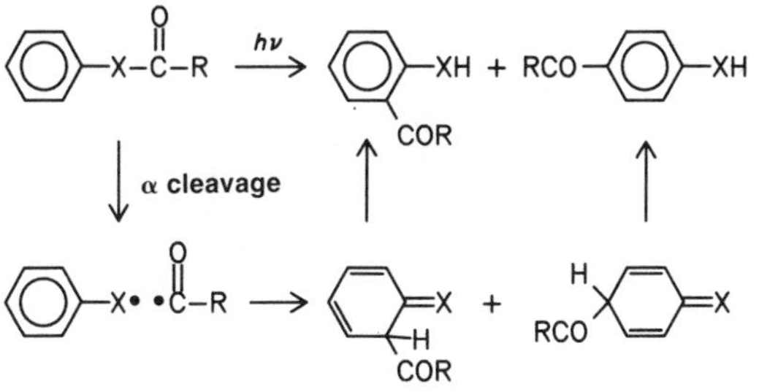

Scheme 4

The rearrangement is almost exclusively intramolecular and proceeds within a solvent cage (Adam et al., 1973). In the case of phenyl acetate (X = O, R = CH_3), the *o*:*p* ratio increases from 1 in the absence of β-cyclodextrine to 6.2 in its presence; the macrocycle provides a cage for the radical pair (Ohara and Watanabe, 1975). However, these reactions are singlet reactions and should be classified as dissociations of the benzylic C—X bond rather than α-cleavage reactions (Grimme and Dreeskamp, 1992), as mentioned above.

7.2.2 N_2 Elimination from Azo Compounds

The lowest excited state of many azo compounds, like that of ketones, is an (n,π^*) state. Photolytic cleavage of a CN bond analogous to the α cleavage of ketones is therefore to be expected:

$$R'-N=N-R \xrightarrow{h\nu} R'-N=N\cdot + \cdot R \longrightarrow \cdot R' + N_2 + \cdot R + \text{secondary products}$$

Another conceivable route would be the concerted elimination of nitrogen:

$$R'-N=N-R \xrightarrow{h\nu} R'\cdot + N_2 + \cdot R$$

Experimental data and theoretical arguments indicate that the concerted path is energetically unfavorable, so in general the two-step mechanism is involved (Engel, 1980).

As a model for this reaction the orbital correlation diagram for the cleavage of one NH bond of *cis*-diimide is shown in Figure 7.17a, and the state correlation diagram derived therefrom is displayed in Figure 7.17b. The sim-

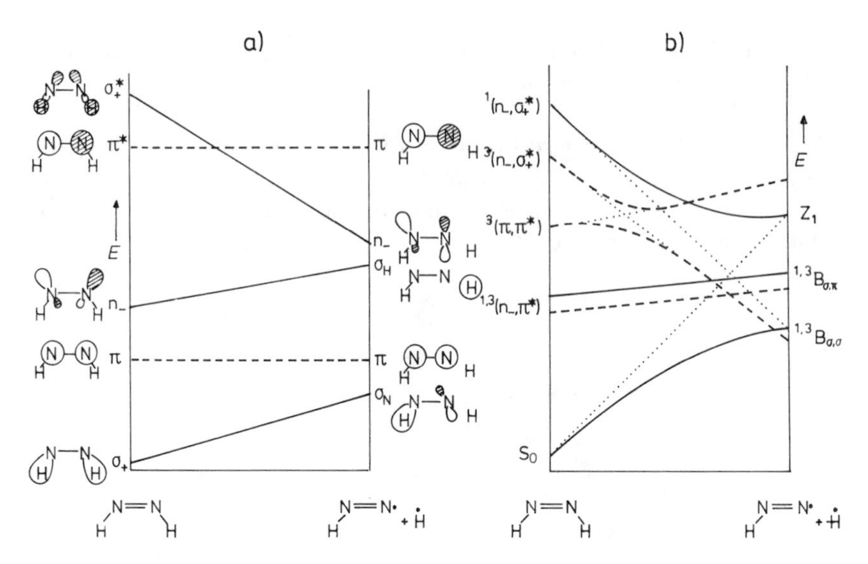

Figure 7.17. Cleavage of one *cis*-diimide NH bond; a) orbital correlation diagram, b) state correlation diagram (adapted from Bigot et al., 1978).

ilarity to the corresponding diagram for the α cleavage of ketones is apparent. The singlet state of the σ,σ biradical correlates with the ground state, and the triplet state with one of the higher excited (n,σ^*) states of the azo compound. The conversion of the ^{1,3}(n,π^*) states of the azo compound to the $^{1,3}B_{\sigma,\pi}$ biradical states is electronically allowed but is expected to be weakly endothermic. The results of quantum chemical calculations shown in Figure 7.18 confirm the expectations based on the correlation diagram.

In comparing these results with experimental data it has to be remembered that in contrast to ketones, azo compounds can also undergo photochemical trans-cis isomerizations. (Cf. Section 7.1.7.) In the gas phase n→π^* excitation results in photodissociation with nearly unit quantum efficiency. At higher pressures, however, and especially in solution, this reaction almost completely disappears and photoisomerization dominates. The latter is observed even at liquid nitrogen temperatures. This is understandable if it is accepted that photodissociation proceeds in the gas phase as a hot ground-state reaction. According to Figure 7.18, it has to overcome a barrier in the excited state and is therefore not observed in solution. For the trans-cis isomerization, on the other hand, no excited-state barrier is to be expected from the results in Section 7.1.7.

For most acyclic azo compounds photodissociation in solution is an indirect process, that is, the photochemical reaction proper is a trans-cis isomerization, and the *cis*-azoalkane formed undergoes thermal decomposition to nitrogen and radicals. This occurs especially if the cis isomer is suffi-

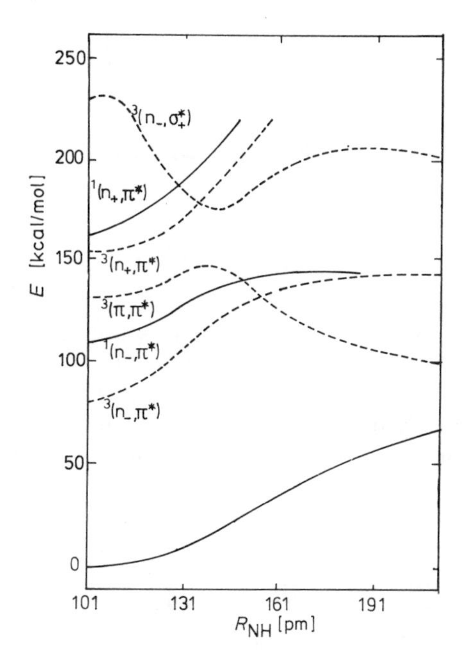

Figure 7.18. Calculated potential energy curves of the ground and low-lying excited states of *cis*-diimide in the one-bond cleavage (by permission from Bigot et al., 1978).

ciently unstable due to the presence of sterically demanding alkyl groups.

The use of triplet sensitizers has shown that the major part of direct photolysis does not involve the triplet state; the extrusion of nitrogen from the triplet state requires an activation energy E_a (Engel, 1980).

Cis-trans isomerization of cyclic azo compounds is only possible for six-membered or larger rings. Otherwise only N_2 loss is observed, and direct irradiation and triplet sensitization can yield different products, as is the case for the cyclic azo compound **17** (Bartlett and Porter, 1968). This is due to the fact that the biradical R ↑ ↑ R formed from the triplet-excited reactant cannot recombine to form a cyclobutane until spin inversion has occurred; the triplet biradical can live long enough to be diverted to the rotamer R ↑ ↑ R′.

According to Scheme 5 cis and trans products are formed; direct photolysis produces **18a** and **18b** in a ratio 10:1, whereas triplet sensitization yields a product ratio of 1.4:1, which is nearly equal to what would be expected for an equilibrium distribution between the different rotameric biradicals. A biradical such as this, which gives different products depending upon its multiplicity, is said to exhibit a *spin-correlation effect*.

Photochemical elimination of N_2 from bicyclic azo compounds produces cyclic *cis*-1,*n*-biradicals followed by stereospecific ring cleavage or cycliza-

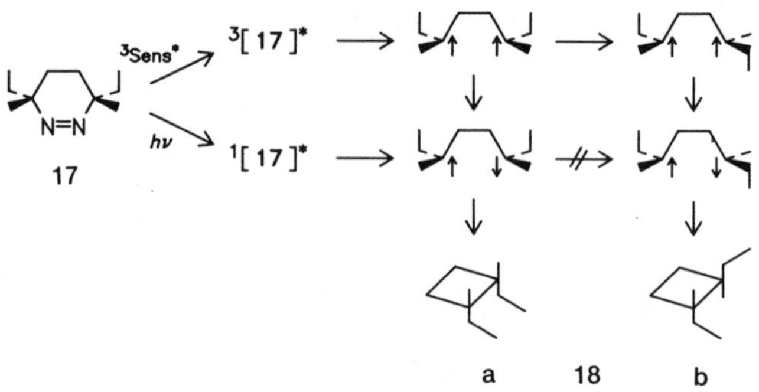

Scheme 5

tion as exemplified in Scheme 6, whereas disproportionation would require a change in conformation and is therefore in general not observed (Cohen and Zand, 1962).

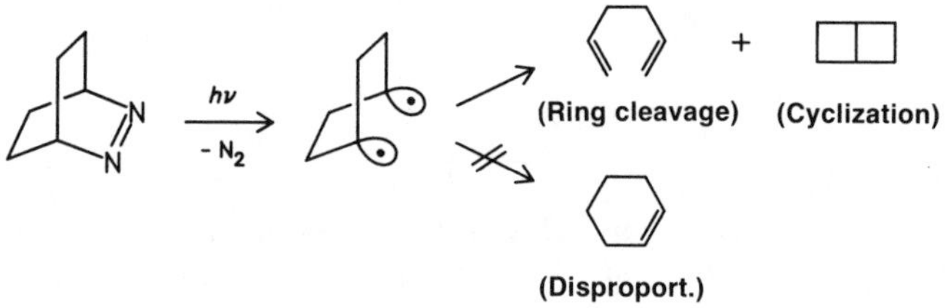

Scheme 6

Example 7.5:
Compound **19**, which yields the valence isomers of benzene on direct irradiation, and diazacyclooctatraene (**20**) as the major product upon triplet sensitization, is an interesting example of the different reactivity of the S_1 and T_1 states of azo compounds (Turro et al., 1977a):

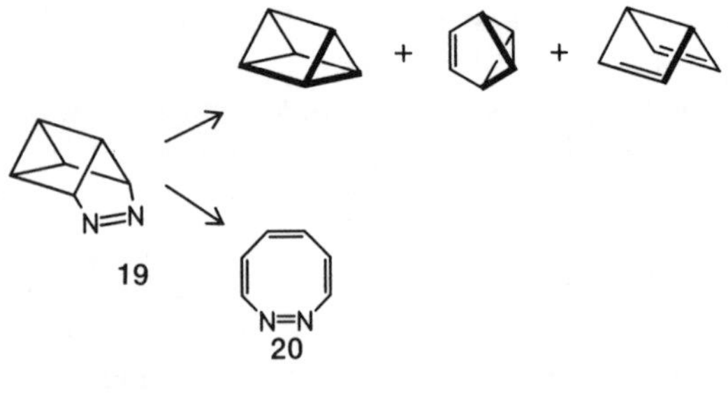

The photochemical parameters for **19** are summarized in Figure 7.19. From these data it is apparent that at low temperatures diazacyclooctatetraene becomes the exclusive product, since the loss of N_2 from the $^1(n,\pi^*)$ state requires an activation energy of ~ 5–6 kcal/mol. Oxygen has a catalytic effect on the

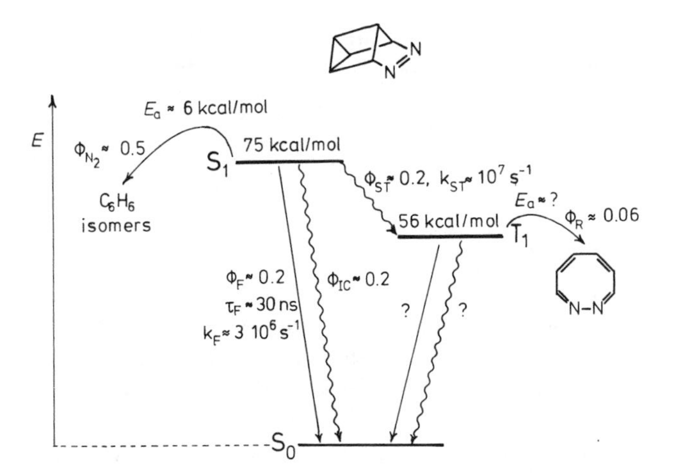

Figure 7.19. Jablonski diagram and photochemical parameters of 7,8-diaza-tetracyclo[3.3.0.0^{2,4}.0^{3,6}]oct-7-ene (by permission from Turro, 1978).

$S_1 \rightarrow T_1$ intersystem crossing and therefore enhances the formation of diazacy-clooctatetraene on direct irradiation.

The photoextrusion of N_2 from cyclic azo compounds is a very useful way of producing strained ring systems such as **21** (Snyder and Dougherty, 1985) or **22** (Lüttke and Schabacker, 1966). Unstable species such as the *o*-quinodimethanes **23** (Flynn and Michl, 1974) and **24** (Gisin and Wirz, 1976), and biradicals such as **25** (Gisin and Wirz, 1976), **26** (Watson et al., 1976), **27** (Platz and Berson, 1977), **28** (Dowd, 1966), and **29** (Roth and Erker, 1973), can also be generated in a matrix by this route and spectroscopically identified.

N=N $\xrightarrow{h\nu}$ **21**

Cl, N, N, Cl $\xrightarrow{h\nu}$ Cl, Cl **22**

23 **24**

25 **26** **27** **28** **29**

Some azo compounds undergo the usual photolysis ($\lambda > 300$ nm) only to a minor degree or not at all and are therefore dubbed "reluctant azoalkanes." These are cyclic azo compounds such as **30**, **31**, and **32**.

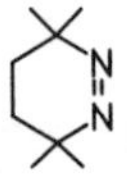

N N N N X N=N

30 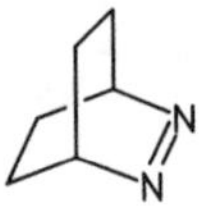 31 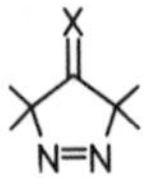 32

Photolysis of such compounds can be accelerated by employing elevated temperatures or by introducing substituents that stabilize the radicals formed. (Cf. Engel et al., 1985.) Short-wavelength irradiation ($\lambda = 185$ nm) also enhances photodissociation. Bridgehead azoalkanes such as **33** are also reluctant compounds and undergo photochemical trans-cis isomerization (Chae et al., 1981). Loss of nitrogen and formation of bridgehead radicals are observed upon excitation to the second singlet state (S_2) of the trans or the cis isomer, with quantum yields of $\Phi_t = 0.3$ and $\Phi_c = 0.16$, respectively (Adam et al., 1983).

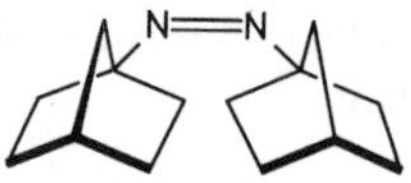

N=N

33

7.2.3 Photofragmentation of Oligosilanes and Polysilanes

Alkylated and arylated oligosilanes and polysilanes, the silicon analogues of alkanes and of polyethylene, have recently attracted considerable attention (Miller and Michl, 1989). Unlike saturated hydrocarbons, these materials absorb in the near UV region. The reasons for this are related to the electropositive nature of silicon and can be understood in simple terms (Michl, 1990). Their excited states bear considerable similarities to those of polyenes, but also exhibit significant differences (Balaji and Michl, 1991). Upon irradiation, oligosilanes (Ishikawa and Kumada, 1986) and particularly polysilanes (Trefonas et al., 1985), readily fragment to lower-molecular-weight species, and polysilanes show promise as photoresists.

Three distinct photochemical processes have been identified as shown in Scheme 7: (1) chain abridgement by silylene extrusion, (2) chain cleavage by silylene elimination, and (3) chain cleavage by homolytic scission (Miller and Michl, 1989).

The two silylene-generating processes are believed to occur in the singlet excited state in pericyclic fashion, while the radical-pair-forming homolytic cleavage process is believed to occur in the triplet state (Michl and Balaji, 1991), as would be expected from Figure 4.5.

According to the definition given in the beginning of this section, the singlet 1,1-elimination (reductive elimination) processes qualify as fragmenta-

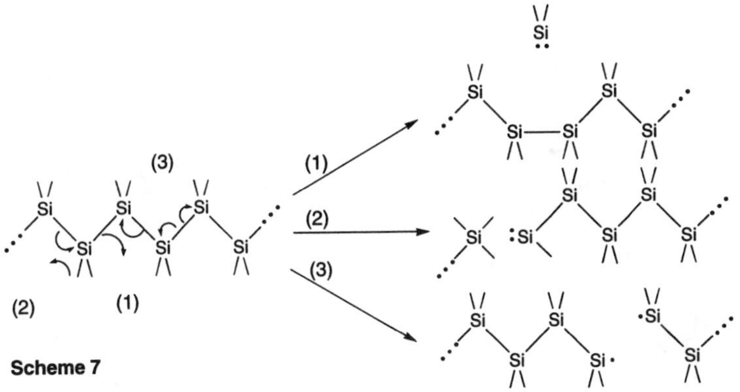

Scheme 7

tion reactions, but not as dissociation reactions, since more than one single or double bond is broken in these processes. They are closely related to the pericyclic processes discussed in Sections 7.4 and 7.5 and are formally isoelectronic with other excited-state-allowed four-electron pericyclic reactions, such as cheletropic elimination of CO from cyclopropanone and disrotatory electrocyclic ring closure in butadiene. The analogy of the chain-abridgement reaction to the latter is illustrated in Figure 7.20, which shows the orbitals involved in the two reactions.

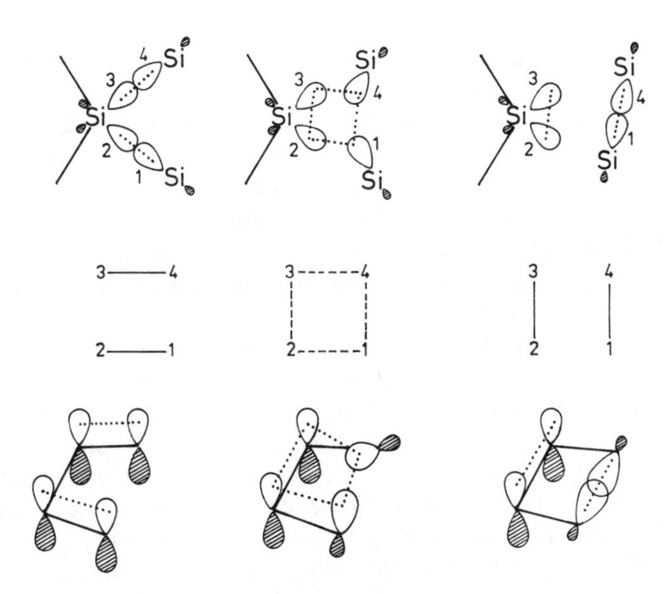

Figure 7.20. Comparison of the AO interactions in the photochemical chain abridgement in a polysilane (top) and in the disrotatory electrocyclic ring closure of butadiene (bottom) (by permission from Michl and Balaji, 1991).

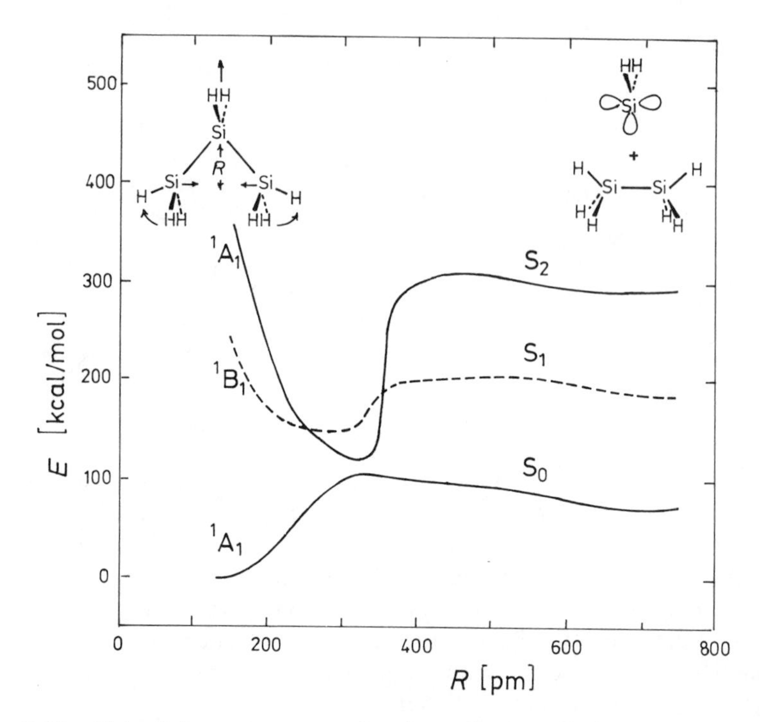

Figure 7.21. Potential energy curves for the reductive elimination of SiH_2 from Si_3H_8, as calculated by an ab initio method (by permission from Michl and Balaji, 1991).

Figure 7.21 shows the behavior of the calculated singlet state energies along an idealized reaction path from trisilane to disilane and silylene (Michl and Balaji, 1991), quite similar to the classical behavior calculated for disrotatory ring closure of butadiene (van der Lugt and Oosterhoff, 1969). The S_0 state of trisilane attempts to correlate with a doubly excited state of the products, in which both electrons of the silylene lone pair are in the Si 3p orbital. The S_0 state of the products originates in a doubly HOMO→LUMO excited state of the starting trisilane, and the would-be crossing is weakly avoided (along a less symmetrical path it may not be avoided at all), resulting in a barrier in the S_0 and a minimum or funnel in the S_1 state ("pericyclic funnel"). The optically strongly allowed HOMO→LUMO excited (σ,σ^*) state correlates with a singly excited state of the products, with silylene in its first excited singlet, and does not impose a barrier in the excited state surface. The reaction presumably proceeds by relaxation into the ground state through the pericyclic funnel, followed by return to the starting geometry for some of the molecules, and dissociation into the products for the remainder.

7.3 Hydrogen Abstraction Reactions

One of the earliest photoreactions to be studied was the photoreduction of benzophenone (Ciamician and Silber, 1900)—that is, the conversion of a carbonyl compound into an alcohol by an intermolecular hydrogen abstraction reaction. Intramolecular hydrogen abstraction by the carbonyl group, usually from the γ site, is referred to as a Norrish type II reaction. Hydrogen abstraction by olefins and heterocycles has also been observed.

7.3.1 Photoreductions

The correlation diagram for hydrogen abstraction by a ketone has been derived in Example 4.5 (Section 4.2.3). From this diagram it can be concluded that no reaction is to be expected from the $^{1,3}(\pi,\pi^*)$ states for the in-plane

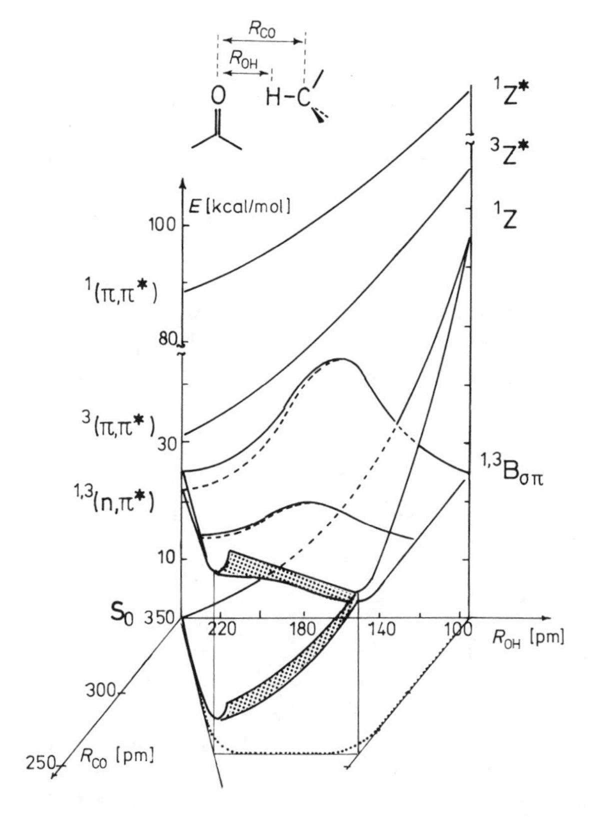

Figure 7.22. Ab initio results for hydrogen abstraction in $CH_2{=}O + CH_4$, as a function of the distance R_{CO} between the carbon in methane and the oxygen in formaldehyde (by permission from Bigot, 1983).

attack; the reaction from the $^{1,3}(n,\pi^*)$ states will have to overcome a barrier that results from a strongly avoided crossing of states if natural orbital correlations are used. (Cf. Figure 6.21.) The results of ab initio calculations for the formaldehyde-methane system depicted in Figure 7.22 show that the barrier decreases with decreasing distance R_{CO} between the methane carbon and carbonyl oxygen. As a consequence, the hydrogen abstraction requires the separation between the reactants to be very small (Bigot, 1983). The crossing between the correlation lines starting at the ground state and the $^1(n,\pi^*)$ state of the reactant shown in Figure 7.22 becomes avoided if the system deviates from a coplanar arrangement. The crossing corresponds to a funnel, which mediates the return both to the product and reactant ground state in the usual way. This reduces the efficiency of the singlet reaction. Since in aryl ketones intersystem crossing is very fast, it is obvious why the reaction is observed only from the $^3(n,\pi^*)$ state.

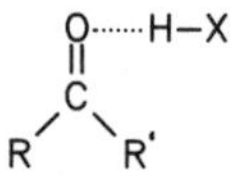

O·······H–X
||
C
R R'

In-plane attack

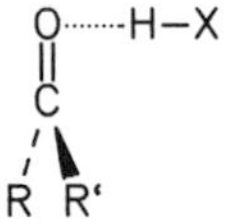

O·······H–X
||
C
R R'

Perpendicular attack

In addition to the reaction between the n_O orbital of the carbonyl group and the σ_{CH} orbital, another conceivable reaction path for hydrogen abstraction is based on an interaction between the π_{CO} and the σ_{CH} orbital. This is referred to as a perpendicular attack.

In the correlation diagram for the perpendicular attack derived in Example 7.6, the $^3(\pi,\pi^*)$ state correlates with the $^3B_{\sigma,\sigma}$ ground state of the primary product. However, the corresponding reaction is expected to show a considerable barrier. The reaction from the $^{1,3}(n,\pi^*)$ states is forbidden in the perpendicular case. These results are also confirmed by formaldehyde-methane calculations.

Example 7.6:

The natural orbital correlation for the perpendicular hydrogen abstraction reaction is shown in Figure 7.23a. The n_O orbital of the ketone is seen to correlate with the n_O orbital of the product, whereas the σ_{CH} orbital turns into the n_C' orbital of the alkyl radical center and the π_{CO} orbital turns into the σ_{OH} orbital. The product ground state is a biradicaloid $^{1,3}B_{\sigma,\sigma}$ state with singly occupied n_C' and n_C orbitals, which are both in the plane of the reacting centers and therefore have σ symmetry. In addition, there is a $^{1,3}B_{\sigma,\pi}$ state with a singly occupied n_O orbital with π symmetry. The resulting configuration correlation diagram is indicated in Figure 7.23b by dotted lines, whereas the state correlation diagram, which takes into account avoided crossings, is shown by full and broken lines for singlets and triplets, respectively. In contrast to the in-plane attack, the perpendicular attack should be most favorable from the $^3(\pi,\pi^*)$ state and

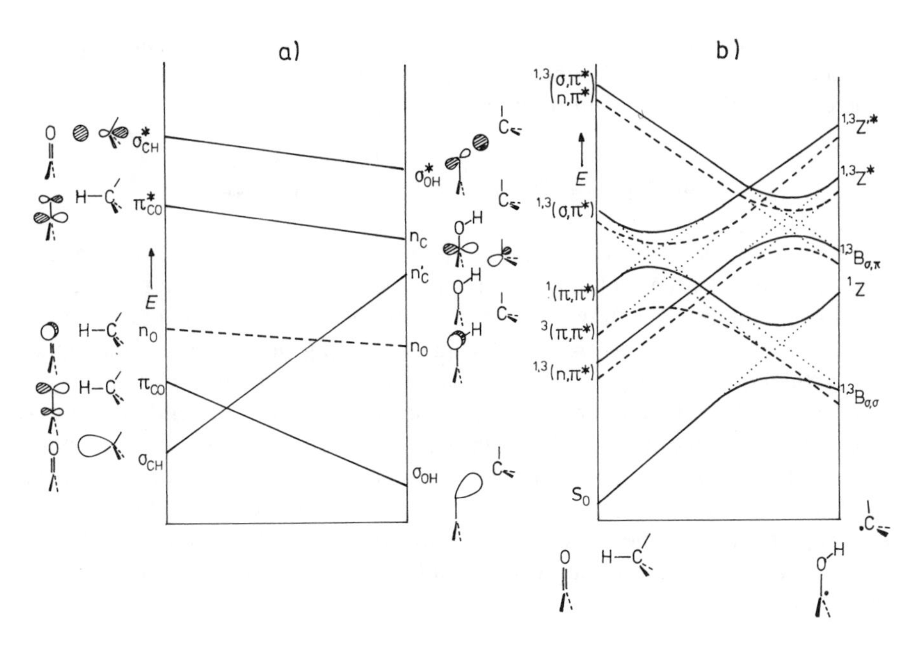

Figure 7.23. Perpendicular hydrogen abstraction reactions: a) the natural orbital correlation diagram and b) the resulting state correlation diagram (adapted from Bigot, 1980).

should therefore be conceivable for molecules such as biphenylyl *t*-butyl ketone (**10**), whose $^3(\pi,\pi^*)$ state lies below the $^3(n,\pi^*)$ state. No perpendicular hydrogen abstraction is observed even with such compounds because of the barrier expected from the natural orbital correlation.

The reaction of benzophenone (**34**) with benzhydrol (**35**) is a representative example of hydrogen abstraction:

$$\underset{\mathbf{34}}{Ph_2C{=}O} + \underset{\mathbf{35}}{Ph_2CHOH} \xrightarrow{h\nu} \underset{\mathbf{36}}{Ph_2\dot{C}{-}OH + Ph_2\dot{C}{-}OH} \longrightarrow \underset{\mathbf{37}}{Ph_2C(OH){-}C(OH)Ph_2}$$

In the first step the spectroscopically detectable ketyl radical **36** is formed, which then recombines to form benzopinacol (**37**) (Weiner, 1971). The same pinacol is obtained by reacting benzophenone with 2-propanol, since the dimethylketyl radical (**38**) produced in the hydrogen abstraction step is a strong reductant and transfers a hydrogen atom to the excess benzophenone to form another molecule of the diphenylketyl radical (**36**).

$$Ph_2C{=}O + \underset{\mathbf{38}}{(CH_3)_2\dot{C}{-}OH} \longrightarrow \underset{\mathbf{36}}{Ph_2\dot{C}{-}OH} + (CH_3)_2C{=}O$$

Thus the quantum yield for the disappearance of benzophenone has a limiting value of $\Phi_R = 2$, since only one photon is needed to convert two molecules of the reactant into one product molecule:

$$2\ Ph_2C{=}O + (CH_3)_2CHOH \xrightarrow{h\nu} Ph_2C(OH){-}C(OH)Ph_2 + (CH_3)_2C{=}O$$

Some ketones such as 2-acetylnaphthalene (**39**) show no photochemical reactivity. In these cases the lowest triplet state is a (π,π^*) state, the reaction of which is inhibited by a barrier which from the correlation diagram in Figure 4.18 is seen to result from an avoided crossing of the levels starting at the vertical (π,π^*) and (n,π^*) states. The efficiency of the hydrogen abstraction by 4-hydroxybenzophenone (**40**) is strongly solvent dependent. In cyclohexane, the quantum yield is high, $\Phi_R = 0.9$, but in the polar 2-propanol solvent, the (n, π^*) state is destabilized to such an extent that it no longer is the lowest triplet state, and the quantum yield drops to $\Phi_R = 0.02$ (Porter and Suppan, 1964). These examples demonstrate the influence of the energy ordering of states: the wrong order results in correlation-induced barriers. (Cf. Section 6.2.4.)

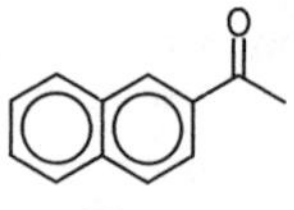

HO

39 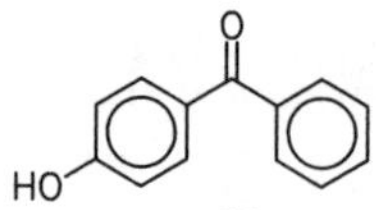 40

Carbonyl compounds can also be reduced by electron transfer instead of hydrogen abstraction (Cohen et al., 1973). For the reduction of benzophenone with diisopropylamine the mechanism can be formulated as indicated in Scheme 8. The intermediate of this reaction again is the ketyl radical **36**, which recombines to form the pinacol **37**:

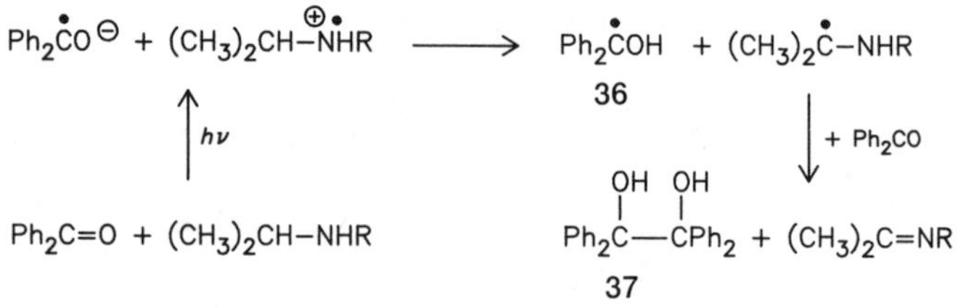

$$Ph_2\dot{C}O^{\ominus} + (CH_3)_2CH{-}\dot{N}^{\oplus}HR \longrightarrow Ph_2\dot{C}OH\ (\mathbf{36}) + (CH_3)_2\dot{C}{-}NHR$$

$$Ph_2C{=}O + (CH_3)_2CH{-}NHR \xrightarrow{h\nu} Ph_2\dot{C}O^{\ominus} + (CH_3)_2CH{-}\dot{N}^{\oplus}HR$$

$$\xrightarrow{+\ Ph_2CO} Ph_2C(OH){-}C(OH)Ph_2\ (\mathbf{37}) + (CH_3)_2C{=}NR$$

Scheme 8

Naphthyl ketones, which are generally unreactive toward photoreduction by 2-propanol, are efficiently reduced by amines. Electron transfer (cf. Section 5.4.4) is considerably faster than hydrogen abstraction. Thus, the reaction cannot be quenched using common triplet quenchers although it proceeds from the triplet state.

Table 7.1 Quantum Yields of Norrish Type II Reactions (Adapted from Horspool, 1976)

Ketone	Solvent	Φ_S	Φ_T
Hexan-2-one	C_6H_6	0.06	0.37
	t-BuOH	0.05	0.80
Octan-2-one	C_6H_6	0.20	0.27
	t-BuOH	0.30	0.64

7.3.2 The Norrish Type II Reaction

A common reaction of aliphatic ketones is intramolecular hydrogen abstraction from the γ position (in rare instances from the δ or even the β position). In addition to regenerating the reactant, the resulting biradical can cleave to give an olefin and an enol, or form a cycloalkanol. Scheme 9 illustrates the most important case of γ hydrogen abstraction.

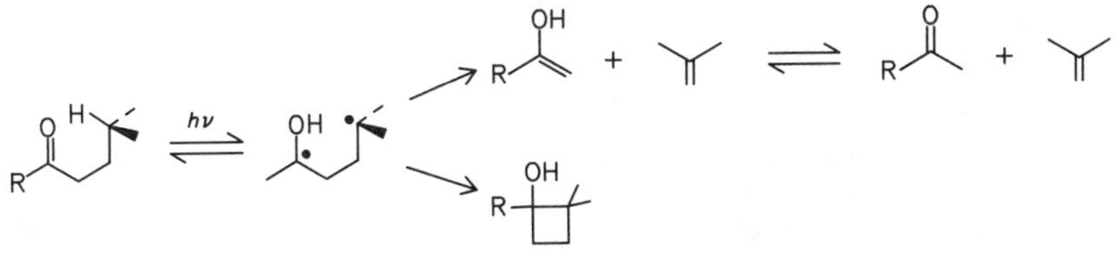

Scheme 9

This reaction is known as Norrish type II reaction. The correlation diagram for the intermolecular hydrogen abstraction discussed in the last section applies equally well to the intramolecular reaction. Accordingly, the in-plane hydrogen abstraction yielding a ketyl radical is allowed from the $^1(n,\pi^*)$ state as well as from the $^3(n,\pi^*)$ state and forbidden from the (π,π^*) states irrespective of their multiplicity. Since singlet states prefer tight biradicaloid geometries, and triplet states prefer loose ones that favor product formation, quantum yields Φ_S of the singlet reaction are generally smaller than those of the triplet reaction (Φ_T), as is exemplified by the data in Table 7.1.

In alkyl aryl ketones spin inversion is so fast that no singlet state reaction is observed (Wagner and Hammond, 1965).

Example 7.7:
MINDO/3 calculations including configuration interaction for the type II reaction of butanal (Dewar and Doubleday, 1978) confirm the conclusions from the correlation diagram. The results are summarized in Figure 7.24: Reactions from the $^1(n,\pi^*)$ state as well as from the $^3(n,\pi^*)$ state proceed over a barrier. The triplet reaction then reaches the $^3B_{\pi,\sigma}$ state of the biradical, which is just

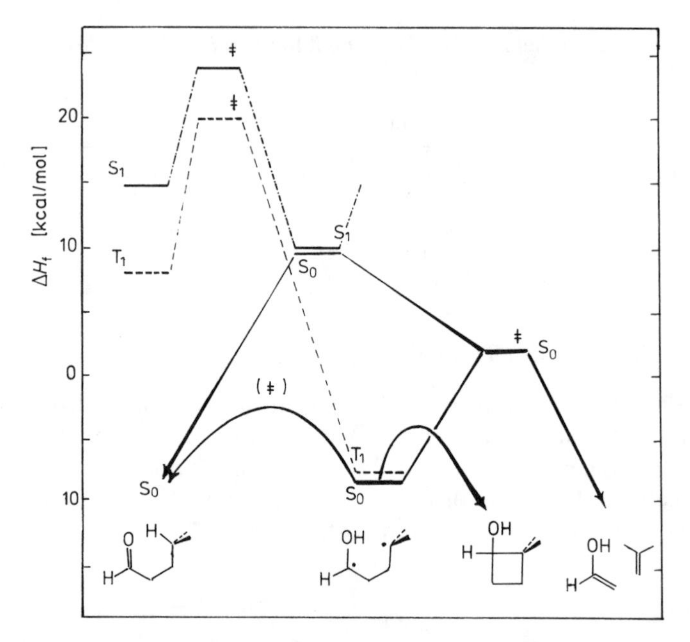

Figure 7.24. Schematic representation of energies of stationary points for the Norrish type II reaction of butanal. The diagram corresponds to a projection of multidimensional potential energy surfaces into a plane. The two energies given for the biradical on the S_0 surface correspond to a geometry optimized for S_0 (front bottom) and optimized for S_1 (middle rear), respectively. A broken line (---) is used for the T_1 surface and a broken-dotted line (-·-·-) for the S_1 surface. The relative energies of T_1, S_0 and S_1 for the geometry of the funnel are not known (by permission from Dewar and Doubleday, 1978).

barely above the ground state. In the correlation diagram, this is apparently the point that corresponds to the crossing of the levels connected to the ground state and to the $^1(n,\pi^*)$ state of the ketone. From this funnel the system can return to the initial geometry or proceed either to the singlet biradical or to the products (cyclobutanol or enol + olefin). The reaction path toward the elimination products passes through a geometry that is effectively the same as that of the transition state for cleavage of the singlet biradical and is stereospecific; that is, the stereochemistry of the initial ketone is preserved in the products. The calculated activation energies of 9 and 12 kcal/mol for the singlet and the triplet reaction, respectively, and of 10 kcal/mol for cleavage of the singlet biradical are presumably too high by 3–5 kcal/mol, but the relative values appear to be correct.

Evidence that the triplet reaction is not concerted, but rather proceeds via the 1,4-biradical, has been obtained from the photoracemization of ketones with a chiral γ-C atom such as **41**, which competes with the hydrogen abstraction (Yang and Elliott 1969):

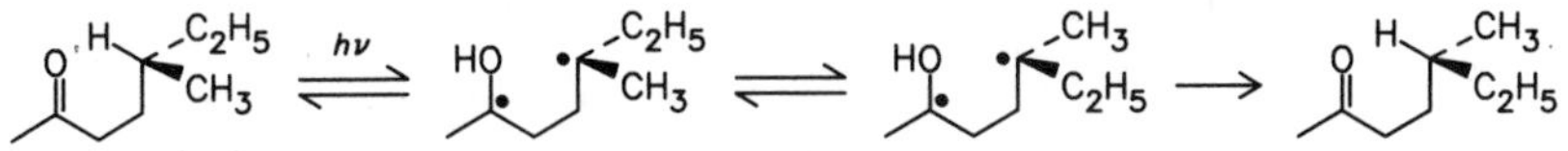

41

The reaction is not concerted and does not yield a triplet olefin, even when this process would be exothermic, as in the case of **42**. Triplet stilbene decays to a 60:40 mixture of *cis*- and *trans*-stilbene, but in the reaction of **42**, 98.6% *trans*-stilbene was observed (Wagner and Kelso, 1969).

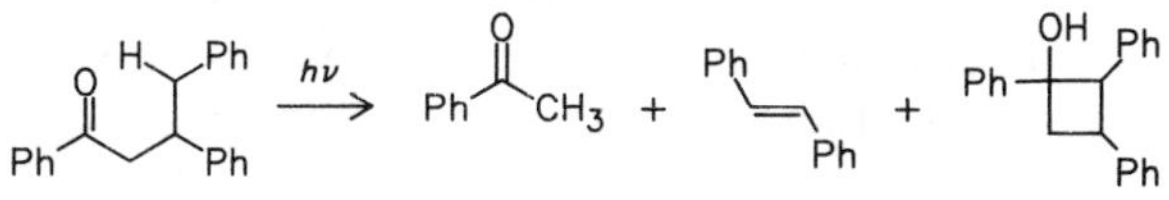

42

Intersystem crossing to the S_0 surface is believed to occur at the 1,4-biradical stage, and to yield one of the three possible types of singlet product (reactant, olefin, cycloalkanol) depending on the geometry at which it occurs, as discussed in Section 4.3.4.

The singlet reaction also proceeds at least partially via a biradical, as was shown indirectly. If the reaction of **43** were concerted, the transfer of H should yield the deuterated cis isomer and the transfer of D the nondeuterated trans isomer:

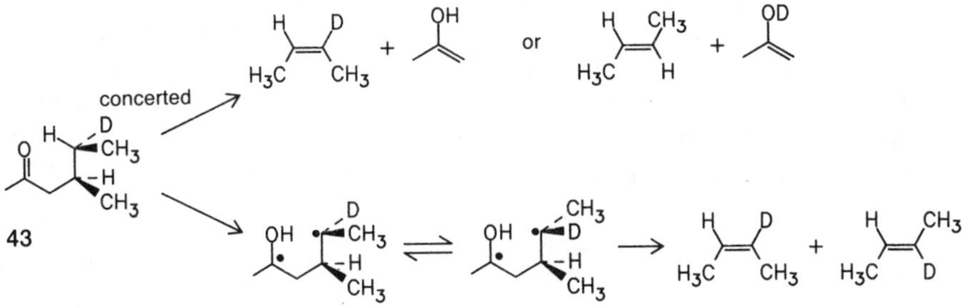

concerted

or

43

With piperylene as the triplet quencher, 10% deuterated trans olefin is found, which must result from rotation about the β,γ bond of the singlet biradical (Casey and Boggs, 1972).

If a molecule has two γ hydrogens available, in the Norrish type II reaction the transfer proceeds over the lower of the two barriers, and according to Scheme 10 a preference for cleavage of the weaker secondary CH bond results (Coxon and Halton, 1974).

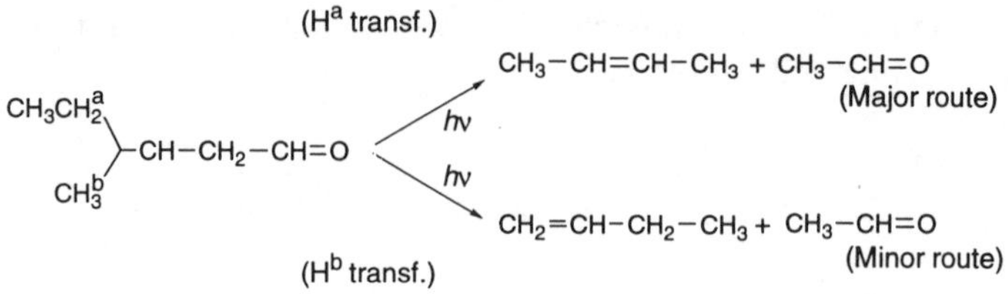

$$CH_3CH_2^{a}(CH_3^{b})CH{-}CH_2{-}CH{=}O \xrightarrow[h\nu]{(H^a\ transf.)} CH_3{-}CH{=}CH{-}CH_3 + CH_3{-}CH{=}O \quad \text{(Major route)}$$

$$CH_3CH_2^{a}(CH_3^{b})CH{-}CH_2{-}CH{=}O \xrightarrow[h\nu]{(H^b\ transf.)} CH_2{=}CH{-}CH_2{-}CH_3 + CH_3{-}CH{=}O \quad \text{(Minor route)}$$

Scheme 10

An increase in temperature or in photon energy reduces the selectivity.

For alkyl aryl ketones electron-releasing substituents in the *p* position decrease the rate constant and quantum yields for type II cleavage. *p*-OH, *p*-NH_2, and *p*-phenyl substituents inhibit the reaction completely. Similarly as in the case of intermolecular hydrogen abstraction, this effect is thought to be a consequence of the $^3(n,\pi^*)$ state no longer being the lowest triplet state, resulting in a larger barrier.

The ratio of olefin to cyclobutanol product yield often depends on substitution. In order to understand this in detail, more would need to be known about the conformational dependence of the spin-orbit coupling matrix element. The present qualitative understanding (Section 4.3.4, Figure 4.26) of the critical intersystem crossing step suggests that it occurs at geometries at which the 2p orbitals of the two radical ends interact through a nonzero resonance integral while their axes lie approximately orthogonal to each other, such that after a 90-degree rotation of one of the orbitals about its center there still is a nonzero resonance integral with the other. According to this analysis a gauche conformation, in which the orbitals interact primarily through space (Figure 7.25a), and an anti conformation, in which they interact primarily through bonds (Figure 7.25b), can both be favorable, provided that the end groups are twisted properly. After intersystem crossing, the former is expected to yield the cyclobutanol, and the latter the fragmentation products, essentially instantaneously. The relative energies of the two types of conformation should be sensitive to the steric demands of substituents.

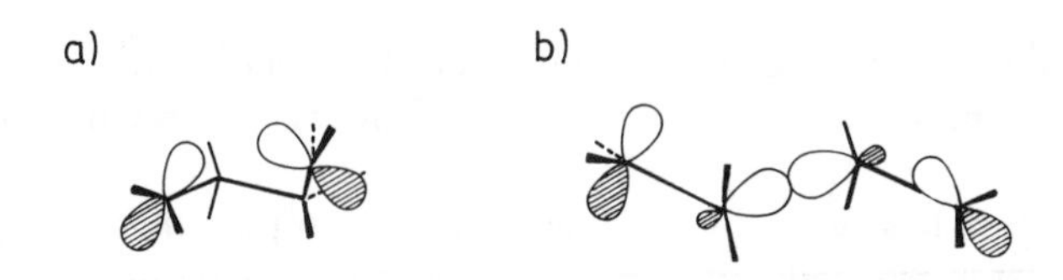

Figure 7.25. Stereoelectronic effects on the Norrish type II reaction. Presumed optimal orbital alignments a) for cyclization and b) for elimination.

The cyclobutanol-forming path is diastereoselective, for example, α-methylbutyrophenone (**44**) and valerophenone (**45**) prefer to place the methyl and the phenyl groups on opposite sides of the four-membered ring (Scheme 11). Such steric discrimination on the T surface may already be present in the open-chain triplet biradical or become felt gradually as the radical ends approach each other and develop the covalent perturbation that leads to intersystem crossing (see Section 4.3.4), and both cases are known and exemplified by the two reactions in Scheme 11. The high stereoselectivity of **44** is believed to be due to a repulsive interaction of the phenyl and the α-methyl groups in the 1,4-biradical (Figure 7.25a). The interaction between the phenyl and the γ-methyl group in the biradical from **45** should be small until after the intersystem crossing has taken place and the 1,4-bond is almost completely formed (Lewis and Hilliard, 1972).

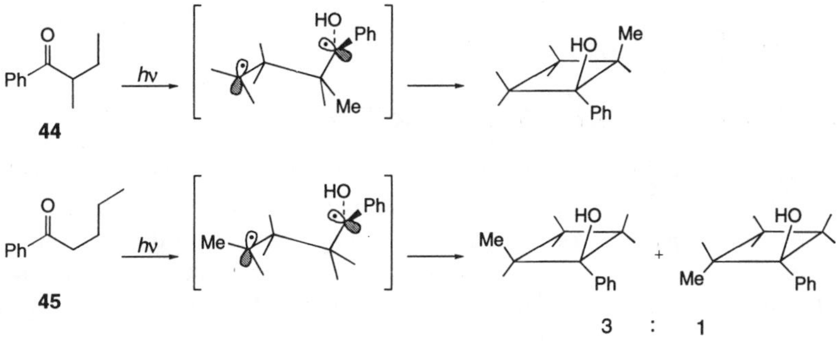

Scheme 11

Abstraction of a δ hydrogen normally competes only when a γ hydrogen is not available, and produces a 1,5-biradical. There are only two choices: return to the starting materials, or cyclization to a cyclopentanol. For instance, α-(o-ethylphenyl)acetophenone (**46**) yields 1-methyl-2-phenyl-2-indanol (**47**):

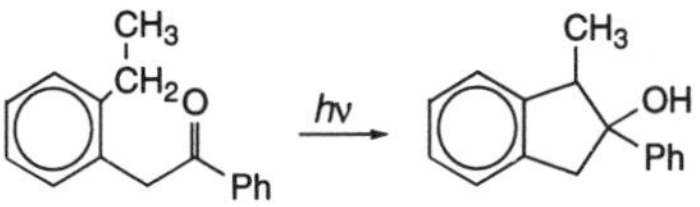

Diastereoselection is again observed and can be understood in terms of the relative energies of the two conformations that are ideally set up for intersystem crossing by spin-orbit coupling (Section 4.3.4). As is seen in Scheme 12, in the favored conformation, a hydroxyl, and in the disfavored conformation, a phenyl, have to be accommodated close to a benzene ring (Wagner et al., 1991).

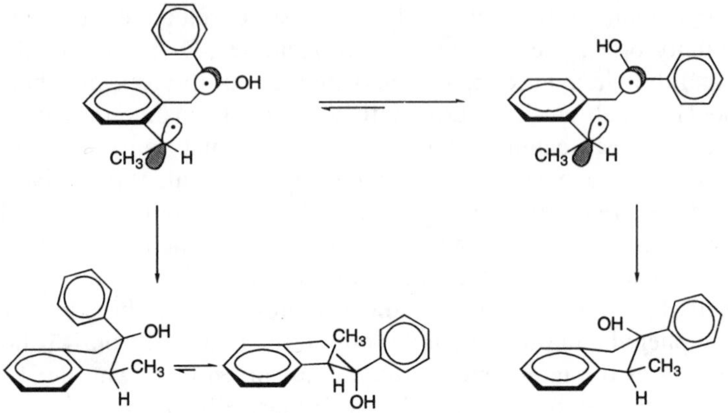

Scheme 12

7.4 Cycloadditions

7.4.1 Photodimerization of Olefins

According to the Woodward–Hoffmann rules, the concerted cycloaddition of two olefins to afford a cyclobutane is allowed photochemically as $[_{\pi}2_{s} + _{\pi}2_{s}]$ reaction and thermally as the $[_{\pi}2_{s} + _{\pi}2_{a}]$ reaction. The different modes of addition give rise to products with different stereochemical structures as indicated in Figure 7.26. If the reaction does not follow a concerted pathway

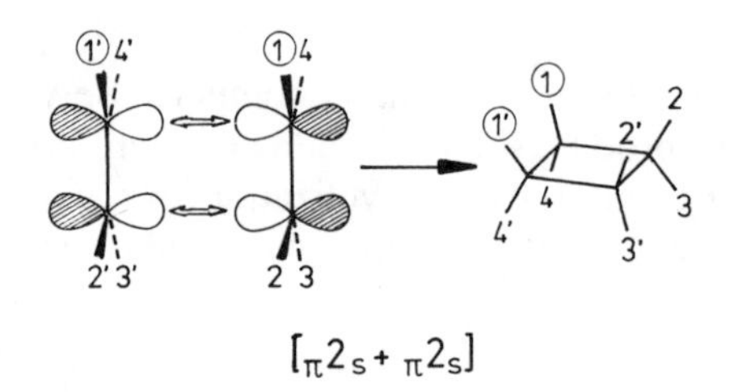

$[_{\pi}2_{s} + _{\pi}2_{a}]$

Figure 7.26. Stereochemical consequences of thermal and photochemical [2 + 2] cycloaddition.

but rather proceeds by a multistep process, and if ring closure of the intermediate is not very rapid compared with bond rotation, rotations about the CC bonds may occur, with subsequent loss of stereospecificity.

Recent calculations for the [2 + 2] photoaddition of two ethylene molecules (Bernardi et al., 1990a,b) demonstrated that the bottom of the pericyclic funnel on the S_1 surface does not lie at the often assumed highly symmetrical rectangular geometry; instead, it is distorted to permit stabilization by diagonal interaction (cf. Section 6.2.1), as suggested by model calculations on H_4 (Gerhartz et al., 1977). Moreover, the calculations show that the S_1–S_0 touching is not even weakly avoided but actually is a conical intersection. These features, a rhomboidal distortion and a conical intersection, are likely to be general for photocycloadditions.

Photodimerization often involves an excimer that can be treated as a supermolecule. (Cf. Section 6.2.3.) Then, the state correlation diagram for the singlet process (Figure 7.27a) ordinarily calls for a two-step return from S_1 to S_0 along the concerted reaction path. First, an excimer intermediate E* is formed. Second, a thermally activated step takes the system to the diagonally distorted pericyclic funnel P* (cf. Section 4.4.1), and the return to S_0 that follows is essentially immediate. The reaction will be stereospecific and concerted in the sense that the new bonds form in concert. However, it will not be concerted in the other sense of the word, in that it involves an intermediate E*.

There may well be systems in which the excimer minimum occurs in the S_2 rather than the S_1 surface (Figure 7.27b). The approach to the pericyclic funnel P* on S_1 may then be barrierless, and an excimer intermediate will

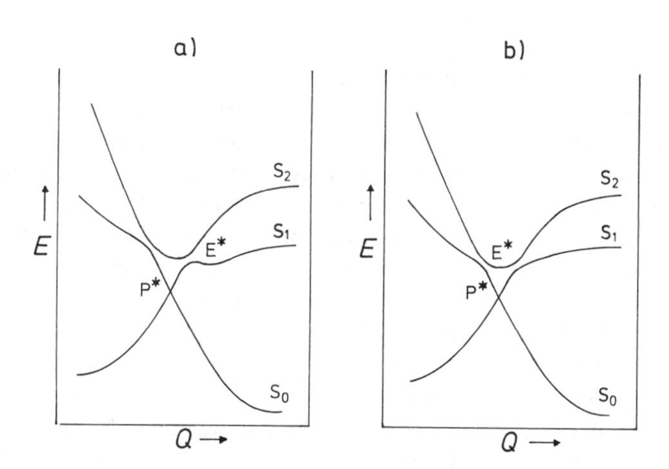

Figure 7.27. Schematic representation of the state correlation diagram for a ground-state–forbidden pericyclic reaction with an excimer minimum E* a) at geometries well before the pericyclic funnel P* is reached, and b) at geometries similar to those of P*.

not be detectable. (For a possible example of the latter case, see Peters et al., 1993.)

These concepts are in very good agreement with experimental findings. There are relatively few examples of photodimerization of simple nonconjugated acyclic olefins because these compounds absorb at very short wavelengths. Irradiation of neat but-2-ene, however, yields tetramethylcyclobutane with a quantum yield $\Phi = 0.04$. For very low conversion, the observed stereochemistry of the adducts is the stereospecific one expected from Scheme 13 for a concerted $[_{\pi}2_s + {}_{\pi}2_s]$ cycloaddition. However, since the major pathway is cis-trans isomerization with a quantum yield $\Phi = 0.5$ (cf. Section 7.1.2), it has been concluded that the molecules that undergo cis-trans isomerization are not involved in photodimerization (Yamazaki et al., 1976).

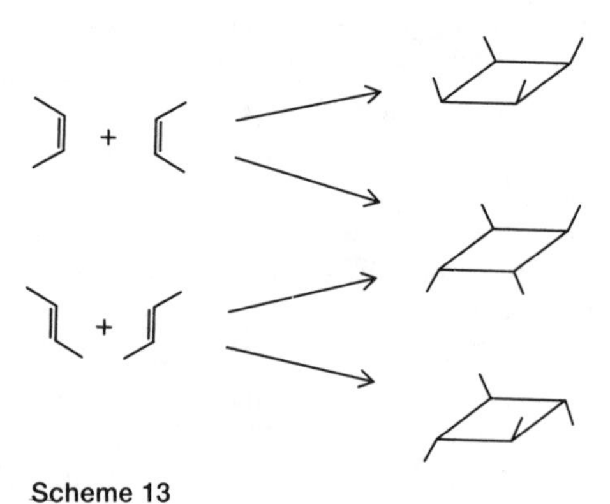

Scheme 13

In simple olefins, direct excitation of the triplet state and intersystem crossing from an excited singlet state to a triplet state do not play an important role. A sensitized reaction of the triplet state is possible and could in principle also be concerted. However, in the triplet state loose geometries with two separate radical centers are energetically more favorable than tight pericyclic geometries with cyclic interaction (cf. Section 6.2.1); it is most probable that one of these favorable minima will be reached prior to return to S_0. This tendency then favors a nonconcerted mechanism with the two new bonds formed in separate reaction steps. Formation of the first bond takes place on the T_1 surface, while the second one will close after the molecule has reached the S_0 surface. The nature and stereochemistry of the product that results from the triplet species upon return to the S_0 surface, in particular cyclobutane formation or back reaction to two olefins, is believed to be dictated by the geometry at which the conversion to S_0 took place. The relative efficiencies are determined by the rates of the various processes, which in turn depend on such factors as the populations of the various conformers in T_1 and the size of the T_1–S_0 spin-orbit coupling matrix element, the geometrical dependence of which was discussed in detail in Section 4.3.4.

Small-ring cyclic alkenes cannot deactivate by cis-trans isomerization. For instance, in contrast to acyclic alkenes, cyclopentene (**48**) therefore undergoes photosensitized [2 + 2] cycloaddition. For cycloalkenes with a six-membered or larger ring, a trans form becomes possible; for molecules such as cyclooctene (**2**), photosensitized cis-trans isomerization is the more efficient reaction path. (Cf. Section 7.1.2.)

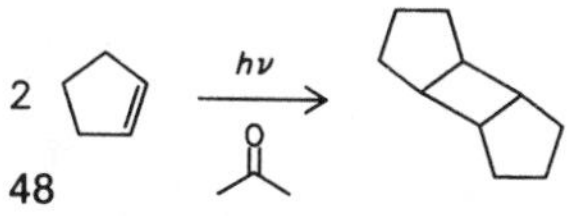

2 48 $h\nu$

Since most simple alkenes have a high triplet energy (E_T = 75–78 kcal/mol), triplet sensitizers have to be chosen accordingly to prevent oxetane formation (see Section 7.4.4), as shown in Scheme 14 for norbornene with acetophenone (E_T = 75 kcal/mol) and benzophenone (E_T = 69 kcal/mol), respectively, as sensitizers (Arnold et al., 1965):

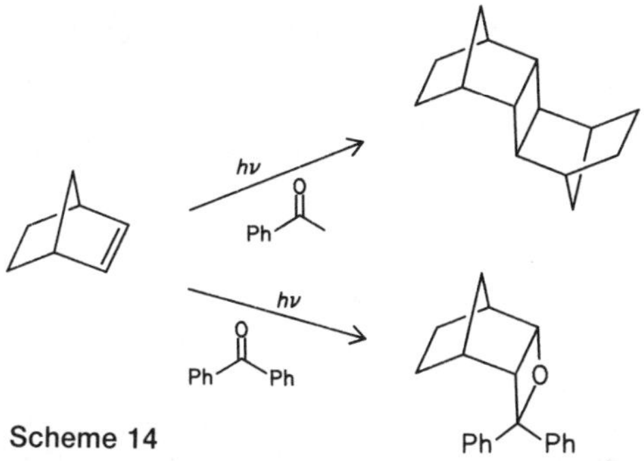

$h\nu$ Ph

$h\nu$ Ph Ph

Scheme 14

O Ph Ph

Finally, it should be remembered that excimer minima E* and pericyclic funnels P* of different energies can result if different mutual arrangements of the reactants can lead to cycloaddition, as indicated in Scheme 3 in Section 6.2.3. The related issues of regio- and stereoselectivity of singlet photocycloaddition are dealt with in Section 7.4.2.

Example 7.8:

Sensitized irradiation of cyclohexenes and cycloheptenes in protic media results in protonation. This phenomenon, which is not shared by other acyclic or cyclic olefins, has been attributed to ground-state protonation of a highly strained *trans*-cycloalkene intermediate. In aprotic media, either direct or triplet-sensitized irradiation of cyclohexene produces a stereoisomeric mixture of [2 + 2] dimers **49–51** as the primary products, with **50** predominating. The reaction apparently involves an initial cis-trans photoisomerization of cyclohexene followed by a nonstereospecific nonconcerted ground-state cycloaddition, promoted by the high degree of strain involved. In contrast, cycloheptene undergoes only a slow addition to the *p*-xylene used as sensitizer, presumably because the trans isomer is not sufficiently strained to undergo the nonconcerted cycloaddition. Copper(I)-catalyzed photodimerization of cyclohexene and cycloheptene affords the product expected for $[2_s + 2_a]$ cycloaddition of the trans isomer to the cis isomer within a *cis,trans* complex **52**. Stereospecificity is high in the former case and complete in the latter. (Kropp et al., 1980.)

hν

49 (31%) **50** (51%) **51** (18%)

hν

52

Irradiation of butadiene in isooctane yields the isomeric 1,2-divinylcyclobutanes **53a** and **53b** and various other products, dependent on reaction conditions. The triplet-sensitized photoreaction yields 4-vinylcyclohexene (**54**) in addition to the divinylcyclobutanes (cf. Example 6.11), and the product ratio depends on the triplet energy of the sensitizer:

2 hν ^{3}Sens* + +

a 53 b 54

When E_T is in excess of 60 kcal/mol, *s-trans*-butadiene (E_T = 60 kcal/mol), which strongly predominates in the thermal equilibrium, is excited and produces mainly divinylcyclobutanes. When the E_T of the sensitizer is not high enough to excite *s-trans*-butadiene, energy transfer to *s-cis*-butadiene (E_T = 54 kcal/mol) occurs instead, yielding vinylcyclohexene. If $E_T < 50$ kcal/mol, nonvertical energy transfer to a twisted diene triplet is believed to occur (Liu et al., 1965).

Intramolecular photoadditions can also occur as so-called $x[2 + 2]$ cycloadditions, as demonstrated for 1,5-hexadiene (**55**): the terminal carbon atom of each double bond adds to the internal carbon atom of the other double bond in such a way the resulting σ bonds cross each other.

hν x[2+2]

55

It is likely that [2 + 2] and $x[2 + 2]$ cycloadditions proceed through the same type of diagonally distorted pericyclic funnel (Section 4.4.1) with a preservation of the diagonal interaction, and eventual production of two diagonal bonds, to yield the $x[2 + 2]$ product in one case; in the other case, there is preservation of perimeter bonding and eventual production of the [2 + 2] product. (Cf. Figure 6.1.5).

Which product will be formed depends on the number of methylene groups between the double bonds. If either one or three CH_2 groups are present, the common [2 + 2] cycloaddition is observed; with two CH_2 groups $x[2 + 2]$ cycloaddition predominates. For cyclic dienes the dependence of the reaction product on the number of CH_2 groups is even more pronounced; 1,5-cyclooctadiene (**56**) yields the product **57** exclusively (Srinivasan, 1963).

hν

56 57

These findings have been rationalized by the *rule of five* (Srinivasan and Carlough, 1967), according to which five-membered cyclic biradicals are preferentially formed, as shown in Scheme 15. (Cf. also the Baldwin rules for radical cyclizations, Baldwin, 1976.)

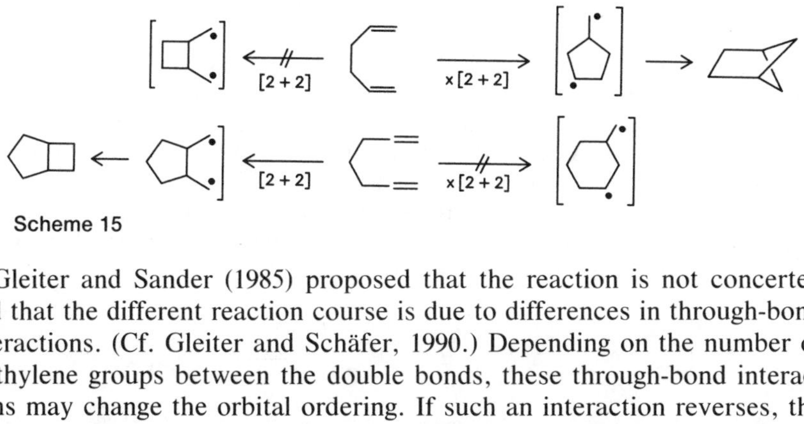

[2 + 2] x[2 + 2]

[2 + 2] x[2 + 2]

Scheme 15

Gleiter and Sander (1985) proposed that the reaction is not concerted and that the different reaction course is due to differences in through-bond interactions. (Cf. Gleiter and Schäfer, 1990.) Depending on the number of methylene groups between the double bonds, these through-bond interactions may change the orbital ordering. If such an interaction reverses, the "natural" ordering of π orbitals imposed by through-space interaction, the excited-state barrier for the $x[2 + 2]$ cycloaddition will be lower than that for the [2 + 2] cycloaddition. Evidence for the deleterious effect of the reversal of orbital order on the [2 + 2] process is provided by cases in which the $x[2 + 2]$ reaction is sterically impossible and the [2 + 2] reaction fails to proceed (Example 7.9). However, it is also possible that the reaction is concerted in its initial stages, and that the difference in the relaxation paths after return to S_0 through a distorted pericyclic funnel is dictated by the steric requirements imposed by the alkane chains.

Example 7.9:

Tricyclo[4.2.0.0^{2,5}]octadiene (**58**) does not undergo [2 + 2] photocyclization to cubane, although its geometry appears to be ideally set up for it. In this mol-

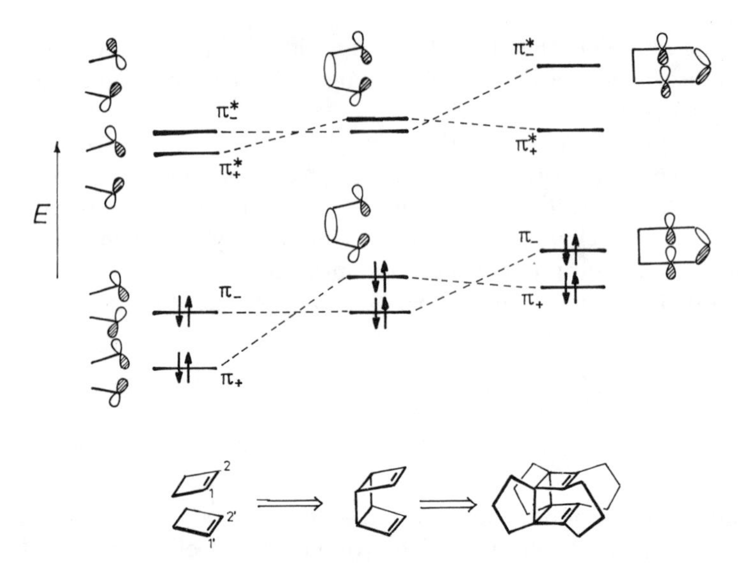

Figure 7.28. Correlation of the π orbitals of two cyclobutene molecules (left) with those of **58** (center) and **59** (right) (by permission from Gleiter, 1992).

ecule, the interaction of the normally more stable in-phase combination of the two π orbitals, $\pi_+ = (\pi_1 + \pi_2)/\sqrt{2}$, with the two doubly allylic σ bond orbitals is very strong and pushes its energy above that of the out-of-phase combination, $\pi_- = (\pi_1 - \pi_2)/\sqrt{2}$, inverting the natural orbital order (Figure 7.28). This converts the normal orbital crossing characteristic of ordinary [2 + 2] cycloadditions into an abnormal orbital crossing, so that the characteristic configuration no longer is the lowest in energy, and a barrier in the potential energy surface results for this reaction path. (See Section 6.2.4.)

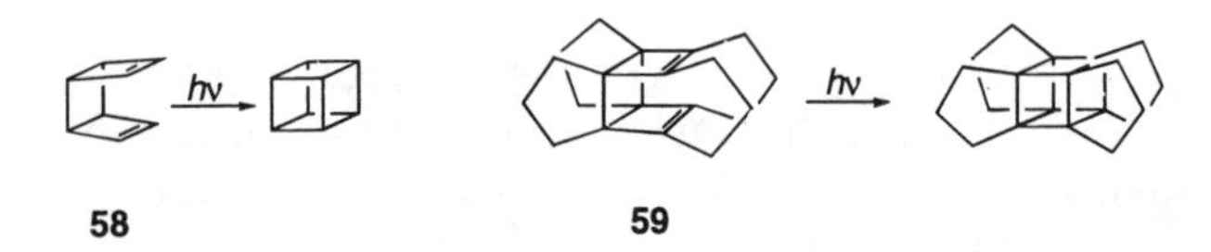

In the quadruply bridged derivative **59**, the [2 + 2] photocycloaddition proceeds. This can be understood as due to the restoration of the natural order of the orbital energies, π_+ below π_-, by the effect of the propano bridges (Figure 7.28). These can be expected to bring the two double bonds closer in space, increasing the through-space interaction, and to introduce additional through-bond interaction opposed in sign to the interaction through the doubly allylic bonds (Gleiter and Karcher, 1988).

Mixed cycloadditions between different olefins are also observed. The regiochemistry follows a pattern that has been rationalized by consideration of the HMO coefficients (Herndon, 1974). Some typical examples of triplet-sensitized mixed cycloadditions are summarized in Scheme 16 (Scharf, 1974).

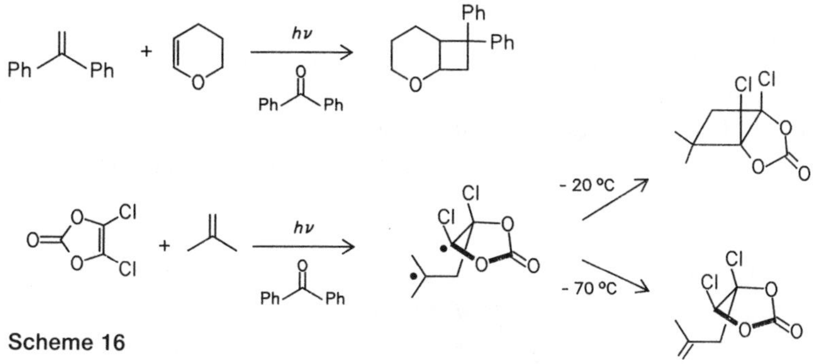

Scheme 16

The addition of *trans*-stilbene to 2,3-dimethylbutene-2 appears to be a singlet reaction:

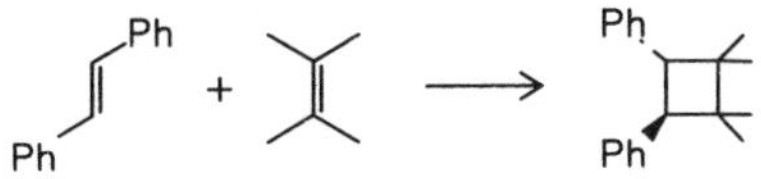

Furthermore, [2 + 2] cycloadditions to N=N and C=N double bonds have been described (Prinzbach et al., 1982; Albert et al., 1984). The regioselectivity of the addition of electron-rich olefins to cyclic ketoiminoethers such as **60** may be rationalized using perturbation theory (Fabian, 1985):

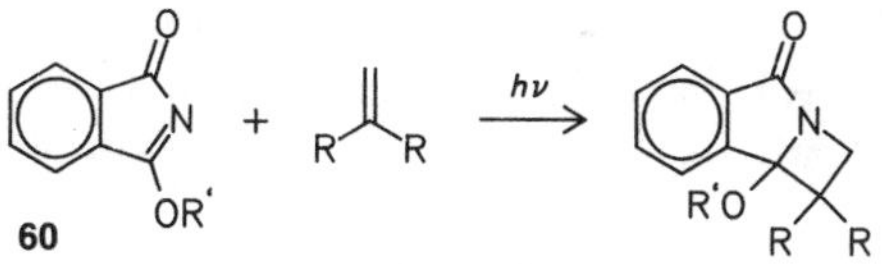

7.4.2 Regiochemistry of Cycloaddition Reactions

An important aspect of photocycloaddition consists of its regio- and stereoselectivity. Thus, dimerization of substituted olefins generally yields head-to-head adducts **61a** and **61b** in a regiospecific reaction, while head-to-tail dimers **63** are obtained from 9-substituted anthracenes (**62**) in polar solutions (Applequist et al., 1959; cf. Kaupp and Teufel, 1980).

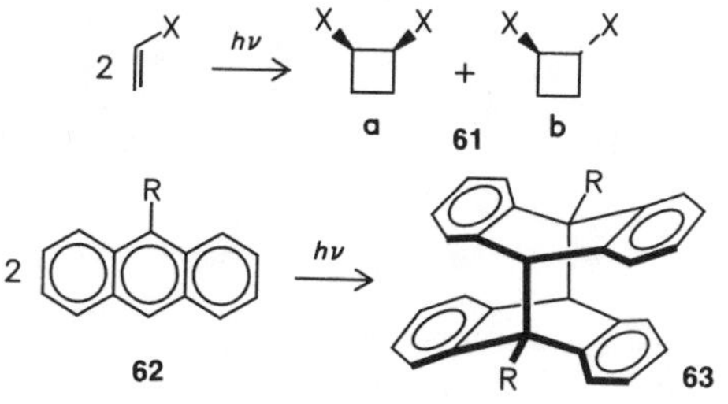

The portion of the head-to-head product increases in nonpolar solvents and in micelles (Wolff et al., 1983). The regioselectivity of 9-methylanthracene is temperature dependent, and a photochemical equilibrium exists between the two isomers (Wolff, 1985).

Singlet acenaphthylene (**64**) stereospecifically gives the syn dimer **65a**, while in the T_1 reaction the anti dimer **65b** predominates. The syn-anti ratio can be influenced by solvents with heavy-atom effect (Cowan and Drisco, 1970b), as well as by micellar solvents (Ramesh and Ramamurthy, 1984).

64 → 65 a + b

S_1: 100 : 0

T_1: 1 : 9

Triplet quenchers such as O_2 or ferrocene inhibit formation of the anti dimer, while the formation of the syn dimer is hardly affected. The syn-anti ratio in the triplet reaction was determined by comparing the outcome of the reaction with and without use of a quencher (Cowan and Drisco, 1970a).

Many factors must be considered to explain these facts, not least the relative stabilities of the various possible excimers or exciplexes and the accessibility of the pericyclic funnels. (Cf. Scheme 3 in Section 6.2.3.) In addition, the geometrical structure at the funnel and the ground-state relaxation pathways originating in the decay region determine which products will be formed.

Excimer and exciplex formation will mainly influence the excited-state reaction path. Simple perturbation theory suggests that in concerted singlet photocycloadditions, electronic factors always favor the most highly symmetric excimer affording head-to-head regiochemistry and syn or cis stereochemistry, since HOMO-HOMO and LUMO-LUMO interactions are decisive in excimer formation according to Section 5.4.2. (Cf. Figure 5.21.) Interaction between centers whose LCAO coefficients are either both large or both small, that is, the head-to-head addition, is then favored:

X X X

X

head-to-head addition head-to-tail addition

Similarly, secondary orbital interactions favor the syn arrangement of two acenaphthylene molecules, as illustrated in Figure 7.29.

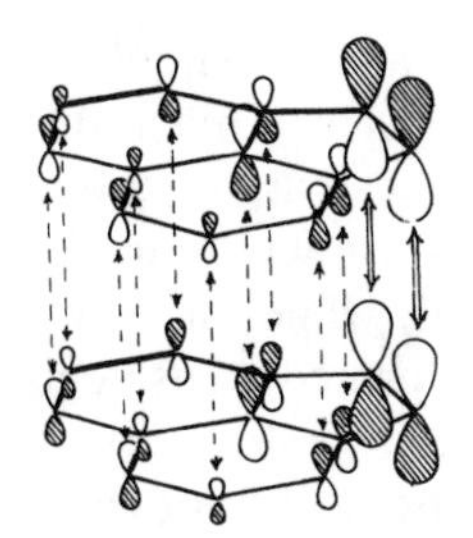

Figure 7.29. Secondary orbital interactions (broken arrows) in the photochemical dimerization of acenaphthylene. The primary orbital interactions are indicated by full double arrows.

The formation of head-to-tail dimers such as those of 9-substituted anthracenes and similar species can be rationalized by the conjecture that in this case it is not the most favorable excimer that determines the stereochemical outcome of the reaction but rather, the most readily reached pericyclic funnel. This can be understood if the effect of substituents on the pericyclic funnel is first considered at the simple two-electron–two-orbital model level (3 × 3 CI, Section 4.3.1). The discussion of cyclodimerization of an olefin is restricted to the four orbitals directly involved in the reaction.

To simplify the presentation, we shall initially ignore the possible effects of a diagonal distortion. Then, the localized nonbonding orbitals at the rectangular pericyclic geometry have equal energies. They correspond to the degenerate π MOs ϕ_2 and ϕ_3 of cyclobutadiene.

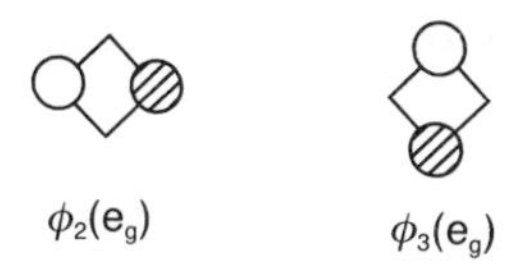

Since they are occupied with a total of two electrons, the system is a perfect biradical ($\delta = 0$) and the energy splitting ΔE between S_0 and S_1 is relatively large. (Cf. Figure 4.19.) Substitution in positions 1,3 or 2,4 removes the degeneracy (Section 6.2.1), and the biradical is converted into a heterosymmetric biradicaloid ($\delta \neq 0$). According to the results in Section 4.3.3, this brings the S_1 and S_0 states closer and is likely to reduce the energy of the S_1 state. For instance, ab initio calculations on perturbed cyclobutadienes indicate that the splitting vanishes almost completely when two CH groups on diagonally opposite corners of the cyclobutadiene (positions 1 and 3) are replaced by isoelectronic $NH^{\oplus}$ groups, yielding a critically heterosymmetric biradicaloid (Bonačić-Koutecký et al., 1989).

As a result of the rectangular geometry, the pericyclic "minimum" at the head-to-tail dimerization path, which corresponds to a 1,3-disubstituted heterosymmetric biradicaloid, is likely to be deeper than that at the head-to-head reaction path. This in turn should reduce the barriers around the minimum and thus make it easier to reach the head-to-tail rectangular "minimum" from the excimer. These arguments would suggest that head-to-tail regiochemistry is favored when access to the pericyclic "minimum" determines the product formation, while the head-to-head product is expected if the energy of the excimer intermediate is decisive for the reaction path (Bonačić-Koutecký et al., 1987).

Next, we consider also the effects of rhomboidal distortions, which permit a diagonal interaction. These may either reinforce or counteract the effect of the substituents, depending on which of the two diagonals has been shortened. For a head-to-tail approach of two substituted ethylenes, there will be two funnels corresponding to 1,3-disubstituted critically heterosymmetric biradicaloids, one at less and one at more diagonally distorted geometry than for the unsubstituted ethylene. This is confirmed by the results of calculations shown in Figure 7.30.

In agreement with expectations from the two-electron–two-orbital model (lower part of Figure 7.30), the effect of donor substituents reinforces the orbital splitting produced by a 1,3-diagonal interaction, which brings the

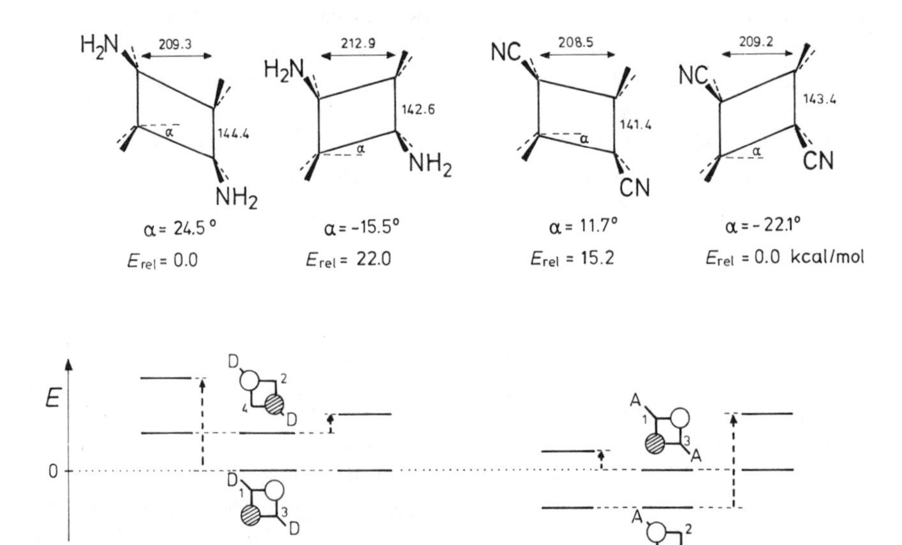

Figure 7.30. Geometries of the conical intersections for the syn head-to-tail [2 + 2] cycloaddition of two aminoethylene molecules (left) and of two acrylonitrile molecules (right), and a schematic representation of the energies of the cyclobutadiene-like "nonbonding" biradicaloid orbitals. The effect of the donor (D) and acceptor (A) substituents on orbital energies is incorporated in the levels shown in the center of each half of the diagram. The effect of the diagonal interaction required to reach the critical value of the orbital energy difference (δ_0, cf. Section 4.3.3) is indicated by broken arrows.

substituted atoms closer to each other, while the effect of acceptor substituents opposes that of the 1,3 interaction. The opposite is true for the 2,4 interaction, which places the substituted atoms apart. As a result, a small rhomboidal distortion is sufficient to reach the "critically biradicaloid" geometry at which S_1 and S_0 are degenerate when the 1,3 diagonal is short for donor substituents and when the 2,4 diagonal is short for acceptor substituents. A large distortion is needed for donor substituents if the short diagonal is 2,4 and for acceptor substituents if it is 1,3. Numerical calculations yield a lower energy for the conical intersection when the substituent effect is opposed to that of the rhomboidal distortion, that is, at the more strongly distorted geometries (Klessinger and Hoinka, 1995).

For small rhomboidal distortions, the peripheral bonding will dominate at the funnel geometry after return to S_0, and the formation of the head-to-tail product is likely to be favored. The diagonal bonding will dominate for large distortions, however, and head-to-head product could possibly be formed by $x[2 + 2]$ cycloaddition—that is, by formation of one diagonal bond and subsequent closure of the other. Thus, the result of the photoaddition will depend on the distorted rhomboidal geometries, that is, on the relative heights of the barriers that lead to the various funnels, typically via an excimer or exciplex minimum. That is where steric effects as well as abnormal orbital crossing effects (cf. Example 7.9) may come into play.

Finally, as we pointed out in Example 6.2, a number of different ground-state trajectories leading to one or the other of the possible photoproducts may emanate from the apex of the cone. This is illustrated schematically in Figure 7.31, which shows energy contour maps for the S_1 and S_0 states of the syn head-to-tail approach of two acrylonitriles. Three valleys starting at the conical intersection are seen on the ground-state surface (Figure 7.31b), one leading to the cyclobutane minimum (B), the other one to the diagonally bonded product (D), and a broad slope back to the reactants (R). The steric outcome of the reaction will depend on the way in which the system enters the funnel region (Figure 7.31a) and on the dynamics of the passage through the cone.

A model for estimating the barrier height between the excimer minimum and the pericyclic minimum for a given geometrical arrangement of the adducts, which ignores the effects just mentioned, was introduced in Section 6.2.3. The resonance integral $\beta(r_c)$ for the bonds formed in the cycloaddition reaction may be estimated according to Equation (6.3) from the singlet and triplet excitation energies and the HMO coefficients of the HOMO and LUMO. A small absolute value will correspond to early surface crossings (cf. Figure 6.17) and thus to low barriers, and high reactivity. Table 7.2 summarizes some results obtained from this model. It is seen that aromatic compounds and olefins undergo facile dimerizations when $|\beta(r_c)| < 20$ kcal/mol, while for $|\beta(r_c)| \approx 20$–24 kcal/mol the reactivity is moderate, and for $|\beta(r_c)| \geq 24$ kcal/mol dimerizations have so far not been reported.

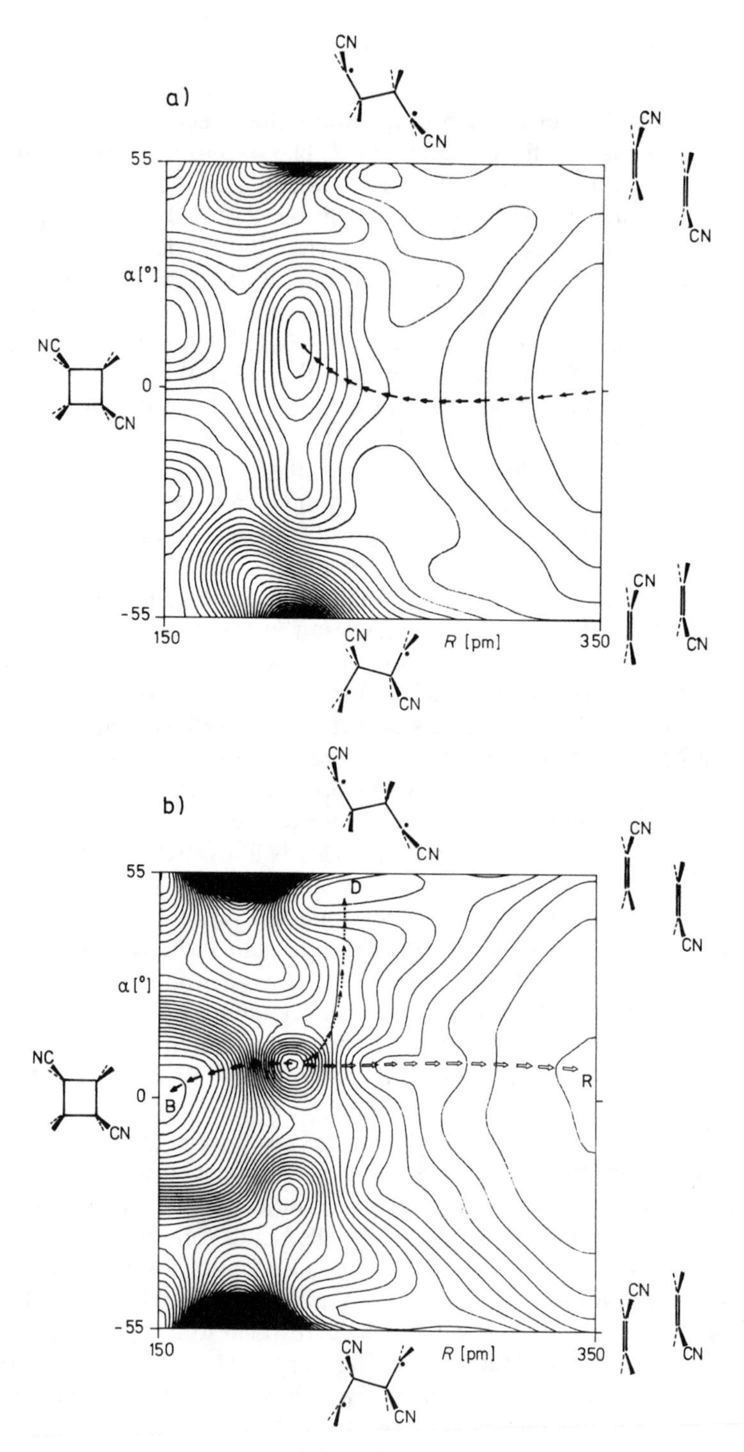

Figure 7.31. Energy contour maps for the face-to-face [2 + 2] cycloaddition of two acrylonitrile molecules in the syn head-to-tail arrangement, showing (a) the lowest excited state S_1 and (b) the ground state S_0. Possible reaction paths via the low-energy conical intersection are indicated by arrows.

Table 7.2 Dimerization of Olefins and Aromatic Compounds: Singlet and Triplet Excitation Energies E_S and E_T, and Interaction Integral $|\beta(r_c)|$ (kcal/mol)

	E_S	E_T	$\sum ab$[a]	$\lvert\beta(r_c)\rvert$	Exp.
Anthracene	75.6	42.6	1.55*	6.2	+
Stilbene	85.3	49	1.55	8.2	+
1,3-Cyclohexadiene	91.3	53	2.88*	5.1	other reactions
			1.98	7.4	
Simple alkenes	124	78	4.04	16	+
Acenaphthylene	61.1	43–47	1.45	17–23	+
Naphthalene	90.9	60.9	1.41*	21.9	±
			0.98	31.7	−
Pyrene	77.0	48.3	0.72	27.2	−
Phenanthrene	82.4	61.7	1.41	29.1	−

[a] $\sum ab = \sum_{\varrho\sigma} (a_{\varrho\text{HO}} b_{\sigma\text{HO}} + a_{\varrho\text{LU}} b_{\sigma\text{LU}})$ for [2 + 2] or [4 + 4] cycloadditions; asterisks indicate results for the latter (adapted from Caldwell, 1980).

In discussing the state correlation diagram of the $H_2 + H_2$ reaction, which can serve as a model for ethylene dimerization, it has been pointed out that the doubly excited state corresponds in the limit to an overall singlet coupling of two initial molecules $^3M^*$ in their triplet-excited states. As a consequence, the pericyclic minimum should be accessible not only from the excimer minimum but also directly through triplet-triplet annihilation. This was demonstrated for anthracene dimerization by determining the quantum yield as a function of the light intensity. The probability of excimer formation by triplet–triplet annihilation was determined experimentally as $p_D' = 0.115$, which is in good agreement with the spin statistical factor 1/9 and is accounted for by noting that apart from the singlet complex, triplet and quintet complexes are also formed upon triplet–triplet encounter. (Cf. Section 5.4.5.5.) Only the singlet species can yield the pericyclic intermediate, whereas the other complexes decay into monomers in the ground state or in a singlet- or triplet-excited state (Saltiel et al., 1981).

It is likely that triplet–triplet annihilation also plays a role in other photochemical reactions. This is especially true at high concentrations and high intensities of the exciting light. The sensitized photodimerization of anthracene presumably proceeds entirely through triplet–triplet annihilation.

7.4.3 Cycloaddition Reaction of Aromatic Compounds

Photodimerizations are observed not only for olefins, but also for aromatic compounds, allenes, and acetylenes. The photodimerization of anthracene, which may be considered to be a ground-state–forbidden $[_{\pi}4_s + _{\pi}4_s]$ cycloaddition, was in fact described as early as in 1867 (Fritsche, 1867).

Example 7.10:
Much information about the detailed mechanism of anthracene dimerization was gained in the study of intramolecular photoreactions of linked anthracenes such as α,ω-bis(9-anthryl)alkanes (**66**). It was shown that luminescence and cycloaddition are competing pathways for the deactivation of excimers. In compounds with sterically demanding substituents R and R′ that impair the cycloaddition reaction, the radiative deactivation is enhanced (H.-D. Becker, 1982).

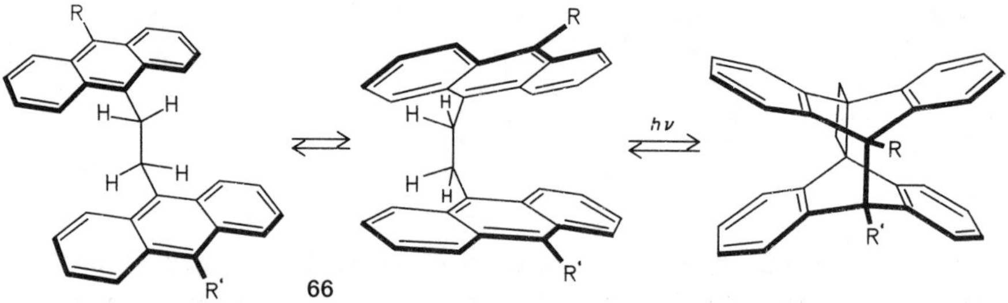

Photocyclization of bis(9-anthryl)methane (**67**) and the corresponding photocycloreversion were shown by picosecond laser spectroscopy to have a common intermediate whose electronic structure is different in polar or nonpolar solvents as judged by different absorption spectra in various solvents (Manring et al., 1985).

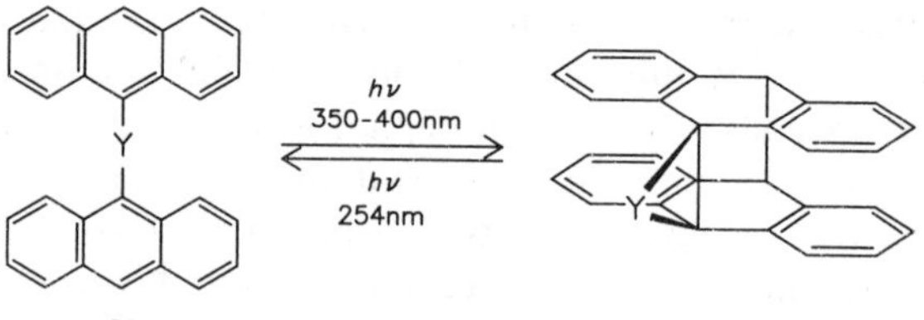

α,α'-Disubstituted bis-9-anthrylmethyl ethers (**68**) are characterized in their meso form by mirror-plane symmetry (σ) and perfectly overlapping anthracene moieties (**68a**). The corresponding racemic diastereomers (**68b**) assume a conformation having a twofold axis of symmetry (C_2). Both the meso and racemic diastereomers cyclomerize. However, whereas the meso compounds are virtually nonfluorescent, the racemic diastereomers deactivate radiatively both from the locally excited state and from the excimer state. Thus, in the emitting excimer state of linked anthracenes the aromatic moieties overlap only partially. Apparently, the formation of luminescent excimers from bichromophoric aromatic compounds is associated with perfectly overlapping π systems only when intramolecular cycloaddition is an inefficient process (H.-D. Becker, 1982).

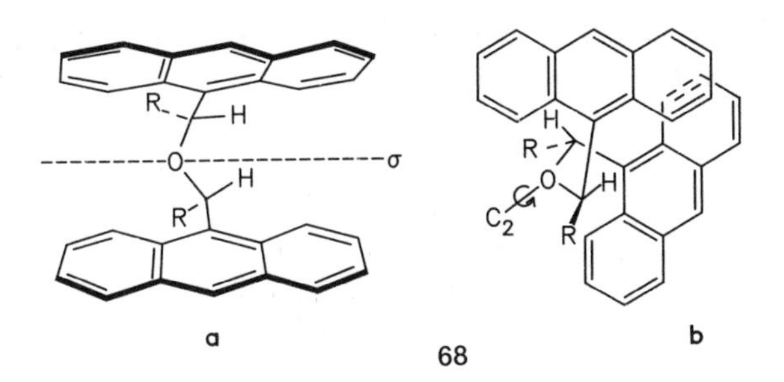

The [2 + 2] cyclodimerization of benzene has been studied theoretically (Engelke et al., 1984) and used practically in the synthesis of polycyclic hydrocarbons such as **69** (Fessner et al., 1983). This represents a first step in the synthesis of the undecacyclic hydrocarbon pagodane (**70**), which in turn can be isomerized to give dodecahedrane (Fessner et al., 1987).

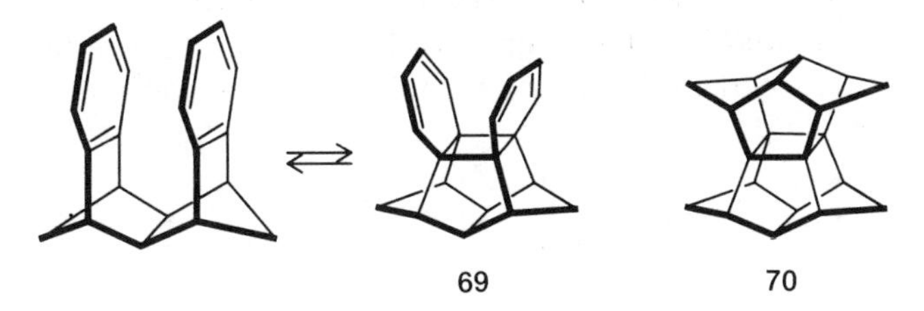

Mixed photocycloadditions of anthracene and conjugated polyenes yield products that correspond to a concerted reaction path, as well as others that are Woodward-Hoffmann–forbidden and presumably result from nonconcerted reactions. For example, the reaction of singlet-excited anthracene with 1,3-cyclohexadiene yields small quantities of the $[_{\pi}4_{s} + {}_{\pi}2_{s}]$ product **72** in addition to the allowed $[_{\pi}4_{s} + {}_{\pi}4_{s}]$ product **71**.

X + → 71 + 72

The yield of **71** increases with increasing polarity of either the solvent or the substituent X in the 9-position of anthracene. This result has been explained by invoking a stabilization of the exciplex, which should reduce the barrier between the exciplex minimum and the pericyclic minimum. The observation that in the presence of methyl iodide **72** becomes the major product for X = H, due to the heavy-atom effect, is compatible with the obvious assumption that the multistep process is a triplet reaction. However, if X = CN, no methyl iodide heavy-atom effect is observed (N. C. Yang et al., 1979, 1981).

Addition of an alkene to benzene can occur in three distinct ways indicated in Scheme 17: ortho, meta, or para.

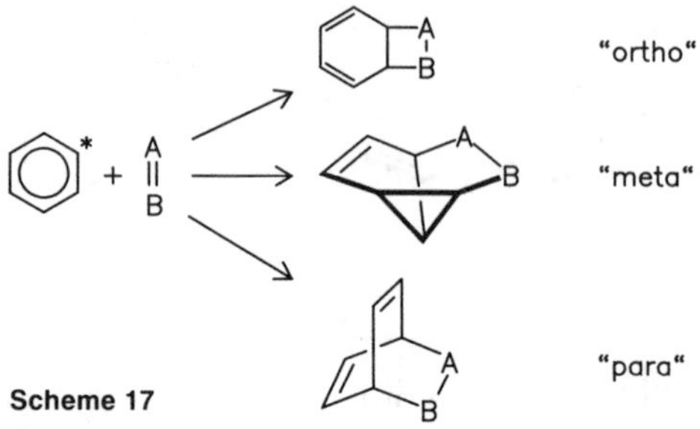

"ortho"

"meta"

"para"

Scheme 17

In general, dienophiles produce the ortho product in addition to small amounts of the para product. Maleic anhydride gives the 1:2 adduct **73** (Gilbert, 1980). Maleimides undergo analogous reactions. Both classes of compounds form CT complexes with benzene and its derivatives, and the reaction can be initiated by irradiation into the CT band (Bryce-Smith, 1973). Alkylethylenes, on the other hand, produce the meta adducts.

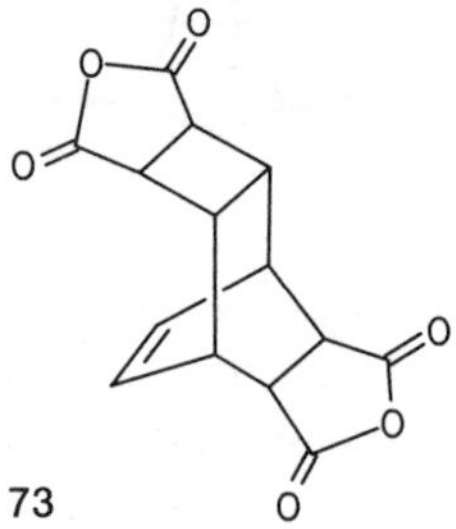

73

The selectivity may be explained by a consideration of the possible interactions of the degenerate benzene HOMO, ϕ_s, ϕ_a, and LUMO, $\phi_{s'}$, $\phi_{a'}$ (cf. Section 2.2.5 and Figure 2.33), with the ethylene π and π^* MOs (Houk, 1982). In the ortho approach, the benzene MOs ϕ_a and $\phi_{a'}$ can interact with the ethylene MOs π and π^*, while in the meta approach, ϕ_s and $\phi_{a'}$ can interact with π and π^*. Therefore, the configuration $\Phi_{a \rightarrow s'}$ is predicted to be energetically favored along the ortho cycloaddition path, while the configuration $\Phi_{s \rightarrow a'}$ is stabilized for the ortho and particularly the meta cycloaddition path. The magnitude of the stabilization depends on orbital overlap and relative orbital energies. The former factor favors the ortho approach and the latter the meta approach of the reactants.

Altogether, one arrives at the prediction that for the addition of ethylene to the S_1 ($^1B_{2u}$) state of benzene, whose wave function can be written as $(\Phi_{s \rightarrow a'} - \Phi_{a \rightarrow s'})/\sqrt{2}$, comparable amounts of ortho and meta adducts should be formed, while the meta addition should dominate whenever the alkene MOs π and π^* are close in energy to the benzene MOs ϕ_s, ϕ_a and $\phi_{s'}$, $\phi_{a'}$.

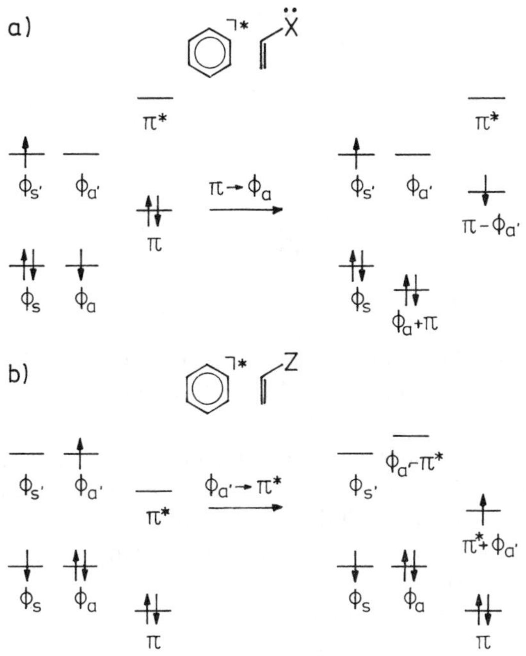

Figure 7.32. Orbital energy diagram for the photocycloaddition of excited benzene to an olefin a) with an electron-donating substituent X and b) with an electron-withdrawing substituent Z (adapted from Houk, 1982).

The presence of substituents alters the situation in that charge transfer between the reaction partners may become significant, and ortho addition should then be preferred over meta addition. For electron-rich alkenes or electron-poor arenes, the ethylene π MO lies at much higher energy than the singly occupied bonding MO of the excited arene (ϕ_a for the ortho path), and thus charge transfer from the alkene to the arene will take place. For an electron-poor alkene or an electron-rich arene, charge transfer from the arene to the alkene will occur (Figure 7.32).

This analysis does not require a specification of the timing of the bond-forming events. This could correspond (1) to a fully concerted, synchronous pathway, (2) to the cyclization of benzene to a biradical referred to as prefulvene followed by addition of the olefin, or (3) to the bonding of the olefin to meta positions and subsequent cyclopropane formation, as indicated for the case of meta cycloaddition in Scheme 18.

Mechanism 2 of Scheme 18, which was first proposed in 1966 (Bryce-Smith et al., 1966), has been discarded because recent experimental evidence excludes the intermediacy of prefulvene (Bryce-Smith et al., 1986). Mechanism 3 is presently favored. Experimental and theoretical results sup-

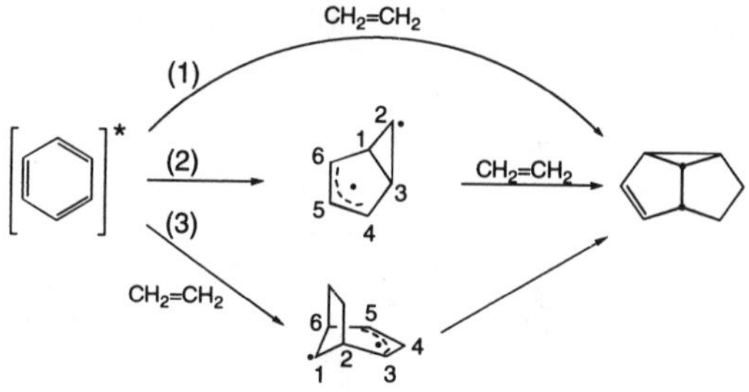

Scheme 18

port the notion that the σ bonds between the reactants are formed during the initial stage of the reaction, while the cyclopropane ring closure is not expected until crossing to the ground-state energy surface (De Vaal et al., 1986; van der Hart et al., 1987). A fully concerted process (mechanism 1 in Scheme 18) is considered unlikely. An argument against it, and in favor of mechanism 3, is provided by independent photochemical generation of the proposed biradical intermediate (Scheme 19), which yields the same product ratio as is obtained from the corresponding arene–alkene photocycloaddition at low conversions (Reedich and Sheridan, 1985).

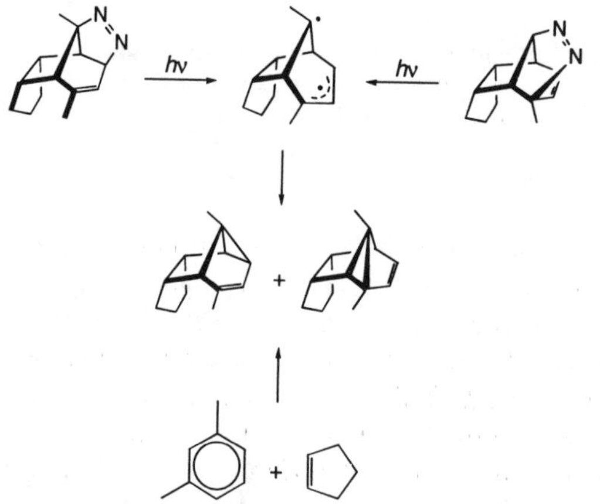

Scheme 19

Furthermore, the occurrence of an exciplex intermediate is assumed; an empirical correlation based on the ΔG_{ET} values calculated from the Weller relation, Equation (5.28), has been established which allows the prediction of mode selectivity for a wide range of arene–alkene photocycloadditions (Mattay, 1987). For negative or very small values of ΔG_{ET}, addition is preferred over cycloaddition. This is the case for the reactions of certain electron-rich alkenes with excited arenes. For positive values of ΔG_{ET}, cycloaddition is expected. Ortho cycloaddition is favored for ΔG_{ET} values up to about 1.4–1.6 eV, and meta cycloaddition for more strongly positive values (Scheme 20).

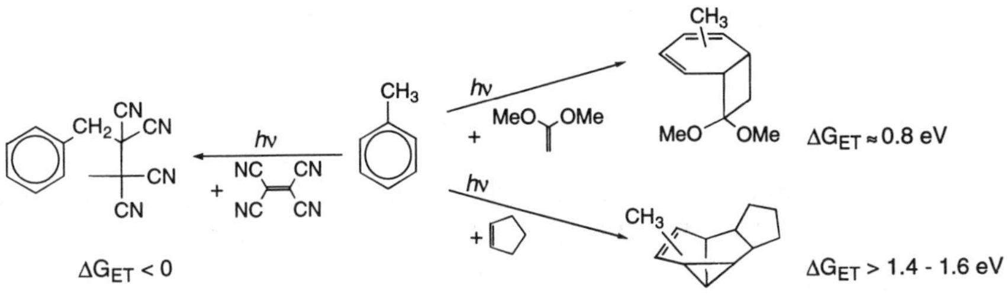

Scheme 20

Regioselectivity in the meta cycloaddition to substituted benzenes has been assumed to depend on charge polarization in the biradical intermediate shown in Scheme 21 (van der Hart et al., 1987). The theoretical calculations mentioned earlier, however, point out that in the early stage of the reaction the excited state has appreciable polar character that disappears as the reaction proceeds; the biradical itself does not have any particular polarity.

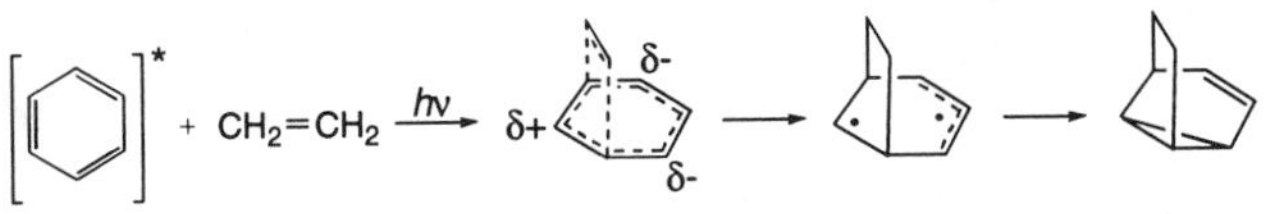

Scheme 21

Irradiation of mixtures of various acetylenes with benzene gives cyclooctatetraenes, presumably via an intermediate ortho adduct **74** (Bryce-Smith et al., 1970). For acetylenes with bulky substituents, the bicyclooctatriene intermediate is sufficiently stable for a subsequent intramolecular photocycloaddition to a tetracyclooctene (**75**) (Tinnemans and Neckers, 1977):

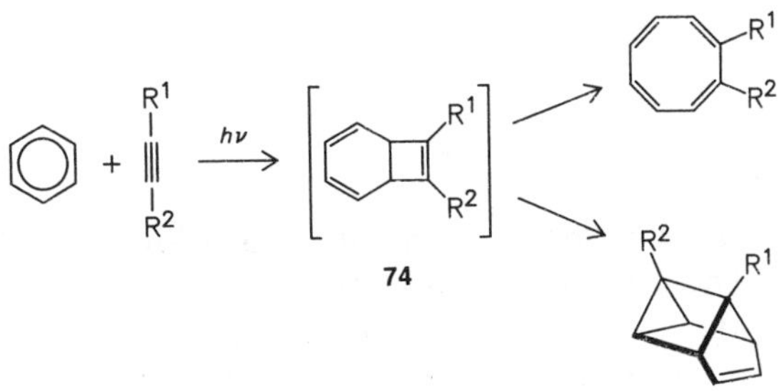

(R^1 = Ph, R^2 = $COOCH_3$)

74

75

7.4.4 Photocycloadditions of the Carbonyl Group

Another photocycloaddition reaction that has been known for a long time is the *Paterno-Büchi reaction,* which involves the formation of oxetanes through the addition of an excited carbonyl compound to olefins:

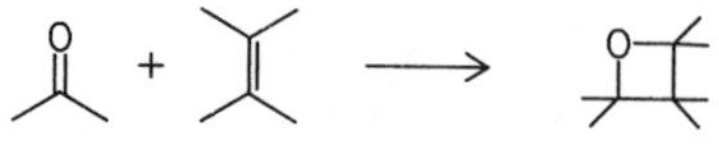

As far as the addition of aromatic carbonyl compounds is concerned, only the triplet state is reactive, and hydrogen abstraction occurs as a side reaction. Another competing reaction path involves energy transfer (cf. Section 7.1.2); efficient oxetane formation is therefore observed only when the triplet energy of the carbonyl compound is not high enough for sensitized triplet excitation of the olefin. As is to be expected from the loose geometry of a triplet biradical intermediate, the reaction is not stereospecific; the same mixture of oxetanes **76** and **77** is produced at low conversion by irradiating *cis*- or *trans*-2-butene with benzophenone (Turro et al., 1972a; Carless, 1973):

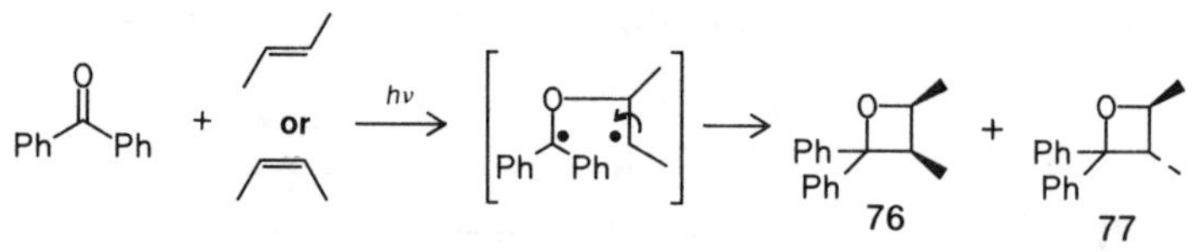

Ph Ph + or $h\nu$ Ph Ph Ph Ph **76** + Ph Ph **77**

That the reaction is partly regiospecific, as indicated in Scheme 22, has been attributed to the differing stabilities of the biradical intermediates (Yang et al., 1964).

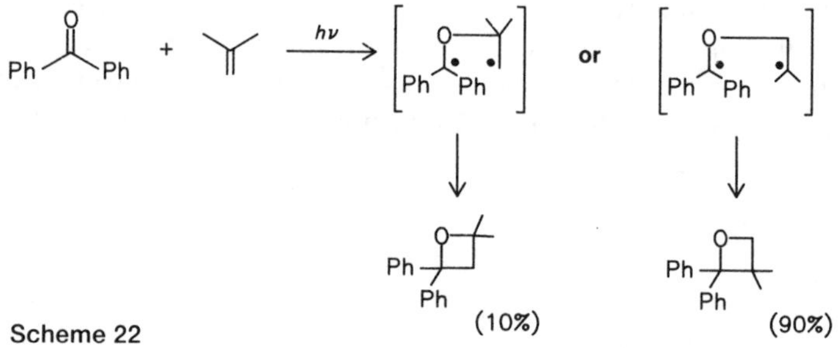

Ph Ph + $h\nu$ Ph Ph or Ph Ph

Ph Ph (10%) Ph Ph (90%)

Scheme 22

The mechanism of the photochemical oxetane formation is summarized in Figure 7.33. The $^3(n,\pi^*)$ excited state of the ketone produced by light absorption and subsequent intersystem crossing (ISC) attacks the olefin to form a triplet 1,4-biradical (k_b). Intersystem crossing (k_{ISC}) to the singlet, either directly or via a contact ion pair (k_{ip}), leads to ring closure (k_c) to form the oxetane, or to β cleavage to re-form the reagents (k_β). The contact ion

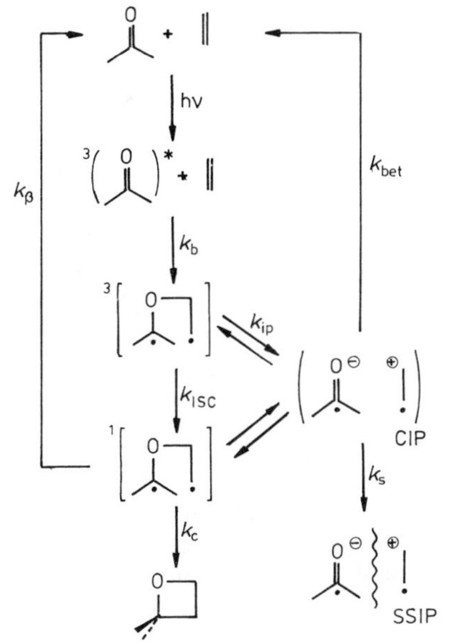

Figure 7.33. Mechanism of the photochemical oxetane formation: k_b, k_{ISC}, and k_c are rate constants of triplet biradical formation, intersystem crossing, and cyclization; k_{ip} and k_s correspond to the formation of a contact ion pair and solvent-separated ion pairs; and k_β and k_{bet} are rate constants of the back reaction of the singlet biradical and back-electron transfer, respectively (by permission from Buschmann et al., 1991).

pair may also dissociate to a solvent-separated ion pair (k_s) or return to the reactants by back-electron transfer (k_{bet}). The triplet biradical intermediate was identified by picosecond spectroscopy (Freilich and Peters, 1985). A mechanism involving the formation of polar exciplexes in the first step, as well as electron-transfer processes, has been invoked in the interpretation of biacetyl emission quenching by electron-rich alkenes (Mattay et al., 1984b; Gersdorf et al., 1987).

Diastereomeric oxetanes are formed from chiral carbonyl compounds such as menthyl phenylglyoxylate (**78**) (Buschmann et al., 1989).

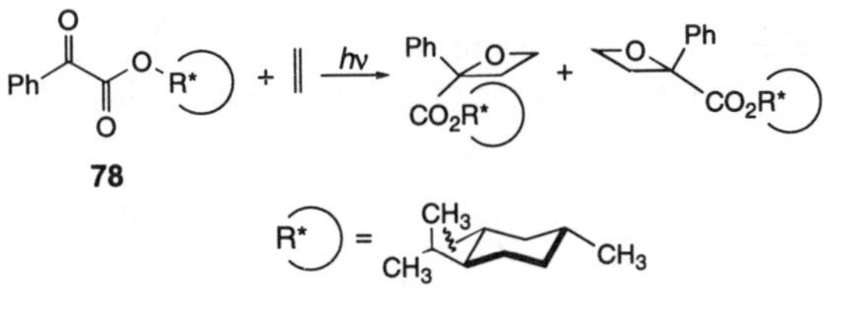

The general kinetic scheme (Fig. 7.34) displays two stages of diastereoselection: (1) a preferred formation of that of the two diastereomeric 1,4-

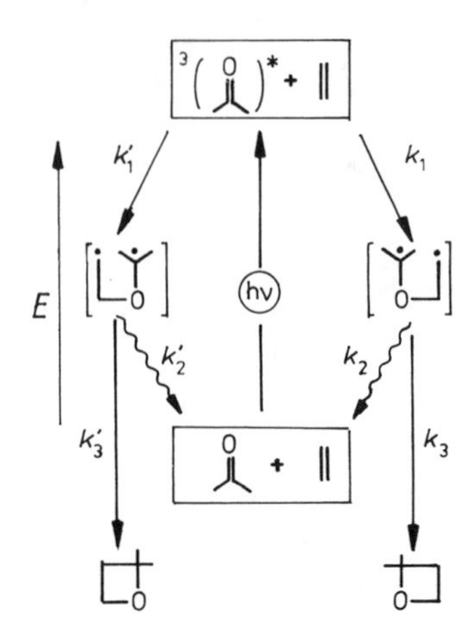

Figure 7.34. A simplified kinetic scheme for the diastereoselective oxetane formation in the Paterno-Büchi reaction (k_2 and k_2' as well as k_3 and k_3' include intersystem crossing steps) (by permission from Buschmann et al., 1989).

biradicals ($k_1 > k_1'$) which leads to the majority oxetane product, and (2) a preferred β cleavage of the biradical that leads to the minority oxetane product ($k_2' > k_2$), or a preferred ring closure of the biradical, which leads to the majority oxetane product ($k_3 > k_3'$). Upon continued irradiation, the β-cleavage products reenter the reaction cycle, establishing a "photon-driven selection pump."

The Arrhenius plot for the diastereoselectivity (cf. Section 6.1.5.2) contains two linear regions, one with a positive and one with a negative slope, changing into each other at the "inversion temperature," T_{inv}. Each region has a different dominant selection step. At temperatures above T_{inv}, selection is driven primarily by enthalpy, as expected for an early transition state in the bond cleavage in the energy-rich biradical intermediate. Below T_{inv}, selection is driven primarily by entropy, as expected for an early transition state in the bond-forming step in the formation of the biradical from the energy-rich reactants (Buschmann et al., 1989).

Example 7.11:

An interesting example of diastereoselectivity is provided by the photocycloaddition of aromatic aldehydes to electron-rich cyclic olefins such as 2,3-dihydrofuran (**79**):

79 + ArCH=O —hν→ "endo" + "exo"

The stereochemistry of the reaction can be accounted for by the conformational dependence of spin-orbit coupling elements discussed in Section 4.3.4.

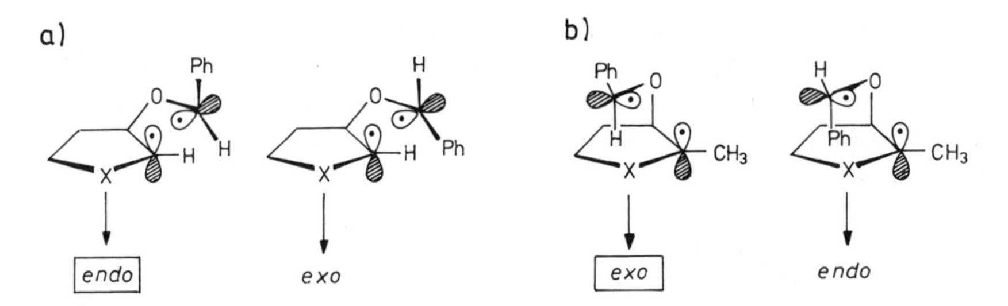

Figure 7.35. Preferred conformations for intersystem crossing during the diastereoselective oxetane formation (adapted from Griesbeck and Stadtmüller, 1991).

Figure 7.35a shows the postulated geometry of favored intersystem crossing in the biradical derived from an unsubstituted cycloalkene; bond formation following spin inversion is faster than conformational changes, and the endo adduct is formed from the less sterically hindered conformation. The exoselectivity observed in 1-substituted cycloalkenes is explained by increasing gauche interactions with the β-alkoxy group that favor the biradical conformations shown in Figure 7.35b (Griesbeck and Stadtmüller, 1991).

The photoaddition of a furan and an aldehyde can serve as a photochemical version of a stereoselective aldol reaction, since the photoadduct can be viewed as a protected aldol, as indicated in Scheme 23 (Schreiber et al., 1983).

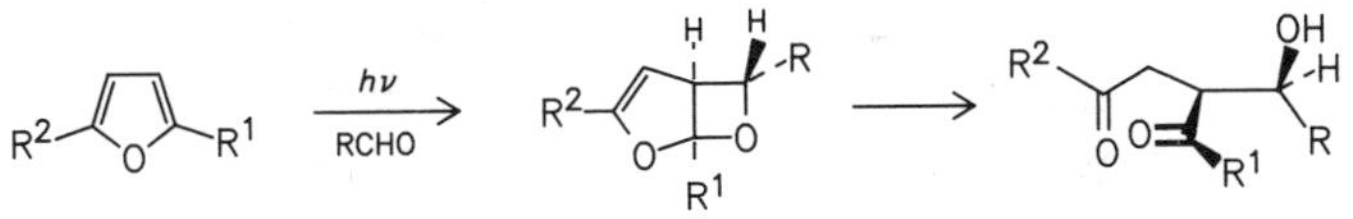

Scheme 23

For aliphatic ketones, the situation is complicated by less efficient intersystem crossing, thus permitting reaction of the $^1(n,\pi^*)$ as well as the $^3(n,\pi^*)$ state of the carbonyl compound, as revealed by the use of triplet quenchers. The S_1 reaction is more stereospecific, presumably because of the tight geometry of the singlet biradical, and yields less cis-trans isomerization by a competing path.

Thus, the stereochemistry of *cis*-1-methoxy-1-butene (**80**) is partially retained when acetone singlets attack the olefin, but it is almost completely scrambled in the reaction of triplet acetone (Turro and Wriede, 1970):

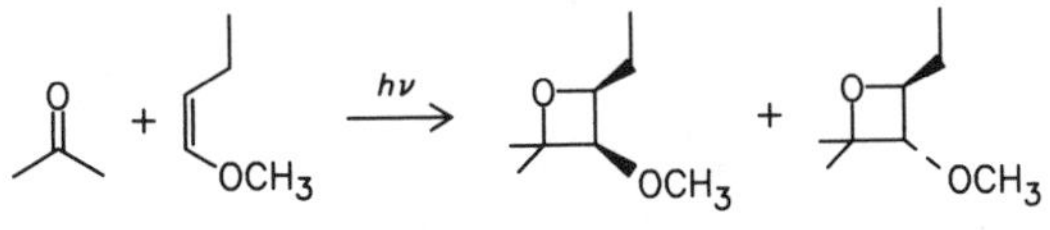

80

S_1 4 : 1

T_1 1 : 1

The situation is different with electron-poor olefins such as 1,2-dicyanoethylene. Only the S_1 state of acetone forms oxetanes, and the reaction is highly stereospecific, as indicated in Scheme 24. The competing cis-trans isomerization of the olefin arises exclusively from the triplet state (Dalton et al., 1970).

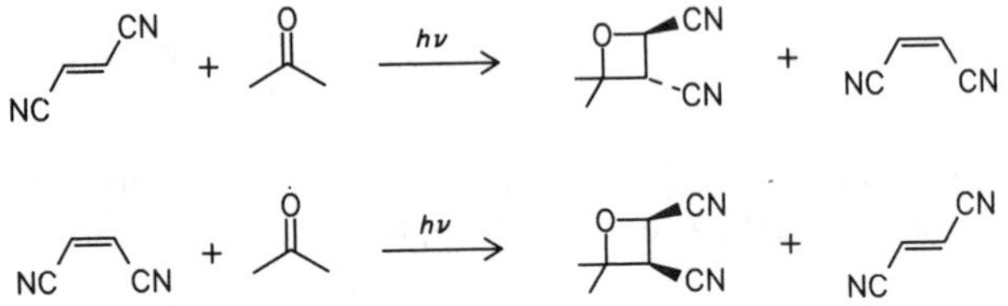

Scheme 24

The specificity of this reaction has been used to *chemically titrate* both the excited-singlet acetone and the triplet acetone produced through thermal decomposition of tetramethyl-1,2-dioxetane. (Cf. Section 7.6.4.) For this purpose the thermolysis was carried out in the presence of *trans*-1,2-dicyanoethylene, and the quantities of singlet and triplet acetone formed were obtained from the yields of dioxetane and *cis*-1,2-dicyanoethylene, respectively (Turro and Lechtken, 1972).

The various reactions of excited carbonyl compounds with olefins may be rationalized on the basis of correlation diagrams. In principle, four different pathways have to be discussed: the perpendicular and the parallel approaches (Figure 7.36) and the initial formation of a CO and a CC bond, yielding a C,C-biradical and C,O-biradical, respectively:

C,C-biradical

C,O-biradical

However, only three out of these four possibilities are realistic, since there is no carbonyl group orbital available for a perpendicular approach and formation of a CC bond. Formation of a C,O-biradical is therefore possible only through a parallel approach.

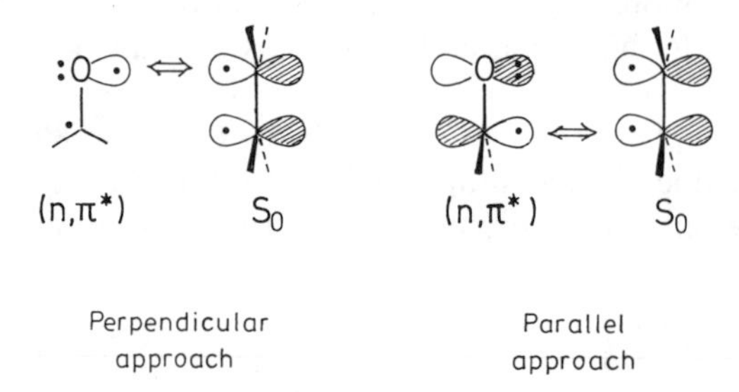

Figure 7.36. The perpendicular and the parallel approach for the interaction of an n→π^* excited ketone with a ground-state olefin.

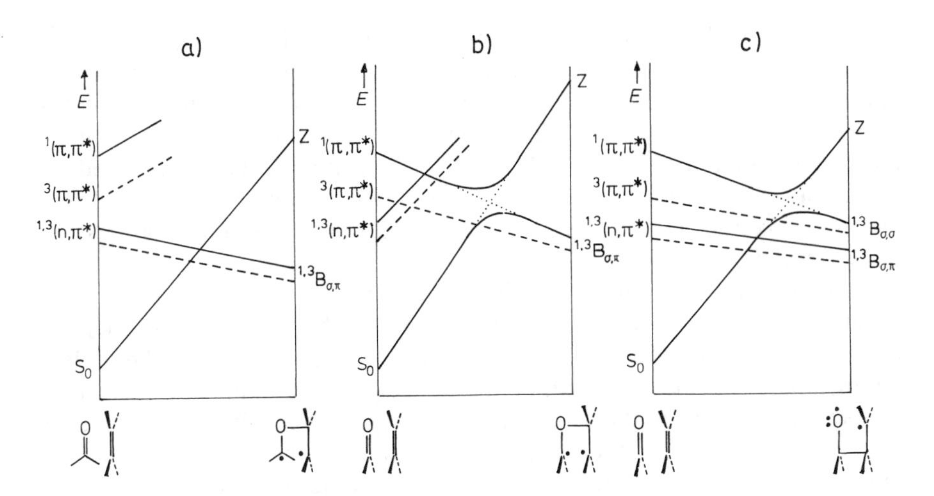

Figure 7.37. Correlation diagram a) for the perpendicular approach of ketone and olefin, and for the parallel approach resulting in b) a C,C-biradical and c) a C,O-biradical.

From Figure 7.37a, the perpendicular approach leading to a C,C-biradical is seen to be electronically allowed. Since, however, the two p AOs of the unpaired electrons are oriented perpendicular to each other, a rotation of the CH_2 group of the ketone is required prior to cyclization both for the singlet biradical and for the triplet biradical, which assume similar geometries and therefore can each give an oxetane. Small rotations in the less sterically hindered direction to optimal geometries for ISC and subsequent reaction may lead to stereospecific oxetane formation from the triplet state. (Cf. Example 7.11.)

Correlation diagrams for the two modes of parallel approach are shown in Figures 7.37b and 7.37c. If the CO bond is formed first, the (n,π^*) excited reactant states correlate with highly excited (n,σ^*_{CO}) states of the products and a correlation-induced barrier results. Hence this reaction is electronically forbidden (Figure 7.37b). If, however, the CC bond is formed first, the unpaired electron at the oxygen can be localized in a p AO with either σ or π symmetry, that is, either in the π^*_{CO} MO or in the n_O orbital. Since the $^{1,3}(n,\pi^*)$ reactant states correlate with the $^{1,3}B_{\sigma,\pi}$ product states, no correlation-induced barrier is to be expected, and the reaction is likely to be exothermic (Figure 7.37c). In contrast to the perpendicular approach, the triplet biradical of the parallel approach will have a loose geometry and should result in cis-trans isomerization of the olefin.

The conclusions from the correlation diagrams have been nicely confirmed by early ab initio calculations for the carbon–oxygen attack of formaldehyde on ethylene (Salem, 1974) and by more recent calculations on the same model reaction considering both modes of attack. (Palmer et al., 1994).

The results for the carbon–oxygen attack are summarized schematically in Figure 7.38. The excited-state branch of the reaction path terminates in a conical intersection point at a CO distance of 177 pm before the biradical is fully formed (cf. Figure 7.37a). Thus the system can evolve back to the reactants or produce a transient C,C-biradical intermediate that is isolated by small barriers (< 3 kcal/mol) to fragmentation (TS_1) or to rotation and ring closure to oxetane (TS_2). The singlet and triplet biradical minima are essentially coincident.

A schematic representation of the surfaces for the carbon–carbon attack is shown in Figure 7.39. The very flat region of the S_0 surface (barriers of the order of 1 kcal/mol) corresponds to the $^{1,3}B_{\sigma,\sigma}$ C,O-biradical. The $^{1,3}B_{\sigma,\pi}$ biradical has a CC bond length of 156 pm and corresponds to a conical intersection geometry in the case of the singlet, and to a minimum in the case of the triplet. Thus for the singlet photochemistry the decay to S_0 occurs close to the products, and the reaction appears to be concerted. Since, however, the formation of the singlet biradical is also possible from the same funnel, a certain fraction of photoexcited reactant can evolve via a nonconcerted route.

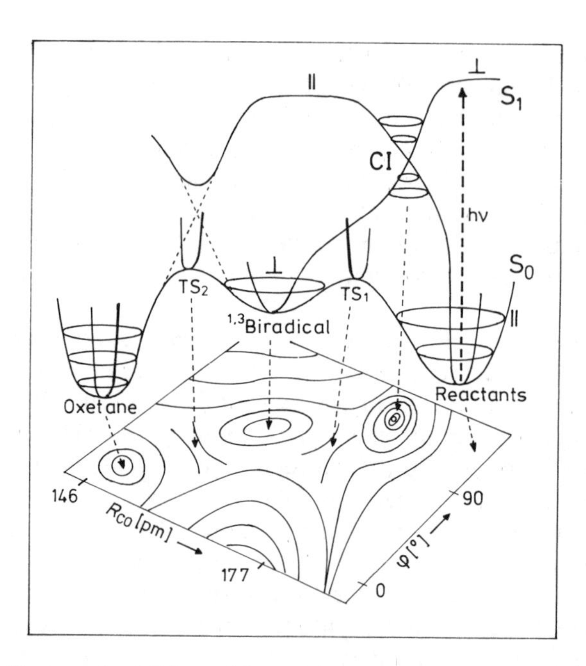

Figure 7.38. Photocycloaddition of formaldehyde and ethylene. Schematic representation of the surfaces involved in the carbon–oxygen attack as a function of the C,O distance R_{CO} and the dihedral angle φ between the formaldehyde and ethylene fragments. $\|$ and $\perp$ denote the parallel and perpendicular approach of the reactants, respectively; CI marks the conical intersection (by permission from Palmer et al., 1994).

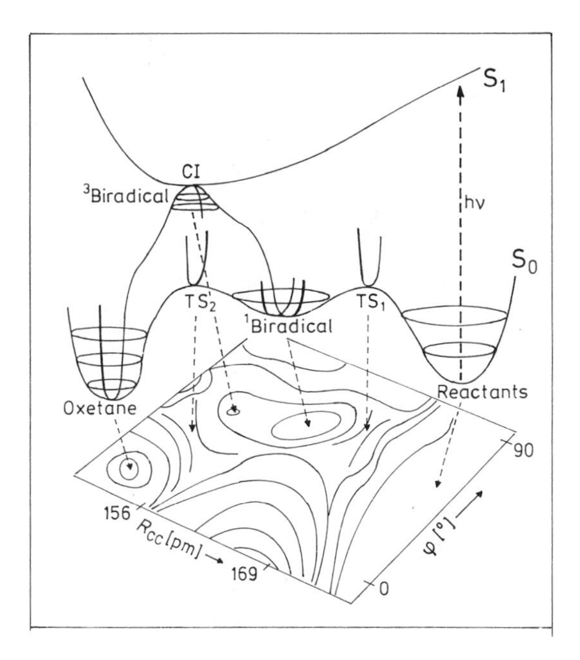

Figure 7.39. Photoaddition of formaldehyde and ethylene. Schematic representation of the carbon–oxygen attack as a function of the C,C distance R_{CC} and the dihedral angle φ between the formaldehyde and ethylene fragments (by permission from Palmer et al., 1994).

The overall nature of the singlet reaction path is determined by whether the system returns to S_0 through an early funnel along the path of carbon–oxygen attack to produce a C,C-biradical intermediate (Figure 7.38) or through a late funnel along the path of carbon–carbon attack to produce oxetane in a direct process (Figure 7.39). The stereochemistry and efficiency of the reaction will depend on which of these two excited-state reaction paths is followed.

Conclusions similar to those obtained from the correlation diagrams can be reached by means of either PMO theory (Herndon, 1974) or simple frontier orbital theory, as indicated in Figure 7.40. Orbital energies may be estimated from the assumption that the olefinic π MOs will be stabilized by electron-withdrawing substituents and destabilized by electron-releasing ones. The HOMO energy of an electron-rich olefin will then be comparable to that of the carbonyl n orbital, whereas the LUMO of electron-poor olefins will be similar in energy to the π^* MO of the carbonyl group. Interactions between orbitals of comparable energy require a parallel approach in the case of electron-poor olefins and result in CC bond making and formation of a C,O-biradical. Perpendicular attack is required for electron-rich olefins and yields a C,C-biradical by CO bond formation.

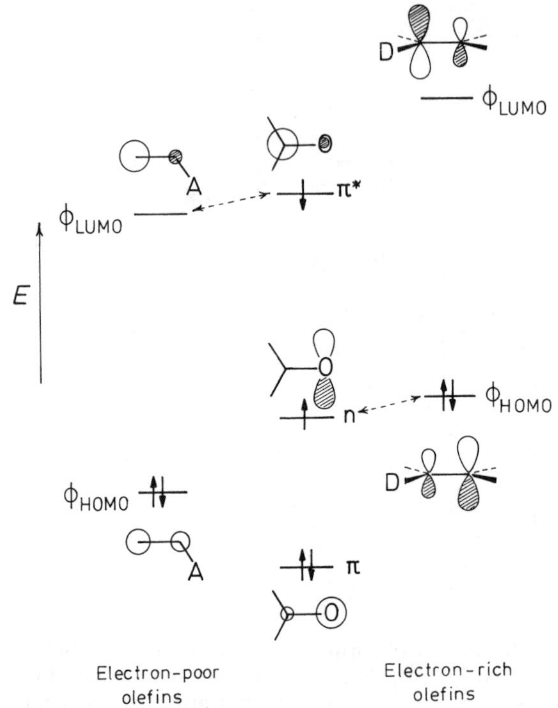

Figure 7.40. Frontier orbitals for oxetane formation. Interaction of one of the half-filled carbonyl orbitals with the LUMO of electron-poor olefins (left) and with the HOMO of electron-rich olefins (right).

The magnitude of the LCAO MO coefficients of the interacting orbitals permits a prediction of the expected oxetane regiochemistry. Both in the HOMO of a donor-substituted olefin and in the LUMO of an acceptor-substituted olefin, the coefficient of the unsubstituted carbon atom is the larger one in absolute value. Therefore, electron-poor olefins regioselectively afford the oxetane with the substituted carbon next to the oxygen. (Cf. Barltrop and Carless, 1972.) In contrast, an electron-rich olefin predominantly yields the oxetane with the unsubstituted carbon next to the oxygen. (Cf. Scheme 22.)

Finally, Figure 7.40 permits the conclusion that oxetane formation can be considered to be either a nucleophilic attack by a ketone on an electron-poor olefin or an electrophilic attack by a ketone on an electron-rich olefin, corresponding to the predominant interaction of a half-filled carbonyl orbital either with the empty π^* MO or with the doubly filled π MO of the olefin.

7.4.5 Photocycloaddition Reactions of α,β-Unsaturated Carbonyl Compounds

Intermolecular photoadditions of α,β-unsaturated carbonyl compounds can take place either at the CC or at the CO double bond. Photodimerizations with formation of a cyclobutane ring are quite common. In cases such as cyclopentenone (**81**), head-to-head as well as head-to-tail dimers are produced.

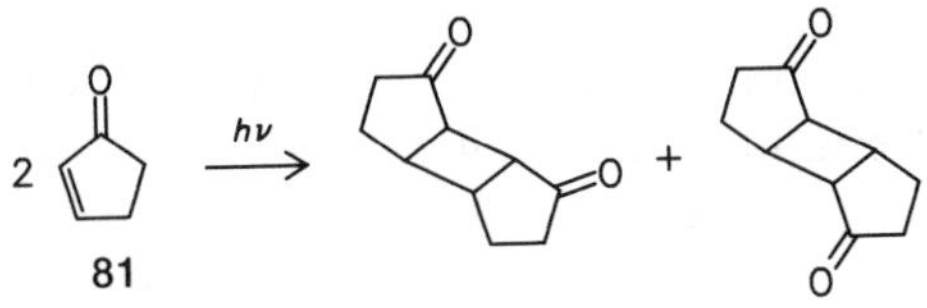

2 **81** $\xrightarrow{h\nu}$ +

While double-bond cis-trans isomerization in β-substituted α,β-enones occurs at both 313 and 254 nm, oxetene forms only when short wavelengths are used (Friedrich and Schuster, 1969, 1972). This differentiation has been explained on the basis of calculations that revealed a S_1–S_0 conical intersection lying 15 and 10 kcal/mol above the $^1(n,\pi^*)$ minimum of *s-trans-* and *s-cis*-acrolein, respectively (Reguero et al., 1994). Oxabicyclobutane, however, has not been detected. This behavior of α,β-enones is in contrast with that of butadiene, where cyclobutene and bicyclobutane are formed simultaneously. (Cf. Sections 6.2.1 and 7.5.1.)

Mixed cycloadditions, for example, between cyclopentenone and ethyl vinyl ether, have also been observed frequently (Schuster, 1989). As indicated in Scheme 25, attack of the $^3(\pi,\pi^*)$ excited state of the enone on the olefin will give a triplet 1,4-biradical that ultimately yields cyclization and disproportionation products (Corey et al., 1964; De Mayo, 1971). The regio-

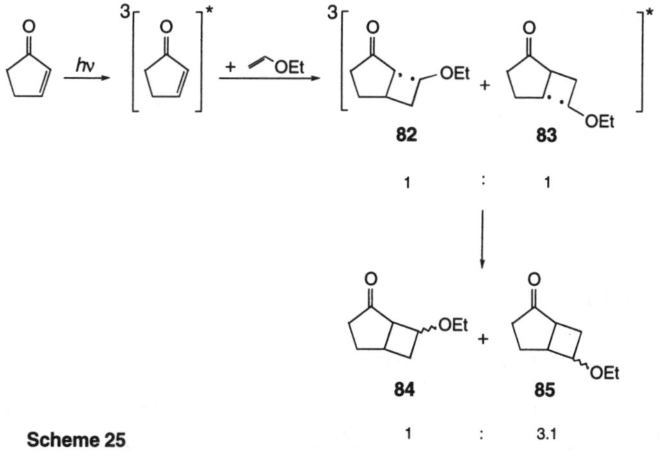

$h\nu$ + OEt

82 OEt + **83** OEt

1 : 1

84 OEt + **85** OEt

1 : 3.1

Scheme 25

selectivity of the reaction, which was explained earlier in terms of an initial interaction of the enone triplet with an alkene to give an exciplex, can alternatively be accounted for by the relative efficiencies with which each of the isomeric biradical intermediates proceeds to the annelation products (as opposed to reverting to the ground-state enone). While the structural isomers **84** and **85** are produced in a ratio 1.0 : 3.1, trapping with hydrogen selenide H_2Se indicates that the biradicals **82** and **83** are formed in the ratio 1.0:1.0 (Hastings and Weedon, 1991).

Kinetic studies indicating that triplet 3-methylcyclohex-2-en-1-one (**86**) reacts with maleic or fumaric dinitrile directly to yield triplet 1,4-biradicals argue against the intermediacy of an exciplex (Schuster et al., 1991a). Dynamical properties of enone $^3(\pi,\pi^*)$ states have been determined and discussed; torsion around the olefinic bond is the principal structural parameter that affects their energies and lifetimes (Schuster et al., 1991b). Twisting around the C═C bond causes an increase in energy on the S_0 surface and a concomitant decrease in energy on the $^3(\pi,\pi^*)$ surface (cf. Figure 4.6). This reduces the T_1–S_0 gap and facilitates T_1–S_0 radiationless decay in a process dominated by spin-orbit coupling (Section 4.3.4). Average energies and lifetimes of the mixture of head-to-head and head-to-tail 1,4-biradicals have been determined by photoacoustic calorimetry (Kaprinidis et al., 1993).

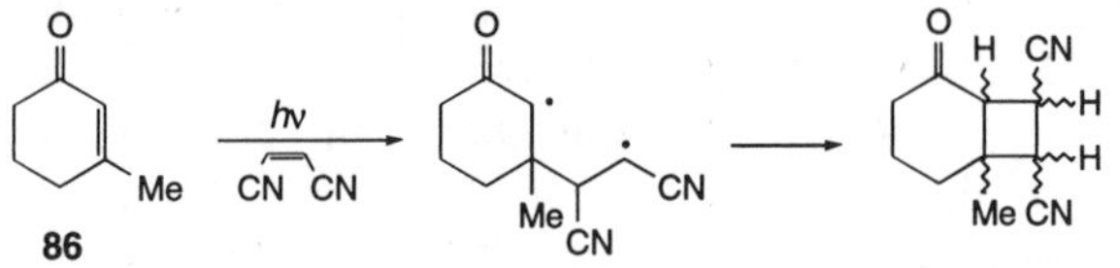

O Me 86 hν CN CN O Me CN CN O H CN H H Me CN

Synthetic applications of [2 + 2] cycloadditions of α,β-unsaturated carbonyl compounds are numerous. The synthesis of cubane (Eaton and Cole, 1964), in which cage formation is achieved by the following photochemical reaction step, is an example:

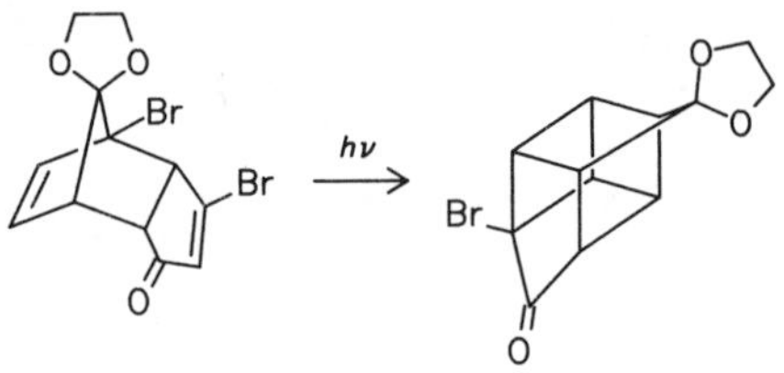

O O Br Br O hν Br O O O

7.5 Rearrangements

7.5.1 Electrocyclic Reactions

Electrocyclic ring-opening and -closure reactions represent another important field in which the Woodward-Hoffmann rules apply. These rules were in fact derived as a rationalization of the chemistry of vitamin D, which is characterized by a number of thermal as well as photochemical electrocyclic reactions and sigmatropic shifts, some of which are as follows:

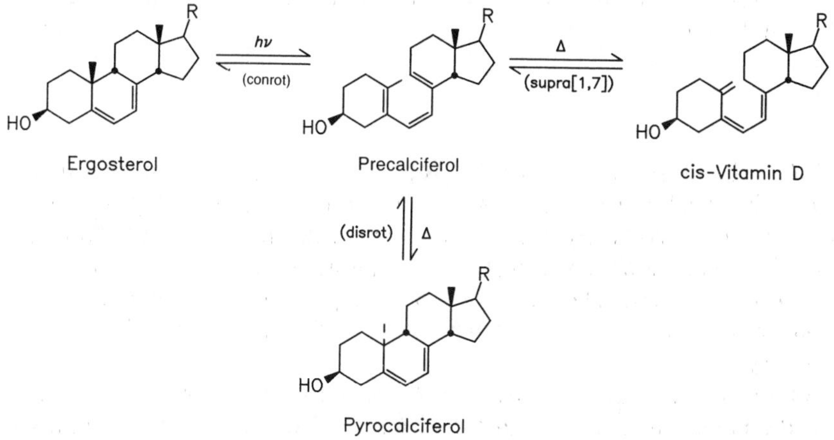

hν (conrot) Δ (supra[1,7]) (disrot) Δ

Ergosterol Precalciferol cis-Vitamin D Pyrocalciferol

The number of photoreactions that proceed in agreement with the Woodward-Hoffmann rules (Table 7.3) is very large. Some examples are collected in Scheme 26.

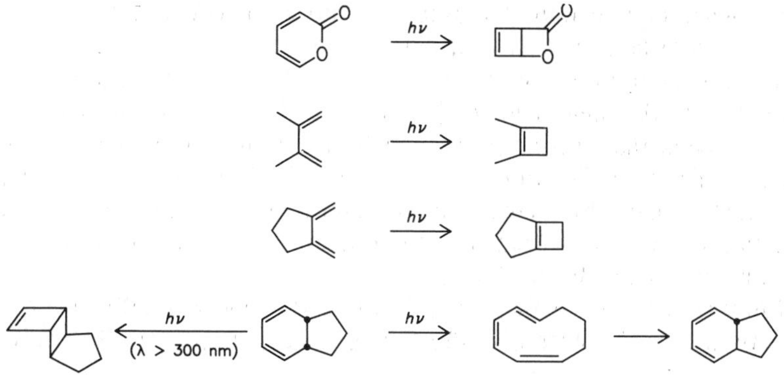

hν (λ > 300 nm)

Scheme 26

Table 7.3 Orbital Symmetry Rules for Thermal (Δ) and Photochemical (*hν*) Electrocyclic Reactions

	Number of π Electrons in Open-Chain Polyene	
Reaction	$4n$	$4n + 2$
Δ	Conrotatory	Disrotation
$h\nu$	Disrotatory	Conrotation

The important role of avoided crossings and the resulting pericyclic minima for the mechanisms of photochemical reactions was first pointed out on the example of the butadiene–cyclobutene conversion (van der Lugt and Oosterhoff, 1969).

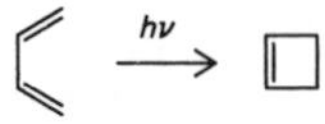

$$\xrightarrow{h\nu}$$

While the Oosterhoff model that follows from the state correlation diagrams discussed in Section 4.2.3 describes the stereochemistry of electrocyclic reactions correctly and in agreement with the Woodward-Hoffmann rules, it is oversimplified in that it does not attempt to actually locate the bottom of the pericyclic minimum and simply assumes a planar carbon framework. It therefore predicts a nonzero S_1–S_0 gap at perfect biradicaloid geometry.

As discussed earlier (cf. Sections 4.4.1, 6.2.1, etc.) symmetry-lowering distortions that remove the exact degeneracy of the nonbonding orbitals at the perfect biradical geometry by introducing a heterosymmetric perturbation δ lower the energy of the S_1 state and reduce the S_1–S_0 gap. Recent calculations on butadiene (Olivucci et al., 1993) have demonstrated that the gap is actually reduced to zero and that the funnel corresponds to a true S_1–S_0 conical intersection at geometries with pericyclic and diagonal interactions. According to these calculations, the ring closure of butadiene to form cyclobutene proceeds through the same funnel as the cis-trans isomerization, and the stereochemical decision is taken mainly on the excited-state branch of the reaction pathway, where the low-energy pathway corresponds to disrotatory motion, as discussed in detail in Section 6.2.1. In fact, an excited-state *cis*-butadiene enters the conical intersection region in a conformation already consistent with the production of only one ground-state cyclobutene isomer (Figure 7.41).

However, not all reactions that follow a Woodward–Hoffmann allowed path are necessarily stereospecific. (Cf. Clark and Leigh, 1987). Irradiation

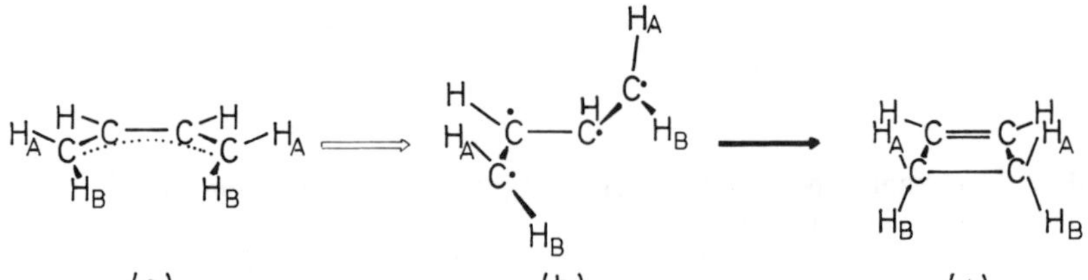

(a) (b) (c)

Figure 7.41. Schematic representation of the geometry changes of excited state *s-cis*-butadiene a) entering the conical intersection region b) and producing ground-state cyclobutene c). Light and dark arrows indicate excited-state and ground-state pathways, respectively (adapted from Olivucci et al., 1993).

of bicyclo[4.2.0]oct-7-enes (**87**) yields the products shown in Scheme 27 (Leigh et al., 1991). Recent calculations suggest that the motion on the S_1 surface is disrotatory, as demanded by the Woodward-Hoffmann rules, and that a loss of steric information results from motion on the ground-state surface after return through the pericyclic funnel (Bernardi et al., 1992c). The mechanism of the fragmentation step (also shown in Scheme 27) has not yet been elucidated in a definitive fashion.

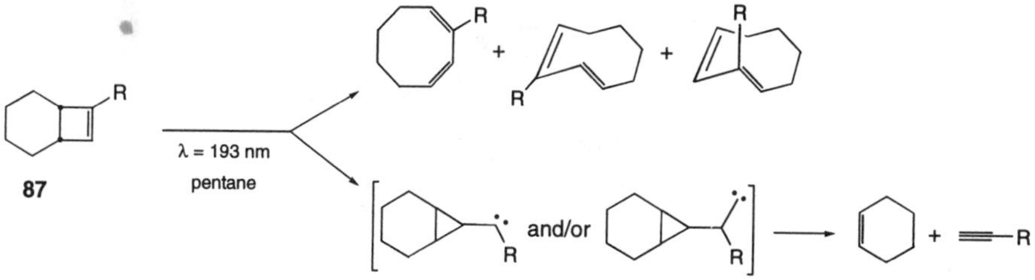

λ = 193 nm
pentane
and/or

Scheme 27

Irradiation of *s-cis*-2,3-dimethylbutadiene yields both electrocyclic ring closure and double-bond isomerization products (Scheme 28). While the ring closure in *s-cis*-butadiene, isoprene, 2-isopropylbutadiene, and 1,3-pentadiene is considerably less efficient than s-cis-s-trans isomerization, in *s-cis*-2,3-dimethylbutadiene it is about 50 times faster (Squillacote and Semple, 1990). This effect is counterintuitive, since the two methyl groups appear to hinder the rotation about the central C—C bond in butadiene.

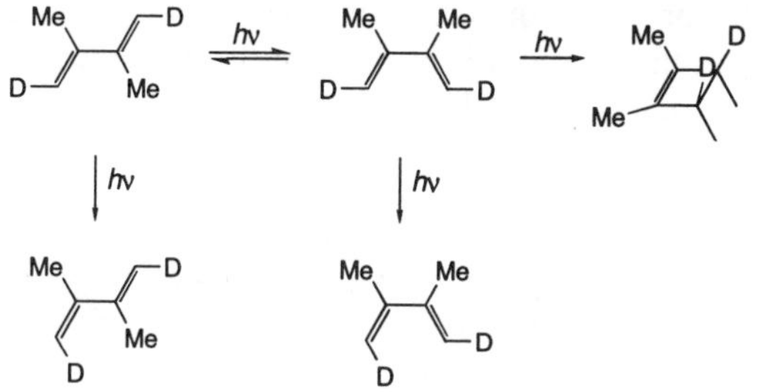

Scheme 28

Example 7.12:
According to ab initio calculations of Olivucci et al., (1994b), the presence of the methyl substituents in *s-cis*-2,3-dimethylbutadiene does not alter the nature of the pathway on S_1 to the pericyclic funnel. In both the parent and the dimethyl derivative, the atoms attached in positions 2 and 3 remain eclipsed. The different behavior could therefore be due to differences in the region of the S_0 surface explored after return through the funnel, or to differences in the kinematics caused by the larger inertia of the methyl groups compared to hydrogen. The latter explanation is favored by Olivucci et al., but dynamical calculations are needed to settle the issue.

At the conical intersection the C_1–C_2–C_3 valence angle ω is close to 90°, suggesting that in strained systems, where this value is hard to reach, the excited-state barrier on the pathway to the conical intersection or the conical intersection itself may be so high in energy that the photoreaction cannot take place. The quantum yields for cyclobutene formation in dimethylene-cycloalkanes summarized in Scheme 29 (Aue and Reynolds, 1973; Leigh and Zheng, 1991) are in agreement with this expectation.

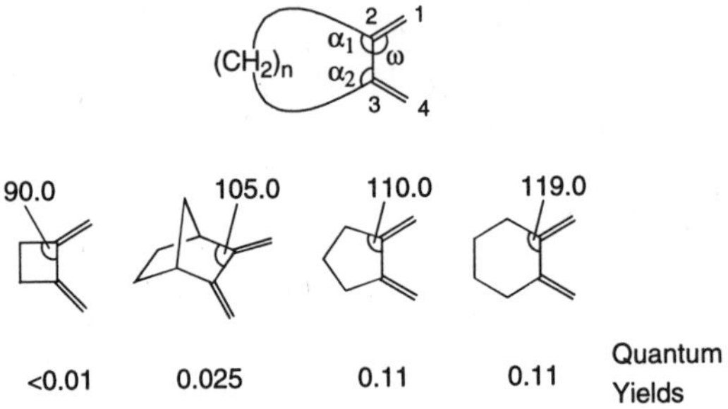

(CH2)n α1 α2 ω 2 1 3 4

90.0 105.0 110.0 119.0

<0.01 0.025 0.11 0.11 Quantum Yields

Scheme 29

Bicyclo[1.1.0]butane is usually a side product of the photocyclization of butadiene to cyclobutene (Srinivasan, 1963); in isooctane, the quantum yield ratio is 1 : 16 (Sonntag and Srinivasan, 1971). It becomes the major product in systems in which the butadiene moiety is constrained near an s-trans conformation and bond formation between the two terminal methylene groups that leads to cyclobutene is disfavored. An example is the substituted diene **88** in Scheme 30, for which the bicyclobutane is the major product; a nearly orthogonal conformation should result from the presence of the 2,3-di-*t*-butyl substituents (Hopf et al., 1994).

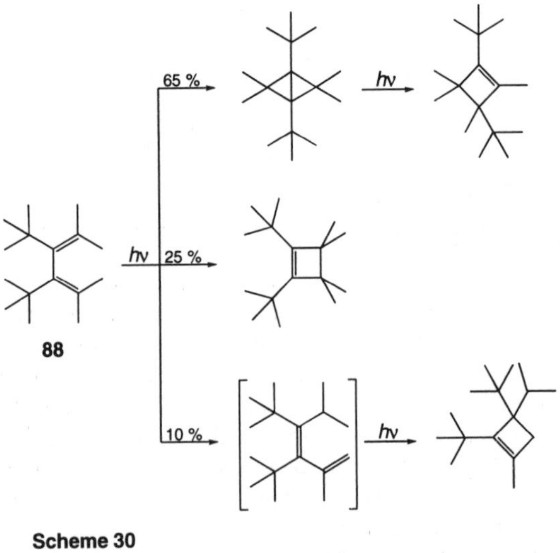

hν 65 % hν 25 % 10 % hν

88

Scheme 30

Example 7.13:
The photochemical reaction network and photoequilibrium involved in the commercial synthesis of vitamin D using the photochemical electrocyclic ring opening of ergosterol to give precalciferol, which subsequently undergoes a thermal [1,7]-H shift to produce vitamin D itself (cf. Jacobs and Havinga, 1979), has been studied using a combination of molecular mechanics and simple 2 × 2 VB methods (Bernardi et al., 1992b). The results are summarized in Figure 7.42. The focal feature of the reaction network centered on precalciferol is the existence of three funnels analogous to those discussed for the cis-trans isomerization of *cis*-hexatriene in Section 7.1.3, except that there are two conical intersections $CI^{+}_{c\text{-}t}$ and $CI^{-}_{c\text{-}t}$ for the s-cis-s-trans isomerization (*c-t*) since the

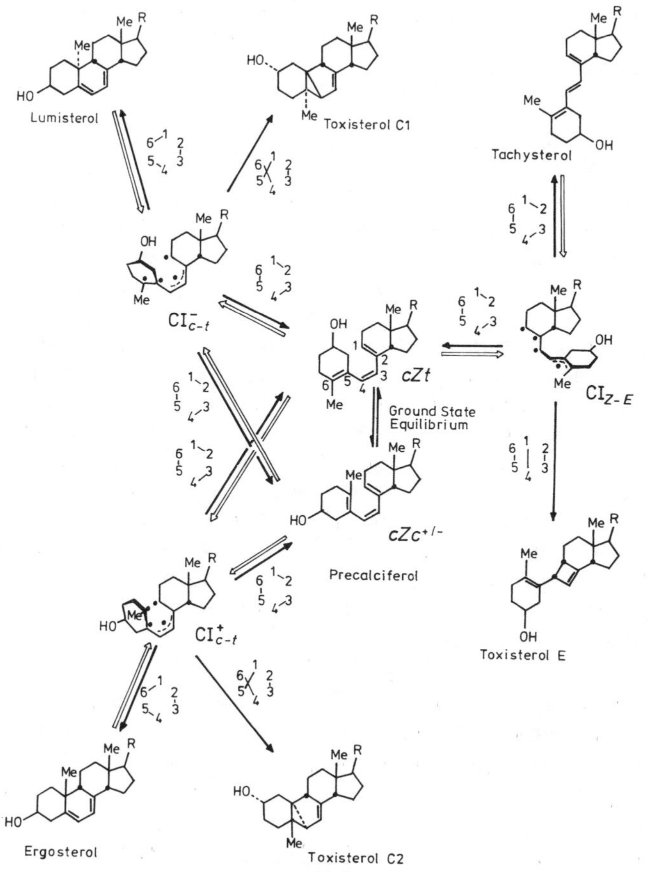

Figure 7.42. Mechanistic scheme for the precalciferol reaction network including the three conical intersection regions $CI^{+}_{c\text{-}t}$, $CI^{-}_{c\text{-}t}$ and $CI_{Z\text{-}E}$. Dark arrows indicate ground-state pathways, while light arrows indicate pathways on the excited-state surface (by permission from Bernardi et al., 1992b).

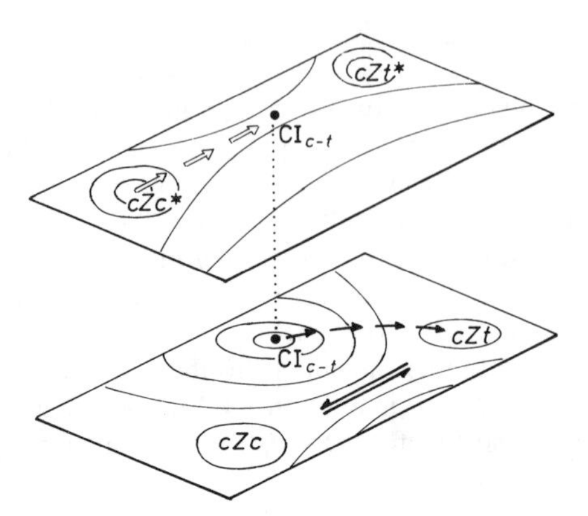

Figure 7.43. Illustration of the NEER principle: interconversion of the excited-state conformers of precalciferol is prohibited by the high probability of decay to the ground state through the conical intersection $CI_{c\text{-}t}$ (by permission from Bernardi et al., 1992b).

s-cis-*Z*-s-cis isomer of precalciferol has two diastereomers, which may be denoted as cZc^+ and cZc^-.

The possibility of bifurcation of the reaction path on the S_0 surface corresponding to the various possible bond-forming schemes (indicated in Figure 7.42 by dark arrows) and the existence of several pathways in the S_1 state (light arrows) are the reasons for the great diversity of photoproducts from precalciferol.

In agreement with the existence of three ground-state conformers cZc^+, cZc^- and cZt of precalciferol, there are three excited-state minima, cZc^{+*}, cZc^{-*}, and cZt^*. Search for transition structures between these excited-state minima led toward the funnels $CI^+_{c\text{-}t}$ and $CI^-_{c\text{-}t}$. This work explains the short singlet lifetime of an excited-state conformer, as well as the inability of excited conformers to interconvert, as stated by the NEER (nonequilibration of excited rotamers) principle. (Jacobs and Havinga, 1979; see also Example 6.9.) The interconverting conformer (e.g., cZc^*) will have high probability of decaying to the ground state through the conical intersection $CI_{c\text{-}t}$ before the barrier for interconverting to cZt^* could be completely overcome. This is illustrated qualitatively in Figure 7.43.

The photocyclization of *cis*-stilbene gives dihydrophenanthrene (**5**), which may be thermally or photochemically converted back to *cis*-stilbene or oxidized to phenanthrene (Moore et al., 1963; Muszkat, 1980).

The electrocyclic ring-closure reaction proceeds exclusively in the S_1 state and yields solely the trans product by a conrotatory mode of reaction, as is to be expected from Table 7.3 for a 6π-electron system. Competing reactions are cis-trans isomerization (cf. Section 7.1.4) and intersystem crossing to T_1. From the T_1 state, generally only cis-trans isomerization is observed. Stilbenes with substituents that enhance spin inversion, such as Br, RCO, and NO_2, do not undergo the cyclization reaction efficiently.

This reaction is not limited to stilbene itself. Whether a diarylethylene may be cyclized, and if so, in which way, can be predicted from the sum $F^*_{\varrho\sigma} = F^*_\varrho + F^*_\sigma$ of the free valence numbers $F^*_\varrho = \sqrt{3} - \Sigma p^*_{\varrho\varrho'}$ calculated for the reacting positions ϱ and σ of the HOMO–LUMO excited reactant. Here, $\Sigma p^*_{\varrho\varrho'}$ is the sum of excited-state π-bond orders for all bonds ϱ—ϱ' originating from atom ϱ. Laarhoven (1983) derived the following rules:

1. Photocyclizations do not occur when $F^*_{\varrho\sigma} < 1.0$.
2. When two or more cyclizations are possible in a particular compound, only one product arises if one $F^*_{\varrho\sigma}$ value is larger by at least 0.1 than all the others.
3. When planar as well as nonplanar products (pentahelicene or higher helicenes) can be formed, the planar aromatic will in general be the main product regardless of rule 2, provided rule 1 is fulfilled for its formation.

The application of these rules is illustrated in the following example.

Example 7.14:
The excited-state π-bond orders of an alternant hydrocarbon are given by

$$p^*_{\varrho\varrho'} = p_{\varrho\varrho'} - 2c_{\varrho\mathrm{HO}}c_{\varrho'\mathrm{HO}}$$

where $p_{\varrho\varrho'}$ is the π-bond order of the bond between atom ϱ and its neighbor ϱ', while $c_{\varrho\mathrm{HO}}$ and $c_{\varrho'\mathrm{HO}}$ are the corresponding LCAO coefficients of the HOMO. Using $F^*_2 = \sqrt{3} - 1.282 = 0.450$, $F^*_4 = 0.620$,and $F^*_{2'} = 0.491$, $F^*_{2,2'} = 0.941$, and $F^*_{4,2'} = 1.111$ are found for compound **89**. From rule 1 only ring closure between positions 4 and 2′ to yield the hydrocarbon **91** is expected, and no formation of **92** by ring closure between positions 2 and 2′ should occur. Similarly, for **90** one has $F^*_{2,10'} = 1.095$ and $F^*_{4,10'} = 1.188$; from rule 1, formation of both **91** and **92** could be expected, but since **91** is nonplanar, **92** should be the main product according to rule 3. Experimental results show that only products allowed by rules 1–3 are formed: **89** gives **91** in 60% yield and **90** yields only the planar compound **92**, in 75% yield (Laarhoven, 1983).

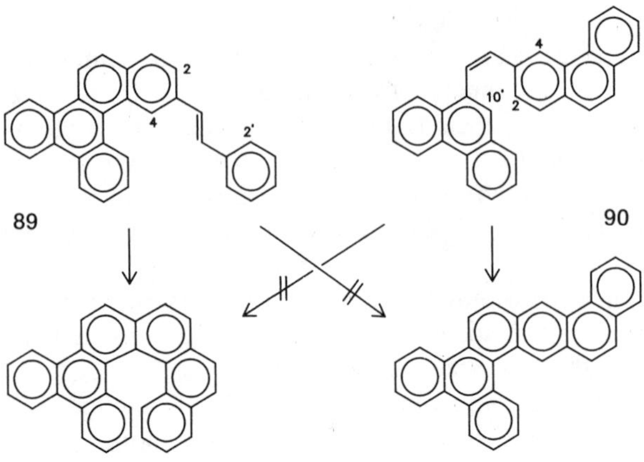

89 90

91 92

The analogous photocyclization of *N*-methyldiphenylamine has been studied in detail (Förster et al., 1973; Grellmann et al., 1981) and utilized synthetically (Schultz, 1983). In contrast to stilbene, the reaction proceeds from the triplet state of the amine by an adiabatic conrotatory ring closure to give a dihydrocarbazole (**93**), in accordance with the Woodward-Hoffmann rules. After return to the ground state, the initial product is oxidized to a carbazole (Scheme 31):

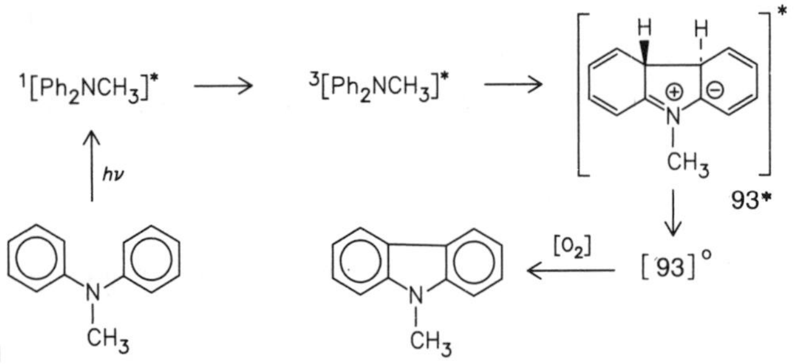

$^1[Ph_2NCH_3]^*$ ⟶ $^3[Ph_2NCH_3]^*$ ⟶ $[93]^*$ ⟶ $[93]^\circ$ $\xrightarrow{[O_2]}$ carbazole

Scheme 31

This reaction is of special interest because it is a stereospecific triplet reaction. Another known example is the ring opening by CC bond cleavage of the heterocyclic three-membered rings aziridine (**94**), oxirane (**95**), etc. (Huisgen, 1977; Padwa and Griffin, 1976). These reactions are believed to proceed through a cyclic antiaromatic triplet minimum which is isoelectronic with $C_3H_3^{\ominus}$; this in turn is an axial biradical with $E(T_1) < E(S_0)$. (Cf. Section 4.3.2.) It has been suggested that the unusual degree of triplet stabilization at the tight pericyclic geometry in axial biradicals is responsible for the stereospecific course of the reaction (Michl and Bonačić-Koutecký, 1990).

NH O

94 **95**

A situation similar to that of stilbene is found for dimethyldihydropyrene (**96**), which is converted into the dimethyl-substituted metacyclophanediene **97** by irradiation with visible light (λ = 465 nm); complete back reaction to the reactants is achieved either with lower wavelength light (λ = 313 nm) or thermally.

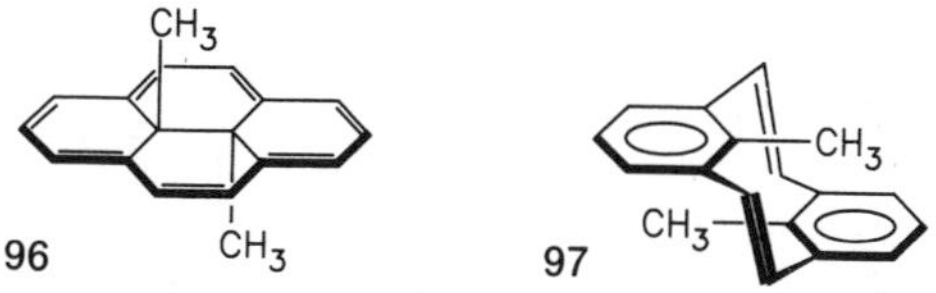

The two systems differ insofar as for stilbene the thermodynamic stability of the "open" form is greater, while for dihydropyrene the "ring-closed" form is more stable. The thermal back reaction **97** → **96**, which requires an activation energy of 22 kcal/mol, is apparently a symmetry-forbidden concerted reaction. An estimate of the ground-state barrier based on EHT calculations yields values that are in good agreement with experimental activation energies (Schmidt, 1971).

In general, electrocyclic ring-closure and ring-opening reactions are singlet processes, believed to proceed via tight biradicaloid geometries. In contrast, triplet excitation frequently yields five-membered rings (see the rule of five, Section 7.4.1) via loose biradicaloid intermediates. For example, singlet cyclohexadiene gives bicyclo[2.2.0]hexene (**98**) and hexatriene, while triplet hexatriene yields bicyclo[3.1.0]hexene (**99**) (Jacobs and Havinga, 1979):

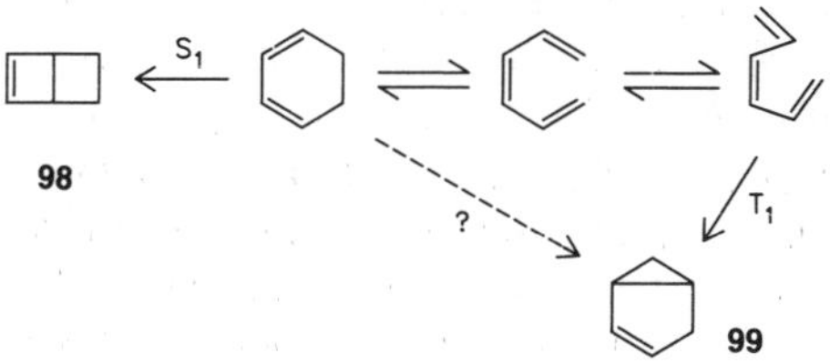

The reactions of butadiene are very typical. As indicated in Scheme 32, direct irradiation yields predominantly cyclobutene; in the presence of Cu(I) or Hg, however, bicyclobutane formed by an x[2 + 2] process is the major product, with minor products formed through [2 + 2] cycloaddition. (Cf. Section 7.4.2.)

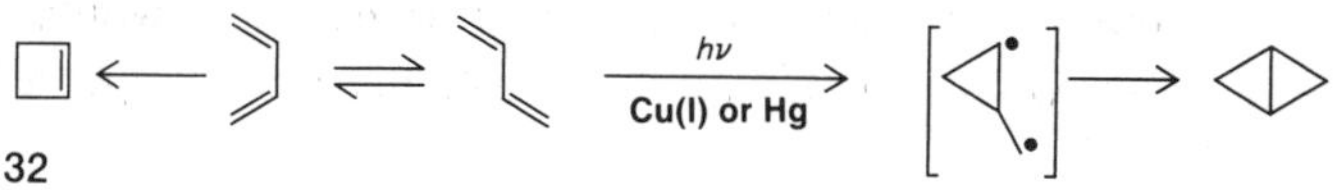

Scheme 32

Example 7.15:
Direct irradiation of myrcene (**100**) gives the cyclobutane derivative **101** by an electrocyclic ring-closure reaction, together with β-pinene (**102**) formed in a [2 + 2] cycloaddition process. The sensitized reaction, however, yields the bicyclic compound **103**.

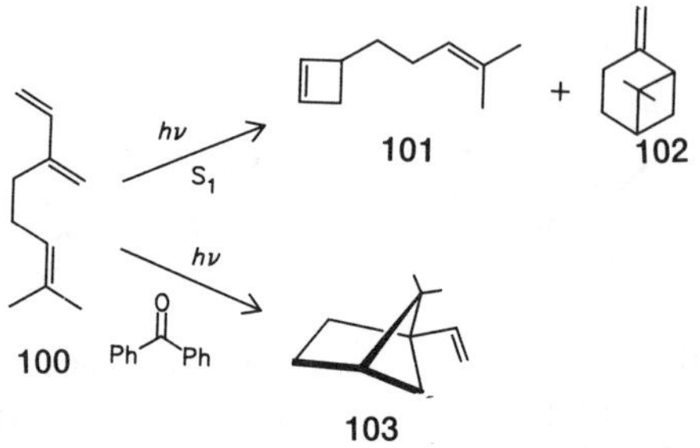

The triplet reaction is assumed to proceed as a two-step process with the most stable five-ring biradical as intermediate according to the "rule of five" (cf. Section 7.4.1) (Liu and Hammond, 1967).

The kinetics of the photochemical ring opening of cyclic dienes and trienes such as 1,3,5-cyclooctatriene (**104**) were determined by picosecond time-resolved UV resonance Raman spectroscopy (Ried et al., 1990) and provide excellent direct support for the Woodward–Hoffmann rules.

The photochemical ring opening of cyclohexa-1,3-diene involves a rapid (10 fs) radiationless decay of the initially excited $1B_2$ spectroscopic state to the lower $2A_1$, presumably through an S_2–S_1 conical intersection reached by a conrotatory motion, in agreement with Woodward-Hoffmann rules (Trulson et al., 1989). According to the latest calculations (Celani et al., 1994), this is followed by further conrotatory motion on the $2A_1$ surface along a pericyclic path distorted in the usual diagonal way from twofold symmetry, to a shallow minimum located most of the way toward the ring-open geometry of *s-cis-Z-s-cis*-hexatriene. As judged by the experimentally observed 6 ps appearance time of the ground-state 1,3,5-hexatriene product (Reid et al., 1993), the molecule has the time to equilibrate vibrationally in this shallow minimum before it decays to the ground state, presumably mostly via thermal activation to a more strongly diagonally distorted pericyclic conical intersection area located only about 1 kcal/mol higher in energy. As is common for pericyclic funnels, this return to S_0 is about equally likely to be followed by relaxation to the starting 1,3-cyclohexadiene and to the product, *s-cis-Z-s-cis*-hexatriene. The calculations assign little if any importance to an alternative ground-state relaxation path that would preserve the diagonal 1,5 interaction and lead to the 5-methylenecyclopent-2-en-1-yl biradical and products of its further transformation. These are indeed not observed.

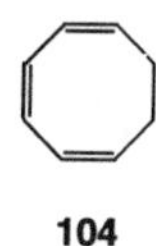

104

7.5.2 Sigmatropic Shifts

Sigmatropic shifts represent another important class of pericyclic reactions to which the Woodward-Hoffmann rules apply. The selection rules for these reactions are best discussed by means of the *Dewar-Evans-Zimmerman rules*. It is then easy to see that a suprafacial [1,3]-hydrogen shift is forbidden in the ground state but allowed in the excited state, since the transition state is isoelectronic with an antiaromatic $4N$-Hückel system (with $n = 1$), in which the signs of the $4N$ AOs can be chosen such that all overlaps are positive. The antarafacial reaction, on the other hand, is thermally allowed, inasmuch as the transition state may be considered as a Möbius system with just one change in phase.

[1,3] hydrogen shift

For sigmatropic shifts of organic groups other than hydrogen, inversion of configuration at the migrating site is possible and introduces another change in phase. No phase change occurs if the configuration is retained. The resulting rules are summarized in Table 7.4.

Table 7.4 Orbital Symmetry Rules for Thermal (Δ) and Photochemical ($h\nu$) Sigmatropic $[i, j]$ Shifts

Reaction	$1 + j = 4n$	$i + j = 4n$	$1 + j = 4n + 2$	$i + j = 4n + 2$
Δ	*supra* + Inv.[a]	*supra-antara*	*supra* + Ret.	*supra-supra*
	antara + Ret.	*antara-supra*	*antara* + Inv.[a]	*antara-antara*
$h\nu$	*supra* + Ret.	*supra-supra*	*supra* + Inv.[a]	*supra-antara*
	antara + Inv.[a]	*antara-antara*	*antara* + Ret.	*antara-supra*

[a] Hydrogen shifts with inversion are not possible.

Interesting examples of sigmatropic shift reactions are provided by the rearrangements of 1,5-dienes, which, as indicated in Scheme 33 for geranonitrile (**105**), undergo a [3,3] shift (Cope rearrangement) in the ground state and a [1,3] shift in the singlet-excited state, while the triplet excited state undergoes an *x*[2 + 2] addition via a biradical, in agreement with the rule of five (Section 7.4.1). Stereospecificity of the singlet reaction, as well as results of deuterium labeling experiments, verify the expectations from the selection rules for concerted photochemical reactions (Cookson, 1968).

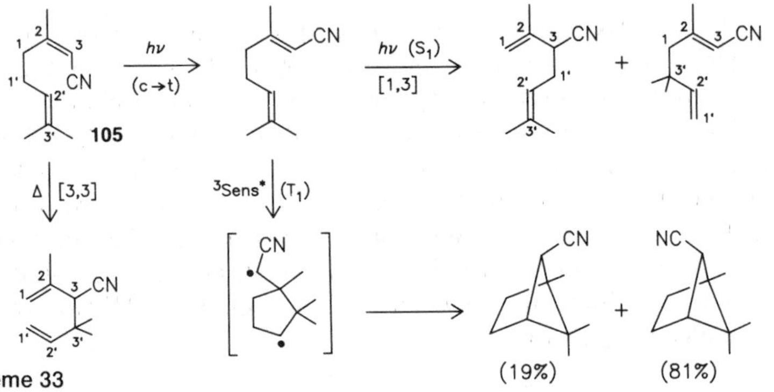

Scheme 33

However, mixtures of [1,2]- and [1.3]-shift products may be obtained by irradiating the same reactant, as shown in Scheme 34 for the direct irradiation of 1,5-hexadienes (Manning and Kropp, 1981). Configuration and state correlation diagrams for the [1,3] shift show that a pericyclic geometry in S_1 is indeed present, making the excited-singlet reaction quite analogous to the disrotatory interconversion of butadiene and cyclobutene discussed in the preceding section. Appropriate diagonal distortions of the pericyclic geometry may again lower the excited-state energy and possibly produce a conical intersection. The [1,2] or pseudosigmatropic shift, on the other hand, involves a cyclic array of an odd number of interacting orbitals that is isoelectronic with a generalized fulvene, and the reaction is always ground-state–forbidden because the product is a biradical.

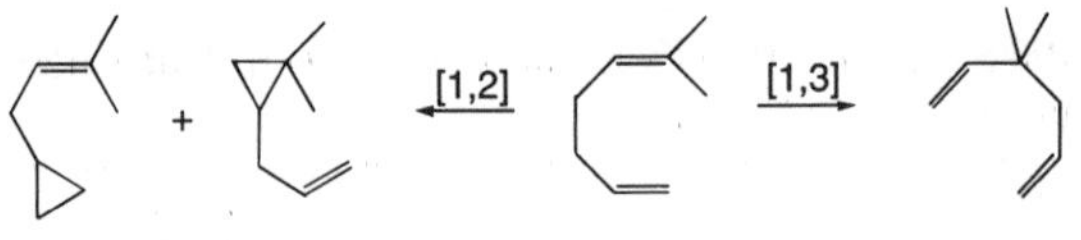

Scheme 34

These expectations have been confirmed by recent theoretical investigations on methyl shifts in but-1-ene (Scheme 35). The results demonstrate the existence of a funnel on the S_1 surface at a geometry from which travel on the ground-state surface can produce either a [1,2] or a [1,3] sigmatropic shift. The geometry and orientation of the migrating CH_3 radical along the

[1,2] and [1,3] pathways is consistent with a suprafacial process and retention of configuration at the migrating carbon (Bernardi et al., 1992a).

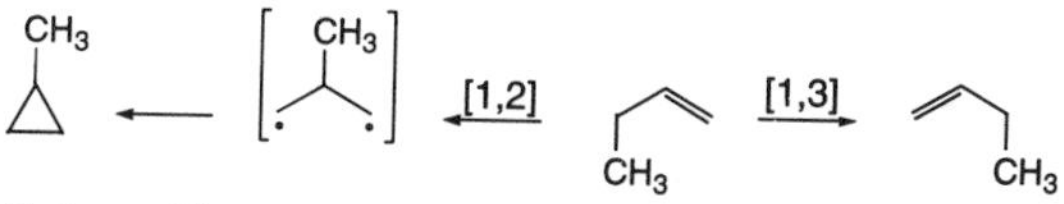

Scheme 35

The irradiation of diisopropylidenecyclobutane (**106**) causes an allowed antarafacial [1,5] shift (Kiefer and Tanna, 1969). For cycloheptatriene (**107**), a suprafacial [1,7] sigmatropic shift (Roth, 1963) and formation of bicyclo[3.2.0]hepta-2,6-diene (**108**) by intramolecular cyclization (Dauben and Cargill, 1961) are observed. Both these reactions originate from the 2A′ excited state, which is reached from the optically prepared 1A″ state via an S_2–S_1 conical intersection, as has been suggested from lifetime measurements (Borell et al., 1987; Reid et al., 1992, 1993) and confirmed by theoretical calculations. The latter also showed that no such conical intersection exists for dibenzosuberene (**109**), and the relaxed S_1 minimum is still of A″ symmetry (Steuhl and Klessinger, 1994). This state exhibits excited-state carbon acid behavior, showing that in fact the HOMO→LUMO excited-singlet state of ground-state antiaromatic carbanions is stabilized compared to the excited state of ground-state aromatic carbanions (Wan and Shukla, 1993).

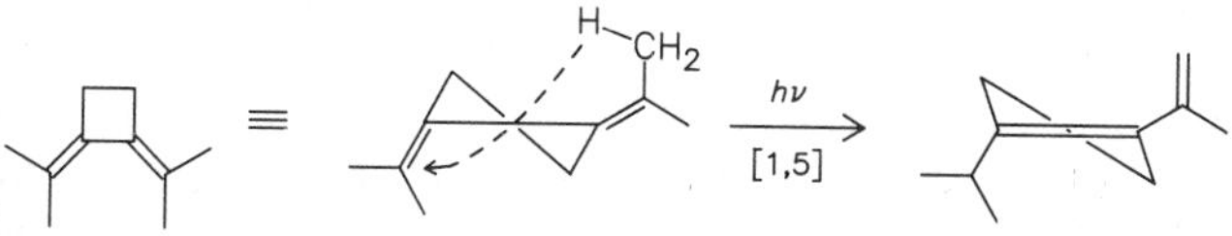

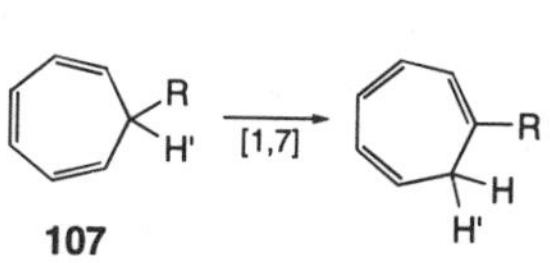

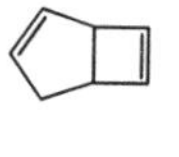

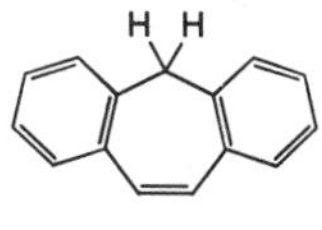

A formal [1,5] hydrogen shift occurs in photoenolization of compounds such as *o*-methylacetophenone (**110**). The reaction proceeds in the triplet state by way of hydrogen abstraction and involves biradical intermediates and an equilibrium of the triplet states of *Z*- and *E*-enol (Haag et al., 1977; Das et al., 1979).

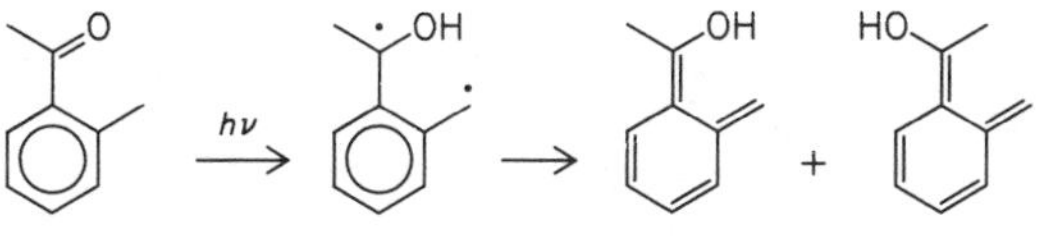

cis-Crotonaldehyde (**111**) has been used as a model for the theoretical investigation of photoenolization reactions (Sevin et al., 1979; Dannenberg and Rayez, 1983).

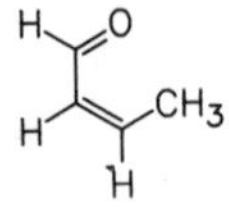

111

Compound **112** represents an example of a photochromic material in which the colored enol is stabilized through hydrogen bonding and by the phenyl groups (Henderson and Ullmann, 1965).

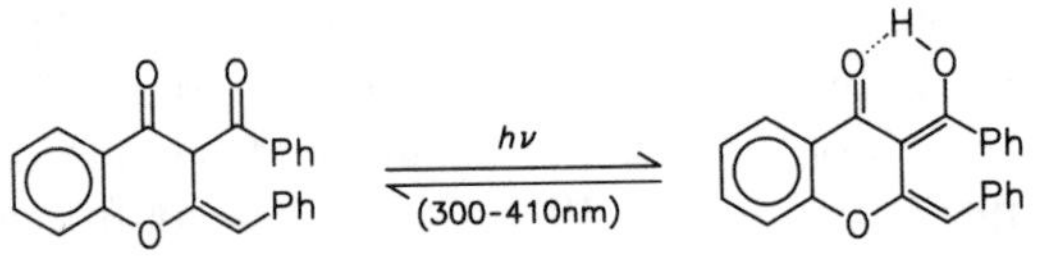

hν (300-410nm)

112

7.5.3 Photoisomerization of Benzene

The photochemical valence isomerizations of benzene summarized in Scheme 36 are wavelength-dependent singlet reactions. The first excited singlet state gives benzvalene (**115**) and fulvene (**116**); prefulvene (**117**), which had been postulated to be involved in the addition of olefins to benzene (cf. Section 7.4.3), was suggested to be an intermediate for both isomers (Bryce-Smith, 1968). The second excited singlet state leads to the formation of dewarbenzene (**113**) in addition to benzvalene (**115**) (Bryce-Smith et al., 1971) and fulvene (**116**); prismane (**114**) is not produced directly from benzene but rather results from a secondary reaction of dewarbenzene. (Cf. Bryce-Smith and Gilbert, 1976, 1980.) The conversion of dewarbenzene into prismane may be treated as a $[_{\pi}2_s + _{\pi}2_s]$ cycloaddition.

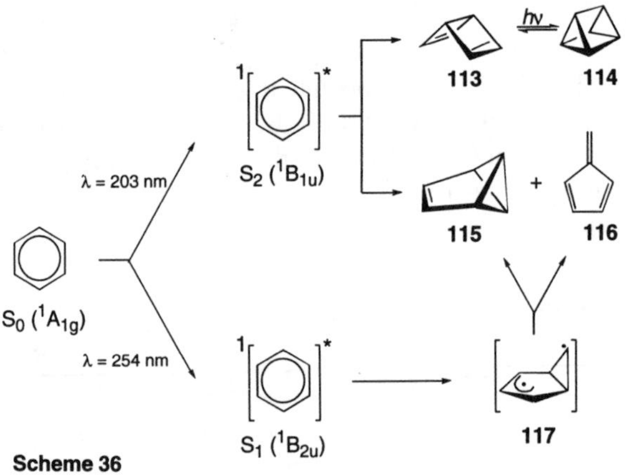

S_0 ($^1A_{1g}$) λ = 203 nm S_2 ($^1B_{1u}$) λ = 254 nm S_1 ($^1B_{2u}$) **113** hν **114** **115** + **116** **117**

Scheme 36

The formation of benzvalene is formally an $x[2 + 2]$ cyclo-addition. The S_1 (B_{2u}) reaction path from benzene toward prefulvene starts at an excited-state minimum with D_{6h} symmetry and proceeds over a transition state to the geometry of prefulvene, where it enters a funnel in S_1 due to an S_1–S_0 conical intersection and continues on the S_0 surface, mostly back to benzene, but in part on to benzvalene (Palmer et al., 1993; Sobolewski et al., 1993). At prefulvene geometries, S_0 has a flat biradicaloid region of high energy with very shallow minima whose exact location depends on calculational details (Kato, 1988; Palmer, et al., 1993, Sobolewski et al., 1993). Fulvene has been proposed to be formed directly from prefulvene or via secondary isomerization of benzvalene (Bryce-Smith and Gilbert, 1976). Calculations support the former pathway with a carbene intermediate (Dreyer and Klessinger, 1995).

The conversion of benzene into dewarbenzene can be formally considered to be a disrotatory 4π ring-closure reaction; thus it is not surprising that it is forbidden in the ground state and allowed in the excited state. The correlation diagrams in Figure 7.44 indicate that the reaction proceeds from the S_2 state (at vertical geometries this is $^1B_{1u}$, but it soon acquires E_{2g} character as motion along the reaction path occurs). (Cf. Example 7.16.) The best currently available calculations (Palmer et al., 1993) suggest that the molecule moves downhill on the S_2 surface to an S_2–S_1 conical intersection and

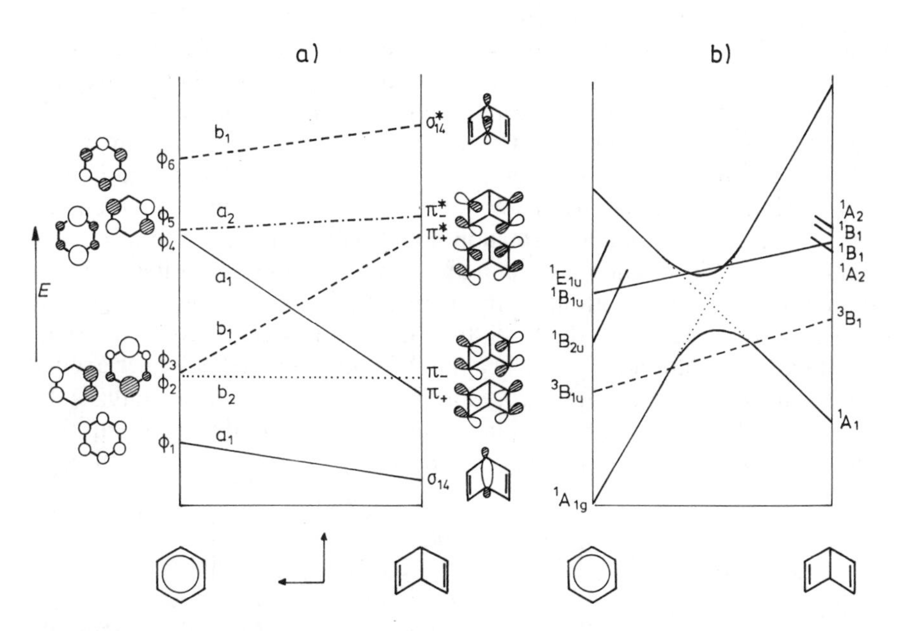

Figure 7.44. Benzene–dewarbenzene valence isomerization; a) orbital correlation diagram (C_{2v} symmetry) and b) state correlation diagram, devised with the help of Table 7.5 and experimental state energies.

reaches the S_1 surface at a geometry where S_1, too, has a funnel. This is caused by an S_1–S_0 conical intersection and provides further immediate conversion to the S_0 state. Thus, there is no opportunity for vibrational equilibration in the S_1 state. The biradicaloid geometry at which these stacked funnels are located is calculated to be such that relaxation in S_0 can produce dewarbenzene, benzvalene, or benzene.

In low-temperature argon matrices, dewarbenzene has been shown to be a primary product of benzene photolysis at 253.7 nm (Johnstone and Sodeau, 1991). Hexafluorobenzene yields hexafluorodewarbenzene upon irradiation in the liquid or vapor phase (Haller, 1967).

Isomerization reactions of benzene that result in "scrambling" of hydrogen and carbon atoms presumably involve a benzvalene intermediate and thermal or photochemical rearomatization; depending on which of the CC bonds indicated in Scheme 37 by a heavy or a broken line is cleaved, either the rearranged product is formed or the reactant is recovered (Wilzbach et al., 1968).

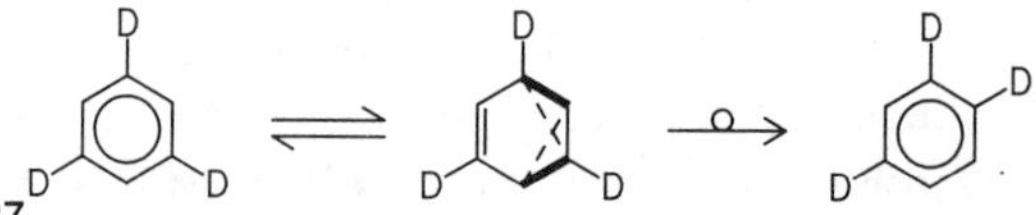

Scheme 37

The major product of the direct irradiation of dewarbenzene is benzene, while some prismane is produced via an intramolecular [2 + 2] cyclization, which involves a funnel that lies between the geometries of dewarbenzene and prismane (Palmer et al., 1992). The electronic factors that control the existence of this funnel are the same as those for the intermolecular [2 + 2] cycloaddition of two ethylenes. (Cf. Sections 6.2.1 and 7.4.1.) An important difference is due to the strong steric strain that arises in the cagelike σ-bond framework as a diagonal distortion is introduced, producing a high-energy ridge along the rhomboidal distortion coordinate. This moves the conical intersection to higher energies, but it apparently remains accessible. Further strain may locate the bottom of the pericyclic funnel at a less diagonally distorted structure, where the S_0–S_1 touching is still weakly avoided.

Triplet sensitization ($E_T > 66$ kcal/mol) exclusively yields triplet excited benzene in an adiabatic reaction. According to Scheme 38, this catalyzes the conversion of dewarbenzene into benzene. The reaction thus proceeds as a chain reaction with quantum yields of up to $\Phi_{lim} \approx 10$ (Turro et al., 1977b).

Scheme 38

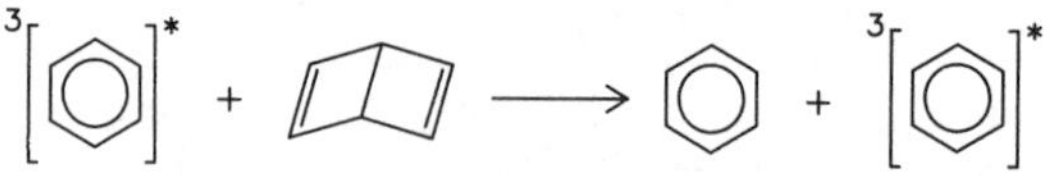

The major reaction to result from direct excitation of benzvalene is a degenerate [1,3] sigmatropic shift as indicated in Scheme 39. The same reaction is observed upon sensitized triplet excitation if the triplet energy of the sensitizer E_T is below 65 kcal/mol. If, however, E_T exceeds 65 kcal/mol, reversion to benzene takes place. The latter is an adiabatic hot triplet reaction, which proceeds from one of the higher excited triplet states (T_2) of benzvalene and produces triplet excited benzene. It results in a chain reaction with $\Phi_{lim} \approx 4$, similar to the reaction of dewarbenzene shown in Scheme 38 (Renner et al., 1975).

Scheme 39

Direct irradiation as well as sensitized triplet excitation of prismane results in rearomatization to produce benzene and in isomerization to form dewarbenzene, which predominates even in the triplet reaction. An adiabatic rearrangement to excited benzene is not observed (Turro et al., 1977b).

Benzene valence isomers undergo thermal rearomatization to form benzene rather easily. Half-times of dewarbenzene and benzvalene at room temperature are approximately 2 and 10 days, respectively. Although all these reactions are strongly exothermic, chemiluminescence (cf. Section 7.6.4) is observed only for the thermal ring opening of dewarbenzene. This reaction produces triplet excited benzene in only very small amounts. While benzene phosphorescence is undetectable in fluid solution near room temperature, triplet-to-singlet energy transfer with 9,10-dibromoanthracene as acceptor, followed by fluorescence, leads to a readily detectable indirect chemiluminescence (Lechtken et al., 1973; Turro et al., 1974).

Example 7.16:
The experimental results on valence isomerization of benzene can be rationalized by means of correlation diagrams that were first discussed for hexafluorobenzene by Haller (1967). The orbital correlation diagram for the conversion of benzene to dewarbenzene (Figure 7.44a) has been constructed on the basis of a classification of the benzene π MOs according to C_{2v} symmetry. The full symmetry of benzene is D_{6h}. The symmetry elements common to both point groups C_{2v} and D_{6h} are E, C_2, and $\sigma_v(x,z) = \sigma_v^{(1)}$, which is the plane through

Table 7.5 Interrelations between Irreducible Representations of Point Groups C_{2v} and D_{6h}

D_{6h}	E	C_2	$\sigma_v = \sigma_v(xy)$	$\sigma_d = \sigma_v'(yz)$	C_{2v}
B_{1u}	1	−1	1	−1	B_1
B_{2u}	1	−1	−1	1	B_2

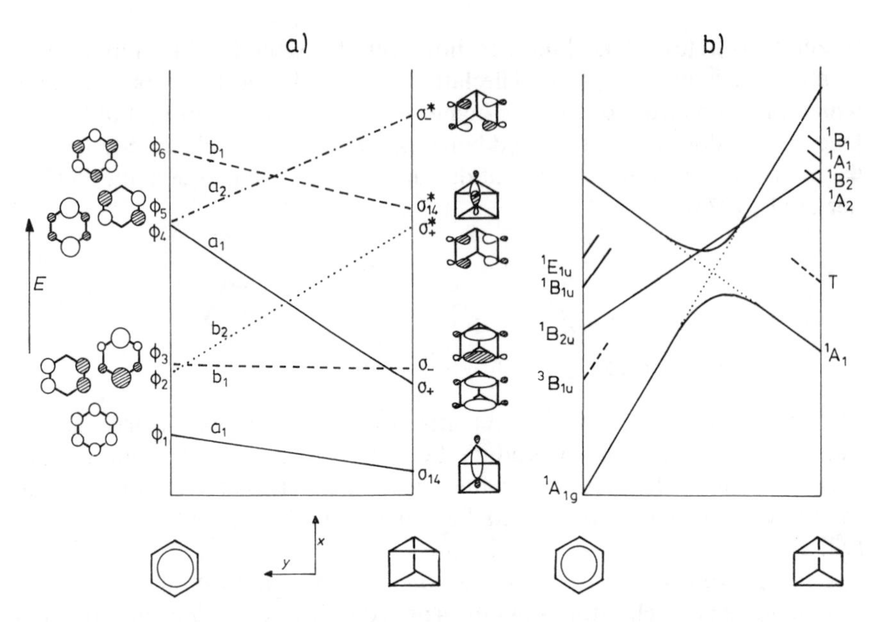

Figure 7.45. Benzene–prismane valence isomerization; a) orbital correlation diagram (C_{2v} symmetry) and b) state correlation diagram, devised with the help of Table 7.5 and experimental state energies.

atoms C-1 and C-4, and $\sigma_v'(yz)$ or $\sigma_d^{(1)}$, perpendicular to the former plane. From symmetry behavior with respect to these elements, interrelations between irreducible representations of the two point groups may be derived as indicated in Table 7.5.

Making use of experimental singlet and triplet excitation energies and of the relative ground-state energies (cf. Turro et al., 1977b), one can obtain the state correlation diagram of the benzene–dewarbenzene interconversion shown in Figure 7.44b; according to Table 7.5 the $^1B_1$ state of dewarbenzene correlates with the $^1B_{1u}$ state of benzene, located 5.96 eV above the ground state. This explains why dewarbenzene is formed only by short-wavelength excitation (Scheme 36). In agreement with the ab initio results discussed earlier, both an S_2–S_1 conical intersection and an S_1–S_0 funnel are easily recognized, although the calculated intersections occur at geometries of lower symmetry, as is to be expected from the discussion in Section 6.2.1. The rearomatization observed on singlet excitation and the adiabatic formation of triplet excited benzene observed for sensitized triplet excitation are apparent from this diagram. It is also seen that correlations can be drawn such that the S_0 and T_1 lines cross, permitting chemiluminescence to occur. (Cf. Section 7.6.4.)

Corresponding correlation diagrams for the benzene–prismane interconversion in Figure 7.45 explain why the direct irradiation of benzene does not produce prismane, while the reverse reaction is quite efficient. Correlation diagrams for the other benzene valence isomerization reactions can be derived in a similar way. (See also Halevi, 1977.)

While the electrocyclic ring opening to *o*-quinodimethanes is the major reaction pathway in the irradiation of substituted benzocyclobutenes (cf. Example 6.14), the irradiation of unsubstituted benzocyclobutene yields 1,2-dihydropentalene (**119**) and 1,5-dihydropentalene (**120**) as major products. The mechanism shown with "prebenzvalene" (**118**) as primary photochemical intermediate has been proposed to explain the formation of the isomeric dihydropentalenes (Turro et al., 1988). Supporting calculations that yield the same mechanism for the benzene-to-fulvene transformation have been published (Dreyer and Klessinger, 1995).

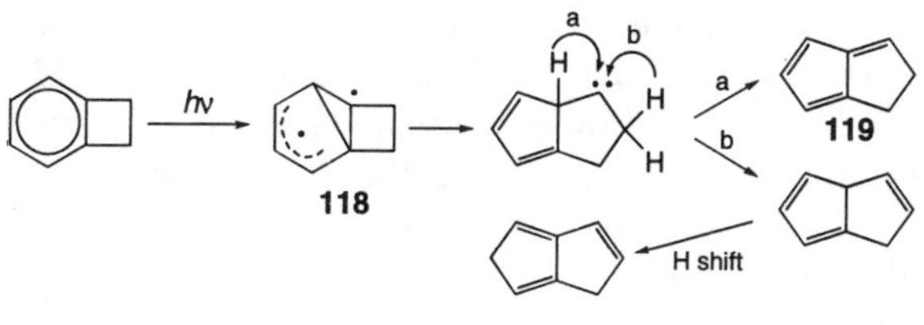

hν
118
a
b
H
H
H
a
b
119
H shift
120

7.5.4 Di-π-methane Rearrangement

1,4-Dienes and related compounds undergo a photochemical rearrangement reaction known as the *di-π-methane* or *Zimmerman rearrangement* (Zimmerman et al., 1966, 1980; Hixson et al., 1973). The reaction occurs also with β,γ-unsaturated carbonyl compounds and is then called the oxa-di-π-methane rearrangement. (See Section 7.5.5.) According to Scheme 40, the reaction formally involves a [1,2] shift, but the second double bond between carbons C-1 and C-2 is apparently also involved in the reaction process. The most favorable structural feature for di-π-methane rearrangement is an interaction of the C-2 and C-4 centers of the 1,4-diene in a way that permits a stepwise reaction involving a 1,4-biradical and a 1,3-biradical, as formulated in Scheme 40. In contrast to experimental results (Paquette and Bay, 1982), ab initio calculations suggested that the 1,4-biradical may be a true intermediate (Quenemoen et al., 1985).

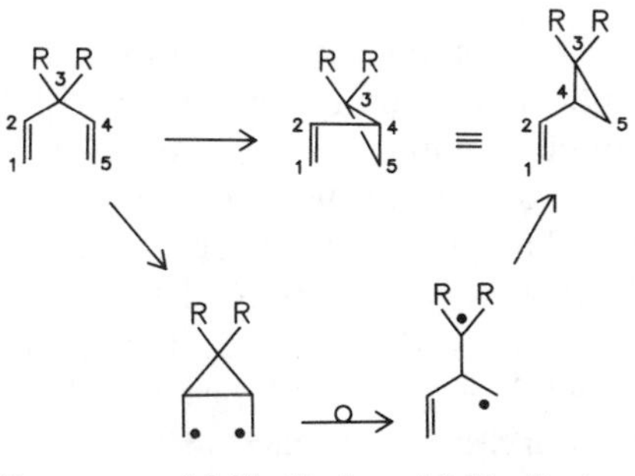

Scheme 40 **1,4-Biradical** **1,3-Biradical**

However, more recent calculations for the excited-singlet reaction of 1,4-pentadiene (Reguero et al., 1993) suggest that the preferred relaxation from S_1 to S_0 occurs via a funnel that corresponds to a 1,2 shift of a vinyl group (see Section 7.5.2) and produces the 1,3-biradical, avoiding the formation of the 1,4-biradical intermediate altogether. The 1,3-biradical is unstable and undergoes a ground-state barrierless ring-closure process to yield the final vinylcyclopropane product. In particular, three funnels have been found: one structure with both double bonds twisted, which would lead only to cis-trans isomerization or back to the starting material (cf. Section 7.1.3); one rhomboidal structure similar to the conical intersection for the ethylene + ethylene $[2_s + 2_s]$ cycloaddition, which would lead to cycloaddition products or to the 1,4-biradical (which arises from an asynchronous [2 + 2] addition, cf. Section 7.4.2); and the lowest-energy conical intersection (23 kcal/mol lower in energy than the first and 31 kcal/mol lower than the second), which corresponds to the sigmatropic [1,2] shift and would produce the 1,3-biradical. A ground state reaction path emanating from the geometry of this funnel leads to vinylcyclopropane.

From the stereochemistry at C-1 and C-5, it has been concluded that the ring closure occurs in a disrotatory way and anti with respect to the migrating vinyl group, as shown in Scheme 41. Only when the anti-disrotatory ring closure is not possible for steric reasons is a stereochemistry that corresponds to a syn-disrotatory mode of reaction observed.

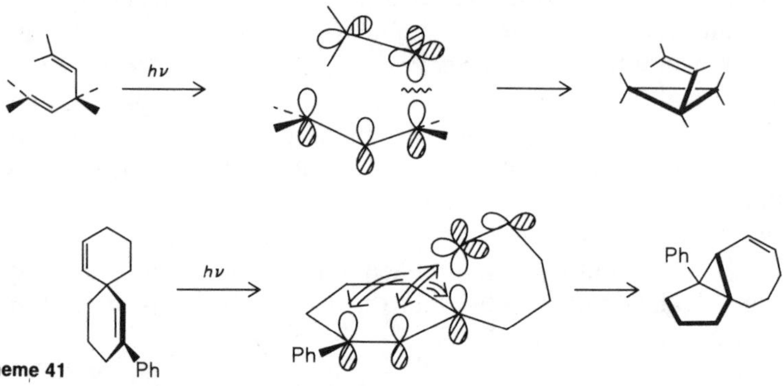

Scheme 41

Both reaction paths that involve breaking of one σ and two π bonds and the making of three new bonds are in agreement with the Woodward-Hoffmann rules. According to Figure 7.46, the disrotatory ring closure between C-3 and C-5 requires the orbitals *D* and *F* to overlap anti or syn with respect to the bond between C-2 and C-3 which is being cleaved; overlap of orbitals *C* and *B* produces the new bond between C-2 and C-4, and orbitals *A* and *E* form the new π bond between C-1 and C-2. The rearrangement may therefore be formulated as a $[_\pi2_a + {_\sigma2_a} + {_\pi2_a}]$, or a $[_\pi2_s + {_\sigma2_s} + {_\pi2_a}]$ reaction, for the anti- or syn-disrotatory mode of reaction, respectively. Both cases correspond to ground-state–forbidden, photochemically allowed reactions. The

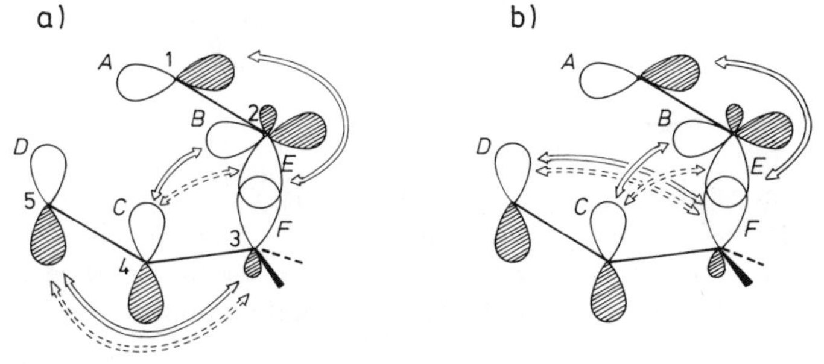

Figure 7.46. Orbital interactions at the di-π-methane rearrangement a) anti-disrotatory mode ($\Leftrightarrow [_{\pi}2_a + {_{\sigma}2_a} + {_{\pi}2_a}]$, $\Leftarrow \Rightarrow [_{\pi}2_a + {_{\sigma}2_a}]$), b) syn-disrotatory mode ($\Leftrightarrow [_{\pi}2_s + {_{\sigma}2_s} + {_{\pi}2_a}]$, $\Leftarrow \Rightarrow [_{\pi}2_s + {_{\sigma}2_s}]$).

same conclusion may be reached by the observation from Figure 7.46 that the relevant orbitals form a $(4N + 2)$-Möbius system, since signs of the six orbitals can be chosen in such a way that just one overlap is negative.

Inversion of configuration at C-3, which is expected for an anti-disrotatory reaction path, has been observed, for instance, for the acyclic chiral 1,4-diene **121**.

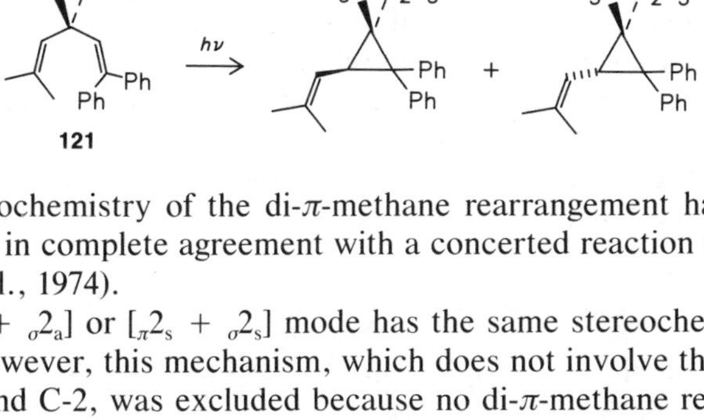

121

The stereochemistry of the di-π-methane rearrangement has thus been shown to be in complete agreement with a concerted reaction course (Zimmerman et al., 1974).

The $[_{\pi}2_a + {_{\sigma}2_a}]$ or $[_{\pi}2_s + {_{\sigma}2_s}]$ mode has the same stereochemical consequences. However, this mechanism, which does not involve the π bond between C-1 and C-2, was excluded because no di-π-methane rearrangement is observed upon irradiation of compound **122**, which contains only one double bond.

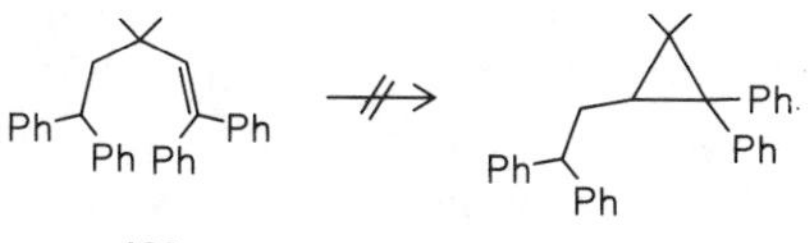

122

For acyclic and monocyclic 1,4-dienes, the di-π-methane rearrangement occurs in general from the singlet excited state, since loose geometries are favored in the triplet state and cis-trans isomerization is the preferred reaction, as shown in Scheme 42.

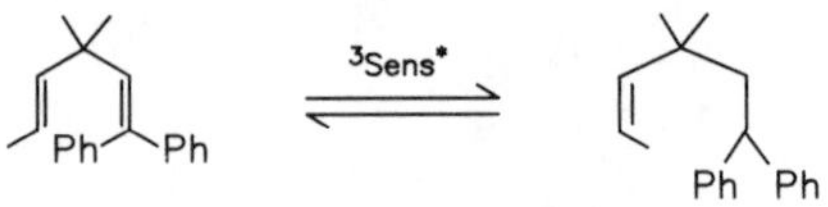

Scheme 42

For singlet reactions, the direction of ring closure is the one that involves the most stable 1,3-biradical, as shown in Scheme 43. This is equivalent to the assumption that in a concerted mode of reaction the bond between centers C-3 and C-5 is only weakly formed during the initial stages of the reaction.

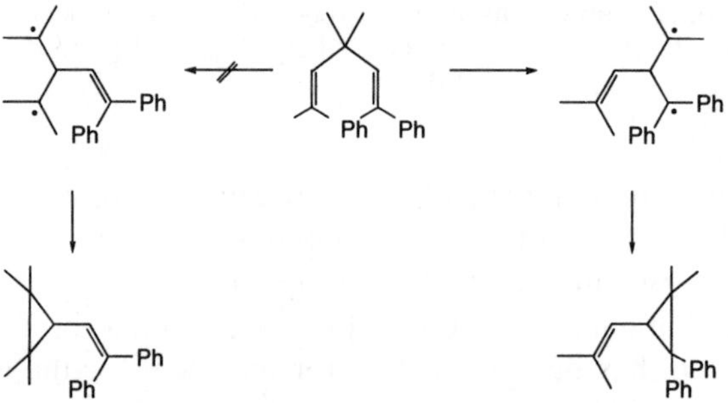

Scheme 43

The di-π-methane rearrangement is not confined to acyclic and monocyclic systems. Bicyclic 1,4-dienes such as barrelene (**123**) also rearrange, but they do so upon triplet sensitization (Zimmerman and Grunewald, 1966). In all probability, the built-in geometry constraint of a bicyclic molecule does not allow for the loose geometries otherwise characteristic of the triplet state, while in the singlet state other reactions such as [2 + 2] cycloaddition predominate.

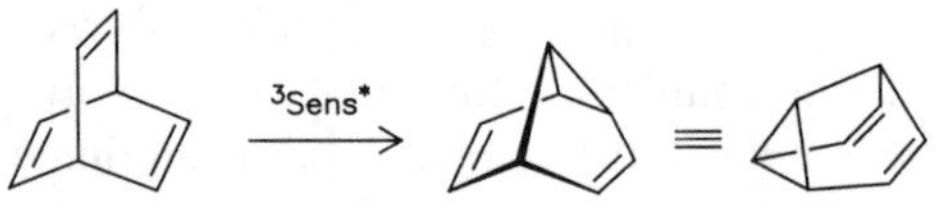

123 Semibullvalene

All these experimental features are consistent with a concerted pathway that passes through a funnel for a [1,2] vinyl shift and leads to the 1,3-biradical region (Reguero et al., 1993). First, the regiospecifity is readily rationalized via the different accessibility of the two conical intersections leading to the 1,3-biradicals (Scheme 43): the one with the most stable substituted radical moiety will be lower in energy, and the vinyl radical without radical stabilizing substituents will migrate. Second, the double bond in the migrating vinyl radical retains its original configuration along the excited-state–ground-state relaxation path. Third, inversion of configuration on C-3 (cf. Scheme 40) occurs during the evolution of the conical intersecction structure toward the 1,3-biradical structure, as was shown by reaction path calculations: the σ bond between the two terminal methylene radical centers is formed via a barrierless process on the side of the —CH_2 fragment opposite to the initial position of attachment of the migrating vinyl radical.

The triplet reaction probably proceeds by a different mechanism, likely to involve a 1,4-biradical intermediate. In an elegant study of the rearrangement of deuterated *m*-cyanodibenzobarrelene (**124**) to the corresponding semibullvalene **125**, it was in fact shown that the 1,4-biradical is the product-determining intermediate (Zimmerman et al., 1993).

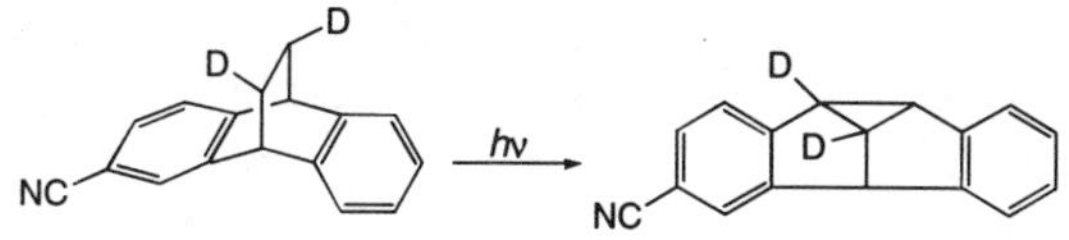

124 → (hν) 125

A large number of rearrangements such as **126** → **127** can be classified as belonging to the di-π-methane type even though the molecules do not formally contain a 1,4-diene unit, since in these cases an aryl ring can take the place of one of the olefinic bonds.

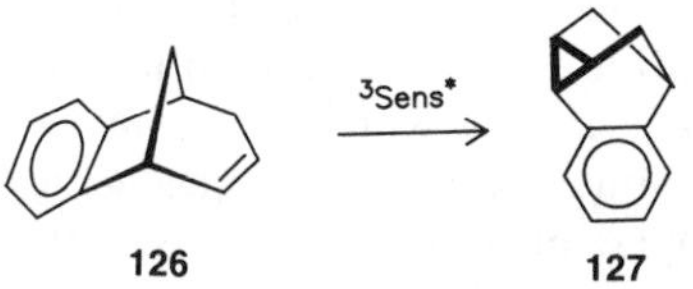

126 → (^{3}Sens*) 127

Substituent effects have been studied in detail and have been rationalized on the basis of the proposed mechanism (Zimmerman, 1980).

Example 7.17:
Donor-substituted benzonorbornadienes **128** generally yield the product of a di-π-methane rearrangement corresponding to *m*-bridging, while acceptor-substituted compounds **129** give the *p*-bridged product:

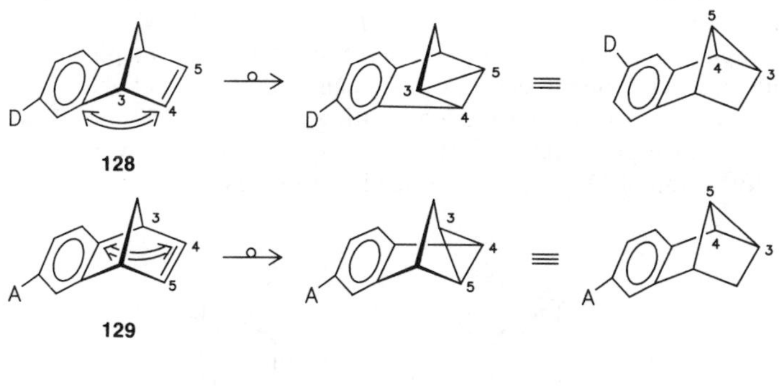

128

129

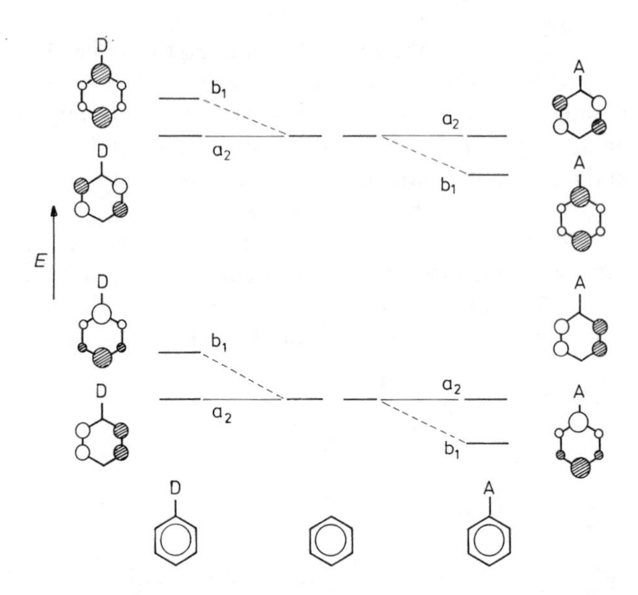

Figure 7.47. Orbital ordering in donor- and acceptor-substituted benzenes.

If the triplet states of **128** and **129** involve mainly single configurations with a half-occupied HOMO and a half-occupied LUMO, the pronounced regioselectivity may be rationalized on the basis of an interaction of the half-occupied LUMO of the aromatic moiety with the vacant LUMO of the ground-state ethylene moiety. Substitution removes the degeneracy of the benzene LUMO, and the product formed (meta or para) depends on the magnitude of the LCAO coefficients of the lower of these MOs. From Example 3.10 the $\phi_{s'}$ MO is known to be the LUMO in acceptor-substituted benzenes, while $\phi_{a'}$ is the LUMO in donor-substituted benzenes. (Cf. Figure 3.18.) This is shown again in Figure 7.47, together with the labeling of the MOs corresponding to C_{2v} symmetry; in the b_1 MO the LCAO coefficient of the para position is largest in absolute value, while in the a_2 MO the meta coefficient is larger than the para coefficient (Santiago and Houk, 1976).

Example 7.18:

The complex mechanism of the rearrangement of metacyclophanediene **130** into dihydrocyclopropapyrene **136** was elucidated by Wirz et al. (1984) and is shown in Figure 7.48. Valence isomerization of the type **130** ↔ **131** are symmetry-allowed ground-state reactions and are extraordinarily fast. The adiabatic formation of **131*** was mentioned in Example 6.7.

130 **136**

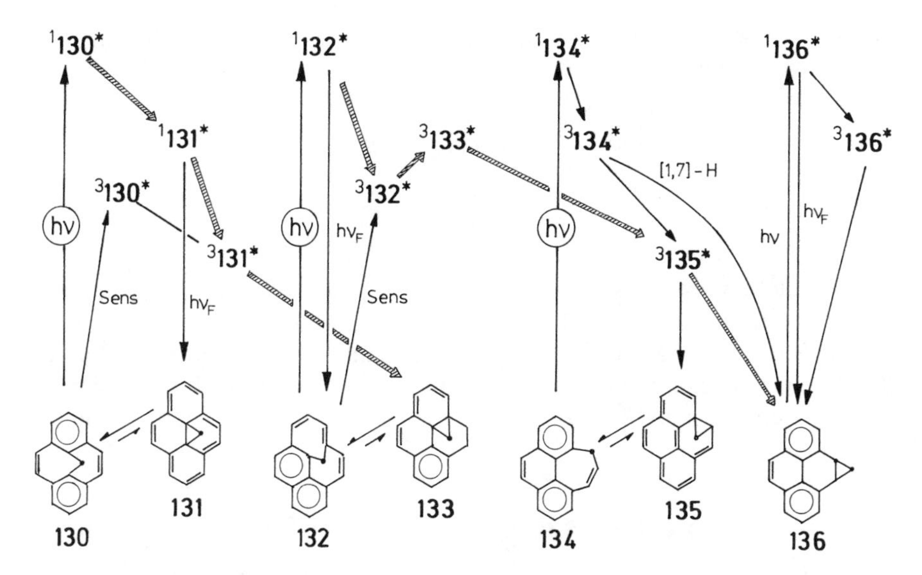

Figure 7.48. Mechanism of the transformation of metacyclophane **130** into dihydrocyclopropapyrene **136** with intermediates **132/133** and **134/135**. The suggested position of the original methano bridge is marked by a bullet. The hatched arrows indicate the two-quantum process that occurs at room temperature and does not involve the ground-state intermediate **134** (adapted from Wirz et al., 1984).

Below $-30°C$ the reaction path consists of a three-quantum process involving two thermally stable, light-sensitive isomers **132** and **134**. The first two steps from **130** to **133** may formally be viewed as di-π-methane rearrangements (cf. Scheme 40, the carbon marked with a bullet in Figure 7.48 corresponds to C-3 in Scheme 40), while the last step from **134** to **136** represents a [1,7] hydrogen shift. At room temperature the reaction proceeds as a two-quantum process, bypassing the ground-state intermediate **134**.

A related reaction, the "bicycle rearrangement," is schematically shown in Scheme 44: A carbon with substituents R^1 and R^2 connected to a π system by two hybrid AOs moves along the molecule as if the hybrid AOs were the wheels and the substituents the handlebars of a bicycle.

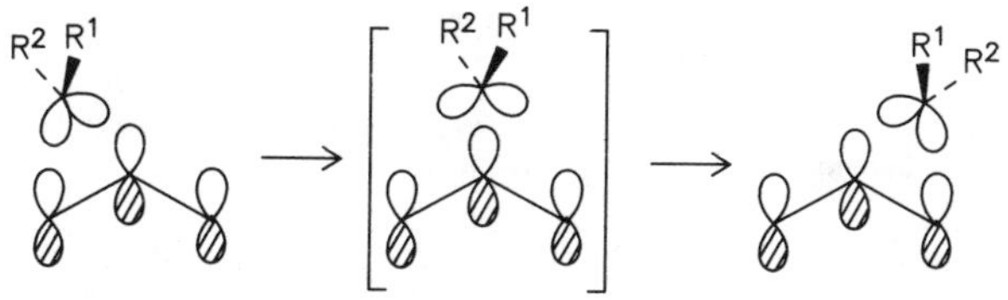

R^2 R^1 $\longrightarrow$ [R^2 R^1] $\longrightarrow$ R^1 R^2

Scheme 44

An example is the rearrangement of 2-methylene-5,6-diphenylbicyclo[3.1.0]hexene (**137**), which yields 1,5-diphenylspiro[2.4]-4,6-heptadiene

(**138**). The shift does not proceed over two bonds as depicted in Scheme 44, but over three bonds (Zimmerman et al., 1971):

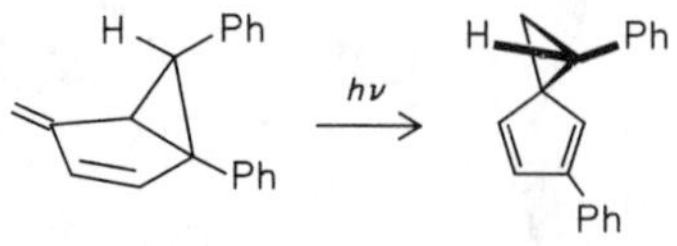

H Ph hν H Ph Ph Ph

137 **138**

The reaction proceeds in the singlet state and is stereospecific in that the configuration of the "handlebars" is retained for migrations of up to three bonds. Both the mechanism of the di-π-methane rearrangement and that of the 2,5-cyclohexadienone rearrangement (dealt with in Section 7.5.5), involve a step that may be formulated as a bicycle rearrangement (Zimmerman, 1982).

7.5.5 Rearrangements of Unsaturated Carbonyl Compounds

In addition to the normal photochemical reactions of saturated ketones, β,γ-unsaturated carbonyl compounds undergo carbonyl migration via a [1,3] shift. Compound **139** in Scheme 45 represents a typical example. Compounds with an alkyl substituent in the β position such as **140** may also undergo a Norrish type II reaction (Kiefer and Carlson, 1967), while for ketones with electron-rich double bonds such as **141**, oxetane formation is also observed (Schexnayder and Engel, 1975).

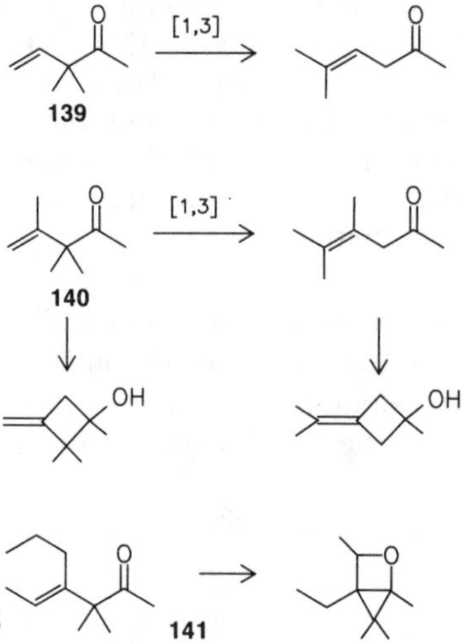

O [1,3] O

139

O [1,3] O

140

OH OH

O O

Scheme 45 **141**

α Cleavage is much faster for β,γ-unsaturated ketones than for saturated ketones. This can be rationalized by the relative stability of the acyl–allyl radical pair, which can experience an additional stabilization by simultaneous bonding interaction of the acyl radical with the γ carbon (Houk, 1976). Whether this interaction is large enough to render the [1,3] shift a concerted

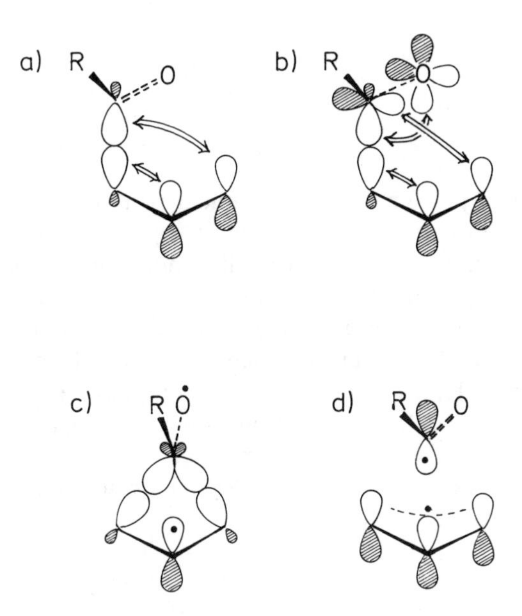

Figure 7.49. Possible mechanisms of the [1,3] shift of β,γ-unsaturated ketones: a) concerted, b) concerted with inclusion of carbonyl orbitals, c) biradical intermediate, and d) radical pair intermediate (adapted from Houk, 1976).

reaction, or whether a mechanism involving α cleavage and subsequent radical recombination prevails, cannot be generally decided.

Four different mechanisms can be envisioned as indicated in Figure 7.49: the [1,3] shift may be concerted and can be classified either as a $[_{\pi}2_s + {_\sigma}2_s]$ reaction (Figure 7.49a) or with inclusion of the π_{CO} and n_O orbitals in the orbital array (Figure 7.49b). Two stepwise mechanisms, one involving the formation of a biradical, the other of a radical pair, can also be formulated (Figure 7.49c,d). Since the [1,3] shift is a ground-state forbidden by orbital symmetry, the ground-state surface will have a saddle point at a geometry approximating the transition state of this reaction, while the S_1 surface will have a minimum or a funnel at a nearby geometry. Thus, whether a biradicaloid intermediate or a radical pair will be involved, or whether the reaction will be concerted, depends on the magnitude of α,γ interaction or on the degree to which the crossing is avoided.

For the $^1(n,\pi^*)$ state of a β,γ-unsaturated ketone to undergo radical α cleavage or a [1,3] shift, the α bond must be approximately parallel to the π orbital, since only then can allyl resonance stabilize the biradical and the carbonyl orbital overlap with the π system at the γ terminus. These geometrical requirements are not fulfilled for the aldehyde **142**, and intersystem crossing dominates α cleavage and [1,3] shift (Baggiolini et al., 1970).

The triplet photochemistry of β,γ-unsaturated carbonyl compounds depends very much on their electronic configuration. The $^3(n,\pi^*)$ state under-

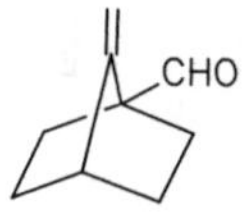

CHO

142

goes a [1,3] shift, presumably via a radical pair. In most β,γ-unsaturated carbonyl compounds, however, the lowest triplet state is the $^3(\pi,\pi^*)$ state, for which a [1,2] shift termed the oxa-di-π-methane rearrangement is characteristic (Schaffner and Demuth, 1986). Concerted mechanisms as well as biradical intermediates may be formulated for this reaction, in a way analogous to the di-π-methane rearrangement. Although the problem has not yet been solved generally, evidence in favor of a stepwise reaction path has been obtained—for instance, for the reaction shown in Scheme 46 (Dauben et al., 1976). This contrasts with the corresponding results for the di-π-methane rearrangement of the hydrocarbon **121**, which occurs in the singlet state.

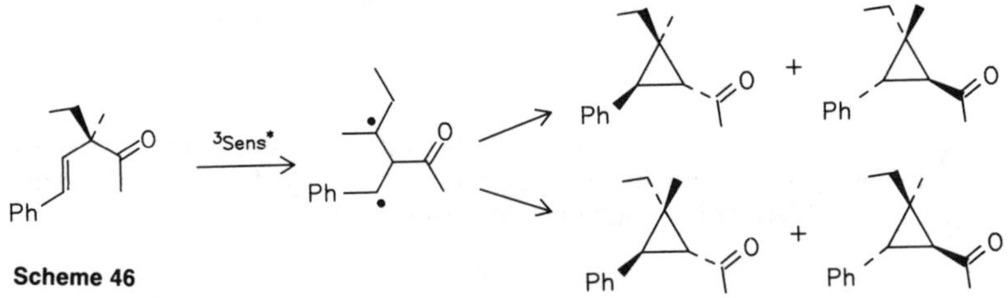

Scheme 46

Because the intersystem crossing probability is low, the triplet reaction can in general be initiated only by sensitization. The difference between singlet and triplet reactions can be utilized synthetically, as indicated in Scheme 47 (Baggiolini et al., 1970; Sadler et al., 1984):

Scheme 47

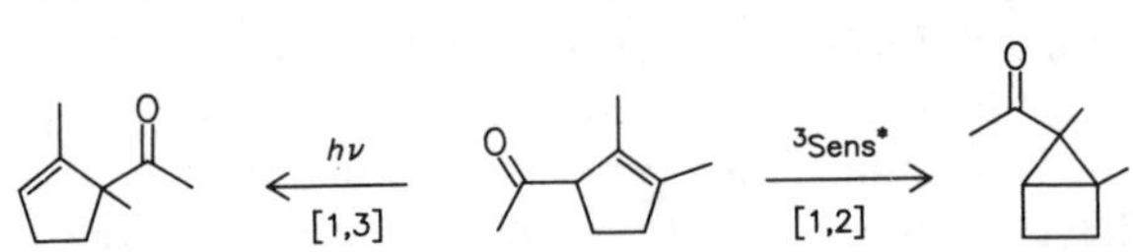

In addition to the reactions discussed in Section 7.4.5, α,β-unsaturated ketones undergo isomerization to give the deconjugated β,γ-unsaturated compound. According to Scheme 48, this reaction may be analyzed as a Norrish type II reaction with cleavage of a π bond instead of a σ bond. However, it may also be interpreted as a photoenolization. (Cf. Section 7.5.2.)

Scheme 48

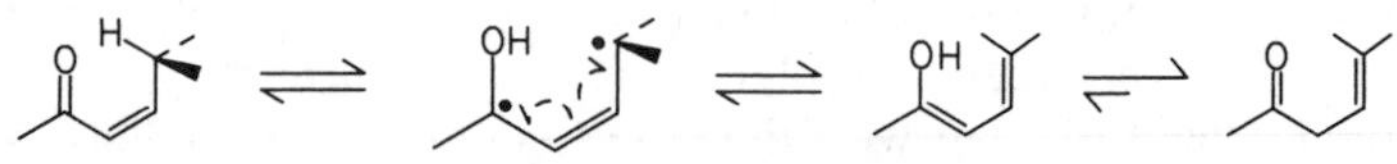

Cyclic enones and dienones undergo a number of photorearrangements. As indicated in Scheme 49, both ring contraction via a [1,2] shift (Zimmerman and Little, 1974) and formation of cyclopropanone derivatives via a

[1,3] shift (Barber et al., 1969) has been observed for cyclopentenones (**143**). Radical stabilizing substituents in the 5-position, as in **144**, favor formation of cyclopropyl ketenes (Agosta et al., 1969).

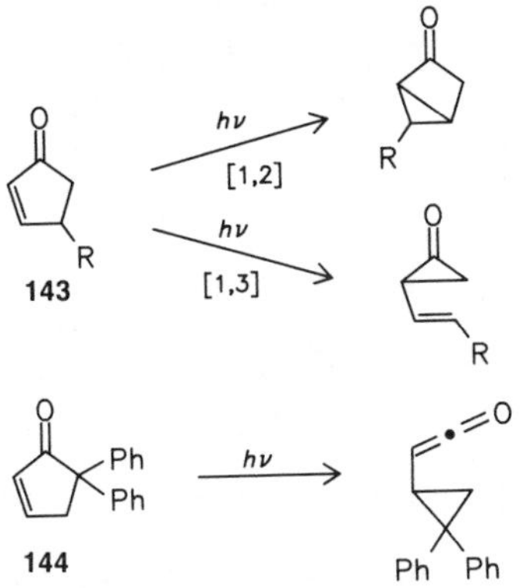

Scheme 49

For cyclohexenones, only [1,2] shifts are typical. According to Scheme 50, two types of [1,2] shift with ring contraction can occur: one through rearrangement of the ring atoms (type A), the other through migration of a ring substituent (type B) (Zimmerman, 1964):

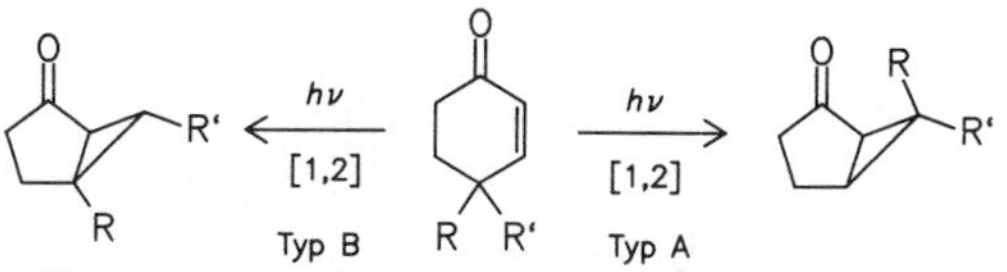

Scheme 50

Photochemical ring opening of linearly conjugated cyclohexadienones affords dienylketenes (**145**), which react in one of the following ways: recyclization to the original or to a stereoisomeric cyclohexadienone, formation of bicyclo[3.1.0]hexenones (**146**), or addition of a protic nucleophile to yield substituted hexadienecarboxylic acids (**147**) (Quinkert et al., 1979).

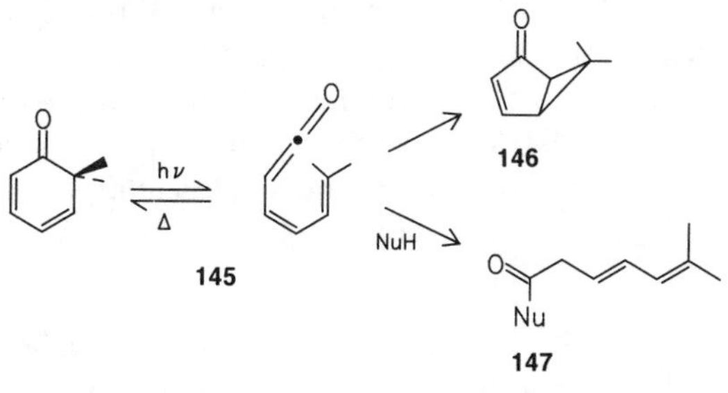

In addition to these reactions typical of the $^1(n,\pi^*)$ state, 6-acetyloxycyclohexadienones (*o*-quinol acetates) form phenols from the $^3(\pi,\pi^*)$ state. These results are summarized in Figure 7.50 (Quinkert et al., 1986). Dienylketenes are produced irrespective of whether irradiation occurs into the $n \rightarrow \pi^*$ or $\pi \rightarrow \pi^*$ band, and phenols can be formed not only through triplet

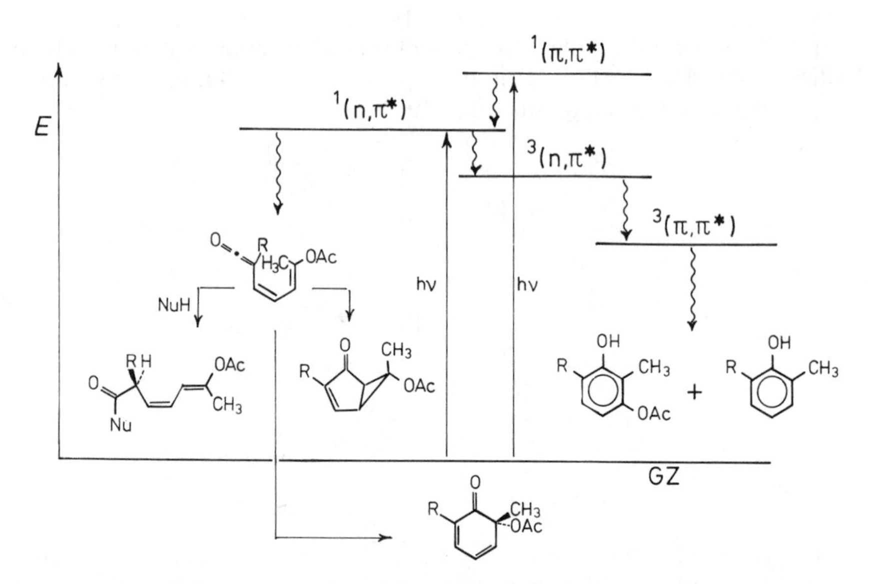

Figure 7.50. Schematic state correlation diagram for *o*-quinol acetate photoreactions. Wavy arrows designate physical as well as chemical radiationless processes (by permission from Quinkert et al., 1986).

sensitization ($E_T > 42$ kcal/mol, corresponding to the $^3(\pi,\pi^*)$ state), but also by extended direct irradiation in aprotic solvents if the quinol acetate is regenerated from the kinetically unstable dienylketene.

7.6 Miscellaneous Photoreactions

7.6.1 Electron-Transfer Reactions

It has become evident only fairly recently that photochemical electron-transfer processes (PET) play an important role in many reactions. In this connection it is of great importance that in an excited state a molecule can be a better oxidant as well as a better reductant than in the ground state. For instance, after HOMO→LUMO excitation the half-occupied HOMO can readily accept another electron, while the single electron from the LUMO can readily be transferred to an acceptor. (Cf. Figure 5.25.)

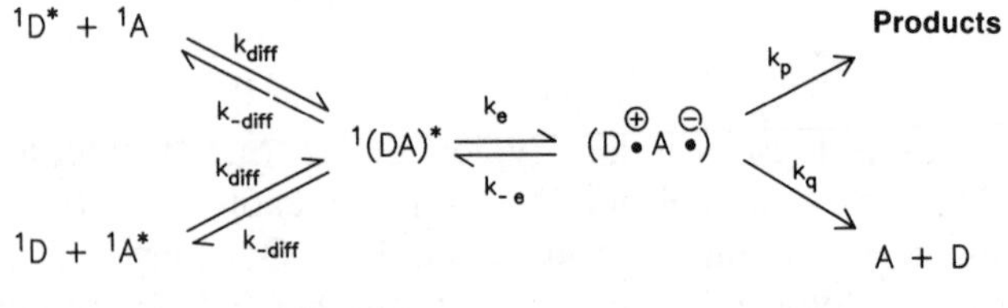

$$
\begin{array}{ccccccc}
{}^1D^* + {}^1A & \underset{k_{-diff}}{\overset{k_{diff}}{\rightleftharpoons}} & & & & \overset{k_p}{\nearrow} & \textbf{Products} \\
 & & {}^1(DA)^* & \underset{k_{-e}}{\overset{k_e}{\rightleftharpoons}} & (D^{\oplus\bullet}\, A^{\ominus\bullet}) & & \\
{}^1D + {}^1A^* & \underset{k_{-diff}}{\overset{k_{diff}}{\rightleftharpoons}} & & & & \underset{k_q}{\searrow} & A + D
\end{array}
$$

Scheme 51

From Scheme 51 it is seen that an exciplex with a certain degree of charge-transfer character can be formed either from an excited donor molecule $^1D^*$ and an acceptor molecule ^1A or from an excited acceptor molecule $^1A^*$ and a donor molecule ^1D. Both alternatives have been observed experimentally. Thus, the use of diethylaniline as donor and either biphenyl or anthracene as acceptor yields exciplexes that can be identified by the typical structureless fluorescence at long wavelengths. This corresponds to the two possibilities depicted in Figure 5.25a and b, excitation of the donor in the first case and of the acceptor in the second case. The charge-transfer character of the complex causes the wavelength of its fluorescence to be highly solvent dependent, and the value of $\mu \approx 10$ D has been derived for the dipole moment (Beens et al., 1967). Electron transfer then gives a radical ion pair ($D^{\oplus\cdot}A^{\ominus\cdot}$), which for $k_q > k_p$ decays to the reactants by back electron transfer; this corresponds overall to an electron-transfer quenching. (Cf. Section 5.4.4.) However, if fast secondary reactions lead to product formation, $k_p \gg k_q$, the process becomes a photoinduced electron-transfer reaction, and this can be utilized synthetically.

Example 7.19:

As an example, photochemical excitation of donor–acceptor complexes may be considered. Irradiation into the CT band of the anthracene–tetracyanoethylene complex leads directly to the radical ion pair, the components of which are identifiable from their UV-visible spectra. The transient absorptions decay in ~60 ps after excitation, as the radical ion pairs undergo rapid back electron transfer to afford the original donor–acceptor complex (Hilinski et al., 1984). With tetranitromethane as acceptor, however, an addition product is obtained in both high quantum and chemical yield. This is due to the fact that the tetranitromethane radical anion undergoes spontaneous fragmentation to a NO_2 radical and a trinitromethyl anion, which is not able to reduce the anthracene radical cation (Masnovi et al., 1985):

$$D + A \rightleftharpoons (DA) \xrightarrow{h\nu_{CT}} (D^{\oplus\cdot} A^{\ominus\cdot}) \xrightarrow{(A = TCNE)} D + A$$

$$\downarrow (A = C(NO_2)_4)$$

$$[D^{\oplus\cdot}, C(NO_2)_3^{\ominus}, \cdot NO_2]$$

$$\downarrow$$

D = (anthracene)

(9-H, 9-C(NO$_2$)$_3$ anthracenyl radical) ⟶ (9-H, 9-C(NO$_2$)$_3$, 10-O$_2$N, 10-H dihydroanthracene)

A detailed study showed that after dissociation of the radical anion a contact ion pair [$D^{\oplus}$ $C(NO_2)_3^{\ominus}$] in a solvent cage is initially formed. It transforms into a solvent-separated ion pair within a few ps, which in turn converts into the free ions within a few ns. Both processes follow first-order kinetics. The free ions then form the adduct in a second-order process (Masnovi et al., 1985).

The photoreduction of carbonyl compounds or aromatic hydrocarbons by amines was one of the early electron-transfer reactions to be studied. Observation of products from primary electron transfer depends on the facility of α deprotonation of the amine, which must be fast compared to back electron transfer. For amines without α hydrogens, quenching by back electron transfer is observed exclusively (Cohen et al., 1973). The solvent plays a quite important role since it determines the yield of radical ion pairs formed from the exciplex (Hirata and Mataga, 1984).

As a typical example the photoreduction of naphthalene by triethylamine (Barltrop, 1973) is shown in Scheme 52. The radicals generated by α deprotonation couple to the products **148** and **149**, and disproportionate to the reduction product **150**.

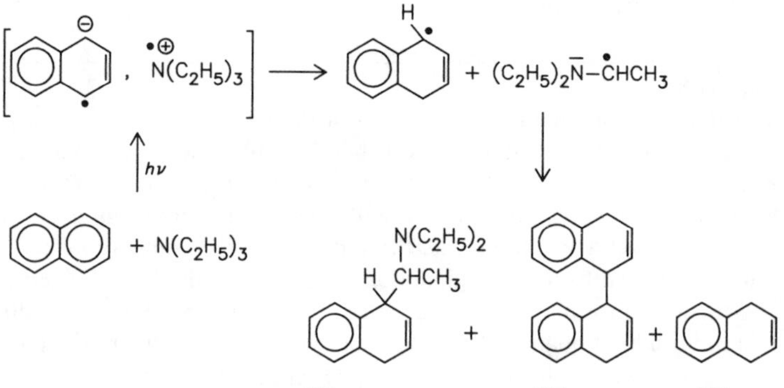

$N(C_2H_5)_3$ $h\nu$ $(C_2H_5)_2\bar{N}-\dot{C}HCH_3$ $N(C_2H_5)_2$ H $CHCH_3$

148 **149** **150**

Scheme 52

With singlet excited *trans*-stilbene (**151**) and tertiary alkyl amines only products characteristic for radical coupling are observed (Lewis et al., 1982).

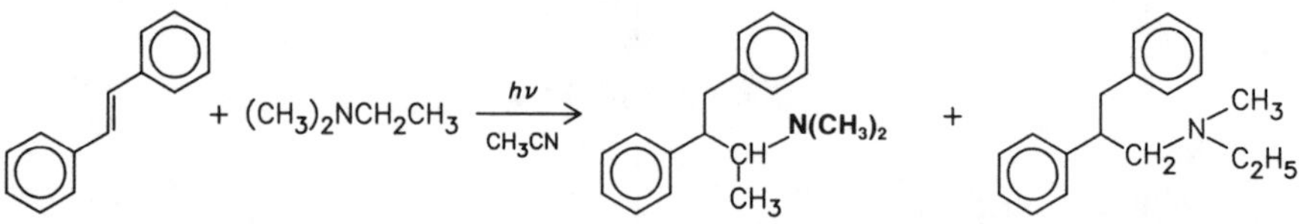

151 + $(CH_3)_2NCH_2CH_3$ $\xrightarrow[CH_3CN]{h\nu}$ $CH-N(CH_3)_2$, CH_3 (16%) + $CH_2-N(CH_3)(C_2H_5)$ (84%)

With unsymmetrical trialkylamines selective formation of the least-substituted α-amino radical is observed. The stereoselectivity is thought to be stereoelectronic in origin, as can be most easily seen in highly substituted amines such as diisopropylmethylamine, where α deprotonation occurs more readily from the conformationally favored transition state **152** rather than from **153** (Lewis et al., 1981):

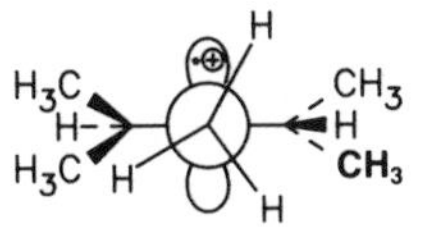

152 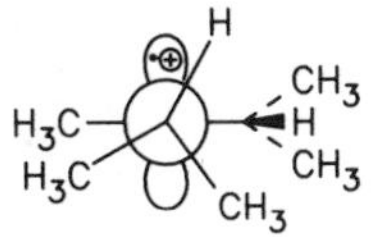 **153**

While electron-transfer reactions of aromatic hydrocarbons are in general reactions of the (π,π^*) state, electron transfer in carbonyl compounds can be into the (n,π^*) as well as into the (π,π^*) state. Correlation diagrams may be constructed for these reactions and are fully equivalent to the corresponding diagrams for hydrogen abstraction. (Cf. Section 7.3.1.) From such diagrams it is, for instance, evident that coplanar electron transfer (attack of the amine within the plane of the carbonyl group) is symmetry allowed in the (n,π^*) state but symmetry forbidden in the (π,π^*) state.

Photoreduction of benzophenones and acetophenones by amines has been studied in detail. A mechanism has been derived for the reaction of benzophenone with *N,N*-dimethylaniline that involves a triplet exciplex in the formation of the radical ion pair, as indicated in Scheme 49.

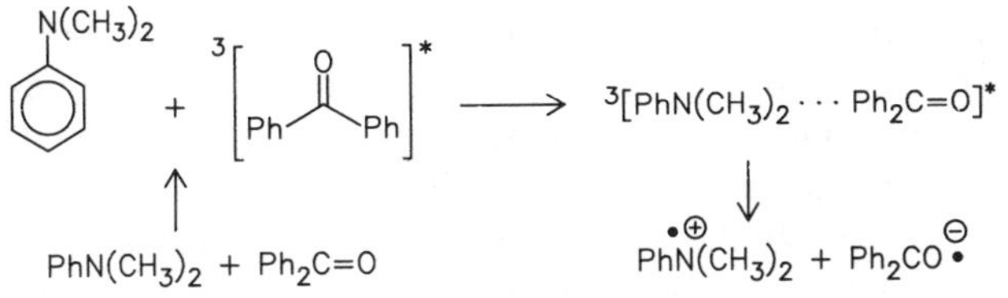

Scheme 53

Subsequent proton transfer yields the ketyl radical and the usual products. (Cf. Section 7.3.1.) This mechanism was confirmed by CIDNP measurements. At the same time it was proven that on excitation of the amine, electron transfer occurs from the singlet excited amine to the ground-state ketone (Hendricks et al., 1979). The exact time scale for those processes that follow the electron transfer was studied by picosecond spectroscopy (Peters et al., 1982).

Photochemical addition reactions may also occur as electron-transfer reactions involving a radical ion pair. An illustrative example is the photochemical reaction of 9-cyanophenanthrene (**154**) with 2,3-dimethyl-2-butene, which, in nonpolar solvents, gives good yields of a [2 + 2] cycloadduct via a singlet exciplex, while in polar solvents radical ions are formed in the primary photochemical process. The olefin radical cation then undergoes deprotonation to yield an allyl radical or suffers nucleophilic attack by the solvent to produce a methoxy alkyl radical. Coupling of these radicals with the aromatic radical anion produces acyclic adducts such as **156** in addition to the cycloaddition and reduction products (Lewis and Devoe, 1982).

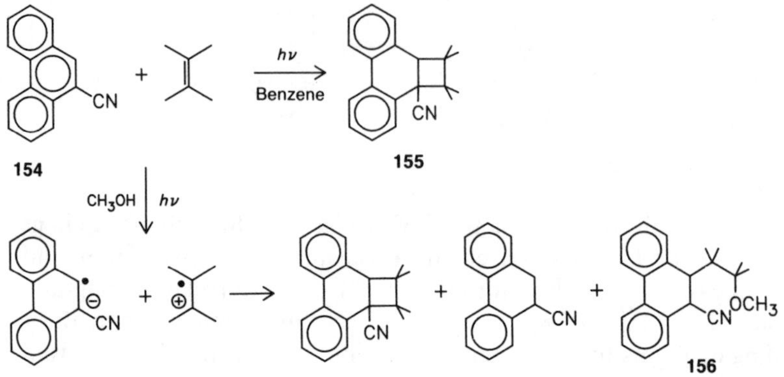

154 + → (hν, Benzene) 155

CH_3OH hν

→ + + 156

Formation of cycloadducts can be completely quenched by conducting the experiment in a nucleophilic solvent. This intercepts radical cations so rapidly that they cannot react with the olefins to yield adducts. In Scheme 54 the regiochemistry of solvent addition to 1-phenylcyclohexene is seen to depend on the oxidizability or reducibility of the electron-transfer sensitizer. With 1-cyanonaphthalene the radical cation of the olefin is generated, and nucleophilic capture then occurs at position 2 to afford the more stable radical. Electron transfer from excited 1,4-dimethoxynaphthalene, however, generates a radical anion. Its protonation in position 2 gives a radical that is oxidized by back electron transfer to the sensitizer radical before being attacked by the nucleophilic solvent in position 1. Thus, by judicious choice of the electron-transfer sensitizer, it is possible to direct the photochemical addition in either a Markovnikov (**157**) or anti-Markovnikov (**158**) fashion (Maroulis and Arnold, 1979).

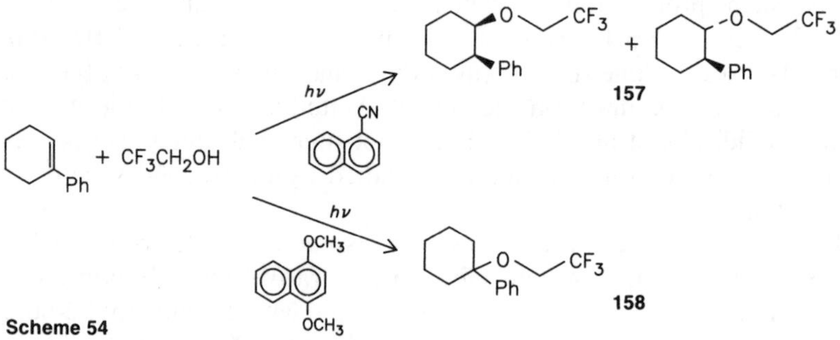

+ CF_3CH_2OH (hν) → 157; (hν) → 158

Scheme 54

Electron transfer can also induce valence isomerizations such as the transformation of hexamethyldewarbenzene (**159**) to hexamethylbenzene. The quantum yield of this reaction is larger than unity; a chain reaction mechanism is therefore assumed (Jones and Chiang, 1981).

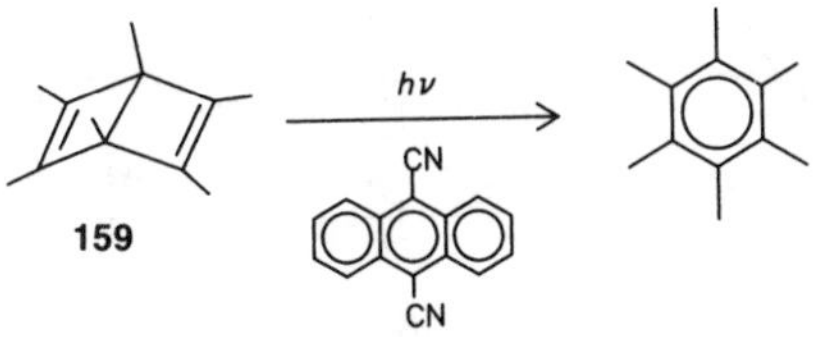

159

Spectroscopic studies with photo-CIDNP techniques revealed the existence of two distinct radical cations generated from hexamethyldewarbenzene, presumably rapidly interconverting. In one of these, the central carbon—carbon bond is significantly stretched and bears the unpaired spin density. In the second, the spin density is confined to one of the olefinic bonds. This example is the first to show conclusively that two different radical ion structures can correspond to a single minimum on the ground-state surface of the neutral (Roth et al., 1984).

Quadricyclanes (**160**) also undergo a valence isomerization to norbornadienes if irradiated in the presence of electron acceptors such as fumaronitrile (Jones and Becker, 1982). Two distinct radical cation structures are observed for the hydrocarbon, corresponding roughly to the bonding patterns of norbornadiene and quadricylane, respectively (Roth et al., 1981).

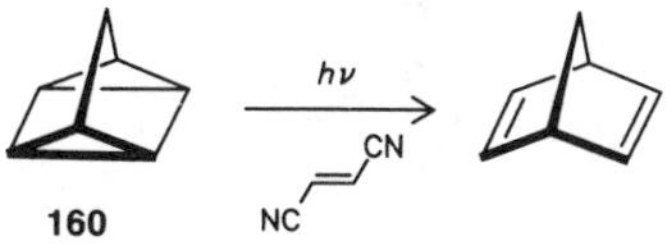

160

Quadricyclane in CsI or KBr matrices, prepared by deposition in the salt under conditions that yield single-molecule isolation, is rapidly converted into norbornadiene under conditions that induce color center formation in the alkali halide: rapid-growth vapor deposition, or UV or X-ray irradiation. The reaction proceeds only at temperatures at which color centers of the "missing electron" type (H center) are mobile. At lower temperatures ($T <$ 90 K), UV irradiation of norbornadiene converts it into quadricyclane in the usual fashion (Kirkor et al., 1990).

The electron-transfer sensitized interconversion of *trans*- (**161**) and *cis*-1,2-diphenylcyclopropane (**162**) (Wong and Arnold, 1979) is thought to proceed via the radical ion pair from which back electron transfer generates a triplet biradical that undergoes geometric isomerization; the corresponding ring-opened radical cations **163** are conformationally stable (Roth and Schilling, 1981).

Cycloadditions of carbonyl compounds to olefins generally involve exciplexes and biradicals. (Cf. Section 7.4.4.) While normal olefins frequently yield a number of products, photoinduced electron transfer may be utilized in the case of electron-rich olefins to influence the regioselectivity. Thus, irradiation of the ketene acetal **165** and biacetyl (**164**) yields exclusively the oxetane **167**. Since the radical cation **166** could be trapped, electron transfer

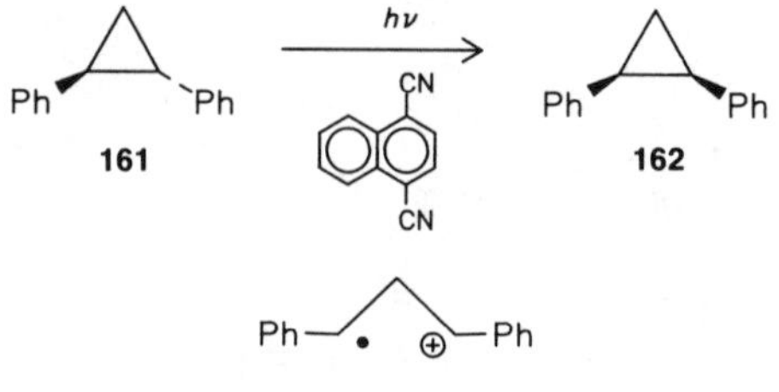

161 162 163

is assumed to be the photochemical primary step. As the regioisomeric oxetane **168** is generated by a thermal process involving a dipolar intermediate, this reaction constitutes a case of reversal ("Umpolung") of the carbonyl reactivity by photoinduced electron transfer (Mattay et al., 1984a).

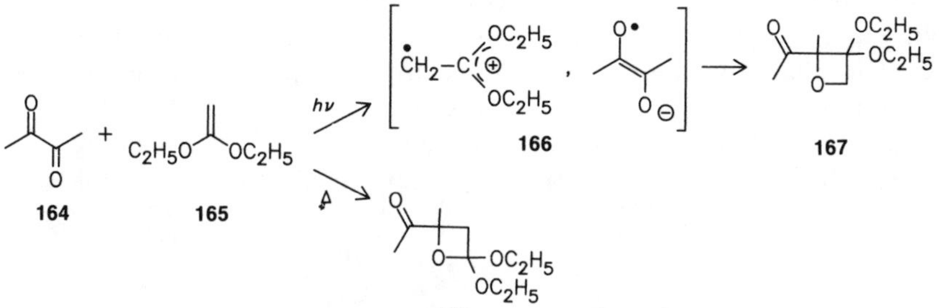

164 165 166 167 168

In the photoinduced singlet dimerization of indene shown in Scheme 55, the endo head-to-head dimer expected from the more favorable excimer geometry is the preferred product. With electron-transfer sensitization, the less sterically hindered exo head-to-head dimer is formed (Farid and Shealer, 1973).

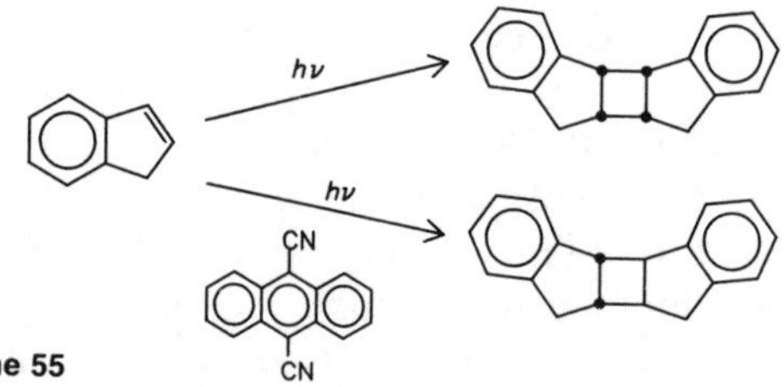

Scheme 55

Besides cyclobutane formation, alternative ring closures are sometimes observed. One example is the 9,10-dicyanoanthracene sensitized dimerization of 1,1-diphenylethylene (**169**). The six-membered ring is formed via the 1,4-radical cation, which results from the addition of the free radical cation to diphenylethylene as indicated in Scheme 56, while the 1,4-biradical generated by back electron transfer from the radical ion pair yields tetraphenylcyclobutane (**170**) (Mattes and Farid, 1983).

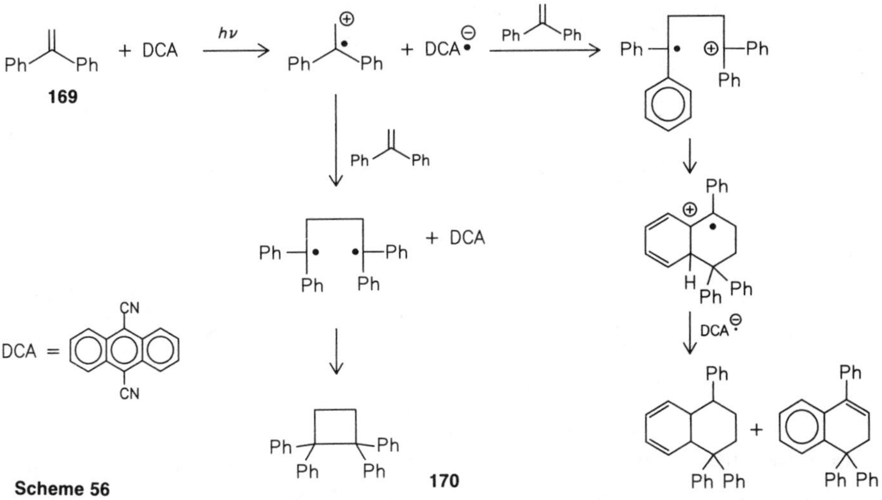

Scheme 56

The [4 + 2] cycloaddition of electron-rich dienes to electron-rich dienophiles in nonpolar solvents can be catalyzed by electron-poor arene sensitizers (Calhoun and Schuster, 1984). The proposed mechanism involves a triplex (ternary complex) formed by the reaction of the diene with an exciplex composed of the sensitizer and the dienophile. Using a chiral sensitizer, (−)-1,1′-bis(2,4-dicyanonaphthyl), an enantioselective cycloaddition was observed as shown in Scheme 57. In the case of 1,3-cyclohexadiene and trans-β-methylstyrene, the enantiomeric excess was 15 ± 3% (Kim and Schuster, 1990). (Cf. the enantioselective cis-trans isomerization via a triplex discussed in Section 7.1.2.)

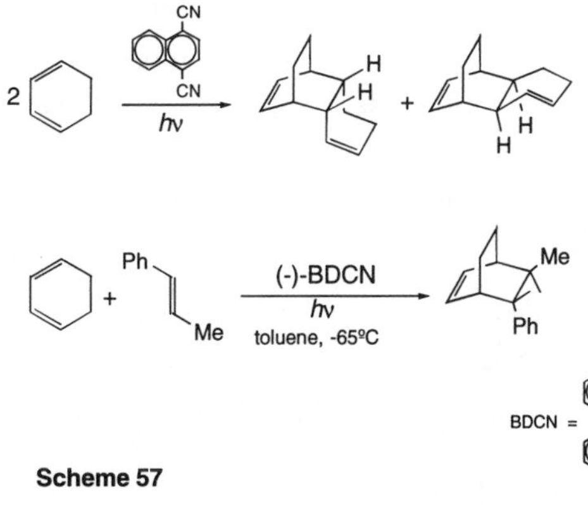

Scheme 57

Products resulting from photolysis of alkyl iodides indicate that according to Scheme 58 homolysis of the C—X bond is followed by electron transfer, which results in an ion pair in a solvent cage (Kropp, 1984). In the case of norbornyl iodide (**171**) and other bridgehead iodides, bridgehead cations result; these are difficult to obtain from solvolytic reactions. In other cases Wagner–Meerwein rearrangements of the ionic intermediates have been observed.

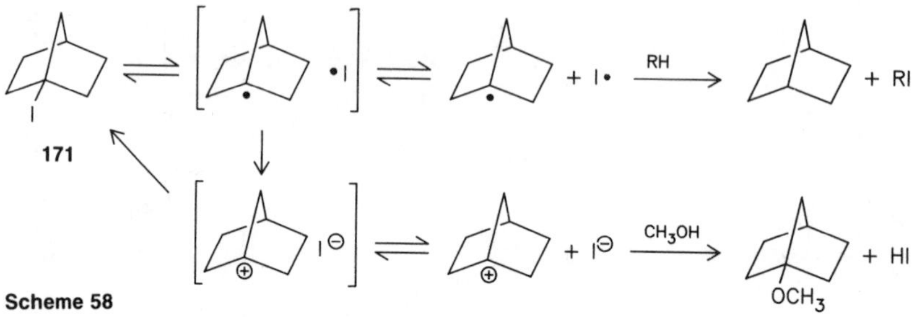

171

Scheme 58

A particularly important photoinduced electron-transfer process occurs in photosynthesis in green plants. The overall process amounts to the splitting of water by sunlight into oxygen and metabolically bound hydrogen, and this forms the basis for the existence of higher organized living systems on earth (Kirmaier and Holten, 1987; Feher et al., 1989; Boxer, 1990).

In contrast to simpler photosynthetic bacteria that have only one photosystem and are not able to oxidize water, two quanta of light are used by plants to split water by means of two photosystems (PS I and PS II). This proceeds via a sequence of redox processes, indicated schematically in Figure 7.51.

Photosynthesis may be formally described as charge separation induced by electron transfer with the positive charge (electron defect) being used for

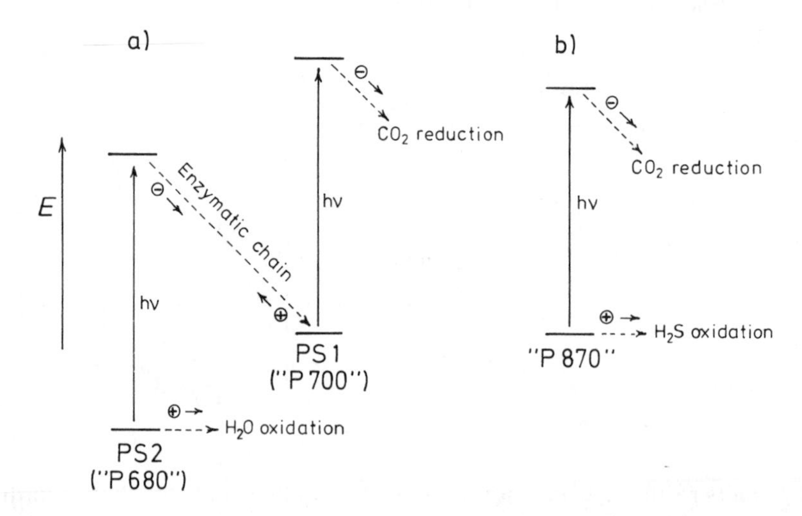

Figure 7.51. Schematic representation of charge separation in the photosynthetic cycle a) in green plants involving photosystems PS II and PS I and b) in photosynthetic active bacteria (by permission from Rettig, 1986).

oxidation of H_2O (or H_2S) and the negative charge for reduction of CO_2, according to the overall equation

$$n\mathrm{CO}_2 + n\mathrm{H}_2\mathrm{O} \xrightarrow{h\nu} (\mathrm{CH}_2\mathrm{O})_n + n\mathrm{O}_2$$

However, the saccharides $(CH_2O)_n$ are not produced by the photoreaction but by a subsequent dark reaction of the photochemically generated hydrogenated nicotinamide adenine dinucleotide phosphate (**172**) ($NADP^{\oplus}\cdot H_2$).

The energy absorbed by pigment–protein complexes in the light-gathering antennae, which also contain carotenoids as triplet quenchers, is transferred to the photochemically active reaction center and produces the excited singlet state of pigment P680 absorbing at 680 nm, which constitutes a special dimer complex of chlorophyll a (**173**) referred to as the *special pair.*

The oxidation potential of excited P680 is sufficient to remove one electron from water; an electron is transferred from the excited special pair, via a primary electron acceptor, possibly a pheophytin a molecule (an Mg-free chlorophyll a), to a plastoquinone (**174**). From Figure 7.52 it is seen that after these two electron-transfer processes the ground state of P680 is regenerated; P680 thus acts as a photocatalyst for the transfer of an electron from water to plastoquinone.

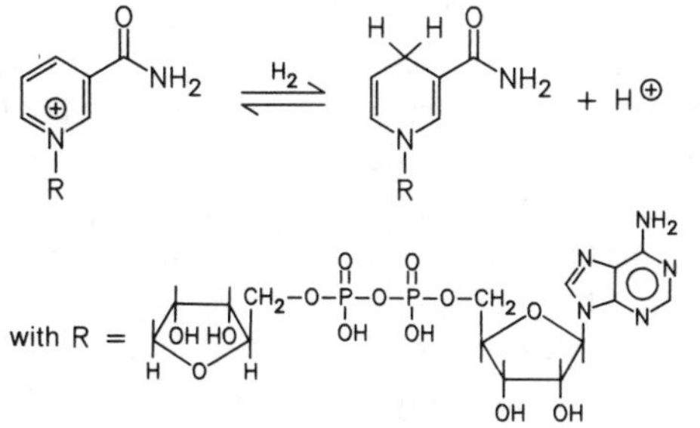

with R =

172

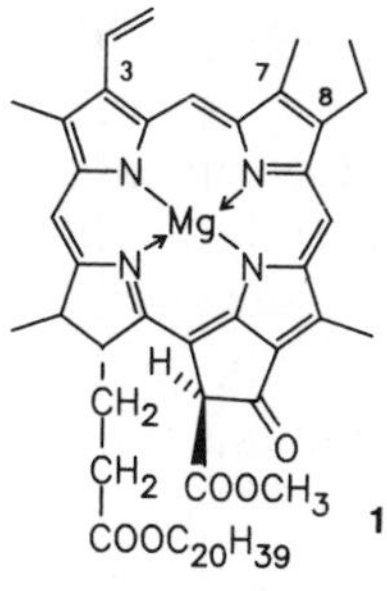

173

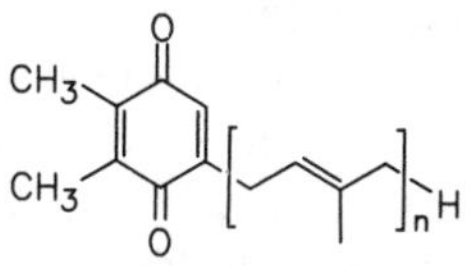

174

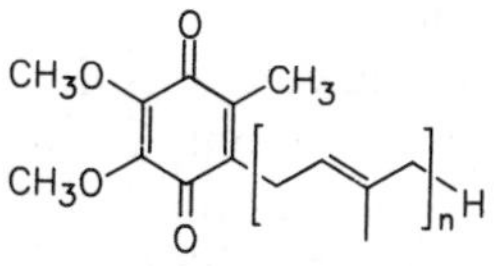

175

(In bacteriochlorophyll C3 carries a $COCH_3$ group and the C7-C8 double bond is hydrogenated)

Plastoquinone in turn is a reductant for excited P700 of photosystem PS I, which operates similarly to the system PS II and has a reduction potential sufficient for an electron transfer to the iron–sulfur complex of ferredoxin and finally to $NADP^{\oplus}$, producing $NADP^{\oplus}\cdot H_2$.

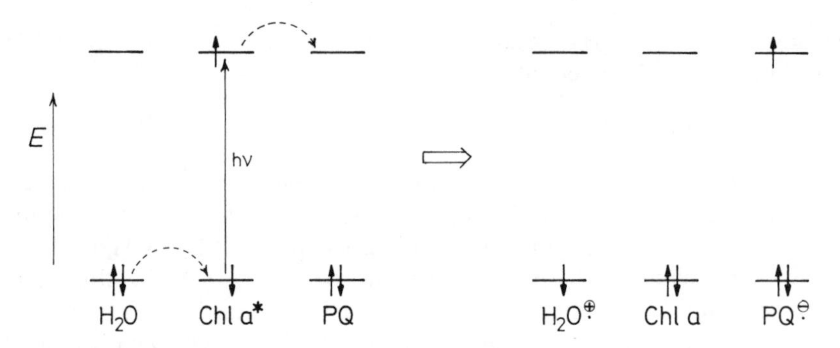

Figure 7.52. Redox process of photosystem PS II during electron transfer from water to plastoquinone.

The protein matrix essentially determines the properties and the π-electron redox chemistry of the reaction center and has to be considered an apoenzyme of this functional unit. The geometrical arrangement of the chromophores in the reaction center of the purple bacterium *Rhodopseudomonas viridis* has been elucidated using X-ray crystallographic analysis (Deisenhofer et al., 1985). On the basis of these structures, several mechanisms have been proposed to explain the primary electron-transfer event (Bixon et al., 1988; Bixon and Jortner, 1989; Marcus, 1988). The special pair in this system consists of a bacteriochlorophyll dimer that transfers an electron from the excited singlet state via a bacteriopheophytin to an ubiquinone (**175**). The high symmetry of this arrangement and the small overlap of the chromophores in the reaction center are apparently essential for the high efficiency of the system, as has been concluded from a comparison with TICT states, or "sudden polarization" states (cf. Section 4.3.3) (Rettig, 1986). In solution, electron transfer from a metal porphyrin to benzoquinone is, in fact, efficient only in the triplet state, and in the singlet state back electron transfer predominates (Huppert et al., 1976). INDO/S calculations demonstrate that the observed charge separation can be reproduced only by including the effects of the protein; utilizing the self-consistent reaction field solvent model dramatically lowers the energy of all the charge-transfer states (Thompson and Zerner, 1991).

7.6.2 Photosubstitutions

Many aromatic compounds undergo heterolytic photosubstitution on irradiation. Nucleophilic aromatic substitutions are particularly frequent. The orientation rules are reversed compared to those known for ground-state reactivity; that is, electron-withdrawing substituents orient incoming groups in meta positions, and strong electron-releasing substituents such as —OCH_3 behave as activating substituents and direct incoming nucleophiles to the ortho and para positions. The photoreactions of the three nitroanisoles with $^{\ominus}CN$ are collected in Scheme 59 to illustrate this behavior (Letsinger and McCain, 1969).

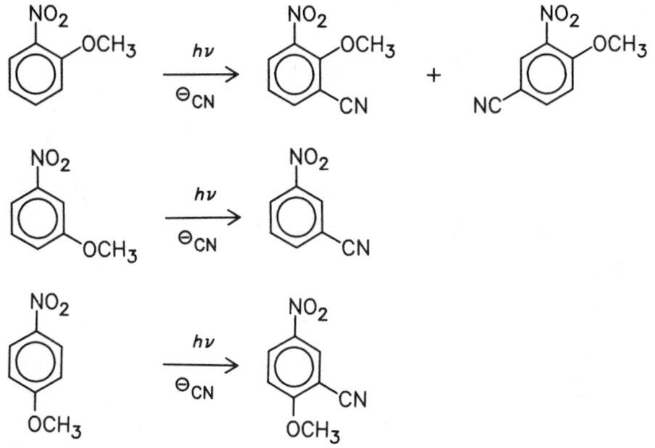

Scheme 59

Aromatic photosubstitution can proceed by various mechanisms. Electron-withdrawing substituents presumably include a $S_N2\ {}^3Ar^*$ mechanism involving a σ complex of the triplet excited aromatic moiety with the nucleophile, as indicated in Scheme 60.

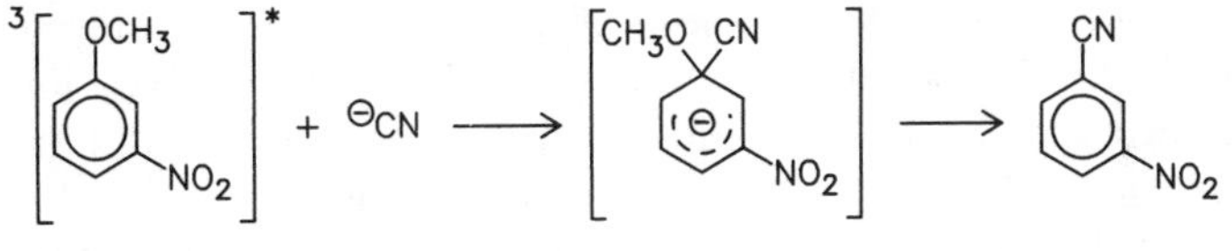

Scheme 60

This mechanism has been proven conclusively for the $CH_3O^{\ominus}$ exchange reaction of nitroanisoles and similar compounds. Attack by the nucleophile occurs at the position with the highest calculated positive charge in the triplet state. The directing effect of the nitro group can also be described by HOMO and LUMO charge densities but the activating influence is difficult to rationalize on the basis of charge densities. This has been explained by the fact that the excited triplet and the ground-state potential energy surfaces are particularly close to each other at the geometry of the σ complex for meta substitution, which is very unfavorable in the ground state (Van Riel et al., 1981).

The activating effect of methoxy groups and similar substituents has been explained by an $S_{R^{\oplus}N}1\ {}^3Ar^*$ mechanism with the formation of a radical cation through electron transfer as the principal step as shown in Scheme 61 (Havinga and Cornelisse, 1976):

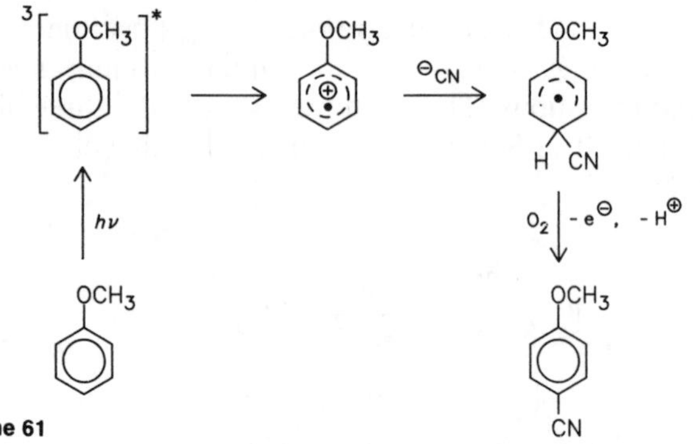

Scheme 61

For halobenzenes an $S_{R^{\ominus}N}1$ Ar* mechanism with the formation of a radical anion by electron transfer as shown in Scheme 62 has been discussed (Bunnett, 1978):

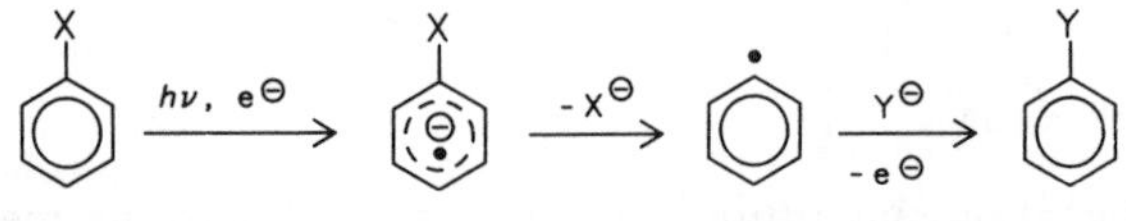

Scheme 62

Aliphatic and alicyclic molecules such as cyclohexane undergo photosubstitution with nitrosyl chloride (Pape, 1967). The reaction is of considerable industrial importance in the synthesis of ε-caprolactam, an intermediate in the manufacture of polyamides (nylon 6). (Cf. Fischer, 1978.) At long wavelengths a cage four-center transition state between alkane and an excited nitrosyl chloride molecule is involved, as indicated in Scheme 63. In contrast to light-induced halogenation, photonitrosation has a quantum yield smaller than unity, and is not a chain reaction.

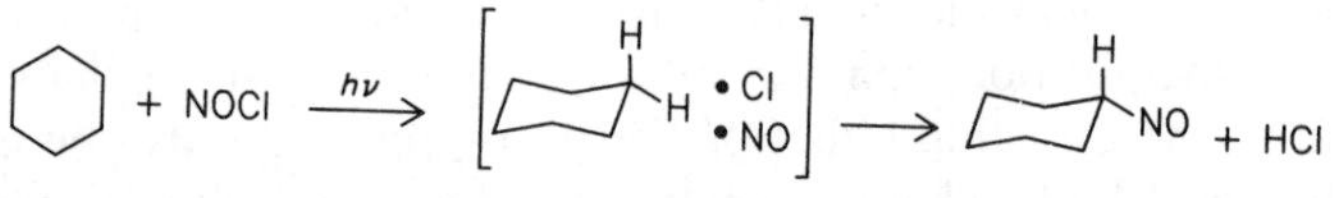

Scheme 63

7.6.3 Photooxidations with Singlet Oxygen

Under photochemical conditions oxygen can insert itself into a substrate with the formation of hydroxyhydroperoxides (**176**), hydroperoxides (**177**), peroxides (**178**), and dioxetanes (**179**). There are essentially two different reaction courses, involving either a photochemically generated radical reacting with ground-state oxygen or a ground-state substrate molecule reacting with singlet oxygen. (Cf. Example 7.20.) Some typical examples are summarized in Scheme 64.

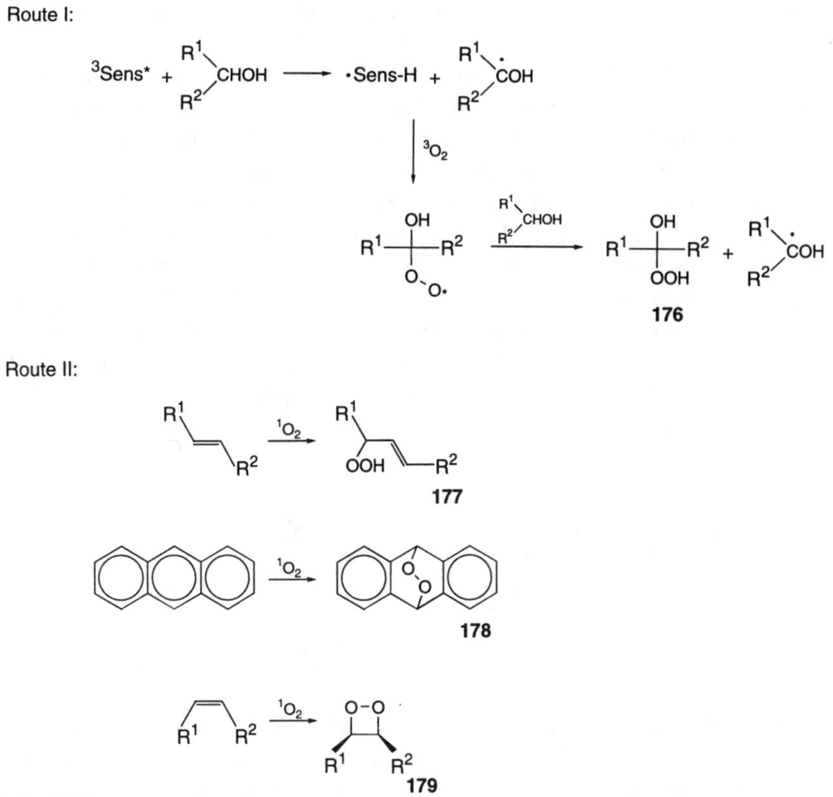

Route I:

^{3}Sens* + R^1R^2CHOH ⟶ ·Sens-H + $R^1R^2\dot{C}$OH

↓ $^3O_2$

R^1R^2C(OH)O–O· —(R^1R^2CHOH)⟶ R^1R^2C(OH)OOH + $R^1R^2\dot{C}$OH

176

Route II:

R^1CH=CHR^2 —$^1O_2$⟶ R^1CH(OOH)CH=CHR^2

177

—$^1O_2$⟶

178

R^1CH=CHR^2 —$^1O_2$⟶ O–O, R^1, R^2

179

Scheme 64

Formation of endoperoxides of type **178** can be interpreted as a $[_{\pi}4_s + _{\pi}2_s]$ cycloaddition that involves singlet oxygen and therefore constitutes a photochemical reaction. Similarly, formation of hydroperoxides from nonconjugated olefins with an allylic hydrogen generally appears to be a concerted ene reaction as indicated in Scheme 65. Other mechanisms have

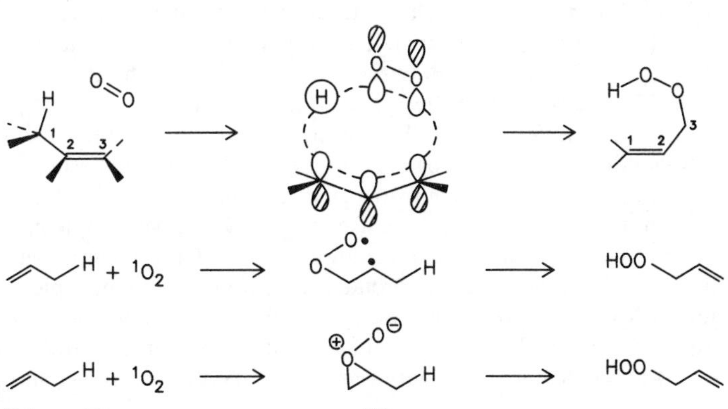

H, O=O, 1, 2, 3 ⟶ (H) ⟶ H–O–O, 1, 2, 3

CH$_2$=CHCH$_2$–H + $^1O_2$ ⟶ ·O–O·, H ⟶ HOO–CH$_2$CH=CH$_2$

CH$_2$=CHCH$_2$–H + $^1O_2$ ⟶ ⊕O–O⊖, H ⟶ HOO–CH$_2$CH=CH$_2$

Scheme 65 180

been proposed, and according to Scheme 65 involve a biradical or a perepoxide **180**.

Ab initio calculations favor the biradical mechanism, at least in the gas phase (Harding and Goddard III, 1980), while semiempirical calculations suggest the perepoxide to be a genuine intermediate (Dewar and Thiel, 1977). Reactions of singlet oxygen are characterized by low activation energies and very fast reaction times. Therefore, a detailed mechanism is in general difficult to establish. Many experimental findings suggest, however, that an interaction with charge-transfer character occurs at the initial stages of reaction, with the stereochemistry given by the HOMO of the olefin as the electron donor and the π^* LUMO of the oxygen (Stephenson et al., 1980). This is exemplified in Scheme 66 for the HOMO-LUMO interaction of 2-butene with oxygen.

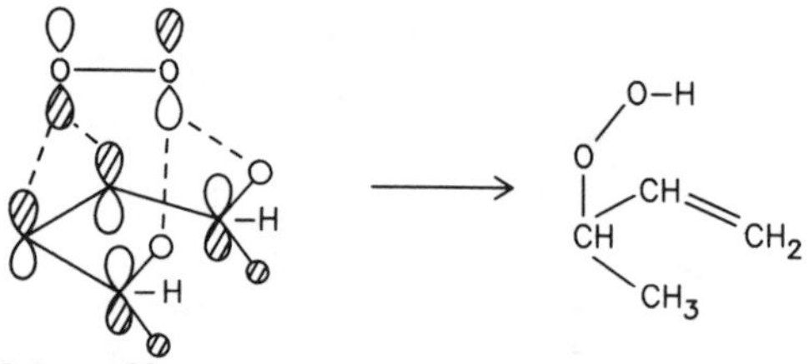

Scheme 66

Some substituted alkenes react with singlet oxygen to form a dioxetane in a $[_{\pi}2_a + _{\pi}2_s]$ cycloaddition reaction. Most dioxetanes readily decompose to carbonyl compounds in an exothermic reaction that is accompanied by a bluish luminescence. The chemiluminescence will be dealt with in more detail in Section 7.6.4.

Example 7.20:
The ground state of molecular oxygen is a ${}^3\Sigma_g^-$ state according to Hund's rule, and the MOs π_x and π_y, which are degenerate by symmetry, are singly occupied by electrons with parallel spin. The corresponding singlet state ${}^1\Sigma_g^+$ is higher in energy by 35 kcal/mol. Between these two Σ states there is a degenerate ${}^1\Delta_g$ state, 22 kcal/mol above the ${}^3\Sigma_g^-$ ground state. This ${}^1\Delta_g$ state is generally referred to as *singlet oxygen*. The description of these states becomes particularly intelligible if it is recognized that the O_2 molecule is a perfect axial biradical, especially if the symmetry-adapted complex π MOs $\pi_+ = (\pi_x + i\pi_y)/\sqrt{2}$ and $\pi_- = (\pi_x - i\pi_y)/\sqrt{2}$ are used. Figure 7.53 gives a schematic representation of the wave functions of the lowest states of molecular oxygen expressed in terms of complex and real orbitals, with MOs doubly occupied in all states not shown. In an axial biradical, all possible real combinations of the degenerate orbitals are localized or delocalized to the same degree. (Cf. Section 4.3.2.) Potential energy curves of molecular oxygen are shown in Figure 7.54.

State	Complex Orbitals	Real Orbitals
$^1\Sigma_g^+$	$[(\uparrow)_+ (\downarrow)_-] - [(\downarrow)_+ (\uparrow)_-]$	$[(\downarrow\uparrow)_x (\ )_y] + [(\ )_x (\downarrow\uparrow)_y]$
$^1\Delta_g$	$[(\downarrow\uparrow)_+ (\ )_-]$	$[(\downarrow\uparrow)_x (\ )_y] - [(\ )_x (\downarrow\uparrow)_y]$
	$[(\ )_+ (\downarrow\uparrow)_-]$	$[(\uparrow)_x (\downarrow)_y] - [(\downarrow)_x (\uparrow)_y]$
$^3\Sigma_g$	$[(\uparrow)_+ (\uparrow)_-]$	$[(\uparrow)_x (\uparrow)_y]$
	$[(\uparrow)_+ (\downarrow)_-] + [(\downarrow)_+ (\uparrow)_-]$	$[(\uparrow)_x (\downarrow)_y] + [(\downarrow)_x (\uparrow)_y]$
	$[(\downarrow)_+ (\downarrow)_-]$	$[(\downarrow)_x (\downarrow)_y]$

Figure 7.53. Spin-orbital diagrams for the lowest molecular oxygen states (by permission from Kasha and Brabham, 1979).

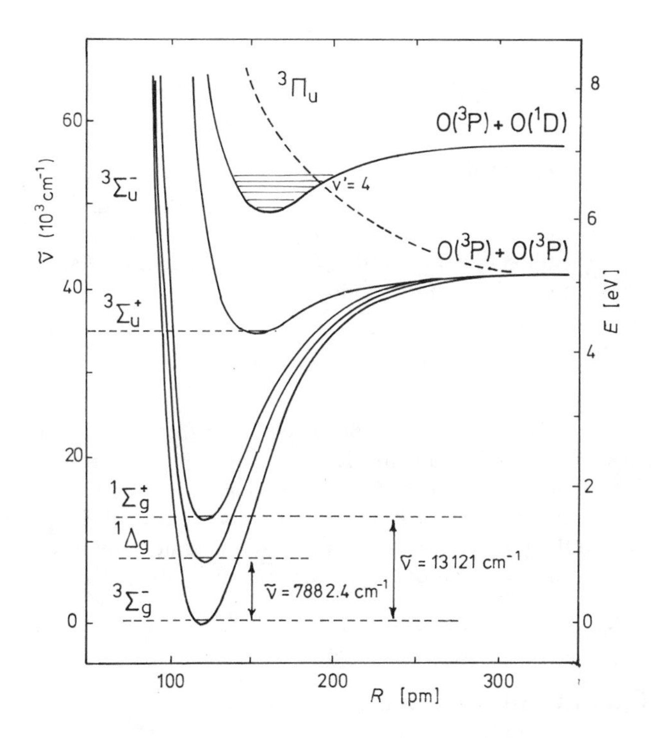

Figure 7.54. Potential energy curves for the lowest molecular oxygen states (adapted from Herzberg, 1950).

Singlet oxygen can be generated either by thermal or by photochemical methods. The most general and synthetically useful method is photosensitization with a strongly absorbing dye such as Rose Bengal or methylene blue, which can be used advantageously as a polymer-bound sensitizer (Schaap et al., 1975). Singlet oxygen is generated by triplet–triplet annihilation according to

$$\text{Sens} + h\nu \rightarrow {}^{1}\text{Sens}^{*} \rightarrow {}^{3}\text{Sens}^{*}$$
$${}^{3}\text{Sens}^{*} + {}^{3}O_2 \rightarrow \text{Sens} + {}^{1}O_2$$

The thermal generation of singlet oxygen from hydrogen peroxide and hypochlorite presumably involves the chloroperoxy anion; other synthetically useful examples involve decomposition of phosphite ozonides or endoperoxides, as indicated in Scheme 67 (cf. Murray, 1979).

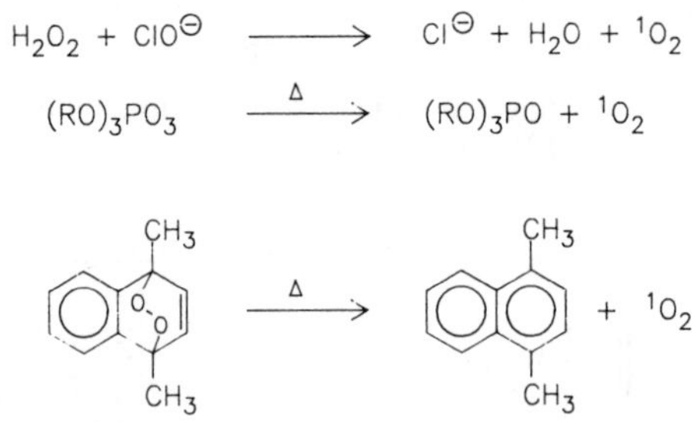

$$H_2O_2 + ClO^{\ominus} \longrightarrow Cl^{\ominus} + H_2O + {}^{1}O_2$$

$$(RO)_3PO_3 \xrightarrow{\Delta} (RO)_3PO + {}^{1}O_2$$

1,4-dimethylnaphthalene endoperoxide $\xrightarrow{\Delta}$ 1,4-dimethylnaphthalene + $^{1}O_2$

Scheme 67

Example 7.21:
The orbital correlation diagram for the [4 + 2] cycloaddition of singlet oxygen is shown in Figure 7.55 for the reaction

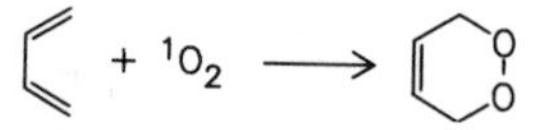

butadiene + $^{1}O_2$ $\longrightarrow$ 3,6-dihydro-1,2-dioxine

The ordering of the reactant orbitals is obtained from ionization potentials of molecular oxygen and butadiene. The reactant configuration with singly occupied MOs π_x^* and π_y^* ($^3O_2$) correlates with a highly excited product configuration, while the configuration $\pi_x^{*2} - \pi_y^{*2}$ with doubly occupied π_x^* or π_y^* MOs correlates with the product peroxide in the ground state. Due to perturbation by the reaction partner, the configuration π_x^{*2} develops as the reaction proceeds.

7.6.4 Chemiluminescence

From the schematic representation in Figure 7.56 it is seen that chemiluminescence can be described as a reverse photochemical reaction. Chemiluminescence is afforded by a transition from the ground-state potential energy surface to an isoenergetic vibrational level of an excited-state surface and

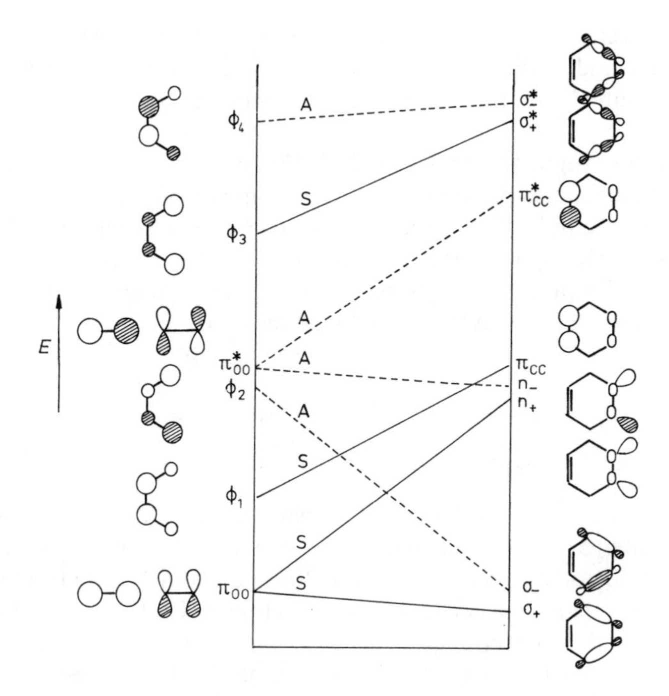

Figure 7.55. Orbital correlation diagram for endoperoxide formation.

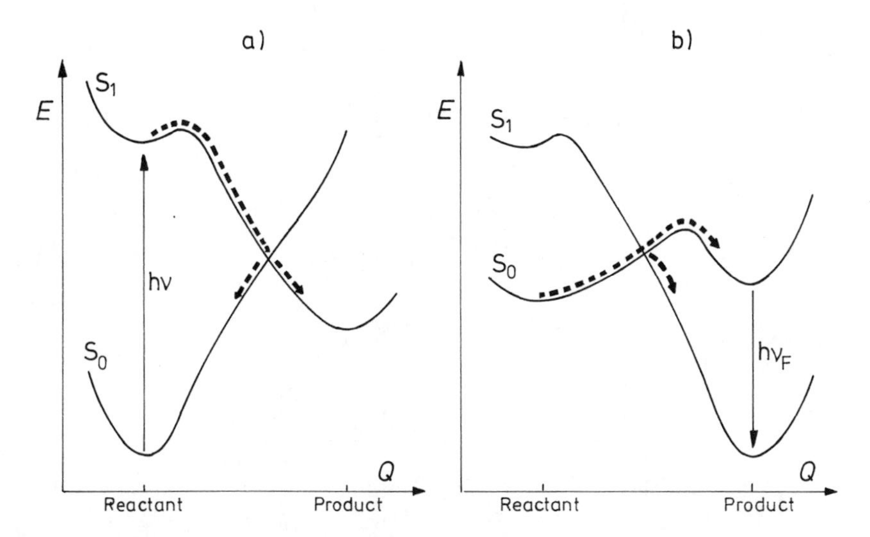

Figure 7.56. Schematic state correlation diagram a) for a singlet photoreaction and b) for a singlet chemiluminescent process.

by escape over a small barrier into a deeper well. If the separation of the ground-state and excited-state surfaces has increased in the process as shown in Figure 7.46b, the resulting excited molecule can reveal its presence by emission.

Oxidation of luminol (see Example 7.22) and thermolysis of endoperoxides or 1,2-dioxetanes provide important examples of chemiluminescent reactions. Tetramethyl-1,2-dioxetane (**181**) has been studied in great detail; the thermolysis is clearly first order and the activation enthalpy in butyl phthalate is $\Delta H^{\ddagger} \approx 27$ kcal/mol. The enthalpy difference between the reactants and ground-state products is $\Delta H_0 = -63$ kcal/mol.

$$\mathbf{181} \xrightarrow[(100\%)]{\Delta} {}^3[\text{acetone}]^* + \text{acetone}$$

181

The sum $-\Delta H_0 + \Delta H^{\ddagger}$ is greater than the excitation energies of the T_1 (n,π^*) state ($\Delta E_T = 78$ kcal/mol) as well as the S_1 (n,π^*) state ($\Delta E_S = 84$ kcal/mol) of acetone. Both states can be formed exothermally from the transition state for thermolysis (Lechtken and Höhne, 1973). It could be shown independently that triplet excited acetone ($^3A^*$) is produced directly with a quantum yield $\Phi \approx 0.5$. Singlet excited acetone ($^1A^*$) is formed with a quantum yield $\Phi = 0.005$.

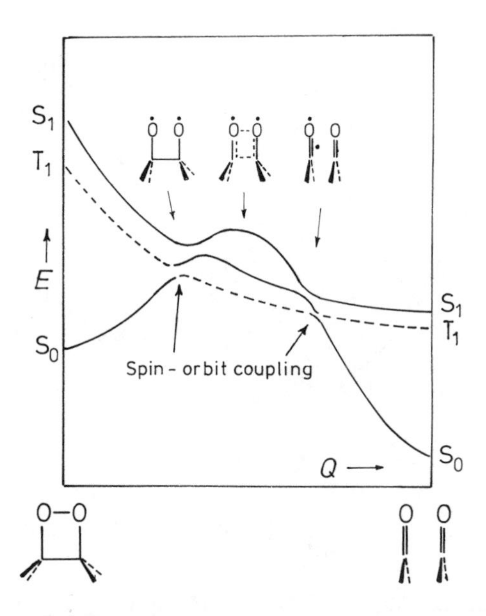

Figure 7.57. State correlation diagram for the chemiluminescence reaction of tetramethyl-1,2-dioxetane (by permission from Turro, 1978).

These results have been rationalized by means of the correlation diagram shown in Figure 7.57 (Turro and Devaquet, 1975). The ground state of tetramethyl-1,2-dioxetane correlates with a doubly $\pi \rightarrow \pi^*$ excited state of the supermolecule consisting of two acetones, while the ground state of the latter correlates with a doubly $\sigma \rightarrow \sigma^*$ excited state of the dioxetane. The (n,π^*) states of acetone will correlate directly with excited dioxetane states of appropriate symmetry, probably with the (n,σ^*) states, which are antisymmetric with respect to the reaction plane. The crossing between S_0 and T_1 will be weakly avoided due to spin-orbit coupling, and a "surface jump" to the T_1 product surface ($^3A^* + A_0$) can occur. The origin of spin-orbit coupling that makes the crossing avoided may be visualized by means of the schematic diagram given in Figure 7.57, which represent the electronic structure of the reaction complex for different stages of the reaction: To the left of the transition state, the O—O bond has lengthened with the "unpaired" electrons in orbitals of σ symmetry; at the product side, however, one of the unpaired electrons is in a π^* MO of one of the carbonyl groups. That a reaction of the type $S_0 \rightarrow T_1 \rightarrow S_0$ is not observed may be due to the fact that the first crossing can be reached many times until spin inversion finally takes place, while the second crossing is passed only once. Probabilities are therefore much lower for the $T_1 \rightarrow S_0$ transition than for the $S_0 \rightarrow T_1$ transition.

The situation is somewhat different in the case of dewarbenzene, which undergoes an electrocyclic ring opening to give triplet excited benzene. (Cf. Example 7.16.) In contrast to 1,2-dioxetanes, this reaction possesses a very low chemiluminescence efficiency. The reason is thought to be the low intersystem crossing probability, which is due to the very weak spin-orbit coupling inherent in hydrocarbon systems. Thus, although the available energy is very favorable for chemiluminescence, the rearrangement proceeds as a ground-state reaction.

Example 7.22:
It is generally agreed that the excitation-producing step in the oxidation of luminol (**182**) is decomposition with loss of nitrogen of the dianion of an azoendoperoxide produced by the action of a base and oxygen:

NH2 O, NH, NH, O — ⊖OH, O2 → NH2 O⊖, O, O, N, N, O⊖ → [NH2 O⊖, O, O, O⊖]* + N2

182

This can be viewed as an allowed $[2_s + 2_s + 2_s]$ pericyclic reaction, and it is not obvious that there should be an avoided or unavoided touching of S_0 with S_1 that could provide an easy "surface jump" to the S_1 surface along the way.

The following alternative mechanism may well provide an explanation of luminol chemiluminescence:

The first step is analogous to the easy retro-Diels-Alder reaction of 1,4-dihydrophthalazine (**183**), whose $\Delta H^{\ddagger}$ is 15 kcal/mol (Flynn and Michl, 1974). The ground state of the resulting energy-rich peroxide dianion has 18 π electrons. Stretching of the O—O bond converts it into a much more stable aminophthalate dianion product, whose ground state, however, has only 16 π electrons. It therefore does not correlate with the ground state, but with a doubly $\pi \rightarrow \sigma^*$ excited state of the peroxide. The 18 π-electron ground state of the peroxide correlates with a doubly n$\rightarrow \pi^*$ excited state of the aminophthalate dianion product. At high symmetries, the crossing of the two states will be avoided. In the first approximation, the splitting will be given by twice the exchange integral between the two orbitals that cross (cf. Section 4.2.3). One of these is a

183

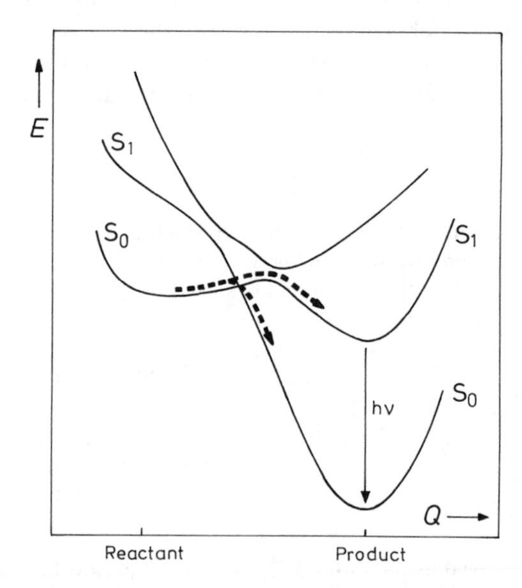

Figure 7.58. Schematic state correlation diagram for the chemiluminescence of a strongly exothermic reaction forbidden in the ground state (adapted from Michl, 1977).

π-type orbital, the other an n-type orbital, and the overlap density and the exchange integral will therefore be only very small. It is likely that an S_1–S_0 conical intersection can be reached when the symmetry is lowered, and this ought to provide a facile radiationless crossing between the two surfaces as indicated in Figure 7.58 (Michl, 1977).

Several additional chemiluminescence mechanisms have been described, which are based on excited-state generation by electron transfer in a radical ion pair according to

$$D^{\oplus\cdot} + A^{\ominus\cdot} \rightarrow D + A^*$$

where $D^{\oplus\cdot} + A^{\ominus\cdot}$ constitutes an excited state of the system D + A. The radical ion pair may be produced either electronically or chemically by electron transfer to a peroxide that subsequently rearranges and loses a neutral molecule. An example of this CIEEL (*chemically induced electron-exchange luminescence*) mechanism is provided by the thermal reaction of diphenoyl peroxide (**184**) (Koo and Schuster, 1977):

+ ACT ⟶ + ACT• ⟶ (- CO2) + ACT• ⟶ + ACT*

184

ACT = activator, for example, aromatic hydrocarbon.

Supplemental Reading

Barltrop, J.A., Coyle, J.D. (1975), *Excited States in Organic Chemistry*; Wiley: London.

Coyle, J.D., Ed. (1986), *Photochemistry in Organic Synthesis,* Special Publ. **57**; The Royal Soc. Chem.: London.

Cowan, D.O., Drisko, R.L. (1976), *Elements of Organic Photochemistry*; Plenum Press: New York.

Coxon, J.M., Halton, B. (1974), *Organic Photochemistry*; Cambridge University Press: Cambridge.

Gilbert, A., Baggott, J. (1991). *Essentials of Molecular Photochemistry*; CRC Press: Boca Raton.

Horspool, W.M., Ed. (1984), *Synthetic Organic Photochemistry*; Plenum Press: New York.

Kagan, J. (1993), *Organic Photochemistry: Principles and Applications;* Academic Press: London.

Kopecky, J. (1992), *Organic Photochemistry: A Visual Approach*; VCH: New York.

Ramamurthy, V., Turro, N.J., Eds. (1993), "Photochemistry," thematic issue of *Chem. Rev.* **93**, 1–724.

Scaiano, J.C., Ed. (1989), *Handbook of Organic Photochemistry*; CRC Press: Boca Raton, Vols. I and II.

Turro, N.J. (1978), *Modern Molecular Photochemistry*; Benjamin/Cummings Publ.: Menlo Park.

Wayne, R.P. (1988), *Principles and Applications of Photochemistry*; Oxford University Press.

Cis-trans Isomerizations

Collin, G.J. (1987), "Photochemistry of Simple Olefins: Chemistry of Electronic Excited States or Hot Ground State?", *Adv. Photochem.* **14**, 135.

Dauben, W.G., McInnis, E.L., Michno, D.M. (1980), "Photochemical Rearrangement in Trienes," in *Rearrangements in Ground and Excited States*, **3**; de Mayo, P., Ed.; Academic Press: New York.

Görner, H., Kuhn, H.J. (1995), "Cis-trans Photoisomerization of Stilbene and Stilbene-Like Molecules," *Adv. Photochem.* (Neckers, D.C., Volman, D.H., von Bünau, G., Eds.) **19**, 1.

Jacobs, H.J.C., Havinga, E. (1979), "Photochemistry of Vitamin D and Its Isomers and of Simple Trienes," *Adv. Photochem.* **11**, 305.

Saltiel, J., Sun, Y.-P. (1990), "Cis-trans Isomerization of C═C Double Bonds," in *Photochromism, Molecules and Systems*; Dürr, H., Bouas-Laurent, H., Eds.; Elsevier: Amsterdam.

Saltiel, J., Sun, Y.-P. (1989), "Application of the Kramers Equation to Stilbene Photoisomerization in *n*-Alkanes Using Translational Diffusion Coefficients to Define Microviscosity," *J. Phys. Chem.* **93**, 8310.

Azoalkanes and Azomethines

Adam, W., De Lucchi, O. (1980), "The Synthesis of Unusual Organic Molecules from Azoalkanes," *Angew. Chem. Int. Ed. Engl.* **19**, 762.

Dürr, H., Ruge, B. (1976), "Triplet States from Azo Compounds," *Topics Curr. Chem.* **66**, 53.

Engel, P.S. (1980), "Mechanism of the Thermal and Photochemical Decomposition of Azoalkanes," *Chem. Rev.* **80**, 99.

Meier, H., Zeller, K.-P. (1977), "Thermal and Photochemical Elimination of Nitrogen," *Angew. Chem. Int. Ed. Engl.* **16**, 835.

Paetzold, R., Reichenbächer, M. Appenroth, K. (1981), "Die Kohlenstoff-Stickstoff-Doppelbindung: Spektren, Struktur, thermische und photochemische *E*/*Z*-Isomerisierung," *Z. Chem.* **12**, 421.

Rau, H. (1990), "Azo Compounds" in *Photochromism, Molecules and Systems*; Dürr, H., Bouas-Laurent, H., Eds.; Elsevier: Amsterdam.

Wentrup, C. (1984), *Reactive Molecules*; Wiley: New York.

Carbonyl Compounds

Cohen, S.G., Parola, A., Parsons Jr., G.H. (1973), "Photoreduction by Amines," *Chem. Rev.* **73**, 141.

Houk, K.N. (1976), "The Photochemistry and Spectroscopy of β,γ-Unsaturated Carbonyl Compounds," *Chem. Rev.* **76**, 1.

Kramer, H.E.A. (1990), "Salicylates, Triazoles, Oxazoles," in *Photochromism, Molecules and Systems*; Dürr, H., Bouas-Laurent, H., Eds.; Elsevier: Amsterdam.

Sammes, P.G. (1986), "Photoenolization," *Acc. Chem. Res.* **4**, 41.

Schaffner, K., Demuth, M. (1980). "Photochemical Rearrangements of Conjugated Cyclic Dienones," in *Rearrangements in Ground and Excited States,* **3**; de Mayo, P., Ed.; Academic Press: New York.

Schuster, D.I. (1980), "Photochemical Rearrangements of Enones," in *Rearrangements in Ground and Excited States,* **3**; de Mayo, P., Ed.; Academic Press: New York.

Turro, N.J., Dalton, J.C., Dawes, K., Farrington, G., Hautala, R., Morton, D., Niemczyk, M., Schore, N. (1972), "Molecular Photochemistry of Alkanones in Solution: α-Cleavage, Hydrogen Abstraction, Cycloaddition, and Sensitization Reactions," *Acc. Chem. Res.* **5**, 92.

Wagner, P.J. (1980), "Photorearrangements via Biradicals of Simple Carbonyl Compounds," in *Rearrangements in Ground and Excited States,* **3**; de Mayo, P., Ed.; Academic Press: New York.

Wagner, P.J., Park, B.-S. (1991), "Photoinduced Hydrogen Atom Abstraction by Carbonyl Compounds," *Org. Photochem.* (Padwa, A., Ed.) **11**, 111.

Photocycloadditions

Arnold, D.R. (1968), "The Photocyloaddition of Carbonyl Compounds to Unsaturated Systems: The Syntheses of Oxetanes," *Adv. Photochem.* **6**, 301.

Becker, H.-D. (1990), "Excited State Reactivity and Molecular Topology Relationships in Chromophorically Substituted Antracenes," *Adv. Photochem.* **15**, 139.

Bernardi, F., Olivucci, M., Robb, M.A. (1990), "Predicting Forbidden and Allowed Cycloaddition Reactions: Potential Surface Topology and Its Rationalization," *Acc. Chem. Res.* **23**, 405.

Bouas-Laurent, H., Desvergne, J.-P. (1990), "Cycloaddition Reactions Involving 4n Electrons: [4 + 4] Cycloaddition Reactions Between Unsaturated Conjugated Systems," in *Photochromism, Molecules and Systems*; Dürr, H., Bouas-Laurent, H., Eds.; Elsevier: Amsterdam.

Caldwell, R.A., Creed, D. (1980), "Exciplex Intermediates in [2 + 2] Photocyloadditions," *Acc. Chem. Res.* **13**, 45.

Cornelisse, J. (1993), "The Meta Photocycloaddition of Arenes to Alkenes," *Chem. Rev.* **93**, 615.

Crimmins, M.T. (1988), "Synthetic Applications of Intramolecular Enone-Olefin Photocycloadditions," *Chem. Rev.* **88**, 1453.

Desvergne, J.-P., Bouas-Laurent, H. (1990), "Cycloaddition Reactions Involving 4n Electrons: [2 + 2] Cycloadditions; Molecules with Multiple Bonds Incorporated in or Linked to Aromatic Systems," in *Photochromism, Molecules and Systems*; Dürr, H., Bouas-Laurent, H., Eds.; Elsevier: Amsterdam.

Jones, II, G. (1990), "Cycloaddition Reactions Involving 4n Electrons: [2 + 2] Cycloadditions; Photochemical Energy Storage Systems Based on Reversible Valence Photoisomerization" in *Photochromism, Molecules and Systems*; Dürr, H., Bouas-Laurent, H., Eds.; Elsevier: Amsterdam.

Laarhoven. W.H. (1987), "Photocyclizations and Intramolecular Cycloadditions of Conjugated Olefins," *Org. Photochem.* (Padwa, A., Ed.) 9, 129.

McCoullough, J.J. (1987), "Photoadditions of Aromatic Compounds," *Chem. Rev.* **87**, 811.

Schuster, D.I. (1993), "New Mechanism for Old Reactions: Enone Photocycloaddition," *Chem. Rev.* **93**, 3.

Wagner-Jauregg, T. (1980), "Thermische und photochemische Additionen von Dienophilen an Arene und deren Vinyloge und Hetero-Analoge," *Synthesis* **165**, 769.

Wender, P.A. (1986), "Alkenes: Cycloaddition," in *Photochemistry in Organic Synthesis,* Special Publ. **57**; Coyle, J.D., Ed.; The Royal Soc. Chem.: London.

Wender, P.A., Siggel, L., Nuss, J.M. (1989), "Arene-Alkene Photocycloaddition Reactions," *Org. Photochem.* (Padwa, A., Ed.) **10**, 356.

Rearrangement Reactions

Arai, T., Tokumaru, K. (1993), "Photochemical One-way Isomerization of Aromatic Olefins," *Chem. Rev.* **93**, 23.

Bryce-Smith, D., Gilbert, A. (1980), "Rearrangements of the Benzene Ring," in *Rearrangements in Ground and Excited States,* **3**; de Mayo, P., Ed.; Academic Press: New York.

Demuth, M. (1991), "Synthetic Aspects of the Oxadi-π-Methane Rearrangement," *Org. Photochem.* (Padwa, A., Ed.) **11**, 37.

Leigh, W.J. (1993), "Techniques and Applications of Far UV Photochemistry. The Photochemistry of the C_3H_4 and C_4H_6 Hydrocarbons," *Chem. Rev.* **93**, 487.

Zimmerman, H.E. (1991), "The Di-π-Methane Rearrangement," *Org. Photochem.* (Padwa, A., Ed.) **11**, 1.

Electron-Transfer Reactions

Davidson, R.S. (1983), "The Chemistry of Excited Complexes: a Survey of Reactions," *Adv. Phys. Org. Chem.* **19**, 1.

Eberson, L. (1987), *Electron Transfer Reactions in Organic Chemistry*; Springer: Berlin.

Fox, M.A., Ed. (1992), "Electron-Transfer Reactions," *Chem. Rev.* **92**, 365.

Fox, M.A. (1986), "Photoinduced Electron Transfer in Organic Systems: Control of the Back Electron Transfer," *Adv. Photochem.* **13**, 237.

Julliard, M., Chanon, M. (1983), "Photoelectron-Transfer Catalysis: Its Connections with Thermal and Electrochemical Analogues," *Chem. Rev.* **83**, 425.

Kavarnos, G.J., Turro, N.J. (1986), "Photosensitation by Reversible Electron Transfer: Theories, Experimental Evidence, and Examples," *Chem. Rev.* **86**, 401.

Mariano, P.S., Stavinoka, J.L. (1984), "Synthetic Aspects of Photochemical Electron Transfer Reactions," in *Synthetic Organic Photochemistry*; Horspool, W.M., Ed.; Plenum Press: New York.

Mattay, J. (1987), "Charge Transfer and Radical Ions in Photochemistry,"*Angew. Chem. Int. Ed. Engl.* **26**, 825.

Yoon, U.C., Mariano, P.S. (1992), "Mechanistic and Synthetic Aspects of Amine-Enone Single Electron Transfer Photochemistry," *Acc. Chem. Res.* **25**, 233.

Photosubstitution

Cornelisse, J., Havinga, E. (1975), "Photosubstitution Reactions of Aromatic Compounds," *Chem. Rev.* **75**, 353.

Davidson, R.S., Gooddin, J.W., Kemp, G. (1984), "The Photochemistry of Aryl Halides and Related Compounds," *Adv. Phys. Org. Chem.* **20**, 191.

Havinga, E., Cornelisse, J. (1976), "Aromatic Photosubstitution Reactions," *Pure Appl. Chem.* **47**, 1.

Kropp, P.J. (1984), "Photobehavior of Alkyl Halides in Solution: Radical, Carbocation, and Carbene Intermediates," *Acc. Chem. Res.* **17**, 131.

Singlet Oxygen

Foote, C.S. (1968), "Photosensitized Oxygenations and the Role of Singlet Oxygen," *Acc. Chem. Res.* **1**, 104.

Lissi, E., Encinas, M.V., Rubio, M.A. (1993), "Singlet Oxygen Bimolecular Photoprocesses," *Chem. Rev.* **93**, 698.

Stephenson, L.M., Grdina, M.J., Orfanopoulos, M. (1980), "Mechanism of the Ene Reaction between Singlet Oxygen and Olefins," *Acc. Chem. Res.* **13**, 419.

Wasserman, H.H., Murray, R.W. (1979), *Singlet Oxygen*; Academic Press: New York.

Chemiluminescence

Gundermann, K.-D., McCapra, F. (1987), *Chemiluminescence in Organic Chemistry*; Springer: Berlin.

Schuster, G.B., Schmidt, S.P. (1982), "Chemiluminescence of Organic Compounds," *Adv. Phys. Org. Chem.* **18**, 187.

Turro, N.J., Ramamurthy, V. (1980), "Chemical Generation of Excited States"; in *Rearrangements in Ground and Excited States*, **3**; de Mayo, P., Ed.; Academic Press: New York.

Various

Balzani, V., Scandola, F. (1991), *Supramolecular Photochemistry*; Ellis Horwood: New York.

Bünau, G. von, Wolff, T. (1988), "Photochemistry in Surfactant Solutions," *Adv. Photochem.* **14**, 273.

Dürr, H., Bouas-Laurent, H., Eds. (1990), *Photochromism, Molecules and Systems*; Elsevier: Amsterdam.

Guillet, J. (1985), *Polymer Photophysics and Photochemistry*; Cambridge University Press: Cambridge.

Platz, M.S., Leyva, E., Haider, K. (1991), "Selected Topics in the Matrix Photochemistry of Nitrenes, Carbenes, and Excited Triplet States," *Org. Photochem.* (Padwa, A., Ed.) **11**, 367.

Turro, N.J., Cox, G.S., Paczkowski, M.A. (1985), "Photochemistry in Micelles," *Top. Curr. Chem.* **129**, 57.

Epilogue

In the preceding seven chapters, we have gradually developed the framework necessary for a qualitative understanding of the photophysical and photochemical behavior of organic molecules in terms of potential energy surfaces. After introducing the basics of electronic spectroscopy, ordinary and chiral, and the fundamental concepts of photophysics in Chapters 1–4, we described the basic notions of organic photophysics and photochemistry in Chapter 5 and 6, and illustrated their utility on a fair number of specific examples in Chapter 7. Throughout, we have attempted to concentrate on the key concepts provided by the quantum theory of molecular structure, and to relate these to experimental observations. In a sense, this text aspires to being a textbook of both theoretical and mechanistic photochemistry but it makes no pretext of providing practical experimental information on light sources and the like.

More than anything else, our goal has been to introduce the reader to a way of thinking about problems in photophysics and photochemistry. Although many additional organic photochemical processes could be added to Chapter 7, we have chosen not to do so. Instead, we hope that the reader will be able to apply the understanding of the material that we have chosen to present as he or she approaches the study of additional reactions.

References

Abrash, S., Repinec, S., Hochstrasser, R.M. (1990), *J. Chem. Phys.* **93**, 1041.

Adam, W., de Sanabia, J.A., Fischer, H. (1973), *J. Org. Chem.* **38**, 2571.

Adam, W., Mazenod, F., Nishizawa, Y., Engel, P.S., Baughman, S.A., Chae, W.-K., Horsey, D.W., Quast, H., Seiferling, B. (1983), *J. Am. Chem. Soc.* **105**, 6141.

Adam, W., Grabowski, S., Wilson, R.M. (1990), *Acc. Chem. Res.* **23**, 165.

Agosta, W.C., Smith, III, A.B., Kende, A.S., Eilermann, R.G., Benham, J. (1969), *Tetrahedron Lett.* **8**, 4517.

Albert, B., Berning, W., Burschka, C., Hünig, S., Prokschy, F. (1984), *Chem. Ber.* **117**, 1465.

Allinger, N.L. (1977), *J. Am. Chem. Soc.* **99**, 8127.

Almgren, M. (1972), *Mol. Photochem.* **4**, 213.

Alves, A.C.P., Christofferson, J., Hollas, J.M. (1971), *Mol. Phys.* **20**, 625.

Angliker, H., Rommel, E., Wirz, J. (1982), *Chem. Phys. Lett.* **87**, 208.

Aoyagi, M., Osamura, Y., Iwata, S. (1985), *J. Chem. Phys.* **83**, 1140.

Applequist, D.E., Litle, R.L., Friedrich, E.C., Wall, R.E. (1959), *J. Am. Chem. Soc.* **81**, 452.

Arnett, J.F., Newkome, G., Mattice, W.L., McGlynn, S.P. (1974), *J. Am. Chem. Soc.* **96**, 4385.

Arnold, B.R., Balaji, V., Michl, J. (1990), *J. Am. Chem. Soc.* **112**, 1808.

Arnold, B.R., Balaji, V., Downing, J.W., Radziszewski, J.G., Fisher, J.J., Michl, J. (1991), *J. Am. Chem. Soc.* **113**, 2910.

Arnold, D.R., Trecker, D.J., Whipple, E.B. (1965), *J. Am. Chem. Soc.* **87**, 2596.

Atchity, G.J., Xantheas S.S., Ruedenberg, K. (1991) *J. Chem. Phys.* **95**, 1862.

Aue, D.H., Reynolds, R.N. (1973), *J. Am. Chem. Soc.* **95**, 2027.

Bachler, V., Polanski, O.E. (1988), *J. Am. Chem. Soc.* **110**, 5972, 5977.

Badger, G.M. (1954), *The Structure and Reactions of Aromatic Compounds*; Cambridge University Press: Cambridge.

Baggiolini, E., Hamlow, H.P., Schaffner, K. (1970), *J. Am. Chem. Soc.* **92**, 4906.

Balaji, V., Michl, J. (1991), *Polyhedron* **10**, 1265.

Baldwin, J.E. (1976), *J.C.S. Chem. Commun.* 734.

Barber, L.L., Chapman, O.L., Lassila, J.D. (1969), *J. Am. Chem. Soc.* **91**, 3664.

Barltrop, J.A. (1973), *Pure Appl. Chem.* **33**, 179.

Barltrop, J.A., Carless, H.A.J. (1972), *J. Am. Chem. Soc.* **94**, 1951.

Bartlett, P.D., Porter, N.A. (1968), *J. Am. Chem. Soc.* **90**, 5317.

Bauschlicher, Jr., C.W., Langhoff, S.R. (1991), *Theor. Chim. Acta* **79**, 93.

Bearpark, M.J., Robb, M.A., Schlegel, H.B. (1994), *Chem. Phys. Lett.* **223**, 269.

Beck, S.M., Liverman, M.G., Monts, D.L., Smalley, R.E. (1979), *J. Chem. Phys.* **70**, 232.

Becker, H.-D. (1982), *Pure Appl. Chem.* **54**, 1589.

Becker, R.S., Freedman, K. (1985), *J. Am. Chem. Soc.* **107**, 1477.

Beckmann, S., Wessel, T., Franck, B., Hönle, W., Borrmann, H., von Schnering, H.G. (1990), *Angew. Chem. Int. Ed. Engl.* **29**, 1395.

Beens, H., Weller, A. (1968), *Chem. Phys. Lett.* **2**, 140.

Beens, H., Knibbe, H., Weller, A. (1967), *J. Chem. Phys.* **47**, 1183.

Beer, M., Longuet-Higgins, H.C. (1955), *J. Chem. Phys.* **23**, 1390.

Bentzien, J., Klessinger, M. (1994), *J. Org. Chem.* **59**, 4887.

Bernardi, F., Olivucci, M., McDonall, J.J.W., Robb, M.A. (1988), *J. Chem. Phys.* **89**, 6365.

Bernardi, F., De, S., Olivucci, M., Robb, M.A. (1990a) *J. Am. Chem. Soc.* **112**, 1737.

Bernardi, F., Olivucci, M., Robb, M.A. (1990b), *Acc. Chem. Res.* **23**, 405.

Bernardi, F., Olivucci, M. Robb, M.A., Tonachini, G. (1992a), *J. Am. Chem. Soc.* **114**, 5805.

Bernardi, F., Olivucci, M., Ragazos, I.N., Robb, M.A. (1992b), *J. Am. Chem. Soc.* **114**, 8211.

Bernardi, F., Olivucci, M., Ragazos, I.N., Robb, M.A. (1992c), *J. Am. Chem. Soc.* **114**, 2752.

Bigot, B. (1980), in *Quantum Theory of Chemical Reactions,* Vol. II; Daudel, R., Pullman, A., Salem, L., Veillard, A., Eds.; Reidel: Dordrecht.

Bigot, B. (1983), *Israel J. Chem.* **23**, 116.

Bigot, B., Sevin, A., Devaquet, A. (1978), *J. Am. Chem. Soc.* **100**, 2639.

Bigwood, M., Boué, S. (1974), *J.C.S. Chem. Commun.* 529.

Birks, J.B. (1970), *Photophysics of Aromatic Molecules*; Wiley: London, New York.

Bischof, H., Baumann, W., Detzer, N., Rotkiewicz, K. (1985), *Chem. Phys. Lett.* **116**, 180.

Bixon, M., Jortner, J. (1968), *J. Chem. Phys.* **48**, 715.

Bixon, M., Jortner, J. (1989), *Chem. Phys. Lett.* **159**, 17.

Bixon, M., Michel-Beyerle, M.E., Jortner, J. (1988), *Israel J. Chem.* **28**, 155.

Blair, J.T., Krogh-Jesperen, K., Levy, R.M. (1989), *J. Am. Chem. Soc.* **111**, 6948.

Blais, N.C., Truhlar, D.G., Mead, C.A. (1988), *J. Chem. Phys.* **89**, 6204.

Blout, E.R., Fields, M. (1948), *J. Am. Chem. Soc.* **70**, 189.

Bonačić-Koutecký, V., Ishimaru, S. (1977), *J. Am. Chem. Soc.* **99**, 8134.

Bonačić-Koutecký, V., Michl, J. (1985a), *Theor. Chim. Acta* **68**, 45.

Bonačić-Koutecký, V., Michl, J. (1985b), *J. Am. Chem. Soc.* **107**, 1765.

Bonačić-Koutecký, V., Bruckmann, P., Hiberty, P., Koutecký, J., Leforestier, C., Salem, L. (1975), *Angew. Chem. Int. Ed. Engl.* **14**, 575.

Bonačić-Koutecký, V., Michl, J., Köhler, J. (1984), *Chem. Phys. Lett.* **104**, 440.

Bonačić-Koutecký, V., Koutecký, J., Michl, J. (1987), *Angew. Chem. Int. Ed. Engl.* **26**, 170.

Bonačić-Koutecký, V., Schöffel, K., Michl, J. (1989), *J. Am. Chem. Soc.* **111**, 6140.

Bonneau, R., Joussot-Dubien, J., Salem, L., Yarwood, A.J. (1976), *J. Am. Chem. Soc.* **98**, 4329.

Borell, P.M., Löhmannsröben, H.-G., Luther, K. (1987), *Chem. Phys. Lett.* **136**, 371.

Boumann, T.D., Lightner, D.A. (1976), *J. Am. Chem. Soc.* **98**, 3145.

Boxer, S.G. (1990), *Annu. Rev. Biophys. Chem.* **19**, 267.

Briegleb, G., Czekalla, J. (1960), *Angew. Chem.* **72**, 401.

Brogli, F., Heilbronner, E., Kobayashi, T. (1972), *Helv. Chim. Acta* **55**, 274.

Brooker, L.G.S., Keyes, G.H., Heseltine, D.W. (1951), *J. Am. Chem. Soc.* **73**, 5350.

Bruckmann, P., Salem, L. (1976), *J. Am. Chem. Soc.* **98**, 5037.

Bryce-Smith, D. (1968), *Pure Appl. Chem.* **16**, 47.

Bryce-Smith, D. (1973), *Pure Appl. Chem.* **34**, 193.

Bryce-Smith, D., Gilbert, A. (1976), *Tetrahedron* **32**, 1309.

Bryce-Smith, D., Gilbert, A. (1980), in *Rearrangements in Ground and Excited States*, Vol. 3; de Mayo, P., Ed.; Academic Press: New York; p. 349.

Bryce-Smith, D., Gilbert, A., Orger, B.H. (1966), *J.C.S., Chem. Commun.* 512.

Bryce-Smith, D., Gilbert, A. Grzonka, J. (1970), *J.C.S. Chem. Commun.* 498.

Bryce-Smith, D., Gilbert, A., Robinson, A.D. (1971), *Angew. Chem. Int. Ed. Engl.* **10**, 745.

Bryce-Smith, D., Gilbert, A., Mattay, J. (1986), *Tetrahedron* **42**, 6011.

Buckingham, A.D., Ramsy, D.A., Tyrell, J. (1970), *Can. J. Phys.* **48**, 1242.

Buenker, R.J., Peyerimhoff, S.D. (1974), *Chem. Rev.* **74**, 127.

Buenker, R.J., Shih, S., Peyerimhoff, S.D. (1976), *Chem. Phys. Lett.* **44**, 385.

Buenker, R.J., Bonačić-Koutecký, V., Poglinai, L. (1980), *J. Chem. Phys.* **73**, 1836.

Buncel, E., Rajagopal, S. (1990), *Acc. Chem. Res.* **23**, 226.

Bunnett, J.F. (1978), *Acc. Chem. Res.* **11**, 413.

Bunton, C.A., Savelli, G. (1986), *Adv. Phys. Org. Chem.* **22**, 213.

Buschmann, H., Scharf, H.-D., Hoffmann, N., Plath, M.W., Runsink, J. (1989), *J. Am. Chem. Soc.* **111**, 5367.

Buschmann, H., Scharf, H.-D., Hoffmann, N., Esser, P. (1991), *Angew. Chem. Int. Ed. Engl.* **30**, 477.

Caldwell, R.A. (1980), *J. Am. Chem. Soc.* **102**, 4004.

Caldwell, R.A., Carlacci, L, Doubleday, Jr., C.E., Furlani, T.R., King, H.F., McIver, Jr., J.W. (1988), *J. Am. Chem. Soc.* **110**, 6901.

Calhoun, G.C., Schuster, G.B. (1984), *J. Am. Chem. Soc.* **106**, 6870.

Callomon, J.H., Dunn, T.M., Mills, I.M. (1966), *Phil. Trans. R. Soc. London* **259A**, 499.

Calzaferri, G., Gugger, H., Leutwyler, S. (1976), *Helv. Chim. Acta* **59**, 1969.

Carlacci, L., Doubleday, Jr., C., Furlani, T.R., King, H.F., McIver, Jr., J.W. (1987), *J. Am. Chem. Soc.* **109**, 5323.

Carless, H.A.J. (1973), *Tetrahedron Lett.* **14**, 3173.

Casey, C.P., Boggs, R.A. (1972), *J. Am. Chem. Soc.* **94**, 6457.

Castellan, A., Michl, J. (1978), *J. Am. Chem. Soc.* **100**, 6824.

Castellan, A., Kolc, J., Michl, J. (1978), *J. Am. Chem. Soc.* **100**, 6687.

Cave, R.J., Davidson, E.R. (1988), *Chem. Phys. Lett.* **148**, 190.

Celani, P., Ottani, S., Olivucci, M., Bernardi, F., Robb, M.A. (1994), *J. Am. Chem. Soc.* **116**, 10141.

Celani, P., Bernardi, F., Olivucci, M., Robb, M.A. (1995), *J. Chem. Phys.* **102**, 5733

Chae, W.-K., Baughman, S.A., Engel, P.S., Bruch, M., Özmeral, C., Szilagyi, S., Timberlake, J.W. (1981), *J. Am. Chem. Soc.* **103**, 4824.

Charlton, J.L., Dabestani, R., Saltiel, J. (1983), *J. Am. Chem. Soc.* **105**, 3473.

Charney, E. (1979), *The Molecular Basis of Optical Acitivity*; Wiley: New York.

Chu, N.Y.C., Kearns, D.R. (1970), *J. Phys. Chem.* **74**, 1255.

Ciamician, G., Silber, P. (1900), *Berchte* **33**, 2911.

Clar, E. (1964), *Polycyclic Hydrocarbons*; Academic Press: London, New York.

Clark, K.B., Leigh, W.J. (1987), *J. Am. Chem. Soc.* **109**, 6086.

Closs, G.L., Miller, J.R. (1981), *J. Am. Chem. Soc.* **103**, 3586.

Closs, G.L., Miller, J.R. (1988), *Science* **240**, 440.

Closs, G.L., Calcaterra, L.T., Green, N.J., Penfield, K.W., Miller, J.R. (1986), *J. Phys. Chem.* **90**, 3673.

Closs, G.L., Johnson, M.D., Miller, J.R., Piotrowiak, P. (1989), *J. Am. Chem. Soc.* **111**, 3751.

Closs, G.L., Forbes, M.D.E., Piotrowiak, P. (1992), *J. Am. Chem. Soc.*, **114**, 3285.

Cohen, S.G., Zand, R. (1962), *J. Am. Chem. Soc.* **84**, 586.

Cohen, S.G., Parola, A., Parsons, Jr., G.H. (1973), *Chem. Rev.* **73**, 141.

Colle, R., Montagnani, R., Riani, P., Salvetti, O. (1978), *Theor. Chim. Acta* **49**, 37.

Condon, E.U. (1928), *Phys. Rev.* **32**, 858.

Constanciel, R. (1972), *Theor. Chim. Acta* **26**, 249.

Cookson, R.C. (1968), *Q. Rev.* **22**, 423.

Corey, E.J., Bass, J.D., LeMahieu, R., Mitra, R.B. (1964), *J. Am. Chem. Soc.* **86**, 5570.

Cotton, F.A. (1971), *Chemical Applications of Group Theory*; Wiley: Chichester.

Cowan, D.O., Drisko, R.L.E. (1970a,b), *J. Am. Chem. Soc.* **92**, 6281, 6286.

Cowan, D.O., Gleiter, R., Hashmall, J.A., Heilbronner, E., Hornung, V. (1971), *Angew. Chem. Int. Ed. Engl.* **10**, 401.

Coxon, J.M., Halton, B. (1974), *Organic Photochemistry*; Cambridge University Press: Cambridge.

Craig, D.P. (1950), *J. Chem. Soc.* (London), 2146.

Dalton, J.C., Wriede, P.A., Turro, N.J. (1970), *J. Am. Chem. Soc.* **92**, 1318.

Dannenberg, J.J., Rayez, J.C. (1983), *J. Org. Chem.* **48**, 4723.

Das, P.K., Becker, R.S. (1978), *J. Phys. Chem.* **82**, 2081.

Das, P.K., Encinas, M.V., Small, Jr., R.D., Scaiano, J.C. (1979), *J. Am. Chem. Soc.* **101**, 6965.

Datta, P., Goldfarb, R.D., Boikess, R.S. (1971), *J. Am. Chem. Soc.* **93**, 5189.

Dauben, W.G., Cargill, R.L. (1961), *Tetrahedron* **12**, 186.

Dauben, W.G., Kellogg, M.S. (1971), *J. Am. Chem. Soc.* **93**, 3805.

Dauben, W.G., Salem, L., Turro, N.J. (1975), *Acc. Chem. Res.* **8**, 41.

Dauben, W.G., Lodder, G., Robbins, J.D. (1976), *J. Am. Chem. Soc.* **98**, 3030.

Day, A.C., Wright, T.R. (1969), *Tetrahedron Lett.*, 1067.

De Mayo, P. (1971), *Acc. Chem. Res.* **4**, 41.

De Mayo, P., Shizuka, H. (1973), *J. Am. Chem. Soc.* **95**, 3942.

De Vaal, P., Lodder, G., Cornelisse, J. (1986), *Tetrahedron* **42**, 6011.

DeBoer, C.D., Herkstroeter, W.G., Marchetti, A.P., Schultz, A.G., Schlessinger, R.H. (1973), *J. Am. Chem. Soc.* **95**, 3963.

Dehareng, D., Chapuisat, X., Lorquet, J.-C., Galloy, C., Raseev, G. (1983), *J. Chem. Phys.* **78**, 1246.

Deisenhofer, J. Epp, O., Miki, K., Huber, R., Michel, H. (1985), *Nature* **318**, 618.

Del Bene, J., Jaffé, H.H. (1968), *J. Chem. Phys.* **48**, 1807.

Devaquet, A., Sevin, A., Bigot, B. (1978), *J. Am. Chem. Soc.* **100**, 2009.

Dewar, M.J.S. (1950), *J. Chem. Soc.* (London), 2329.

Dewar, M.J.S. (1969), *The Molecular Orbital Theory of Organic Chemistry*; McGraw-Hill: New York.

Dewar, M.J.S., Doubleday, C. (1978), *J. Am. Chem. Soc.* **100**, 4935.

Dewar, M.J.S., Dougherty, R.C. (1975), *The PMO Theory of Organic Chemistry*; Plenum Press: New York, London.

Dewar, M.J.S., Thiel, W. (1977), *J. Am. Chem. Soc.* **99**, 2338.

Dewar, M.J.S., Thompson, Jr., C.C. (1966), *Tetrahedron Suppl.* **7**, 97.

Dexter, D.L. (1953), *J. Chem. Phys.* **21**, 836.

Dick, B., Nickel, B. (1983), *Chem. Phys.* **78**, 1.

Dick, B., Gonska, H., Hohlneicher, G. (1981), *Ber. Bunsenges. Phys. Chem.* **85**, 746.

DMS-UV-Atlas organischer Verbindungen (1966–1971); Butterworths: London; Verlag Chemie: Weinheim.

Doany, F.E., Hochstrasser, R.M., Greene, B.I., Millard, R.R. (1985), *Chem. Phys. Lett.* **118**, 1.

Döhnert, D., Koutecký, J. (1980), *J. Am. Chem. Soc.* **102**, 1789.

Dörr, F. (1966), in *Optische Anregung organischer Systeme*; Foerst, W., Ed.; Verlag Chemie: Weinheim.

Dörr, F. Held, M. (1960), *Angew. Chem.* **72**, 287.

Doubleday, Jr., C., McIver, Jr., J.W., Page, M. (1982), *J. Am. Chem. Soc.* **104**, 6533.

Doubleday, Jr., C., McIver, Jr., J.W., Page, M. (1985), *J. Am. Chem. Soc.* **107**, 7904.

Doubleday, Jr., C., Turro, N.J., Wang, J.-F. (1989), *Acc. Chem. Res.* **22**, 199.

Dougherty, T.J. (1992), *Adv. Photochemistry* **17**, 275.

Dowd, P. (1966), *J. Am. Chem. Soc.* **88**, 2587.

Downing, J.W., Michl, J., Jørgensen, P., Thulstrup, E.W. (1974), *Theor. Chim. Acta* **32**, 203.

Dreeskamp, H., Koch, E., Zander, M. (1975), *Chem. Phys. Lett.* **31**, 251.

Dreyer, J., Klessinger, M. (1994), *J. Chem. Phys.* **101**, 10655.

Dreyer, J., Klessinger, M. (1995), *Chemistry Eur. J.* (in press).

Dubois, J.T., van Hemert, R.L. (1964), *J. Chem. Phys.* **40**, 923.

Dubois, J.T., Wilkinson, F. (1963), *J. Chem. Phys.* **39**, 899.

Dunning, Jr., T.H., Hosteny, R.P., Shavitt, I. (1973), *J. Am. Chem. Soc.* **95**, 5067.

Dvorak, V., Michl, J. (1976), *J. Am. Chem. Soc.* **98**, 1080.

Eaton, P.E., Cole, Jr., T.W. (1964), *J. Am. Chem. Soc.* **86**, 3158.

Eberson, L. (1982), *Adv. Phys. Org. Chem.* **18**, 79.

Eckert, R., Kuhn H. (1960), *Z. Elektrochem.* **64**, 356.

El-Sayed, M.A. (1963), *J. Chem. Phys.* **38**, 2834.

Engel, P.S. (1980), *Chem. Rev.* **80**, 99.

Engel, P.S., Keys, D.E., Kitamura, A. (1985), *J. Am. Chem. Soc.* **107**, 4964.

Engelke, R., Hay, P.J., Kleier, D.A., Wadt, W.R. (1984), *J. Am. Chem. Soc.* **106**, 5439.

Englman, R., Jortner, J. (1970), *Mol. Phys.* **18**, 145.

Eriksson, M., Nordén, B., Lycksell, P.-O., Gräslund, A., Jernström, B. (1985), *J.C.S. Chem. Commun.* 1300.

Evleth, E.M., Kassab, E. (1978), *J. Am. Chem. Soc.* **110**, 7859.

Eyring, H., Walter, J., Kimball, G. (1944), *Quantum Chemistry;* Wiley: New York.

Fabian, W. (1985), *Z. Naturforsch.* **40a**, 279.

Farid, S., Shealer, S.E. (1973), *J.C.S. Chem. Commun.* 677.

Feher, G., Allen, J.P., Okamura, M.Y., Rees, D.C. (1989), *Nature* **339**, 111.

Fessner, W.-D., Prinzbach, H., Rihs, G. (1983), *Tetrahedron Lett.* **24**, 5857.

Fessner, W.-D., Sedelmeier, G., Spurr, P.R., Rihs, G., Prinzbach, H. (1987), *J. Am. Chem. Soc.* **109**, 4626.

Fieser, L.F., Fieser, M., Rajagopalan, S. (1948), *J. Org. Chem.* **13**, 800.

Fischer, M. (1978), *Angew. Chem. Int. Ed. Engl.* **17**, 16.

Fischer, S.F., van Duyne, R.P. (1977), *Chem. Phys.* **26**, 9.

Flynn, C.R., Michl, J. (1974), *J. Am. Chem. Soc.* **96**, 3280.

Förster, E.W., Grellmann, K.H., Linschitz, H. (1973), *J. Am. Chem. Soc.* **95**, 3108.

Förster, T. (1950), *Ber. Bunsenges. Phys. Chem.* **54**, 531.

Förster, T. (1951), *Fluoreszenz organischer Verbindungen*; Vandenhoek und Ruprecht: Göttingen.

Förster, T. (1959), *Disc. Faraday Soc.* **27**, 7.

Förster, T. (1970), *Pure Appl. Chem.* **24**, 443.

Förster, T., Kasper, K. (1955), *Z. Elektrochem.* **59**, 976.

Förster, T., Seidel, H.-P. (1965), *Z. Phys. Chem. N.F.* **45**, 58.

Forbes, M.D.E., Schulz. G.R. (1994), *J. Am. Chem. Soc.* **116**, 10174.

Franck, J. (1926), *Trans. Faraday Soc.* **21**, 536.

Freilich, S.C., Peters, K.S. (1985), *J. Am. Chem. Soc.* **107**, 3819.

Freund, L., Klessinger, M. (1995), to be published.

Friedrich, L.E., Schuster, G.B. (1969), *J. Am. Chem. Soc.* **91**, 7204.

Friedrich, L.E., Schuster, G.B. (1972), *J. Am. Chem. Soc.* **94**, 1193.

Friedrich, J., Metz, F., Dörr, F. (1974), *Ber. Bunsenges. Phys. Chem.* **78**, 1214.

Fritsche, J. (1867), *J. Prakt. Chem.* [1] **101**, 333.

Frölich, W., Dewey, H.J., Deger, H., Dick, B., Klingensmith, K.A., Püttmann, W., Vogel, E., Hohlneicher, G., Michl, J. (1983), *J. Am. Chem. Soc.* **105**, 6211.

Geldof, P.A. , Rettschnick, R.P.H., Hoytink, G.J. (1969), *Chem. Phys. Lett.* **4**, 59.

Gerhartz, W., Poshusta, R.D., Michl, J. (1976), *J. Am. Chem. Soc.* **98**, 6427.

Gerhartz, W., Poshusta, R.D., Michl, J. (1977), *J. Am. Chem. Soc.* **99**, 4263.

Gersdorf, J., Mattay, J., Görner, H. (1987), *J. Am. Chem. Soc.* **109**, 1203.

Gilbert, A. (1980), *Pure Appl. Chem.* **52**, 2669.

Gisin, M., Wirz, J. (1976), *Helv. Chim. Acta* **59**, 2273.

Gleiter, R. (1992), *Angew. Chem. Int. Ed. Engl.* **31**, 27.

Gleiter, R., Karcher, M. (1988), *Angew. Chem. Int. Ed. Engl.* **27**, 840.

Gleiter, R., Sander, W. (1985), *Angew. Chem. Int. Ed. Engl.* **24**, 566.

Gleiter, R., Schäfer, W. (1990), *Acc. Chem. Res.* **23**, 369.

Gleiter, R., Schang, P., Bloch, M., Heilbronner, E., Bünzli, J.-C., Frost, D.C., Weiler, L. (1985), *Chem. Ber.* **118**, 2127.

Godfrey, M., Murrell, J.N. (1964), *Proc. R. Soc. London* **A278**, 57.

Goldbeck, R.A. (1988), *Acc. Chem. Res.* **21**, 95.

Gould, I.R., Moser, J.E., Ege, D., Farid, S. (1988), *J. Am. Chem. Soc.* **110**, 1991.

Gouterman, M. (1961), *J. Mol. Spectrosc.* **6**, 138.

Grabowski, Z.R., Dobkowski, J. (1983), *Pure Appl. Chem.* **55**, 245.

Granville, M.F., Holtom, G.R., Kohler, B.E. (1980), *J. Chem. Phys.* **72**, 4671.

Greene, B.I., Farrow, R.C. (1983), *J. Chem. Phys.* **78**, 3336.

Grellmann, K.-H., Kühnle, W., Weller, H., Wolff, T. (1981), *J. Am. Chem. Soc.* **103**, 6889.

Griesbeck, A.G., Stadtmüller, S. (1991), *J. Am. Chem. Soc.* **113**, 6923.

Griffiths, J. (1972), *Chem. Soc. Rev.* **1**, 481.

Griffiths, J. (1976), *Colour and Constitution of Organic Molecules*; Academic Press: London, New York.

Grimbert, D., Segal, G., Devaquet, A. (1975), *J. Am. Chem. Soc.* **97**, 6629.

Grimme, S., Dreeskamp, H. (1992), *J. Photochem. Photobiol. A: Chem.* **65**, 371.

Grinter, R., Heilbronner, E. (1962), *Helv. Chim. Acta* **45**, 2496.

Gustav, K., Sühnel, J. (1980), *Z. Chem.* **8**, 283.

Gustav, K., Kempka, U., Sühnel, J. (1980), *Chem. Phys. Lett.* **71**, 280.

Haag, R., Wirz, J., Wagner, P.J. (1977), *Helv. Chim. Acta* **60**, 2595.

Halevi, E.A. (1977), *Nouv. J. Chim.* **1**, 229.

Haller, I. (1967), *J. Chem. Phys.* **47**, 1117.

Ham, J.S. (1953), *J. Chem. Phys.* **21**, 756.

Hammond, G.S., Saltiel, J. (1963), *J. Am. Chem. Soc.* **85**, 2516.

Hammond, G.S., Saltiel, J., Lamola, A.A., Turro, N.J., Bradshaw, J.S., Cowan, D.O., Counsell, R.C., Vogt, V., Dalton, C. (1964), *J. Am. Chem. Soc.* **86**, 3197.

Hansen, A.E. (1967), *Mol. Phys.* **13**, 425.

Harada, N., Nakanishi, K. (1972), *Acc. Chem. Res.* **5**, 257.

Harding, L.B., Goddard III, W.A. (1980), *J. Am. Chem. Soc.* **102**, 439.

Haselbach, E., Heilbronner, E. (1970), *Helv. Chim. Acta* **53**, 684.

Hassoon, S., Lustig, H., Rubin, M.B., Speiser, S. (1984), *J. Phys. Chem.* **88**, 6367.

Hastings, D.J., Weedon, A.C. (1991), *J. Am. Chem. Soc.* **113**, 8525.

Hautala, R.R., Dawes, K., Turro, N.J. (1972), *Tetrahedron Lett.* **13**, 1229.

Havinga, E. (1973), *Experientia* **29**, 1181.

Havinga, E., Cornelisse, J. (1976), *Pure Appl. Chem.* **47**, 1.

Hayes, J.M. (1987), *Chem. Rev.* **87**, 745.

Heilbronner, E. (1963), *Tetrahedron* **19**, Suppl. 2., 289.

Heilbronner, E. (1966), in *Optische Anregung organischer Systeme*; Foerst, W., Ed.; Verlag Chemie: Weinheim.

Heilbronner, E., Bock, H. (1968), *Das HMO-Modell und seine Anwendung*; Verlag Chemie: Weinheim.

Heilbronner, E., Murrell, J.N. (1963), *Mol. Phys.* **6**, 1.

Henderson, Jr., W.A., Ullman, E.F. (1965), *J. Am. Chem. Soc.* **87**, 5424.

Hendriks, B.M.P., Walter, R.I., Fischer, H. (1979), *J. Am. Chem. Soc.* **101**, 2378.

Herman, M.F. (1984), *J. Chem. Phys.* **81**, 754.

Herndon, W.C. (1971), *Tetrahedron Lett.* **12**, 125.

Herndon, W.C. (1974), *Fortschr. Chem. Forsch.* **46**, 141.

Herzberg, G. (1950), *Molecular Spectra and Molecular Structure*; Van Nostrand: Princeton.

Herzberg, G., Longuet-Higgins, H.C. (1963), *Trans. Faraday Soc.* **35**, 77.

Herzberg, G., Teller, E. (1947), *Rev. Mod. Phys.* **13**, 75.

Hilinski, E.F., Masnovi, J.M., Kochi, J.K., Rentzepis, P.M. (1984), *J. Am. Chem. Soc.* **106**, 8071.

Hirata, Y., Mataga, N. (1984), *J. Phys. Chem.* **88**, 3091.

Hixson, S.S., Mariano, P.S., Zimmerman, H.E. (1973), *Chem. Rev.* **73**, 531.

Hochstrasser, R.M., Noe, L.J. (1971), *J. Mol. Spectrosc.* **38**, 175.

Hohlneicher, G., Dick, B. (1984), *J. Photochem.* **27**, 215.

Hopf, H., Lipka, H., Traetteberg, M. (1994), *Angew. Chem. Int. Ed. Engl.* **33**, 204.

Horspool, W.H. (1976), *Aspects of Organic Photochemistry*; Academic Press: London.

Hosteny, R.P., Dunning, Jr., T.H., Gilman, R.R., Pipano, A., Shavitt, I. (1975), *J. Chem. Phys.* **62**, 4764.

Houk, K.N. (1976), *Chem. Rev.* **76**, 1.

Houk, K.N. (1982), *Pure Appl. Chem.* **54**, 1633.

Höweler, U., Michl, J. (1995), to be published.

Höweler, U., Chatterjee, P.S., Klingensmith, K.A., Waluk, J., Michl, J. (1989), *Pure Appl. Chem.* **61**, 2117.

Hudson, B.S., Kohler, B.E. (1972), *Chem. Phys. Lett.* **14**, 299.

Hudson, B.S., Kohler, B.E. (1973), *J. Chem. Phys.* **59**, 4984.

Hudson, B.S., Kohler, B.E., Schulten, K. (1982), in *Excited States*, Vol. 6; Lim, E.C., Ed.; Academic Press: New York.

Huisgen, R. (1977), *Angew. Chem. Int. Ed. Engl.* **16**, 572.

Huppert, D., Rentzepis, P.M., Tollin, G. (1976), *Biochim. Biophys. Acta* **440**, 356.

Imamura, A., Hoffmann, R. (1968), *J. Am. Chem. Soc.* **90**, 5379.

Innes, K.K. (1975) in *Excited States,* Vol. 2; Lim, E.C., Ed.; Academic Press: New York.

Inoue, Y., Takamuku, S., Sakurai, H. (1977), *J. Phys. Chem.* **81**, 7.

Inoue, Y., Yamasaki, N., Yokoyama, T., Tai, A. (1993), *J. Org. Chem.* **58**, 1011.

Ireland, J.F., Wyatt, P.A.H. (1973), *J.C.S. Faraday Trans, I* **69**, 161.

Ishikawa, M., Kumada, M. (1986), *Adv. Organomet. Chem.* **19**, 51.

Jacobs, H.J.C., Havinga, E. (1979), *Adv. Photochem,* **11**, 305

Jaffé, H.H., Orchin, M. (1962), *Theory and Applications of Ultraviolet Spectroscopy*; Wiley: New York, London.

Jensen, E., Keller, J.S., Waschewsky, G.C.G., Stevens, J.E., Graham, R.L., Freed, K.F., Butler, L.J. (1993), *J. Chem. Phys.* **98**, 2882.

Job, V.A., Sethuraman, V., Innes, K.K. (1969), *J. Mol. Spectrosc.* **30**, 365.

Johnson, D.R., Kirchhoff, W.H., Lovas, F.J. (1972), *J. Phys. Chem. Ref. Data* **1**, 1011.

Johnson, P.M., Albrecht, A.C. (1968), *J. Chem. Phys.* **48**, 851.

Johnston, L.S., Scaiano, J.C. (1989), *Chem. Rev.* **89**, 521.

Johnstone, D.E., Sodeau, J.R. (1991), *J. Phys. Chem.* **95**, 165.

Jones, II, G., Becker, W.G. (1982), *Chem. Phys. Lett.* **85**, 271.

Jones, II, G., Chiang, S.-H. (1981), *Tetrahedron* **37**, 3397.

Jones, L.B., Jones, V.K. (1969), *Fortschr. Chem. Forsch.* **13**, 307.

Joran, A.D., Leland, B.A., Geller, G.G., Hopfield, J.J., Dervan, P.B. (1984), *J. Am. Chem. Soc.* **106**, 6090.

Jørgensen, N.H., Pedersen, P.B., Thulstrup, E.W., Michl, J. (1978), *Int. J. Quantum Chem.* **S12**, 419.

Jug, K., Bredow, T. (1991), *J. Phys. Chem.* **95**, 9242.

De Kanter, F.J.J., Kaptein, R. (1982), *J. Am. Chem. Soc.* **104**, 4759.

De Kanter, F.J.J., den Hollander, J.A., Huizer, A.H., Kaptein, R. (1977), *Mol. Phys.* **34**, 857.

Kaplan, L., Wilzbach, K.E. (1968), *J. Am. Chem. Soc.* **90**, 3291.

Kaprinidis, N.A., Lem, G., Courtney, S.H., Schuster, D.I. (1993), *J. Am. Chem. Soc.* **115**, 3324.

Karelson, M., Zerner, M.C. (1990), *J. Am. Chem. Soc.* **112**, 9405.

Karelson, M., Zerner, M.C. (1992), *J. Phys. Chem.,* **96**, 6949.

Karwowski, J. (1973), *Chem. Phys. Lett.* **18**, 47.

Kasha, M. (1950), *Disc. Faraday Soc.* **9**, 14.

Kasha, M., Brabham, D.E. (1979), in *Singlet Oxygen*; Wasserman, H.H., Murray, R.W., Eds.; Academic Press: New York.

Kato, S. (1988), *J. Chem. Phys.* **88**, 3045.

Kaupp, G., Teufel, E. (1980), *Chem. Ber.* **113**, 3669.

Kearns, D.R., Case, W.A. (1966), *J. Am. Chem. Soc.* **88**, 5087.

Kearvell, A., Wilkinson, F. (1969), *20ème Réunion Soc. Chim. Phys.*, 125.

Keller, J.S., Kash, P.W., Jensen, E., Butler, L.J. (1992), *J. Chem. Phys.* **96**, 4324.

Kestner, N.R., Logan, J., Jortner, J. (1974), *J. Phys. Chem.* **78**, 2148.

Khudyakov, I.V., Serebrennikov, Y.A., Turro, N.J. (1993), *Chem. Rev.* **93**, 537.

Kiefer, E.F., Carlson, D.A. (1967), *Tetrahedron Lett.* **8**, 1617.

Kiefer, E.F., Tanna, C.H. (1969), *J. Am. Chem. Soc.* **91**, 4478.

Kim, J.-I., Schuster, G.B. (1990), *J. Am. Chem. Soc.* **112**, 9635.

Kiprianov, A.I., Mikhailenko, F.A. (1961), *Zh. Obshch. Khim,* **31**, 1334.

Kirkor, E.S., Maloney, V.M., Michl, J. (1990), *J. Am. Chem. Soc.* **112**, 148.

Kirmaier, C., Holten, D. (1987), *Photosynth. Res.* **13**, 225.

Klessinger, M. (1968), *Fortschr. Chem. Forsch.* **9**, 354.

Klessinger, M. (1978), *Chem. uns. Zeit* **12**, 1.

Klessinger, M. (1995), *Angew. Chem. Int. Ed. Engl.* **34**, 549.

Klessinger, M., Hoinka, C. (1995), to be published.

Klessinger, M., Pötter, T., van Wüllen, C. (1991), *Theor. Chim. Acta* **80**, 1.

Klingensmith, K.A., Püttmann, W., Vogel, E., Michl, J. (1983), *J. Am. Chem. Soc.* **105**, 3375.

Kobayashi, T., Nagakura, S. (1972), *Bull Chem. Soc. Jpn.* **47**, 2563.

Koga, N., Sameshima, K., Morokuma, K. (1993), *J. Phys. Chem.* **97**, 13117.

Kohler, B.E. (1993), *Chem. Rev.* **93**, 41.

Kolc, J., Michl, J. (1976), *J. Am. Chem. Soc.* **98**, 4540.

Kollmar, H., Staemmler, V. (1978), *Theor. Chim. Acta* **48**, 223.

Koo, J., Schuster, G.B. (1977), *J. Am. Chem. Soc.* **99**, 6107.

Kosower, E.M. (1958), *J. Am. Chem. Soc.* **80**, 3253, 3261, 3267.

Koutecký, J. (1965), in *Modern Quantum Chemistry,* Part 1; Sinanoglu, O., Ed.; Academic Press: New York, London.

Koutecký, J. (1966), *J. Chem. Phys.* **44**, 3702.

Koutecký, J. (1967), *J. Chem. Phys.* **47**, 1501.

Koutecký, J., Čižek, J., Dubský, J., Hlavatý, K. (1964), *Theor. Chim. Acta* **2**, 462.

Kroon, J., Verhoeven, J.W., Paddon-Row, M.N., Oliver, A.M. (1991), *Angew. Chem. Int. Ed. Engl.* **30**, 1358.

Kropp, P.J. (1984), *Acc. Chem. Res.* **17**, 131.

Kropp, P.J., Reardon, Jr., E.J., Gaibel, Z.L.F., Williard, K.F., Hattaway, Jr., J.H. (1973), *J. Am. Chem. Soc.* **95**, 7058.

Kropp, P.J., Snyder, J.J., Rawlings, P.C., Fravel, Jr., H.G. (1980), *J. Org. Chem.* **45**, 4471.

Kubota, Y., Motoda, Y., Shigemune, Y., Fujisaki, Y. (1979), *Photochem. Photobiol.* **19**, 1099.

Kuhn, H. (1949), *J. Chem. Phys.* **17**, 1198.

Kuppermann, A., Flicker, W.M., Mosher, O.A. (1979), *Chem. Rev.* **79**, 77.

Laarhoven, W.H. (1983), *Recl. Trav. Chim. Pays-Bas* **102**, 185.

Labhart, H. (1966), *Experientia* **22**, 65.

Labhart, H., Wagnière, G. (1959), *Helv. Chim. Acta* **62**, 2219.

Lakowicz, J.R. (1983), *Principles of Fluroescence Spectroscopy*; Plenum Press; New York, London.

Lamola, A.A. (1968), *Photochem. Photobiol.* **8**, 126.

Landau, L. (1932), *Phys. Z. Sowjet.* **2**, 46.

Langkilde, F.W., Thulstrup, E.W., Michl, J. (1983a), *J. Chem. Phys.* **78**, 3372.

Langkilde, F.W., Gisin, M., Thulstrup, E.W., Michl, J. (1983b), *J. Phys. Chem.* **87**, 2901.

Laposa, J.D., Lim, E.C., Kellogg, R.E. (1965), *J. Chem. Phys.* **42**, 3025.

Lauer, G., Schäfer, W., Schweig, A. (1975), *Chem. Phys. Lett.* **33**, 312.

Lechtken, P., Höhne, G. (1973), *Angew. Chem. Int. Ed. Engl.* **12**, 772.

Lechtken, P., Breslow, R., Schmidt, A.H., Turro, N.J. (1973), *J. Am. Chem. Soc.* **95**, 3025.

Lee, Y.S., Freed, K.F., Sun, H., Yeager, D.L. (1983), *J. Chem. Phys.* **79**, 3862.

Leigh, W.J. (1993), *Can. J. Chem.* **71**, 147.

Leigh, W.J., Zheng, K. (1991), *J. Am. Chem. Soc.* **113**, 4019; Errata, *ibid.* **114** (1992) 796.

Leigh, W.J., Zheng, K., Nguyen, N., Werstink, N.H., Ma, J. (1991), *J. Am. Chem. Soc.* **113**, 4993.

Letsinger, R.L., McCain, J.H. (1969), *J. Am. Chem. Soc.* **91**, 6425.

Levy, R.M., Kitchen, D.B., Blair, J.T., Krogh-Jespersen, K. (1990), *J. Phys. Chem.* **94**, 4470.

Lewis, F.D., DeVoe, R.J. (1982), *Tetrahedron* **38**, 1069.

Lewis, F.D., Hilliard, T.A. (1972), *J. Am. Chem. Soc.* **94**, 3852.

Lewis, F.D., Magyar, J.G. (1972), *J. Org. Chem.* **37**, 2102.

Lewis, F.D., Ho, T.-I., Simpson, J.T. (1981), *J. Org. Chem.* **46**, 1077.

Lewis, F.D., Ho, T.-I., Simpson, J.T. (1982), *J. Am. Chem. Soc.* **104**, 1924.

Li, R., Lim, E.C. (1972), *J. Chem. Phys.* **57**, 605.

Libman, J. (1975), *J. Am. Chem. Soc.* **97**, 4139.

Lin, S.H., Fujimura, Y., Neusser, H.J., Schlag, E.W. (1984), *Multiphoton Spectroscopy of Molecules*; Academic Press: New York.

Linder, R.E., Morrill, K., Dixon, J.S., Barth, G., Bunnenberg, E., Djerassi, C., Seamans, L., Moscowitz, A. (1977), *J. Am. Chem. Soc.* **99**, 727.

Linderberg, J. (1967), *Chem. Phys. Lett.* **1**, 39.

Linderberg, J., Michl, J. (1970), *J. Am. Chem. Soc.* **92**, 2619.

Lippert, E. (1966), in *Optische Anregung organischer Systeme*; Foerst, W., Ed.; Verlag Chemie: Weinheim.

Lippert, E. (1969), *Acc. Chem. Res.* **3**, 74.

Lippert, E., Rettig, W., Bonačić-Koutecký, V., Heisel, F., Miehé, J.A. (1987), *Adv. Chem. Phys.* **68**, 1.

Liptay, W. (1963), *Z. Naturforsch.* **18a**, 705.

Liptay, W. (1966), in *Optische Anregung organischer Systeme*; Foerst, W., Ed.; Verlag Chemie: Weinheim.

Liptay, W. (1969), *Angew. Chem. Int. Ed. Engl.* **8**, 177.

Liptay, W., Wortmann, R., Schaffrin, H., Burkhard, O., Reitinger, W., Detzer, N. (1988a), *Chem. Phys.* **120**, 429.

Liptay, W., Wortmann, R., Böhm, R., Detzer, N. (1988b), *Chem. Phys.* **120**, 439.

Liu, R.S.H., Hammond, G.S. (1967), *J. Am. Chem. Soc.* **89**, 4936.

Liu, R.S.H., Turro, N.J., Hammond, G.S. (1965), *J. Am. Chem. Soc.* **87**, 3406.

Longuet-Higgins, J.C., Abrahamson, E.W. (1965), *J. Am. Chem. Soc.* **87**, 2045.

Longuet-Higgins, H.C., Murrell, J.N. (1955), *Proc. Phys. Soc.* **A 68**, 601.

Lorquet, J.C., Lorquet, A.J., Desouter-Lecomte, M. (1981), in *Quantum Theory of Chemical Reactions*, Vol. II; Daudel, R., Pullman, A., Salem, L., Veillard, A., Eds; Reidel: Dordrecht; p. 241.

Lüttke, W., Schabacker, V. (1966), *Liebigs Ann. Chem.* **698**, 86.

Luzhkov, V., Warshel, A. (1991), *J. Am. Chem. Soc.* **113**, 4491.

Maciejewski, A., Steer, R.D. (1993), *Chem. Rev.* **93**, 67.

Maier, G. (1986), *Pure Appl. Chem.* **58**, 95.

Maier, G., Pfriem, S., Schäfer, U., Malsch, K.-D., Matusch, R. (1981), *Chem. Ber.* **114**, 3965.

Maier, J.P., Seilmeier, A., Laubereau, A., Kaiser, W. (1977), *Chem. Phys. Lett.* **46**, 527.

Majewski, W.A., Plusquellic, D.F., Pratt, D.W. (1989), *J. Chem. Phys.* **90**, 1362.

Malhotra, S.S., Whiting, M.C. (1959), *J. Chem. Soc.* (London), 3812.

Malmqvist, P.-A., Roos, B.O. (1992), *Theor. Chim. Acta* **83**, 191.

Manning, T.D.R., Kropp, P.J. (1981), *J. Am. Chem. Soc.* **103**, 889.

Manring, L.E., Peters, K.S., Jones, II, G., Bergmark, W.R. (1985), *J. Am. Chem. Soc.* **107**, 1485.

Manthe, U., Köppel, H. (1990), *J. Chem. Phys.* **93**, 1658.

Marchetti, A.P., Kearns, D.R. (1967), *J. Am. Chem. Soc.* **89**, 768.

Marcus, R.A. (1964), *Annu. Rev. Phys. Chem.* **15**, 155.

Marcus, R.A. (1988), *Israel J. Chem.* **28**, 205.

Maroulis, A.J., Arnold, D.R. (1979), *Synthesis*, 819.

Martin, H.-D., Schiwek, H.-J., Spanget-Larsen, J., Gleiter, R. (1978), *Chem. Ber.* **111**, 2557.

Martin, R.L., Wadt, W.R. (1982), *J. Phys. Chem.* **86**, 2382.

Masnovi, J.M., Kochi, J.K., Hilinski, E.F., Rentzepis, P.M. (1986), *J. Am. Chem. Soc.* **108**, 1126.

Mason, S.F. (1962), *J. Chem. Soc.* (London), 493.

Mason, S.F. (1981), *Advances in IR and Raman Spectroscopy,* Vol. 8, Chapter 5; Clark, R.J.H., Hester, R.E., Eds.; Heyden: London.

Mason, S.F., Philp, J., Smith, B.E. (1968), *J. Chem. Soc.* (London), A 3051.

Mataga, N. (1984), *Pure Appl. Chem.* **56**, 1255.

Mataga, N., Nishimoto, K. (1957), *Z. Phys. Chem. N.F.* **13**, 140.

Mataga, N., Karen, A., Tadashi, O., Nishitani, S., Kurata, N., Sakata, Y., Misumi, S. (1984), *J. Phys. Chem.* **88**, 5138.

Mattay, J. (1987), *J. Photochem.* **37**, 167.

Mattay, J., Gersdorf, J., Freudenberg, U. (1984a), *Tetrahedron Lett.* **25**, 817.

Mattay, J., Gersdorf, J., Leismann, H., Steenken, S. (1984b), *Angew. Chem. Int. Ed. Engl.* **23**, 249.

Mattes, S.L., Farid, S. (1983), *J. Am. Chem. Soc.* **105**, 1386.

McDiarmid, R. (1976), *J. Chem. Phys.* **64**, 514.

McDiarmid, R., Doering, J.P. (1980), *J. Chem. Phys.* **73**, 4192.

McGlynn, S.P., Azumi, T., Kinoshita, M. (1969), *Molecular Spectroscopy of the Triplet State*; Prentice-Hall: Englewood Cliffs.

McLachlan, A.D. (1959), *Mol. Phys.* **2**, 271.

McWeeny, R. (1989), *Methods of Molecular Quantum Mechanics*; Academic Press: London.

Meinwald, J., Samuelson, G.E., Ikeda, M. (1970), *J. Am. Chem. Soc.* **92**, 7604.

Melander, L.C.S., Saunders, Jr., W.H. (1980), *Reaction Rates of Isotopic Molecules*; Wiley: New York.

Merer, A.J., Mulliken, R.S. (1969), *Chem. Rev.* **69**, 639.

Meth-Cohn, O., Moore, C., van Rooyen, P.H. (1985), *J.C.S. Perkin Trans. I,* 1793.

Michl, J. (1972), *Mol. Photochem.* **4**, 243, 257, 287.

Michl, J. (1973), in *Physical Chemistry,* Vol. VII; Eyring, H., Henderson, D., Jost, W., Eds.; Academic Press: New York.

Michl, J. (1974a), *Top. Curr. Chem.* **46**, 1.

Michl, J. (1974b), in *Chemical Reactivity and Reaction Paths,* Chapter 8; Klopman, G., Ed.; Wiley: New York.

Michl, J. (1974c), *J. Chem. Phys.* **61**, 4270.

Michl, J. (1977), *Photochem. Photobiol.* **25**, 141.

Michl, J. (1978), *J. Am. Chem. Soc.* **100**, 6801, 6812, 6819.

Michl, J. (1984), *Tetrahedron* **40**, 3845.

Michl, J. (1988), *Tetrahedron* **44**, 7559.

Michl, J. (1990), *Acc. Chem. Res.* **23**, 127.

Michl, J. (1991), in *Theoretical and Computational Models for Organic Chemistry*; Formosinho, S.J, Csizmadia, I.G., Arnaut, L.G., Eds.; Kluwer: Dordrecht.

Michl, J. (1992), *J. Mol. Struct. (Theochem)* **260**, 299.

Michl, J., Balaji, V. (1991), in *Computational Advances in Organic Chemistry: Molecular Structure and Reactivity*; Ögretir, C., Csizmadia, I.G., Eds.; Kluwer: Dordrecht.

Michl, J., Bonačić-Koutecký, V. (1990), *Electronic Aspects of Organic Photochemistry*; Wiley: New York.

Michl, J., Kolc, J. (1970), *J. Am. Chem. Soc.* **92**, 4148.

Michl, J., Thulstrup, E.W. (1976), *Tetrahedron* **32**, 205.

Michl, J. Thulstrup, E.W. (1986), *Spectroscopy with Polarized Light*; VCH Publishers.: Deerfield Beach.

Michl, J., West, R. (1980), in *Oxocarbons*; West, R., Ed.; Academic Press: New York.

Miller, R.D., Michl, J. (1989), *Chem. Rev.* **89**, 1359.

Moffitt, W. (1954a), *J. Chem. Phys.* **22**, 320.

Moffitt, W. (1954b), *J. Chem. Phys.* **22**, 1820.

Moffitt, W., Moscowitz, A. (1959), *J. Chem. Phys.* **30**, 648.

Moffitt, W., Woodward, R.B., Moscowitz, A., Klyne, W., Djerassi, C. (1961), *J. Am. Chem. Soc.* **83**, 4013.

Momicchioli, F., Baraldi, I., Berthier, G. (1988), *Chem. Phys.* **123**, 103.

Moore, W.M., Morgan, D.D., Stermitz, F.R. (1963), *J. Am. Chem. Soc.* **85**, 829.

Mosher, O.A., Flicker, W.M., Kuppermann, A. (1973), *J. Chem. Phys.* **59**, 6502.

Moule, D.C., Walsh, A.D. (1975), *Chem. Rev.* **75**, 67.

Mulder, J.J.C. (1980), *Nouv. J. Chim.* **4**, 283.

Mulliken, R.S. (1939), *J. Chem. Phys.* **7**, 20.

Mulliken, R.S. (1952), *J. Am. Chem. Soc.* **74**, 811.

Mulliken, R.S. (1977), *J. Chem. Phys.* **66**, 2448.

Murray, R.W. (1979), in *Singlet Oxygen,* Chapter 3; Wassermann, H.H., Murray, R.W., Eds.; Academic Press: New York.

Murrell, J.N. (1963), *The Theory of the Electronic Spectra of Organic Molecules*; Methuen: London.

Murrell, J.N., Tanaka, J. (1963), *Mol. Phys.* **7**, 363.

Muszkat, K.A. (1980), *Top. Curr. Chem.* **88**, 91.

Nayler, P., Whiting, M.C. (1955), *J. Chem. Soc.* (London), 3037.

Neumann, J. von, Wigner, E. (1929), *Z. Phys.* **30**, 467.

Neusser, H.J., Schlag, E.W. (1992), *Angew. Chem. Int. Ed. Engl.* **31**, 263.

Nickel, B., Roden, G. (1982), *Chem. Phys.* **66**, 365.

Niephaus, H., Schleker, W., Fleischhauer, J. (1985), *Z. Naturforsch.* **40a**, 1304.

Ohara, M., Watanabe, K. (1975), *Angew. Chem. Int. Ed. Engl.* **14**, 820.

Ohmine, I. (1985), *J. Chem. Phys.* **83**, 2348.

Olivucci, M., Ragazos, I.N., Bernardi, F., Robb, M.A. (1993), *J. Am. Chem. Soc.* **115**, 3710.

Olivucci, M., Bernardi, F., Celani, P., Ragazos, I.N., Robb, M.A. (1994a), *J. Am. Chem. Soc.* **116**, 1077.

Olivucci, M., Bernardi, F., Ottani, S., Robb, M.A. (1994b), *J. Am. Chem. Soc.* **116**, 2034.

Onsager, L. (1936), *J. Am. Chem. Soc.* **58**, 1486.

Orlandi, G., Siebrand, W. (1975), *Chem. Phys. Lett.* **30**, 352.

Padwa, A., Griffin, G. (1976), in *Photochemistry of Heterocyclic Compounds*; Burchardt, O.; Ed.; Wiley: New York.

Paetzold, R., Reichenbächer, M., Appenroth, K. (1981), *Z. Chem.* **21**, 421.

Paldus, J. (1976), in *Theoretical Chemistry, Advances and Perspectives,* Vol. 2; Eyring, H., Henderson, D., Eds.; Academic Press: New York.

Palmer, I.J., Olivucci, M., Bernardi, F., Robb, M.A. (1992), *J. Org. Chem.* **57**, 5081.

Palmer, I.J., Ragazos, I.N., Bernardi, F., Olivucci, M., Robb, M.A. (1993), *J. Am. Chem. Soc.* **115**, 673.

Palmer, I.J., Ragazos, I.N., Bernardi, F., Olivucci, M., Robb, M.A. (1994), *J. Am. Chem. Soc.* **116**, 2121.

Pancir, J., Zahradnik, R. (1973), *J. Phys. Chem.* **77**, 107.

Pape, M. (1967), *Fortschr. Chem. Forsch.* **7**, 559.

Paquette, L.A., Bay, E. (1982), *J. Org. Chem.* **47**, 4597.

Pariser, R. (1956), *J. Chem. Phys.* **24**, 250.

Pariser, R., Parr, R.G. (1953), *J. Chem. Phys.* **21**, 466.

Parker, C.A. (1964), *Adv. Photochem.* **2**, 306.

Pasman, P., Mes, G.F., Koper, N.W., Verhoeven, J.W. (1985), *J. Am. Chem. Soc.* **107**, 5839.

Petek, H., Bell, A.J., Christensen, R.L., Yoshihara, K. (1992), *J. Chem. Phys.* **96**, 2412.

Peters, K.S., Pang, E., Rudzki, J. (1982), *J. Am. Chem. Soc.* **104**, 5535.

Peters, K.S., Li, B., Lee, J. (1993), *J. Am. Chem. Soc.* **115**, 11119.

Peyerimhoff, S.D., Buenker, F.J. (1973), *NATO Adv. Study Inst. Ser. C* **8**, 257.

Peyerimhoff, S.D., Buenker, F.J. (1975), *Adv. Quantum Chem.* **9**, 69.

Philips, D., Salisbury, K. (1976), in *Spectroscopy,* Vol. 3; Straughan, B.P., Walker, S., Eds.; Chapman and Hall: London.

Pichko, V.A., Simkin, B.Ya., Minkin, V.I. (1991), *J. Mol. Struct. (Theochem)* **235**, 107.

Pickard, S.T., Smith, H.E. (1990), *J. Am. Chem. Soc.* **112**, 5741.

Platt, J.R. (1949), *J. Chem. Phys.* **17**, 484.

Platt, J.R. (1962), *J. Mol. Spectrosc.* **9**, 288.

Platz, M.S., Berson, J.A. (1977), *J. Am. Chem. Soc.* **99**, 5178.

Pople, J.A. (1953), *Trans. Faraday Soc.* **49**, 1375.

Pople, J.A. (1955), *Proc. Phys. Soc.* **A 68**, 81.

Porter, G., Strachan, W. (1958), *Spectrochim. Acta* **12**, 299.

Porter, G., Suppan, P. (1964), *Pure Appl. Chem.* **9**, 499.

Prinzbach, H., Fischer, G., Rihs, G., Sedelmeier, G., Heilbronner, E., Yang, Z. (1982), *Tetrahedron Lett.* **23**, 1251.

Quenemoen, K., Borden, W.T., Davidson, E.R., Feller, D. (1985), *J. Am. Chem. Soc.* **107**, 5054.

Quinkert, G., Finke, M., Palmowski, J., Wiersdorff, W.-W. (1969), *Mol. Photochem.* **1**, 433.

Quinkert, G., Cech, F., Kleiner, E., Rehm, D. (1979), *Angew. Chem. Int. Ed. Engl.* **18**, 557.

Quinkert, G., Kleiner, E., Freitag, B.-J., Glenneberg, J., Billhardt, U.-M., Cech, F., Schmieder, K.R., Schudok, C., Steinmetzer, H.-C., Bats, J.W., Zimmermann, G., Dürner, G., Rehm, D., Paulus, E.F. (1986), *Helv. Chim. Acta* **69**, 469.

Rabek, J.F. (1982), *Experimental Methods in Photochemistry and Photophysics,* Parts 1 and 2; Wiley: Chichester.

Rademacher, P. (1987), *Strukturen organischer Moleküle;* Verlag Chemie: Weinheim.

Radziszewski, J.G., Burkhalter, F.A., Michl, J. (1987), *J. Am. Chem. Soc.* **109**, 61.

Radziszewski, J.G., Waluk, J., Michl, J. (1989), *Chem. Phys.* **136**, 165.

Radziszewski, J.G., Waluk, J., Nepras, M., Michl, J. (1991), *J. Phys. Chem.* **95**, 1963.

Ragazos, I.N., Robb, M.A., Bernardi, F., Olivucci, M. (1992), *Chem. Phys. Lett.* **197**, 217.

Ramesh, V., Ramamurthy, V. (1984), *J. Photochem.* **24**, 395.

Ramunni, G., Salem, L. (1976), *Z. Phys. Chem. N.F.* **101**, 123.

Ransom, B.D., Innes, K.K., McDiarmid, R. (1978), *J. Chem. Phys.* **68**, 2007.

Rau, H. (1984), *J. Photochem.* **26**, 221.

Reedich, D.E., Sheridan, R.S. (1985), *J. Am. Chem. Soc.* **107**, 3360.

Reguero, M., Bernardi, F., Jones, H., Olivucci, M., Ragazos, I.N., Robb, M.A. (1993), *J. Am. Chem. Soc.* **115**, 2073.

Reguero, M., Olivucci, M., Bernardi, F., Robb, M.A. (1994), *J. Am. Chem. Soc.* **116**, 2103.

Rehm, D., Weller, A. (1970a), *Z. Phys. Chem.* **69**, 83.

Rehm, D., Weller, A. (1970b), *Israel J. Chem.* **8**, 259.

Reichardt, C., Dimroth, K. (1968), *Fortschr. Chem. Forsch.* **11**, 1.

Reichardt, C., Harbusch-Görnert, E. (1983), *Liebigs Ann. Chem.,* 721.

Reid, P.J., Doig, S.J., Mathies, R.A. (1990), *J. Phys. Chem.* **94**, 8396.

Reid, P.J., Wickham, S.D., Mathies, R.A. (1992), *J. Phys. Chem.* **96**, 5720.

Reid, P.J., Doig, S.J., Wickham, S.D., Mathies, R.A. (1993), *J. Am. Chem. Soc.* **115**, 4754.

Reinsch, M., Klessinger, M. (1990), *Phys. Org. Chem.* **3**, 81.

Reinsch, M., Höweler, U., Klessinger, M. (1987), *Angew. Chem. Int. Ed. Engl.* **26**, 238.

Reinsch, M., Höweler, U., Klessinger, M. (1988), *J. Mol. Struct. (Theochem)* **167**, 301.

Renner, C.A., Katz, T.J., Pouliquen, J., Turro, N.J., Waddell, W.H. (1975), *J. Am. Chem. Soc.* **97**, 2568.

Rentzepis, P.M. (1970), *Science* **169**, 239.

Rettig, W. (1986), *Angew. Chem. Int. Ed. Engl.* **25**, 971.

Rettig, W., Majenz, W., Lapouyade, R., Haucke, G. (1992), *J. Photochem. Photobiol. A: Chem.* **62**, 415.

Ridley, J., Zerner, M. (1973), *Theor. Chim. Acta* **32**, 111.

Ridley, J., Zerner, M. (1976), *Theor. Chim. Acta* **42**, 223.

Riedle, E., Weber, T., Schubert, E., Neusser, H.J., Schlag, E.W. (1990), *J. Chem. Phys.* **93**, 967.

Roos, B.O., Taylor, R.P., Siegbahn, P.E.M. (1980), *Chem. Phys.* **48**, 157.

Rosenfeld (1928), *Z. Phys.* **52**, 161.

Roth, H.D., Schilling, M.L.M. (1980), *J. Am. Chem. Soc.* **102**, 4303.

Roth, H.D., Schilling, M.L.M. (1981), *J. Am. Chem. Soc.* **103**, 7210.

Roth, H.D., Schilling, M.L.M., Jones, II, G. (1981), *J. Am. Chem. Soc.* **103**, 1246.

Roth, H.D., Schilling, M.L.M., Raghavachari, K. (1984), *J. Am. Chem. Soc.* **106**, 253.

Roth, W.R. (1963), *Angew. Chem.* **75**, 921.

Roth, W.R., Erker, G. (1973), *Angew. Chem. Int. Ed. Engl.* **12**, 503.

Sadler, D.E., Wendler, J., Olbrich, G., Schaffner, K. (1984), *J. Am. Chem. Soc.* **106**, 2064.

Salem, L. (1974), *J. Am. Chem. Soc.* **96**, 3486.

Salem, L. (1982), *Electrons in Chemical Reactions: First Principles*; Wiley: New York.

Salem, L., Rowland, C. (1972), *Angew. Chem. Int. Ed. Engl.* **11**, 92.

Salem, L., Leforestier, C., Segal, G., Wetmore, R. (1975), *J. Am. Chem. Soc.* **97**, 479.

Salikhov, K.M., Molin, Y.N., Sagdeev, R.Z., Buchachenko, A.L. (1984), *Magnetic and Spin Effects in Chemical Reactions*; Elsevier: Amsterdam.

Saltiel, J., Charlton, J.L. (1980), in *Rearrangements in Ground and Excited States,* Vol. 3; de Mayo, P., Ed.; Academic Press: New York.

Saltiel, J., Zafiriou, O.C., Megarity, E.D., Lamola, A.A. (1968), *J. Am. Chem. Soc.* **90**, 4759.

Saltiel, J., Metts, S., Wrighton, M. (1969), *J. Am. Chem. Soc.* **91**, 5684.

Saltiel, J., Metts, L., Wrighton, M. (1970), *J. Am. Chem. Soc.* **92**, 3227.

Saltiel, J., D'Agostino, J., Megarity, E.D., Metts, L., Neuberger, K.R., Wrighton, M., Zafiriou, O.C. (1973), *Org. Photochem.* **3**, 1.

Saltiel, J., Marchand, G.R., Smothers, W.K., Stout, S.A., Charlton, J.L. (1981), *J. Am. Chem. Soc.* **103**, 7159.

Saltiel, J., Sun, Y.-R. (1990), in *Photochromism, Molecules and Systems*; Dörr, H., Bouas-Laurent, H., Eds.; Elsevier: Amsterdam.

Sandros, K., Bäckström, H.L.J. (1962), *Acta Chem. Scand.* **16**, 958.

Santiago, C., Houk, K.N. (1976), *J. Am. Chem. Soc.* **98**, 3380.

Scaiano, J.C. (1982), *Tetrahedron* **38**, 819.

Schaap, A.P., Thayer, A.L., Blossey, E.C., Neckers, D.C. (1975), *J. Am. Chem. Soc.* **97**, 3741.

Schafer, O., Allan, M., Szeimies, G., Sanktjohanser, M. (1992), *J. Am. Chem. Soc.* **114**, 8180.

Schaffner, K., Demuth, M. (1986), in *Modern Synthetic Methods,* Vol. 4; Scheffold, R., Ed.; Springer: Berlin, Heidelberg.

Scharf, H.-D. (1974), *Angew. Chem. Int. Ed. Engl.* **13**, 520.

Schenck, G.O., Steinmetz, R. (1962), *Bull. Soc. Chim. Belg.* **71**, 781.

Schexnayder, M.A., Engel, P.S. (1975), *J. Am. Chem. Soc.* **97**, 4825.

Schmidt, W. (1971), *Helv. Chim. Acta* **54**, 862.

Scholz, M., Köhler, H.-J. (1981), *Quantenchemische Näherungsverfahren und ihre Anwendung in der organischen Chemie*; Hüthig: Heidelberg.

Schreiber, S.L., Hoveyda, A.H., Wu, H.-J. (1983), *J. Am. Chem. Soc.* **105**, 660.

Schulten, K., Ohmine, I., Karplus, M. (1976a), *J. Chem. Phys.* **64**, 4422.

Schulten, K., Staerk, H., Weller, A., Werner, H.-J., Nickel, B. (1976b), *Z. Phys. Chem. N.F.* **101**, 371.

Schultz, A.G. (1983), *Acc. Chem. Res.* **16**, 210.

Schuster, D.I. (1989), in *The Chemistry of Enones*; Patai, S., Rappoport, Z., Eds.; Wiley: New York.

Schuster, D.I., Heibel, G.E., Woning, J. (1991a), *Angew. Chem. Int. Ed. Engl.* **30**, 1345.

Schuster, D.I, Dunn, D.A., Heibel, G.E., Brown, P.B., Rao, J.M., Woning, J., Bonneau, R. (1991b), *J. Am. Chem. Soc.* **113**, 6245.

Schwartz, S.E. (1973), *J. Chem. Educ.* **50**, 608.

Scott, A.I. (1964), *Interpretation of the Ultraviolet Spectra of Natural Products*; Pergamon Press: Oxford.

Seamans, L., Moscowitz, A., Linder, R.E., Morill, K., Dixon, J.S., Barth, G., Bunnenberg, E., Djerassi, C. (1977), *J. Am. Chem. Soc.* **99**, 724.

Sevin, A., Bigot, B., Pfau, M. (1979), *Helv. Chim. Acta* **62**, 699.

Serrano-Andrés, L., Merchan, M., Nebot-Gil, I., Lindh, R., Roos, B.O. (1993), *J. Chem. Phys.* **98**, 3151.

Sharafy, S., Muszkat, K.A. (1971), *J. Am. Chem. Soc.* **93**, 4119.

Siebrand, W. (1966), *J. Chem. Phys.* **44**, 4055.

Siebrand, W. (1967), *J. Chem. Phys.* **46**, 440, 2411.

Sklar, A.L. (1942), *J. Chem. Phys.* **10**, 135.

Snatzke, G. (1982), *Chem. Zeit* **16**, 160.

Snyder, G.J., Dougherty, D.A. (1985), *J. Am. Chem. Soc.* **107**, 1774.

Sobolewski, A.L., Woywod, C., Domcke, W. (1993), *J. Chem. Phys.* **98**, 5627.

Sonntag, F.J., Srinivasan, R. (1971), *Org. Photochem. Synth.* **1**, 39.

Souto, M.A., Wallace, S.L., Michl, J. (1980), *Tetrahedron* **36**, 1521.

Squillacote, M., Semple, T.C. (1990), *J. Am. Chem. Soc.* **112**, 5546.

Squillacote, M.E., Sheridan, R.S., Chapman, O.L., Anet, F.A.L. (1979), *J. Am. Chem. Soc.* **101**, 3657.

Squillacote, M., Bergman, A., De Felippis, J. (1989), *Tetrahedron Lett.* **30**, 6805.

Srinivasan, R. (1963), *J. Am. Chem. Soc.* **85**, 819.

Srinivasan, R. (1968), *J. Am. Chem. Soc.* **90**, 4498.

Srinivasan, R., Carlough, K.H. (1967), *J. Am. Chem. Soc.* **89**, 4932.

Staab, H.A., Herz, C.P., Krieger, C., Rentea, M. (1983), *Chem. Ber.* **116**, 3813.

Steiner, R.P., Michl, J. (1978), *J. Am. Chem. Soc.* **100**, 6861.

Steiner, U.E., Ulrich, T. (1989), *Chem. Rev.* **89**, 51.

Stephenson, L.M., Grdina, M.J., Orfanopoulos, M. (1980), *Acc. Chem. Res.* **13**, 419.

Stern, O., Volmer, M. (1919), *Z. Phys.* **20**, 183.

Steuhl, H.-M., Klessinger, M. (1994), *Angew. Chem. Int. Ed. Engl.* **33**, 2431.

Stevens, B., Algar, B.E. (1967), *Chem. Phys. Lett.* **1**, 219.

Stevens, B., Ban, M.I. (1964), *Trans. Faraday Soc.* **60**, 1515.

Stohrer, W.-D., Jacobs, P., Kaiser, K.H., Wiech, G., Quinkert, G. (1974), *Top. Curr. Chem.* **46**, 181.

Strickler, S.J., Berg, R.A. (1962), *J. Chem. Phys.* **37**, 814.

Struve, W.S., Rentzepis, P.M., Jortner, J. (1973), *J. Chem. Phys.* **59**, 5014.

Sugihara, Y., Wakabayashi, S., Murata, I., Jinguji, M., Nakazawa, T., Persy, G., Wirz, J. (1985), *J. Am. Chem. Soc.* **107**, 5894.

Sun, Y.-P., Wallraff, G.M., Miller, R.D., Michl, J. (1992a), *J. Photochem. Photobiol.* **62**, 333.

Sun, Y.-P., Hamada, Y., Huang, L.-M., Maxka, J., Hsiao, J.-S., West, R., Michl, J. (1992b), *J. Am. Chem. Soc.* **114**, 6301.

Sun, S., Saigusa, H., Lim, E.C. (1993), *J. Phys. Chem.* **97**, 11635.

Tapia, O. (1982), in *Molecular Interactions,* Vol. 3; Orville-Thomas, J.W., Ed.; Wiley: New York.

Taube, H. (1970), *Electron Transfer Reactions of Complex Ions in Solution*; Academic Press: New York, London.

Taylor, G.N. (1971), *Chem. Phys. Lett.* **10**, 355.

Taylor, P.R. (1982), *J. Am. Chem. Soc.* **104**, 5248.

Teller, E. (1937), *J. Phys. Chem.* **41**, 109.

Terenin, A., Ermolaev, V. (1956), *Trans. Faraday Soc.* **52**, 1042.

Thiel, W. (1981), *J. Am. Chem. Soc.* **103**, 1413.

Thompson, A.M., Goswami, P.C., Zimmerman, G.L. (1979), *J. Phys. Chem.* **83**, 314.

Thompson, M.A., Zerner, M.C. (1991), *J. Am. Chem. Soc.* **113**, 8210.

Thulstrup, E.W., Michl, J. (1976), *J. Am. Chem. Soc.* **98**, 4533.

Thulstrup, E.W., Michl, J., Eggers, J.H. (1970), *J. Phys. Chem.* **74**, 3868.

Tinnemans, A.H.A., Neckers, D.C. (1977), *J. Am. Chem. Soc.* **99**, 6459.

Trefonas, III, P., West, R., Miller, R.D. (1985), *J. Am. Chem. Soc.* **107**, 2737.

Troe, J., Weitzel, K.-M. (1988), *J. Chem. Phys.* **88**, 7030.

Trulson, M.O., Mathies, R.A. (1990), *J. Phys. Chem.* **94**, 5741.

Trulson, M.O., Dollinger, G.D., Mathies, R.A. (1989), *J. Chem. Phys.* **90**, 4274.

Tseng, K.L., Michl, J. (1976), *J. Am. Chem. Soc.* **98**, 6138.

Turner, D.W., Baker, C., Baker, A.D., Brundle, C.R. (1970), *Molecular Photoelectron Spectroscopy*; Wiley-Interscience: London, New York.

Turro, N.J. (1978), *Modern Molecular Photochemistry*; Benjamin: Menlo Park.

Turro, N.J., Devaquet, A. (1975), *J. Am. Chem. Soc.* **97**, 3859.

Turro, N.J., Kraeutler, B. (1978), *J. Am. Chem. Soc.* **100**, 7432.

Turro, N.J., Lechtken, P. (1972), *J. Am. Chem. Soc.* **94**, 2886.

Turro, N.J., Wriede, P.A. (1970), *J. Am. Chem. Soc.* **92**, 320.

Turro, N.J., Dalton, J.C., Dawes, K., Farrington, G., Hautala, R., Morton, D., Niemczyk, M., Schore, N. (1972a), *Acc. Chem. Res.* **5**, 92.

Turro, N.J., Kavarnos, G., Fung, V., Lyons, Jr., A.L., Cole, Jr., T. (1972b), *J. Am. Chem. Soc.* **94**, 1392.

Turro, N.J., Schuster, G., Pouliquen, J., Pettit, R., Mauldin, C. (1974), *J. Am. Chem. Soc.* **96**, 6797.

Turro, N.J., Farneth, W.E., Devaquet, A. (1976), *J. Am. Chem. Soc.* **98**, 7425.

Turro, N.J., Cherry, W.R., Mirbach, M.F., Mirbach, M.J. (1977a), *J. Am. Chem. Soc.* **99**, 7388.

Turro, N.J., Ramamurthy, V., Katz, T.J. (1977b), *Nouv. J. Chim.* **1**, 363.

Turro, N.J., Liu, K.-C., Show, M.-F., Lee, P. (1978), *Photochem. Photobiol.* **27**, 523.

Turro, N.J., Aikawa, M., Butcher, Jr., J.A., Griffin, G.W. (1980a), *J. Am. Chem. Soc.* **102**, 5127.

Turro, N.J., Anderson, D.R., Kraeutler, B. (1980b), *Tetrahedron Lett.* **21**, 3.

Turro, N.J., Cox, G.S., Paczkowski, M.A. (1985), *Top. Curr. Chem.* **129**, 57.

Turro, N.J., Zhang, Z., Trahanovsky, W.S., Chou, C.-H. (1988), *Tetrahedron Lett.* **29**, 2543.

Vala, Jr., M.T., Haebig, J., Rice, S.A. (1965), *J. Chem. Phys.* **43**, 886.

van der Auweraer, M., Grabowski, Z.R., Rettig, W. (1991) *J. Phys. Chem.* **95**, 2083.

van der Hart, J.A., Mulder, J.J.C., Cornelisse, J. (1987), *J. Mol. Struct. (Theochem)* **151**, 1.

van der Lugt, W.T.A.M., Oosterhoff, L.J. (1969), *J. Am. Chem. Soc.* **91**, 6042.

van Riel, H.C.H.A., Lodder, G., Havinga, E. (1981), *J. Am. Chem. Soc.* **103**, 7257.

Vogel, E., Jux, N., Rodriguez-Val, E., Lex, J., Schmickler, H. (1990), *Angew. Chem. Int. Ed. Engl.* **29**, 1387.

Wagner, P.J. (1989), *Acc. Chem. Res* **22**, 83.

Wagner, P.J., Hammond, G.S. (1965), *J. Am. Chem. Soc.* **87**, 1245.

Wagner, P.J., Kelso, P.A. (1969), *Tetrahedron Lett.*, 4151.

Wagner, P.J., Kochevar, I. (1968), *J. Am. Chem. Soc.* **90**, 2232.

Wagner, P.J., Meador, M.A., Zhou, B., Park, B.-S. (1991), *J. Am. Chem. Soc.* **113**, 9630.

Wagniere, G. (1966), *J. Am. Chem. Soc.* **88**, 3937.

Wallace, S.L., Michl, J. (1983), *Photochem. Photobiol.*; Proc. Int. Conf. Alexandria; Zewail, A.H., Ed.; Vol. 2, p. 1191.

Wallraff, G.M., Michl, J. (1986), *J. Org. Chem.* **51**, 1794.

Waluk, J., Michl, J. (1991), *J. Org. Chem.* **56**, 2729.

Waluk, J., Michl, J. (1991), *J. Org. Chem.* **56**, 2729.

Waluk, J., Mueller, M., Swiderek, P., Koecker, M., Vogel, E., Hohlneicher, G., Michl, J. (1991), *J. Am. Chem. Soc.* **113**, 5511.

Wan, P., Shukla, D. (1993), *Chem. Rev.* **93**, 571.

Wasielewski, M.R. (1992), *Chem. Rev.* **92**, 435.

Watkins, A.R. (1972), *J.C.S. Faraday Trans, I* **68**, 28.

Watson, Jr., C.R., Pagni, R.M., Dodd, J.R., Bloor, J.E. (1976), *J. Am. Chem. Soc.* **98**, 2551.

Weber, T., von Bargen, A., Riedle, E., Neusser, H.J. (1990), *J. Chem. Phys.* **92**, 90.

Weeks, G.H., Adcock, W., Klingensmith, K.A., Waluk, J.W., West, R., Vašak, M., Downing, J., Michl, J. (1986), *Pure Appl. Chem.* **58**, 39.

Weiner, S.A. (1971), *J. Am. Chem. Soc.* **93**, 425.

Weller, A. (1958), *Z. Phys. Chem. N.F.* **15**, 438.

Weller, A. (1968), *Pure Appl. Chem.* **16**, 115.

Weller, A. (1982a), *Z. Phys. Chem. NF* **133**, 93.

Weller, A. (1982b), *Pure Appl. Chem.* **54**, 1885.

West, R., Downing, J.W., Inagaki, S., Michl, J. (1981), *J. Am. Chem. Soc.* **103**, 5073.

Wigner, E., Wittmer, E.E. (1928), *Z. Phys.* **51**, 859.

Wilkinson, F. (1964), *Adv. Photochem.* **3**, 241.

Wilkinson, F. (1968), in *Luminescence,* Chapter 8; Bowen, E.J., Ed.; Van Nostrand, London.

Wilzbach, K.E., Harkness, A.L., Kaplan, L. (1968), *J. Am. Chem. Soc.* **90**, 1116.

Winkelhofer, G., Janoschek, R., Fratev, F., v. R. Schleyer, P. (1983), *Croat. Chim. Acta* **56**, 509.

Wirz, J., Persy, G., Rommel, E., Murata, I., Nakasuji, K. (1984), *Helv. Chim. Acta* **67**, 305.

Wolff, T. (1985), *Z. Naturforsch.* **40a**, 1105.

Wolff, T., Müller, N., von Bünau, G. (1983), *J. Photochem.* **22**, 61.

Wong, P.C., Arnold, D.R. (1979), *Tetrahedron Lett.* **23**, 2101.

Woodward, R.B. (1942), *J. Am. Chem. Soc.* **64**, 72.

Woodward, R.B., Hoffmann, R. (1969), *Angew. Chem. Int. Ed. Engl.* **8**, 781.

Xantheas, S.S., Atchity, G.J., Elbert, S.T., Ruedenberg, K. (1991), *J. Chem. Phys.* **94**, 8054.

Yadav, J.S., Goddard, J.D. (1986), *J. Chem. Phys.* **84**, 2682.

Yamazaki, H., Cvetanović, R.J. (1969), *J. Am. Chem. Soc.* **91**, 520.

Yamazaki, H., Cvetanović, R.J., Irwin, R.S. (1976), *J. Am. Chem. Soc.* **98**, 2198.

Yang, K.H. (1976), *Ann. Phys. NY* **101**, 62, 97.

Yang, K.H. (1982), *J. Phys. A* **15**, 437.

Yang, N.C., Elliott, S.P. (1969), *J. Am. Chem. Soc.* **91**, 7550.

Yang, N.C., Feit, E.D., Hui, M.H., Turro, N.J., Dalton, J.C. (1970), *J. Am. Chem. Soc.* **92**, 6974.

Yang, N.C., Nussim, M., Jorgenson, M.J., Murov, S. (1964), *Tetrahedron Lett.*, 3657.

Yang, N.C., Loeschen, R., Mitchell, D. (1967), *J. Am. Chem. Soc.* **89**, 5465.

Yang, N.C., Yates, R.L., Masnovi, J., Shold, D.M., Chiang, W. (1979), *Pure Appl. Chem.* **51**, 173.

Yang, N.C., Masnovi, J., Chiang, W., Wang, T., Shou, H., Yang, D.H. (1981), *Tetrahedron* **37**, 3285.

Yang, N.-C., Noh, T., Gan, H., Halfon, S., Hrnjez, B.J. (1988), *J. Am. Chem. Soc.* **110**, 5919.

Yarkony, D.R. (1990), *J. Chem. Phys.* **92**, 2457.

Yarkony, D.R. (1994), *J. Chem. Phys.* **100**, 3639.

Zechmeister, L. (1960), *Fortschr. Chem. Org. Naturst.* **18**, 223.

Zener, C. (1932), *Proc. R. Soc. London* **A137**, 696.

Zimmerman, H.E. (1964), *Pure Appl. Chem.* **9**, 493.

Zimmerman, H.E. (1966), *J. Am. Chem. Soc.* **88**, 1564, 1566.

Zimmerman, H.E. (1969), *Angew. Chem. Int. Ed. Engl.* **8**, 1.

Zimmerman, H.E. (1980), in *Rearrangements in Ground and Excited States,* Vol. 3; de Mayo, P., Ed.; Academic Press: New York; p. 131.

Zimmerman, H.E. (1982), *Chimia* **36**, 423.

Zimmerman, H.E., Grunewald, G.L. (1966), *J. Am. Chem. Soc.* **88**, 183.

Zimmerman, H.E., Little, R.D. (1974), *J. Am. Chem. Soc.* **96**, 4623.

Zimmerman, H.E., Hackett, P., Juers, D.F., McCall, J.M., Schröder, B. (1971), *J. Am. Chem. Soc.* **93**, 3653.

Zimmerman, H.E., Robbins, J.D., McKelvey, R.D., Samuel, C.J., Sousa, L.R. (1974), *J. Am. Chem. Soc.* **96**, 1974.

Zimmermann, H.E., Penn, J.H., Johnson, M.R. (1981), *Proc. Natl. Acad. Sci. U.S.A.* **78**, 2021.

Zimmerman, H.E., Sulzbach, H.M., Tollefson, M.B. (1993), *J. Am. Chem. Soc.* **115**, 6548.

Index